Edwin Zondervan, Cristhian Almeida-Rivera, and Kyle Vincent Camarda

Product-Driven Process Design

Also of Interest

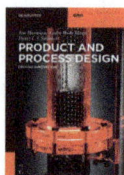

Product and Process Design.
Driving Innovation
Harmsen, de Haan, Swinkels, 2018
ISBN 978-3-11-046772-7, e-ISBN 978-3-11-046774-1

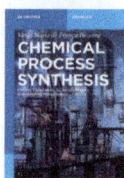

Chemical Process Synthesis.
Connecting Chemical with Systems Engineering Procedures
Bezerra, 2021
ISBN 978-3-11-046825-0, e-ISBN 978-3-11-046826-7

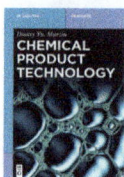

Chemical Product Technology
Murzin, 2018
ISBN 978-3-11-047531-9, e-ISBN 978-3-11-047552-4

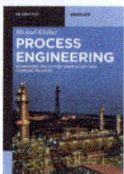

Process Engineering.
Addressing the Gap between Study and Chemical Industry
Kleiber, 2016
ISBN 978-3-11-031209-6, e-ISBN 978-3-11-031211-9

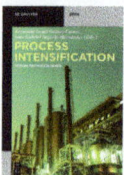

Process Intensification.
Design Methodologies
Gómez-Castro, Segovia-Hernández (Eds.), 2019
ISBN 978-3-11-059607-6, e-ISBN 978-3-11-059612-0

Edwin Zondervan, Cristhian Almeida-Rivera,
and Kyle Vincent Camarda

Product-Driven Process Design

From Molecule to Enterprise

DE GRUYTER

Authors
Prof. dr. ir. Edwin Zondervan
Dept. of Production Engineering
Bremen University
Leobener Str. 7
28359 Bremen
Germany
edwin.zondervan@uni-bremen.de

Prof. Dr. Kyle Vincent Camarda
School of Engineering
Chemical & Petroleum Engineering
University of Kansas
1530 West 15th Street
Lawrence 66045, KS
United States of America
camarda@ku.edu

Dr. Cristhian Almeida-Rivera
Organisation for the Prohibition of Chemical
Weapons (OPCW)
Johan de Wittlaan 32
2517 JR The Hague
The Netherlands
cristhian.almeidarivera@opcw.org

ISBN 978-3-11-057011-3
e-ISBN (PDF) 978-3-11-057013-7
e-ISBN (EPUB) 978-3-11-057019-9

Library of Congress Control Number: 2019950402

Bibliographic information published by the Deutsche Nationalbibliothek
The Deutsche Nationalbibliothek lists this publication in the Deutsche Nationalbibliografie;
detailed bibliographic data are available on the Internet at http://dnb.dnb.de.

© 2020 Walter de Gruyter GmbH, Berlin/Boston
Cover image: gettyimages/iStock/TeamOktopus
Typesetting: VTeX UAB, Lithuania
Printing and binding: CPI books GmbH, Leck

www.degruyter.com

To Prof. Dr. ir. Peter M. M. Bongers (†),
with whom we shared the passion for
product-driven process synthesis
but unfortunately cannot be with us today
to witness the result of our vision

To all people who guided, excited and contributed to the creation of this work:
- All students and colleagues who provided content through discussions!
- My son, Victor Haim Zondervan, who is the driving force of all activity in my life!

— EZO

To all the people who helped and inspired me throughout this journey:
- Prof. em. ir. Johan Grievink, my mentor and former supervisor, from whom I learned the most important lessons of process design,
- Patricia, my wife and life partner, and my two precious peas Lucia and Cristina, who are my reasons for being,
- my parents, who raised me to become persistent and to stay true to my beliefs, and
- the design practitioners out there, who are committed to redesigning this world and making it a better place for everyone.

— CAR

To all the people who supported and inspired me: my family, my colleagues and my students

— KVC

Foreword

The chemical process industry is subject to a rapidly changing environment, characterized by slim profit margins and fierce competitiveness. Being able to operate effectively at reduced costs and with increasingly shorter times-to-market is the common denominator of successful companies and corporations. However, attaining this performance level is not a straightforward or trivial issue. Success is dependent on coping effectively with dynamic environments, short process development and design times and, last but not least, continuously changing consumer (or end-user) needs or expectations.

It is now well established by industry and academia that the focus of the chemical process industry has shifted from a process-centered orientation to a product-centered one. In fact, during the last decades we have experienced how the commodity chemical business gradually released its dominating role towards higher-added value products, such as specialty chemicals, pharmaceuticals and consumer products. These products are not only characterized by their composition and purities, but also by their performance or functionality against the satisfaction of a need or desire of the consumer.

Manufacturing such products in a responsible and sustainable way imposes technical challenges and requires the development of research and development capabilities.

Addressing the dynamic challenges of a consumer-centered market requires the development of an appropriate design methodology. Such methodology should provide the right tools and framework to solve the multi-dimensional and multi-disciplinary nature of design problems, where multiple stakeholders, objectives and types of resources characterize the design space on various time and length scales.

Aiming at a more structured approach toward the synthesis of a consumer-centered product and their manufacturing processes, in this book we propose a methodology that exploits the synergy of combining product and process synthesis workstreams. This methodology, the Product-driven Process Synthesis (PDPS) method, is supported by decomposing the problem into a hierarchy of design levels of increasing refinement, where complex and emerging decisions are made to proceed from one level to another.

This book is structured in two distinctive sections. The first section follows the structure of the proposed methodology, focusing on one level in each chapter, while the second section provides a comprehensive and up-to-date review of the technical areas that reside in the core of chemical and process engineering.

The contents of all the chapters of this book are support by a set of examples and case studies. By addressing industrially relevant and real-life cases the reader (e. g., design practitioners or designers-in-the-make) is exposed to the challenge of design,

https://doi.org/10.1515/9783110570137-201

which hovers over multiple design spaces and with broad time and space scales...
From molecule to enterprise.

Enjoy!

Prof. dr. ir. Edwin Zondervan
Dept. of Production Engineering
Bremen University Germany
edwin.zondervan@uni-bremen.de

Dr. ir. Cristhian Almeida-Rivera
Organisation for the Prohibition of Chemical Weapons (OPCW)
The Netherlands
cristhian.almeidarivera@opcw.org

Prof. Dr. Kyle Vincent Camarda
School of Engineering
Chemical & Petroleum Engineering
University of Kansas
United States of America
camarda@ku.edu

Contents

Part II: Process Design Principles

Part I: **Process Design**

1 Introduction

1.1 The current setting of process industry

"We cannot solve our problems with the same thinking we used when we created them." (Albert Einstein, 1879–1955)

The current chemical process industry (CPI) is subject to a rapidly changing environment, characterized by slim profit margins and fierce competitiveness. Rapid changes are not exclusively found in the demands of society for new, high quality, safe, clean and environmentally benign products (Herder, 1999). They can also be found in the dynamics of business operations, which include, *inter alia*, global operations, competition and strategic alliances mapping. Being able to operate effectively at reduced costs and with increasingly shorter times-to-market is the common denominator of successful companies and corporations. However, attaining this performance level is not a straightforward or trivial issue. Success is dependent on coping effectively with dynamic environments, short process development and design times and, last but not least, continuously changing consumer (or end-user) needs or expectations.

It is now well established by industry and academia that the focus of the CPI has shifted from a process-centered orientation to a product-centered one (Hill, 2004; Bagajewicz, 2007). In recent decades chemical industry has experienced a fundamental transformation in its structure and requirements (Edwards, 2006). In fact, during the last decades we have experienced how the commodity chemical business is gradually releasing its dominating role towards higher-added value products, such as specialty chemicals, pharmaceuticals and consumer products. Chemical industry has been gradually concentrating its activities and streamlining its assets towards value creation and growth by engaging in the manufacture of high-value chemical products. These products are not only characterized by their composition and purities, but by their performance or functionality against the satisfaction of a need or desire of the consumer.

Among process industries, those engaged in the manufacture and marketing of Fast-Moving Consumer Goods are subjected to an even more increasing pressure. The market is not only extremely dynamic and changing, but is also characterized by the predominant role of consumers, which have become more critical, demanding and conscious about their choices since the 1980s. Today's customers are more connected, sceptical about the value-for-money proposition they receive and sophisticated in their demands with ever-increasing expectations of their product and service providers (Deloitte, 2014; Nielsen, 2018).

In this particular market segment, all leading FMCG companies are rapidly transforming from general manufacturing hubs of loosely connected products to companies delivering health, wellness and nutrition in a sustainable manner. Manufacturing in a responsible and sustainable way products within those strategic areas imposes

https://doi.org/10.1515/9783110570137-001

technical challenges to work on and requires the development of research and development capabilities. These challenges and capabilities have the clear aim of delivering a product with all associated benefits (financial performance, environmental and societal impacts) at a short time-to-market and at a reduced manufacturing expenditure. Finding and creating business opportunities to be brought successfully to the market (Verloop, 2004) is the response of leading companies to this rapidly changing environment. In fact, from a Design Thinking perspective, true innovations reside at the intersection of consumer desirability, technology feasibility and financial viability. A key activity in this innovation-driven space is the actual creation of the conversion or manufacturing system (*i. e.*, feasibility), which considers simultaneously the other two dimensions. Addressing the dynamic challenges of the FMCG sector requires the development of an appropriate design methodology. Such methodology should provide the right tools and framework to solve the multi-dimensional and multi-desciplinary nature of the design problems, where multiple stakeholders, objectives and types of resources characterize the design space through various time and length scales.

1.2 The development of the PDPS approach

"The task is not so much to see what no one has yet seen, but to think what nobody yet has thought about that which everybody sees." (Arthur Schopenhauer, 1788–1860)

Aiming at a more structured approach toward the synthesis of consumer-centered products and their manufacturing processes, in this book we propose a methodology that exploits the synergy of combining product and process synthesis workstreams. This methodology, the Product-driven Process Synthesis (PDPS) method, is supported by decomposing the problem into a hierarchy of design levels of increasing refinement, where complex and emerging decisions are made to proceed from one level to another. The PDPS approach is one of the vehicles to embrace most of the design challenges in the FMCG sector, where key players are coping effectively with the dynamic environment of short process development and design times.

By exploiting the synergy of combining product and process synthesis workstreams, PDPS is based on the systems engineering strategy. Each level in the PDPS methodology features the same, uniform sequence of activities, which have been derived from the pioneering work of Douglas (1988) and Siirola (1996) and further extended by Bermingham (2003), Almeida-Rivera et al. (2004), Almeida-Rivera (2005) and the research by Meeuse et al. (1999), Stappen-vander (2005), Meeuse (2005, 2007).

As the methodology has evolved and matured over the years, several areas have included within the scope over which the methodology gravitates (Bongers and Almeida-Rivera, 2008, 2009, 2012; Bongers, 2009). Current efforts in the development of the methodology have been focusing on broadening the design scope to consumer preferences (Almeida-Rivera et al., 2007), product attributes (Bongers, 2008), process variables (Almeida-Rivera et al., 2007) and supply chain considerations and financial factors (de Ridder et al., 2008). The industrial (and academic) relevance of the approach has been validated by the successful application to several cases involving the recovery of high-value compounds from waste streams (Jankowiak et al., 2012, 2015), synthesis of structured matrices with specific consumer-related attributes (Gupta and Bongers, 2011), generation of new-generation manufacturing technologies (Almeida-Rivera et al., 2017) and extraction of high-value compounds from natural matrices (Zderic et al., 2014, 2015, 2016; Zderic and Zondervan, 2016, 2017; Zondervan et al., 2015), among other non-disclosed cases. The applicability of the approach to enterprise-wide scales has been demonstrated by studying the synthesis of biorefinery networks (Zondervan et al., 2011; Zderic et al., 2019; Kiskini et al., 2016, 2015).

1.3 Structure of the book

"Order and simplification are the first steps toward the mastery of a subject." (Thomas Mann, 1875–1955)

This book is structured in two distinctive sections. The first section follows the structure of the proposed methodology, focusing on one level in each chapter, while the second section provides a comprehensive and up-to-date review of the technical areas, that reside in the core of chemical and process engineering.

Chapter 2 addresses the significant role of functional or performance products in our lives. By identifying the current trends derived from a globalized environment, this chapters sets up the conditions for revising available process design approaches. Chapter 3 introduces the PDPS approach. As identified as one of the challenges of the FMCG sector, the development of the approach is justified by the gap analysis of current design approaches, for both products and processes. Chapter 4 is concerned with the relevance of the consumer in current CPI. Identifying consumers' needs, wants and desires – from a Design Thinking standpoint – is a pillar of the approach. Next to the understanding of the needs, this chapter focuses on the translation of those needs – expressed in the language of the consumer – to product attributes. The further translation of those attributes into properties (or technical specifications) is addressed in Chapter 5. This chapter, additionally, comprises the development or generation of product ideas or prototypes and the generation of an input–output structure of the design problem. Chapter 6 is concerned with the creative activity of identifying the set of fundamental tasks that allows the conversion of inputs into outputs. These tasks are integrated in a network, which will be the actual backbone of the process. This chapter also addresses the selection of the mechanism for each task. Chapter 7 looks after the sizing of the units, which are based on the mechanisms previously identified. Chapter 8 focuses on the specific case of multiple products within the design scope and on the plant-wide integration of units and production planning. Chapter 9 addresses the current formulations and algorithms of computer-aided molecular design (CAMD). Polymer design is considered as a tutorial example to highlight the potential of CAMD in the search for novel molecules and environmentally benign materials.

The second section of the book addresses core design fundamentals within chemical and process engineering. The topics are covered with a refreshed view, focusing on the current requirements of designers or design practitioners. The areas addressed include process synthesis (Chapter 10), process simulation (Chapter 11), reactor design (Chapter 12), batch process design (Chapter 13), separation train design (Chapter 14), plant-wide control (Chapter 15), heat integration (Chapter 16), process economics and safety (Chapter 17), design for sustainability (Chapter 18), optimization (Chapter 19) and enterprise-wide optimization (Chapter 20).

The contents of all the chapters of this book are support by a set of examples and case studies. By addressing industrially relevant and real-life cases the reader (*e. g.*, design practitioners or designers-in-the-make) is exposed to the challenge of design, which hovers over multiple design spaces and with broad time and space scales. Enjoy your design journey... From molecule to enterprise.

Bibliography

Almeida-Rivera, C. P. (2005). *Designing reactive distillation processes with improved efficiency*. Thesis/dissertation, Delft University of Technology.

Almeida-Rivera, C. P., Bongers, P., and Zondervan, E. (2017). A structured approach for product-driven process synthesis in foods manufacture. In: M. Martin, M. Eden, and N. Chemmangattuvalappil, eds., *Tools For Chemical Product Design: From Consumer Products to Biomedicine*. Elsevier, pp. 417–441. Chapter 15.

Almeida-Rivera, C. P., Jain, P., Bruin, S., and Bongers, P. (2007). Integrated product and process design approach for rationalization of food products. *Computer-Aided Chemical Engineering*, 24, pp. 449–454.

Almeida-Rivera, C. P., Swinkels, P. L. J., and Grievink, J. (2004). Designing reactive distillation processes: present and future. *Computers and Chemical Engineering*, 28(10), pp. 1997–2020.

Bagajewicz, M. (2007). On the role of microeconomics, planning, and finances in product design. *AIChE Journal*, 53(12), pp. 3155–3170.

Bermingham, S. (2003). *A design procedure and predictive models for solution crystallisation processes*. Thesis/dissertation, Delft University of Technology.

Bongers, P. (2008). Model of the product properties for process synthesis. *Computer-Aided Chemical Engineering*, 25, pp. 1–6.

Bongers, P. (2009). *Intertwine product and process design*. Inaugural lecture. Eindhoven University of Technology. ISBN: 978-90-386-2124-1.

Bongers, P. and Almeida-Rivera, C. P. (2008). Product driven process design methodology. In: *AIChE Symposia Proceedings, Proceedings of the 100th Annual Meeting*. American Institute of Chemical Engineers, p. 623A.

Bongers, P. M. M. and Almeida-Rivera, C. (2009). Product driven process synthesis methodology. *Computer-Aided Chemical Engineering*, 26, pp. 231–236.

Bongers, P. M. M. and Almeida-Rivera, C. (2012). Product driven process design method. *Computer-Aided Chemical Engineering*, 31, pp. 195–199.

de Ridder, K., Almeida-Rivera, C. P., Bongers, P., Bruin, S., and Flapper, S. D. (2008). Multi-criteria decision making in product-driven process synthesis. *Computer-Aided Chemical Engineering*, 25, pp. 1021–1026.

Deloitte (2014). The deloitte consumer review: The growing power of consumers. Technical report, Deloitte.

Douglas, J. (1988). *Conceptual Design of Chemical Process*. New York: McGraw-Hill.

Edwards, M. F. (2006). Product engineering: Some challenges for chemical engineers. *Transactions of the Institute of Chemical Engineers – Part A*, 84(A4), pp. 255–260.

Gupta, S. and Bongers, P. (2011). Bouillon cube process design by applying product driven process synthesis. *Chemical Engineering and Processing*, 50, pp. 9–15.

Herder, P. (1999). *Process Design in a changing environment*. Thesis/dissertation, Delft University of Technology.

Hill, M. (2004). Product and process design for structured products. *AIChE Journal*, 50(8), pp. 1656–1661.

Jankowiak, L., Mendez, D., Boom, R., Ottens, M., Zondervan, E., and van-der Groot, A. (2015). A process synthesis approach for isolation of isoflavons from okara. *Industrial and Engineering Chemistry Research*, 54(2), pp. 691–699.

Jankowiak, L., van-der Groot, A. J., Trifunovic, O., Bongers, P., and Boom, R. (2012). Applicability of product-driven process synthesis to separation processes in food. *Computer-Aided Chemical Engineering*, 31, pp. 210–214.

Kiskini, A., Zondervan, E., Wierenga, P., Poiesz, E., and Gruppen, H. (2015). Using product driven process synthesis in the biorefinery. *Computer-Aided Chemical Engineering*, 37, pp. 1253–1258.

Kiskini, A., Zondervan, E., Wierenga, P. A., Poiesz, E., and Gruppen, H. (2016). Using product driven process synthesis in the biorefinery. *Computers and Chemical Engineering*, 91, pp. 257–268.

Meeuse, F. M. (2005). Process synthesis applied to the food industry. *Computer-Aided Chemical Engineering*, 15(20), pp. 937–942.

Meeuse, M. (2007). Process synthesis for structured food products. In: K. Ng, R. Gani, and K. Dam-Johansen, eds., *Chemical Product Design: Towards a Perspective through Case Studies*, vol. 6. Elsevier, pp. 167–179. Chapter 1.

Meeuse, F. M., Grievink, J., Verheijen, P. J. T., and Stappen-vander, M. L. M. (1999). Conceptual design of processes for structured products. In: M. F. Malone, ed., *Fifth conference on Foundations of Computer Aided Process Design*. AIChE.

Nielsen (2018). What is next in emerging markets. Technical report, Nielsen Company.

Siirola, J. J. (1996). Industrial applications of chemical process synthesis. In: J. Anderson, ed., *Advances in Chemical Engineering. Process Synthesis*. Academic Press, pp. 1–61. Chapter 23.

Stappen-vander, M. L. M. (2005). Process synthesis methodology for structured (food) products. *NPT Procestechnologie*, 6, pp. 22–24.

Verloop, J. (2004). *Insight in Innovation – Managing Innovation by Understanding the Laws of Innovation*. Amsterdam: Elsevier.

Zderic, A. and Zondervan, E. (2016). Polyphenol extraction from fresh tea leaves by pulsed electric field: A study of mechanisms. *Chemical Engineering Research and Design*, 109, pp. 586–592.

Zderic, A. and Zondervan, E. (2017). Product-driven process synthesis: Extraction of polyphenols from tea. *Journal of Food Engineering*, 196, pp. 113–122.

Zderic, A., Tarakci, T., Hosshyar, N., Zondervan, E., and Meuldijk, J. (2014). Process design for extraction of soybean oil bodies by applying the product driven process synthesis methodology. *Computer-Aided Chemical Engineering*, 33, pp. 193–198.

Zderic, A., Zondervan, E., and Meuldijk, J. (2015). Product-driven process synthesis for the extraction of polyphenols from fresh tea leaves. *Chemical Engineering Transactions*, 43, pp. 157–162.

Zderic, A., Almeida-Rivera, C., Bongers, P., and Zondervan, E. (2016). Product-driven process synthesis for the extraction of oil bodies from soybeans. *Journal of Food Engineering*, 185, pp. 26–34.

Zderic, A., Kiskini, A., Tsakas, E., Almeida-Rivera, C. P., and Zondervan, E. (2019). Giving added value to products from biomass: the role of mathematical programming in the product-driven process synthesis framework. *Computer-Aided Chemical Engineering*, 46, pp. 1591–1596.

Zondervan, E., Nawaz, M., Hahn, A., Woodley, J., and Gani, R. (2011). Optimal design of a multi-product biorefinery system. *Computers and Chemical Engineering*, 35, pp. 1752–1766.

Zondervan, E., Monsanto, M., and Meuldijk, J. (2015). Product driven process synthesis for the recovery of vitality ingredients from plant materials. *Chemical Engineering Transactions*, 43, pp. 61–66.

2 Performance products in a challenging environment

2.1 A challenging environment for the process industry

The process industry is subject to a rapidly changing environment, characterized by slim profit margins and fierce competitiveness. Rapid changes are not exclusively found in the demands of society for new, high quality, safe, clean and environmentally benign products (Herder, 1999). They can also be found in the dynamics of business operations, which include global operations and competition and strategic alliances mapping. Being able to operate effectively at reduced costs and with increasingly shorter times-to-market is the common denominator of successful companies and corporations. However, attaining this performance level is not a straightforward or trivial issue. Success is dependent on coping effectively with dynamic environments, short process development and design times and, last but not least, continuously changing consumer (or end-user) needs.

Moreover, taking into account life span considerations of products and processes is becoming essential for development and production activities. Special attention needs to be paid to the potential losses of resources over the process life span. Since such resources are of a different nature (*e. g.*, raw materials, capital and labor) with different degrees of depletion and rates of replenishment, the implementation of a life span perspective becomes a challenge for the process industry.

2.1.1 Environmental and energy challenges

As a result of the continuous stress under which the environment is, the pattern of resources and population (Worldwatch-Institute-Staff, 2014) is simply not comparable to that of the 2010. The world's forest, for instance, shrank by 1.3 % (*ca.* 520,000 km^2 and at a rate of 52,000 km^2 y^{-1}) from 2000 to 2010, deforestation being the key contributor and a mayor challenge in a number of countries and regions. Because of changes in international energy prices and domestic pricing policies and demand, fossil fuel consumption subsidies have declined approximately 44 % during the 2000s. This subsidies fall has favored the competitiveness and use of cleaner, alternative energy sources. As subsidies for newer energy technologies tend to be higher than for established technologies (5 ¢ kWh for renewable sources against 0.8 ¢ kWh for fossil fuel) (Worldwatch-Institute-Staff, 2014), it is expected that by 2020 a cut in global primary energy demand of *ca.* 5 % would happen. It is widely accepted that a gradual shift from fossil fuels to renewables is essential for decarbonization of the global energy systems, leading eventually to a reduction in the greenhouse gas (GHG) emissions (*ca.* 5.8 % of carbon dioxide emissions by 2020) and an increase in **green** new jobs (Worldwatch-Institute-Staff, 2014). According to the widely cited Steiner report (Stern,

https://doi.org/10.1515/9783110570137-002

2007), climate change needs to be considered as a serious threat for our planet with serious and disastrous impacts on the world's growth and development.

In fact, human activities have been changing the composition of the atmosphere and its properties. As reported in Stern (2007), since preindustrial times (*ca.* 1750), carbon dioxide concentrations have increased by just over one-third, from 280 ppm to 380 ppm in 2007 and reaching 400 ppm in 2016 (Figure 2.1), predominantly as a result of burning fossil fuels, deforestation and other changes in land use. Current levels of GHG emissions are higher than at any point in time in at least the past 650,000 years.

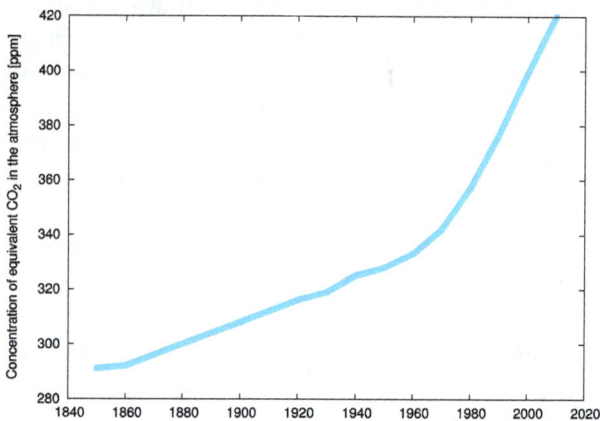

Figure 2.1: Increase in levels of GHG emissions in the atmosphere since the 1840s (Stern, 2007). Preindustrial level: *ca.* 280 ppm; 2019 level: *ca.* 380 ppm; increase rate: *ca.* 2.3 ppm per year.

Because of the inertia in the climate system, any corrective action and its associated cost (or investment) that takes place now and over the coming decades will be only noticeable after 10–20 years. Similarly, any investment in the next 10–20 years will have a profound effect on the climate in the second half of this century and in the next. If no action is undertaken to reduce GHG emissions – which include next to CO_2 also methane and NO_x – the overall costs and risks of climate change will be equivalent to losing at least 5 % of global GDP each year, now and forever. If a wider range of risks and impacts is taken into account, the estimates of damage could rise to 20 % of GDP or more. Moreover, the concentration of GHGs in the atmosphere could reach double its preindustrial level as early as 2035, virtually committing us to a global average temperature rise of over 2 °C. Such a level of global warming will lead to a radical change in the physical geography of the world and, subsequently, to major changes in the way people live their lives. Specifically, climate change will affect basic elements of life for people around the world, like access to water, food production, health and the environment. The impact of climate change will include hundreds of millions of people suffering hunger, water shortages and coastal flooding.

The stock of GHGs in the atmosphere comprises both energy and non-energy emissions, which are related to different human activities. Power, transport, buildings and industry activities are responsible for energy emissions, mainly driven by fossil fuel combustion for energy purposes and amounting to *ca.* 26.2 GtCO$_2$ in 2004. Non-energy emissions originate from land use and agriculture activities, specifically via deforestation, and account for an aggregate of *ca.* 32 % of GHGs in the atmosphere.

As depicted in Figure 2.2, approximately one-quarter of all global GHGs comes from the generation of power and heat, which is mostly used in domestic and commercial buildings, and by industry, including petroleum refineries, gas works and coal mines.

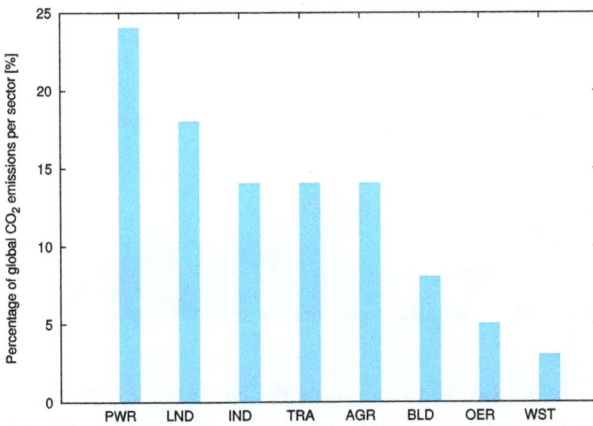

Figure 2.2: Main sources of emissions of GHGs to the atmosphere (Stern, 2007). Legend: PWR, power; LND, land use; IND, industry; TRA, transport; AGR, agriculture; BLD, buildings; OER, other energy related; WST, waste. Note: total emissions: 42 GtCO$_2$.

While land use is the second source of GHG emissions, these are entirely related to deforestation activities and restricted to specific locations. The agriculture sector is the third largest source of emissions, jointly with industry and transport. These non-CO$_2$ emissions are solely related to the use of fertilizers, management of livestock and manure. Furthermore, agriculture activities are indirectly responsible for the energy emissions associated with deforestation, manufacture of fertilizers and movement of goods. GHG emissions originating from the transportation sector comprise road, aviation, rail and shipping activities and amount to *ca.* 14 % of the global GHG stock. With a comparable level of emission, the industry sector accounts for 14 % of total direct emissions of GHG, with 10 % related to CO$_2$ emissions from combustion of fossil fuels in manufacturing and construction activities and 3 % are CO$_2$ and non-CO$_2$ emissions from industrial processes. Direct combustion of fossil fuels and biomass in commercial and residential buildings comprises the GHG emissions of building activities (8 % of the total), which are mostly related to heating and cooking purposes.

The global emissions of CO_2 have consistently increased since 1850 according to reasonable historical data (Stern, 2007). In the period 1950–2000 a step increase rate of approximately 3 % per year is evident (Figure 2.3), mainly due to energy emissions in North America and Europe during that period. In fact, these two geographical regions have produced around 70 % of the CO_2 since 1850, while developing countries account for less than one-quarter of cumulative emissions.

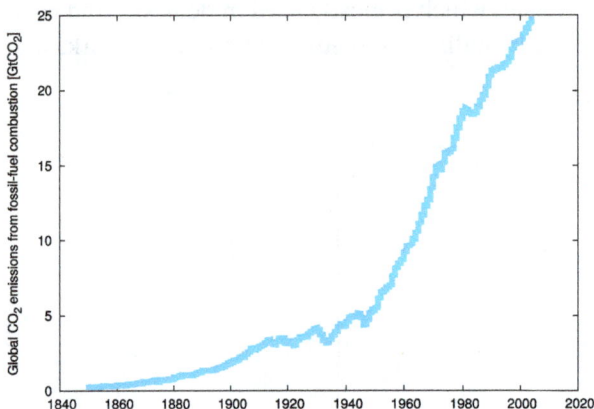

Figure 2.3: Global CO_2 emissions from fossil fuel combustion over the last decades (Stern, 2007).

As previously mentioned, industry is directly responsible for *ca.* 14 % of GHG emissions, or for *ca.* 20 % if upstream emissions from the power sector are included. GHG emissions from industry arise from four main sources: (i) direct fossil fuel combustion in manufacturing and construction, (ii) direct emissions of CO_2 and non-CO_2 emissions from the chemical processes involved in producing various chemicals and metals, (iii) upstream emissions from the power sector and (iv) indirect sources of emissions from the transport sector for the movement of goods.

Two-thirds of the industry-related emissions are coming from direct fossil fuel combustion (point (i)), specifically from the iron and steel, non-metallic minerals and chemicals and petrochemicals sectors (Table 2.1). Direct CO_2 emissions from manufacturing and construction are expected to reach 5.6 $GtCO_2$ in 2030 and 6.5 $GtCO_2$ in 2050, driven by emissions growth in developing countries.

The chemical industry – as one of the key contributors to GHG emissions within manufacturing activities – has seen a tremendous shift in its complexity, business dynamics and even skills' requirement for operators. As mentioned by Costa *et al.* (2014), two generations ago the chemical industry focused on roughly 50 commodities, while today's portfolio spans over thousands of commodity chemicals manufactured in highly optimized facilities. Moreover, the current set of skills expected in chemical industry operators includes flexibility, a wider vision, involvement in non-traditional areas and multi-disciplinary approaches. The change of the skill base has its origin

Table 2.1: Contribution to direct CO_2 emissions from fossil fuel combustion in manufacturing and construction activities (Stern, 2007). Note: total emissions from fossil fuel consumption: 4.3 $GtCO_2$.

Sector	Contribution [%]
Non-metallic minerals	25 %
Non-ferrous metals	2 %
Chemical and petrochemicals	18 %
Other	29 %
Iron and steel	26 %

in society's demand of the so-called specialty or functional products. This subset of high-value-added chemical products are not generally judged by their value based on technical criteria (such as purity), but rather on their performance against requirements defined and valued *by the consumer* (Section 2.2). In fact, because of the need to cope with the market's expectation and – at the same time – guarantee business competitiveness, a new mindset to view chemical industry activities – and by extension chemical engineering – is required.

As mentioned by Stankiewicz and Moulijn (2002), Hill (2009) and Costa *et al.* (2014), two socially accepted paradigms have been historically acknowledged and embraced by practitioners: the unit operations paradigm (1920s–1960s) and the transport phenomena paradigm (1960s–2000s). A third paradigm has been recently suggested to favor creativity, support the invention of new process and to assist the practitioner in the design and manufacture of high-added-value products to meet the needs of the market/consumer. This third paradigm resides in the field of chemical product engineering (Hill, 2009; Costa *et al.*, 2014), which provides the appropriate mindset for the engineers to be able to solve problems where both specialty or functional products and their manufacturing processes must be identified. It is worth mentioning that this third paradigm to be embraced needs to be socially accepted and – more importantly – must combine developments in product engineering and in process engineering, such as process intensification (PI) (Stankiewicz and Moulijn, 2002) and process safety management.

2.1.2 Changes in consumer needs and requirements

Since the 1980s the marketplace has experienced a fundamental change towards conditions in which supply exceeds demand, and where the competition among producers/sellers/service providers is steadily increasing (Kiran, 2017). This situation has turned consumers more connected, critical, demanding, conscious about their choices and sophisticated in their demands with ever-increasing expectations of their product and service providers. In this buyers' market, if the requirements are not met, consumers are simply not content with what they receive and most likely will opt for other alternatives. As a result of this demanding scenario, providers are expected to improve the quality of both their production and their services in order to remain active in the marketplace. Unfortunately, as Henry Ford once mentioned, a market rarely gets saturated with good products, but it very quickly gets saturated with bad ones.

Failing in not meeting customers' requirements involves not being compliant with the perception that consumers have of performance (or fitness for purpose), esthetics, durability, serviceability, warranty, reputation and price (Kiran, 2017). These elements are defined in Table 2.2.

Table 2.2: Elements of the consumers' perception of value or quality (Kiran, 2017).

Element	Description
Performance	or fitness for purpose; product's primary functional characteristics; functional ability and reliability to serve the customers' purpose
Esthetics	attributes related to the external appearance of the product
Durability	period for which the product would serve without failures
Serviceability	speed, courtesy and competence of the after sales service
Warranty	quality assurance, reliability and durability for a minimum length of time
Reputation	confidence created by the company and the brand name in maintenance of quality, reliability and esthetics
Price	monetary value of the product that the consumer is willing to pay in exchange for the satisfaction of their requirements

It is a well-known fact that consumers are becoming more informed, critical and demanding when it comes to the selection of a product or service (Deloitte, 2014), especially over the 2010s. Embracing effectively the requirements of the consumers within the design activity of a service/product contributes to a large extent to its success in the marketplace. The other contributing dimensions to the success of a product or service include the technological feasibility and business viability. In the intersection of these three dimensions resides a true innovation (Figure 2.4) and is the cornerstone of Design Thinking.

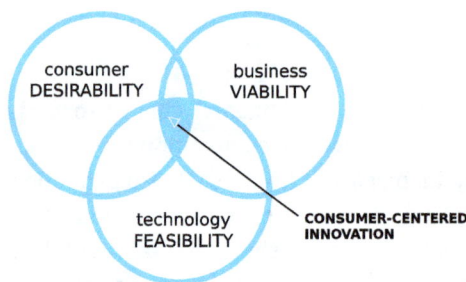

Figure 2.4: Consumer desirability, technological feasibility and business viability in the context of Design Thinking-based innovation.

The term Design Thinking – introduced in 2008 (Brown, 2008) and widely used since then – describes the process of creating solutions using creative problem solving techniques. It has been regarded as a discipline that uses the designer's sensibility and methods to match people's needs with what is technologically feasible and what a viable business strategy can convert into customer value and market opportunity. As mentioned by Widen (2019), Design Thinking is about looking at a problem from the inside out, rather than the other way around. Addressing consumer desirability involves understanding that there is an underserved market, characterized by unmet consumer needs. A central activity of this dimension implies the development of empathy for the consumer. See their world through their eyes. Next to the desirability dimension, truly successful innovations need to be technically feasible. The solution to the design problem needs to be technically feasible with the current resources and technology base. The third dimension is given by the business viability of the design solution, which implies that the solution to the design problem delivers a sustainable profit to the organization, with which the whole investment associated with the development and implementation can be paid back. Failing in properly addressing, solving or simply considering one of these dimensions will result in a product or service that will underperform in the missing dimension. It will be translated into a product that is not a desire from the market, has a weak or unhealthy business case or is not feasible to manufacture.

Building consumer empathy is essential to address the desirability dimension and involves three steps: (i) immersion, where the designer experiences what the user experiences; (ii) observation, where the designer views the behavior of the user in the context of their lives; and (iii) engagement, where the designer interacts with the user (*e. g.*, by interviewing them). Building empathy allows the designer to understand the problem and the people to know what they think and feel, and ultimately to experience what they do and say.

Design Thinking comprises a three-phase approach: EXPLORE ▷ CREATE ▷ IMPLEMENT. As mentioned by Brown (2019), projects may loop back through these steps more than once in an iterative and non-linear way. The exploratory nature of Design Thinking explains this iterative journey. A successful and systematic implementation of Design Thinking requires the identification of the activities to be carried out within each phase (Table 2.3).

The term consumer embraces the person or an organization that could or does receive a product or a service (BSI, 2015). This player, which could be a client, end-user, retailer or receiver, is capable of formulating a set of performance requirements, which are expected to be fulfilled by the product or service. The design of a consumer-centered product should respond to the need of satisfying a set of consumer needs, which are translated into product requirements. Commonly occurring performance requirements involve structure-related, functional and sensorial aspects of the product. In this context, product is defined as the output of an organization that can be produced without any transaction taking place between the organization and the customer. The dominant element of a product is that it is generally tangible. On the other hand, a service is regarded as the output of an organization with

Table 2.3: Activities at each stage of Design Thinking (Ulrich and Eppinger, 2012).

Phase	Purpose/scope	Activity
Explore	involves building empathy, finding out what is wanted, identifying needs, establishing target specifications	benchmarking field work observation
Create	involves making what is wanted, considering the entire design space; creating and creating by modeling	brainstorming prototyping iterations
Implement	involves selecting concepts, further tuning, validating, improving and deploying	prototypes through detailed design testing validation

at least one activity necessarily performed between the organization and the customer. The dominant elements of a service are generally intangible and can often involve activities at the interface with the customer to establish customer requirements as well as upon delivery of the service.

2.1.3 Changes in the network of raw materials and their availability

The dynamic environment of today's world is reflected in the network of current raw materials or feedstocks and their availability to CPI applications. The intricate network of (raw) material resources in the context of the industrial sector is highlighted in the chemical product tree (Gani, 2004a). As depicted in Figure 2.5, this simple and yet powerful representation implies that a countless number of products can be manufactured from a relatively reduced set of raw materials. Climbing along the tree increases not only the number of potential products, but also their molecular complexity and commercial value. Moreover, the availability of predictive models (*e.g.*, for the prediction of physical properties) decreases as we move upwards. As mentioned by Silla (2003), the types of products that can be harvested from this tree might be classified in ten generic categories: chemical intermediates, energy chemicals, foods, food additives, waste treatment chemicals, pharmaceuticals, materials (polymer, metallurgical), personal products, explosives and fertilizers. Chemical intermediates are those chemicals used to synthesize other chemicals (*e.g.*, ethylene, which is produced from hydrocarbons by cracking natural gas either thermally using steam or catalytically and is used to produce polyethylene (45 %), ethylene oxide (10 %), vinyl chloride (15 %), styrene (10 %) and other uses (20 %)). Although these are not sold to the public, they are tremendously relevant to society, as they constitute the building blocks upon which all commercial products are built. Examples of the other processes that result in products of other categories include production of fuels from petroleum or electricity in a steam power plant, biochemical synthesis of bread, synthesis of aspirin or vitamin C, removal of organic matter in municipal and industrial wastewater streams by activated sludge, manufacture of cosmetics, synthesis of thermoplastics (*e. g.*, HDPE), synthesis of nitrocellulose and synthesis of ammonia.

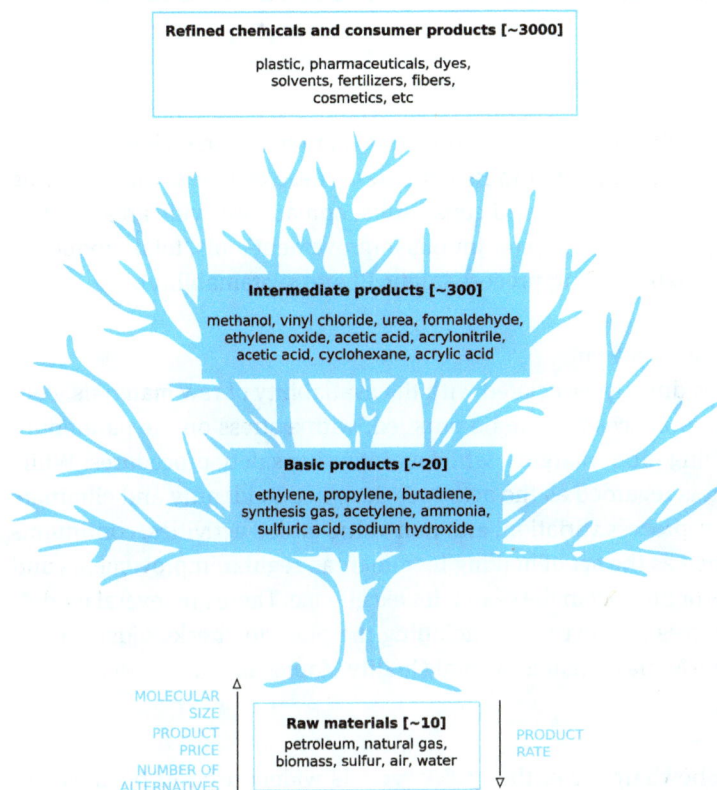

Figure 2.5: Chemical tree of products. Adapted from Gani (2004a). Number of chemicals is indicative.

It is worth noting that the foundation of the chemical tree determines the products and processes to be obtained at the upper ends of the tree. Basic products are commodity of bulk chemicals, produced in large quantities and purchased on the basis of chemical composition, purity and price (Smith, 2015; Costa *et al.*, 2014). Intermediate products, also referred to as fine chemicals, are manufactured in relatively small volumes and – similarly to basic products – are purchased based on their composition, purity and price. The refined chemicals and consumer products, however, are purchased based on their performance rather than their chemical composition. These high-added-value products are commonly produced in small quantities ($<1,000$ t y^{-1}), like perfumes and flavorings, but sometimes also in large volumes, like high-performance polymers and agrochemicals.

The term chemical product used throughout this book embraces all sections of the chemical tree, including base chemicals, intermediate chemicals and refined and consumer products. While this scope rules out mechanical, electronic and electromechanical devices (Bagajewicz, 2007), the extension of design approaches to *all* products is a matter of time.

2.1.4 Technological and operational changes

Over the last few years the chemical industry has experienced large changes in the technologies and approaches related to its activities. These changes, which are the response of current market demand and constraints, spread over wide areas involving manufacturing approaches (*e. g.*, continuous improvement tools, total productive maintenance, quality management, process safety, PI and sustainability).

2.1.4.1 Continuous improvement

Next to considerations directly associated with the availability of raw materials, continuous improvement practices need to be pursued, with a stress on the paramount importance of stretching profit margins, while maintaining safety procedures. While process improvement is regarded as the activities designed to identify and eliminate causes of poor quality, process variation, and non-value-added activities, continuous improvement is defined as the act of making incremental, regular improvements and upgrades to a process or product in the search for excellence. There are several models and strategies for process improvement, including the plan-do-check-adjust (PDCA) cycle, Total Productive Management and Total Quality Management.

PDCA cycle

Also known as the Shewhart cycle, the PDCA cycle is widely used in business for the control and continuous improvement of processes and products (Kiran, 2017). This four-step management method is regarded as an effective problem solving tool with applications in the exploration of new solutions to problems by trying them out and improving them in a controlled way before selecting one for full implementation. Moreover, the PDCA cycle can be applied to planning data collection and analysis, verification and prioritization of problems or root causes and minimization of large-scale waste of resources, among other situations.

The actions of the PDCA cycle in continuous improvement
Plan: In this first step, the practitioner defines what the problem is, devises a systematic way to measure it, gathers data, defines the current process and ensures that the team is working on the correct problem and not merely the symptom of the problem.
Do: This step involves getting to know the probable root cause of the problem. The practitioner determines the reasons how the process could fail and then develops appropriate procedures, or countermeasures, to keep the quality problems from actually happening. After these procedures are devised, education and training should occur with the appropriate members of the process. An implementation

plan should be developed and implemented, and then data should be collected to make sure that the fixes are actually working.

Check: This step is designed to check the results of the procedures and processes that were put into place to correct the initial problem. If the countermeasures are working, the degree of success is measured. If the countermeasures are not working, then new processes need to be designed.

Act: This phase of the PDCA cycle is twofold. If the countermeasures are working, then standardization should occur throughout the entire organization, including the appropriate training and education. If there are any lessons learned from the team, they can be used on the next problem. If the countermeasures have not been successful, the entire process must begin again. It is important to remember that no process stays fixed forever and that you must constantly measure and strive to continually improve the processes to hold your gains and successes.

An illustrative PCDA example is given in Table A.1.

Total Productive Maintenance (TPM)

Total Productive Maintenance (TPM) is a somehow inaccurate translation of the Japanese term Kaizen for improvement (Section 4.3.4), which initially was concerned about maintaining production in an efficient and safe manner by the effective use of people and equipment. TPM can be regarded as a manufacturing improvement strategy within continuous improvement that will drive the business to meet its targets effectively. It aims at restructuring the corporate culture through improvement of human resources, plant, equipment and procedures. By building an organization which prevents every type of loss (*i. e.*, by ensuring zero accidents, zero defects and zero failures) for the life of the production system, TPM's Key Performance Indicators include several areas, as listed in Table 2.4.

Table 2.4: Common Key Performance Indicators of TPM in industry. Legend: OEE, overall equipment effectiveness; SHE, safety, hygiene, environment; LTA, lost time accidents; RWC, restricted work cases.

Area	KPI
Production	OEE [%] Productivity (h/t or h/1000 units) Number of breakdowns (1/y)
Quality	Consumer complaints (1/year) Market auditing index (–)
Cost	Conversion cost (€/t) Loss and gains (%) Maintenance cost (€/t)
Delivery	Comfort to plan (%) Set-up time per product type (min)
SHE	LTA frequency rate (–) RWC frequency rate (–) Volume of effluent (m^3/t)
Morale	One-point lessons (–/y) Trained people (1/y) High-school employees (%)

Total Quality Management (TQM)

The environment in which organizations operate is profoundly different from recent decades. As reported by BSI (2015), today's challenging environment is characterized by accelerated change, globalization of markets and the emergence of knowledge as a principal resource. As a result, society has become better educated and more demanding, making interested parties increasingly more influential. By focusing on quality, an organization promotes a culture that results in the behavior, attitudes, activities and processes that deliver value through fulfilling the needs and expectations of customers and other relevant interested parties. In this context, quality can be understood as the degree to which a set of inherent characteristics of an object fulfills requirements. In fact, the quality of an organization's products and services is determined by the ability to satisfy customers and the intended and unintended impact on relevant interested parties. The quality of products and services includes not only their intended function and performance, but also their perceived value and benefit to the customer. While a wide range of definitions of quality are available, the one by BSI (2015) is globally accepted and adopted within industry. A compendium of relevant definitions is given in Table 2.5.

The impact of quality extends beyond customer satisfaction, as it can also have a direct impact on the organization's reputation. In order to achieve satisfactory quality, the practitioner must be concerned with all three stages of the product or service cycle (Kiran, 2017): (i) the definition of needs; (ii) the product design and conformance; and (iii) the product support throughout its lifetime.

Total Quality Management (TQM) is a formal approach to documenting structure, responsibilities and procedures that will create the ability to achieve effective quality management. It is a method to improve quality while reducing costs and involving the entire organization in continuous improvement. While many definitions exist, TQM can be regarded as a management approach to long-term success through customer satisfaction, based on the participation of all members of an organization striving for improving processes, goods, services and the culture in which they work.

TQM systems are a form of continuous quality improvement and can include problem solving tools, costs of quality, Quality Function Deployment (Section 4.3.4), employee empowerment, Six Sigma, understanding variability, process capability and control and benchmarking. Understanding the needs of the customer is extremely important to drive TQM success. If the practitioners understand what their external customer ultimately wants from the product or service, they can focus on those issues and create and deliver the right type of quality.

> **ℹ** Six Sigma: Motorola developed Six Sigma in the 1980s as an approach to process improvement by reducing variation in its processes. Part of Motorola's approach involved realizing that customers who received a product that had been manufactured correctly the first time were more satisfied than customers who received a product that had internal rework. Six Sigma is currently regarded by many organizations as a method to improve customer satisfaction by reducing variation in the process. Sigma

Table 2.5: Relevant terms as defined by BSI (2015). Figures refer to the section in the ISO 9000:2015 standard.

Term	Definition
Customer (3.2.4)	person or organization that could or does receive a product or a service that is intended for or required by this person or organization
Organization (3.2.1)	person or group of people that has its own functions with responsibilities, authorities and relationships to achieve its objectives
Process (3.4.1)	set of interrelated or interacting activities that use inputs to deliver an intended result
Quality (3.6.2)	degree to which a set of inherent characteristics of an object fulfills requirements
Innovation (3.6.15)	new or changed object realizing or redistributing value
Product (3.7.6)	output of an organization that can be produced without any transaction taking place between the organization and the customer; production of a product is achieved without any transaction necessarily taking place between provider (3.2.5) and customer, but can often involve this service element upon its delivery to the customer; the dominant element of a product is that it is generally tangible; hardware is tangible and its amount is a countable characteristic (*e. g.*, tires); processed materials are tangible and their amount is a continuous characteristic (*e. g.*, fuel and soft drinks)
Service (3.7.7)	output of an organization with at least one activity necessarily performed between the organization and the customer; the dominant elements of a service are generally intangible; service often involves activities at the interface with the customer to establish customer requirements as well as upon delivery of the service and can involve a continuing relationship such as banks, accountancies or public organizations, *e. g.*, schools or hospitals; provision of a service can involve, for example, the following: (i) an activity performed on a customer-supplied tangible product (*e. g.*, a car to be repaired); (ii) an activity performed on a customer-supplied intangible product (*e. g.*, the income statement needed to prepare a tax return); (iii) the delivery of an intangible product (*e. g.*, the delivery of information (3.8.2) in the context of knowledge transmission); or (iv) the creation of ambience for the customer (*e. g.*, in hotels and restaurants)
Improvement (3.3.1)	activity to enhance performance
Continual Improvement (3.3.2)	recurring activity to enhance performance; the process of establishing objectives and finding opportunities for improvement is a continual process through the use of audit findings and audit conclusions, analysis of data, management reviews or other means and generally leads to corrective action or preventive action
Object (3.6.1)	anything perceivable or conceivable
Requirement (3.6.4)	need or expectation that is a stated, generally implied or obligatory
Characteristics (3.10.1)	characteristic distinguishing feature
Defect (3.6.10)	non-conformity related to an intended or specified use; the distinction between the concepts defect and non-conformity is important as it has legal connotations, particularly those associated with product and service liability issues.
Non-conformity (3.6.9)	non-fulfillment of a requirement

stands for standard deviation and is a statistical measure of the performance of a process or product. If a process delivers Six Sigma performance, there will be one defect in 3.4 million opportunities for that defect to occur. Within Six Sigma, a defect is anything that is produced outside of customer specification, and any opportunity for actually touching a part is considered the total chance for a defect. Six Sigma implementations create a system of measurement to achieve world-class performance through process improvement and striving for a near perfection goal. Six Sigma implementation involves the following steps (referred to as DMAIC):

Define: includes identifying and measuring the customer's problems and processes. Improvement goal measurements must be defined and set at this point. Identification of factors critical to the quality of the project will be determined.

Measure: includes gathering the data and information needed to understand the process being studied and the desired output per the customer requirements.

Analyze: uses the data accumulated in the measure phase to determine the causes and effects of the problems that have been discovered.

Improve: involves the development and implementation of solutions to reduce the amount of variation causing the output problems in the process being studied.

Control: involves the definition of provisions to sustain the improvements made in the prior steps. Two additional steps may be added to the control phase, including standardizing the improvements by further documentation and training and ensuring that these steps are built into the process improvement.

According to BSI (2015), Quality Management System (QMS) is a dynamic system that evolves over time through periods of improvement. Every organization has quality management activities, whether they have been formally planned or not. The International Standard ISO 9000:2015 provides guidance on how to develop a formal system to manage these activities, and more importantly to determine whether the activities which already exist in the organization are suitable regarding the context of the organization. A QMS comprises activities by which the organization identifies its objectives and determines the processes and resources required to achieve desired results. Moreover, the QMS manages the interacting processes and resources required to provide value and realize results for relevant interested parties. By embracing QMS in an organization, top management is able to effectively use resources considering the long- and short-term consequences of their decision. A QMS provides the means to identify actions to address intended and unintended consequences in providing products and services.

2.1.4.2 Process safety

Closely related to operational improvement, risk minimization, another process aspect, must also be taken into consideration. In today's world, processes and products must be safe for their complete life span. Major incidents such as Flixborough (1974) with 28 casualties and Bhopal (1984) with 20,000+ casualties may irreversibly affect society's perception of the chemical industry and should stay as a bitter memory of human failure.

The incident of Bhopal and how it could have been prevented

On the 3rd of December 1984 in Bhopal, India the most severe case of fatal release of chemicals took place, involving the toxic chemical methyl isocyanate (Figure 2.6), which was used in the formulation of a specific commercial pesticide.

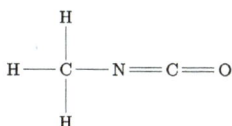

```
      H
      |
H —— C —— N === C === O
      |
      H
```

Figure 2.6: Chemical structure of methyl isocyanate. CASRN: 556-61-6.

The adverse health effects of methyl isocyanate exposure are numerous, as listed in Table 2.6 (Broughton, 2005).

Table 2.6: Health effects of the Bhopal methyl isocyanate gas leak exposure (Broughton, 2005).

Route of exposure/ nature of effects	Effects
Early effects (0–6 months)	
Ocular	Chemosis, redness, watering, ulcers, photophobia
Respiratory	Distress, pulmonary edema, pneumonitis, pneumothorax
Gastrointestinal	Persistent diarrhea, anorexia, persistent abdominal pain
Genetic	Increased chromosomal abnormalities
Psychological	Neuroses, anxiety states, adjustment reactions
Neurobehavioral	Impaired audio and visual memory, impaired vigilance attention and response time, impaired reasoning and spatial ability, impaired psychomotor coordination
Late effects (6 months and onwards)	
Ocular	Persistent watering, corneal opacities, chronic conjunctivitis
Respiratory	Obstructive and restrictive airway disease, decreased lung function
Reproductive	Increased pregnancy loss, increased infant mortality, decreased placental/fetal weight
Genetic	Increased chromosomal abnormalities
Psychological	Neuroses, anxiety states, adjustment reactions
Neurobehavioral	Impaired associate learning, motor speed, precision

In the vicinity of a factory where this extremely toxic chemical was manufactured, an accidental release of about 40 tonnes to the atmosphere took place. The cloud generated by the release of the toxic chemical covered an area of *ca.* 300 ha, exposing nearby inhabitants to concentrations – in most cases – significantly above the lethal thresholds (PEL: TWA 0.02 ppm [0.05 mg/m^3] [skin]; REL: TWA 0.02 ppm [0.05 mg/m^3] [skin], IDLH: 3 ppm, as reported in NIOSH, 2012). Although there is a big discrepancy in the number of fatalities and long-term ill casualties stated in lawsuits, technical reports, disputed claims and the final settlement, the world's worst industrial disaster is responsible for a death toll of up to 20,000 (Lerner and Lerner, 2004; Varma and Varma, 2005), and more than 550,000 others were seriously injured (Broughton, 2005).

Professor Robert M. Solow, Nobel Prize laureate in Economic Sciences in 1987, stated that the rate of economic growth is mainly dependent on the rate of technological progress, despite the potential resource scarcity. This theory is certainly of relevance to chemical industry growth as its practices have been gradually shifted towards more complex manufacturing technologies, involving chemicals with increasing reactivity under more extreme operating conditions (Crowl and Louvar, 2011).

An effective embracement of chemical process safety is fundamentally relevant to cope with the increasing complexity of chemical process technology and its associated challenges. Examples of technical advances in this area include the development of dispersion models to simulate the spread of toxic vapor after release (Figure 2.7) and the development of source models and probability-based models for equipment and process failure.

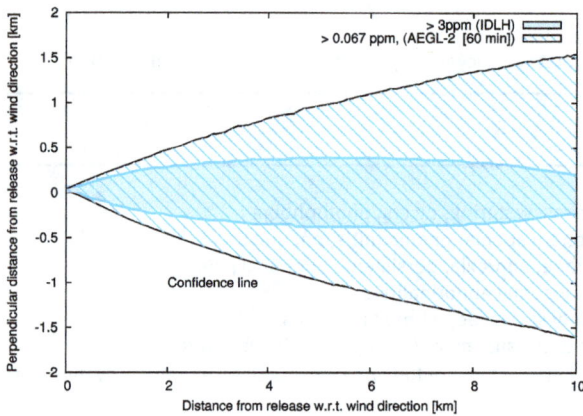

Figure 2.7: Simulation of the release of a cloud of toxic methyl isocyanate (CASRN: 556-61-6) generated from the release of 40 tonnes. Data: wind speed, 10 m/s East-Southeast; ambient temperature, 25 °C. Legend: IDLH, immediately dangerous to life and health concentration; AEGL-2, acute exposure guideline level, above which it is predicted that the general population, including susceptible individuals, could experience irreversible or other serious, long-lasting adverse health effects or an impaired ability to escape.

In the context of this section, the term safety is considered to be interchangeable with the term loss prevention, which includes not only the strategy of accident prevention through personal protective equipment, rules and regulations, but also the identification of hazards and the technical evaluation and design of new engineering features to prevent loss (Crowl and Louvar, 2011). The definitions of relevant concepts are included in Table 2.7.

In accordance with the Engineering Ethics Statement, practitioners within an enterprise are responsible for performing services that minimize losses, provide a safe and secure environment for the enterprise's employees and by doing so maintain and improve the company's bottom line.

Table 2.7: Relevant safety-related definitions (Crowl and Louvar, 2011; Hughes and Ferret, 2010).

Term	Definition
Safety and loss prevention	the prevention of accidents through the use of appropriate technologies to identify the hazard of a chemical plant and eliminate them before an accident occurs
Hazard	the potential of a substance, person, activity or process to cause harm or damage to people, property or the environment. Hazards normally found in chemical plants include mechanical hazards and chemical hazards (fire, explosion, reactive and toxic hazards)
Risk	the likelihood of a substance, activity or process to cause harm and its resulting severity or magnitude of the loss or injury; harm could include human injury, environmental damage or economic loss. Risk is the product of the probability of a release or exposure and the consequences of the exposure
Consequence	a measure of the expected effects of the results of an incident

Contrary to the societal perception on chemical process safety, chemical plants are among the safest of all manufacturing facilities. As shown in Table 2.8, the recent statistics of fatal injuries and rate of fatal injuries in the chemical industry in UK and USA are relatively comparable with the average of all manufacturing activities and substantially lower than fatalities in construction, transportation and agriculture-related activities.

Table 2.8: Number and rate of work-related fatalities in USA in 2017 and UK in 2017 (BLS, 2016; HSE, 2017). Note: Fatal injury rates are per 100,000 full-time equivalent workers.

Activity or sector	USA: number	USA:rate	UK:number	UK: rate
Agriculture/forestry/fishing/hunting	581	23.0	27	7.61
Transportation	882	15.1	14	0.88
All production industries	221	2.6	135	0.43
Chemical industry activities	112	–	18	0.63

The reason why there is a societal concern on safety within the chemical industry, despite the favorable statistics, resides on the industry's potential for high numbers of casualties. In fact, with the increase in the use of more reactive chemicals and in the scale of operations (*i. e.*, production capacity of chemical plants), the potential of fatal injuries per event has increased. As reported by Crowl and Louvar (2011), despite all efforts by the chemical industry to improve safety, the estimated financial loss related to safety incidents increases steadily, doubling every decade.

Achieving an appropriate level of safety requires investment. As with any feature associated with products or services, there is a minimum level beyond which the implementation of the feature is noticeable. Beyond a minimum level of investment in safety, losses associated with safety incidents decrease, increasing the overall enterprise's financial performance. This expenditure is transferred to the manufacturing cost of the product or service. Further increases in safety expenditure, on the other hand, could turn the pricing of the product or service uncompetitive. Safety implementation resides on gauging properly the risk to be accepted and the financial consequences of the expenditure to achieve such risk. Tools such as QRA and LOPA are widely used to support the practitioner in the assessment of the level of risk to be accepted while maintaining healthy profitability.

---

i **Tools to assess the level of acceptable risk and unacceptable risk** (Crowl and Louvar, 2011) QRA: Quantitative Risk Analysis. This method allows the identification of areas where operations, engineering or management systems can be modified to reduce risk. It provides practitioners with a tool to evaluate the overall risk of a process. The key steps of a QRA study include: (i) define potential event sequences and incidents; (ii) evaluate incident consequences; (iii) estimate potential incident frequencies using event trees and fault trees; (iv) estimate incident impacts on people, environment and property; and (v) estimate the risk by combining the impacts and frequencies.

LOPA: Layer of Protection Analysis. This method is semi-quantitative in nature and includes simplified methods to characterize the consequences and estimate the frequencies of an incident. The primary purpose is to determine whether there are sufficient layers of protection against a specific accident scenario. The key steps of a LOPA study include: (i) identification of a single consequence; (ii) identification of an accident scenario and a cause associated with the consequence; (iii) identification of the initiating event for the scenario and estimation of the initiating event frequency; (iv) identification of the protection layers available and estimation of the probability of failure on demand for each protection layer; (v) plot of the consequence *versus* the consequence frequency to estimate the risk; and (vi) evaluation of the risk for acceptability. The consequences of incidents, mainly derived from loss of containment of hazardous materials in the CPI, can be estimated using the following methods: (i) semi-quantitative approach without the direct reference to human harm; (ii) qualitative estimates with human harm; and (iii) quantitative estimates with human harm. The LOPA tool involves the addition of various levels of protection to a process. These layers may include inherently safer concepts, basic process control systems, safety instrumented functions, passive devices, active devices and human intervention (Figure 2.8). The combined effects of the protection layers and the consequences are then compared against some risk tolerance criteria. The consequences and effects are approximated by categories, the frequencies are estimated and the effectiveness of the protection layers is approximated. The criteria to be used to establish the boundary between acceptable and unacceptable risk include frequency of fatalities, frequency of fires, maximum frequency of a specific category of a consequence and the required number of independent layers of protection for a specific consequence category.

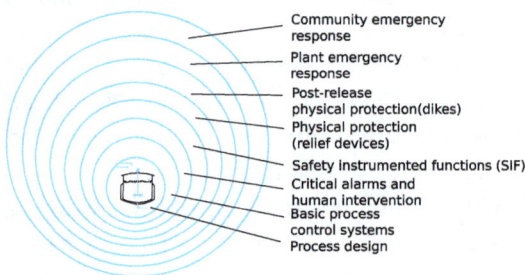

Community emergency response
Plant emergency response
Post-release physical protection(dikes)
Physical protection (relief devices)
Safety instrumented functions (SIF)
Critical alarms and human intervention
Basic process control systems
Process design

Figure 2.8: Layers of protection to lower the frequency of a specific accident scenario.

Generally, there are three types of accidents that occur in the chemical industry (Table 2.9): fire, explosions and toxic release. As mentioned in the report The 100 Largest Losses 1974–2015 (MARSH, 2016), explosion accidents are responsible for the largest business interruption loss since 1974, accounting for nearly two-thirds of the total losses (*ca.* US$33 billion). These are typically vapor-cloud explosions that occur following the loss of containment of light hydrocarbons. Fire accounts for *ca.* 13 % of the total property damage, while toxic release represents *ca.* 1 %.

Table 2.9: Key types of accidents in the chemical process industry, their probability of occurrence and their potential impact (Crowl and Louvar, 2011).

Type of accident	Probability of occurrence	Fatalities	Economic loss
Fire	High	Low	Intermediate
Explosion	Intermediate	Intermediate	High
Toxic release	Low	High	Low

While estimating the probability and consequence of a hazard in an industrial operation is crucial, the first question to be asked is whether the hazard can be eliminated in the first place (Kletz, 1998). Addressing safety considerations during the early phases of process design is fundamentally relevant in view of the implications for society, environment and property. It is at this early phase that designers and practitioners have at their disposal sufficient degrees of freedom in the specifications of operational windows, driving forces, equipment sizing and operations to accommodate for departures from ideal performance. In this context, the term performance entails equipment reliability and operators behavior. As mentioned by Siirola (1996), the decisions made at the conceptual design phase of chemical plants, which account for *ca.* 2 %–3 % of the project costs, fix approximately 80 % of the combined capital and operational costs of the final plant. Considering safety requirements during the design phase will result in plants which are inherently safe, rather than relying on control systems, interlock or special operating procedures to prevent accidents effectively (Crowl and Louvar, 2011). Avoiding hazards – rather than controlling them – can be carried out by adopting one or more of the main strategies shown in Table 2.10.

Referring to the disaster in Bhopal in 1984, PI (Section 2.1.4.3) could have offered an elegant strategy to minimize the inventory of hazardous materials without fundamental changes in the manufacturing technology (Stankiewicz and Moulijn, 2002, 2004). In fact, Stankiewicz and Moulijn (2004) refer to a recent study that showed that methyl isocyanate could be generated and immediately converted to final products in continuous reactors that contained a total inventory of less than 10 kg of the insecticide intermediate. Intensification strategy, when applicable to a design case, does not only lead to inherently safer alternatives, but also results in significant reductions in cost by miniaturizing the required facilities.

Table 2.10: Key strategies to avoid hazards in chemical process industry (Crowl and Louvar, 2011; Kletz, 1998).

Strategy	Approach
Intensification or minimization	reduce quantities of hazardous materials in reactors, separation drums and storage vessels; produce and consume *in-situ* hazardous intermediates; and reduce the inventory of hazardous materials
Substitution	use a safer material in place of the hazardous materials, by exploring alternative chemistry
Attenuation or moderation	use hazardous materials under less hazardous conditions by dilution, refrigeration, agglomeration or milder operational window
Limitation of effect by simplification	keep system neat, easy to follow and maintain

As mentioned by Kletz (1998), the concept of inherently safer design needs to be regarded in conjunction with other desirable criteria such as cost effectiveness, energy usage and process complexity. One of the challenges of the chemical industry for the years to come is the active embracement of technological measures to assure a satisfaction of all four criteria (safety/cost/energy/complexity). Other strategies to be adopted towards inherently safer design include avoidance of knock-on effects, unfeasibility of incorrect assembly and ease of control (Kletz, 1998).

2.1.4.3 Process intensification

The concept of process intensification (PI) has awaken increasing interest since its introduction in the chemical engineering community. While several definitions have been suggested since the 2000s (Table 2.11), it is evident that the scope of the concept has been widened far beyond the miniaturization hallmark. In fact, PI comprises novel equipment, process techniques and process development methods that, when compared to conventional ones, offer substantial improvements in chemical manufacturing and processing.

While initially restricted to the potential benefits in terms of the reduction of plant or equipment sizes, PI has demonstrated unparallel value in the reduction of process costs, prevention of runaway reactions, better mass, heat and momentum transfer, improved process safety and minimized waste production. The reduction of process costs is driven by the reduction of investment costs as a result from equipment compactness, increase in yield and selectivity, enhancement of energy efficiency and reduced waste processing. From a process safety standpoint, PI anticipates an elegant strategy to minimize the inventory of hazardous, toxic and flammable materials. As suggested by Stankiewicz and Moulijn (2000), PI can be seen as divided into two frequently overlapping areas:
- PI equipment, including novel reactors and devices for intensive mixing, heat transfer and mass transfer;

Table 2.11: Selected definitions of PI as summarized by Tsaoulidis (2015), Gerven-van and Stankiewicz (2009) and Lutze and Górak (2016).

Reference	Definition
Ramshaw (1983)	devising exceedingly compact plants which reduce both the main plant item and the installations costs
Cross and Ramshaw (2000)	strategy of reducing the size of chemical plant needed to achieve a given production objective
Stankiewicz and Moulijn (2000)	development of innovative apparatuses and techniques that offer drastic improvements in chemical manufacturing and processing, substantially decreasing equipment volume, energy consumption or waste formation, and ultimately leading to cheaper, safer, sustainable technologies
Tsouris and Porcelli (2003)	technologies that replace large, expensive, energy-intensive equipment or processes with ones that are smaller, less costly, more efficient or that combine multiple operations into fewer equipment
Becht *et al.* (2009)	an integrated approach for process and product innovation in chemical research and development and chemical engineering in order to sustain profitability even in the presence of increasing uncertainties

–   PI methods, including multi-functional reactors, techniques using alternative energy sources (*e. g.*, light, ultrasound, EM energy) and new process-control methods.

Since its introduction to and embracement by the chemical process industry (CPI), PI strategy has resulted in the creation of *smart* units currently available in a wide range of manufacturing activities. In all instances, the intensified unit is characterized by drastic improvements – sometimes of an order of magnitude – in production cost, process safety, controllability, waste or byproducts generation, time-to-market and societal acceptance (Stankiewicz, 2003; Stankiewicz and Moulijn, 2002). Examples of such intensified units are given in Table 2.12.

On the other hand, the methods derived from PI can be classified in three main areas (Table 2.13): multi-functional reactors, hybrid separations and the use of alternative sources of energy. It is worth mentioning that current technological advances have resulted in the emergence of other methods within PI. These approaches have been commercially proven and include, for instance, the use of supercritical fluids in the processing of natural products (Khalloufi *et al.*, 2010; Almeida-Rivera *et al.*, 2010b), cryogenic separation techniques and dynamic operation of chemical reactors.

The challenges in the design of intensified processes are numerous and exciting. The drive for PI and more sustainable plant operations can lead to more intricate geometric structures (*e. g.*, packings, dividing walls, distributed heat exchange elements) and to more refined models, which couple mass and energy transfer in a more realistic way. Aiming at cheaper, cleaner, safer and simpler manufacturing of products, PI

Table 2.12: Selected units originating from the PI strategy (Stankiewicz and Moulijn, 2000).

Equipment	Key features	Applications
Static-mixer reactors	geometric mixing elements fixed within a pipe; advantages: more size- and energy-efficient methods for mixing or contacting fluids; disadvantage: high sensitivity to clogging by solids	nitration and neutralization reactions
Monolithic reactors	metallic and non-metallic bodies with a multitude of straight narrow channels of defined uniform cross-sectional shapes; advantages: very low pressure in single- and two-phase flow, high geometrical areas per reactor volume, high catalytic efficiency, relatively high performance, low investment costs, compact plant layout, multiple-feed distribution, near-to-plug-flow behavior; disadvantages: heat removal for gas-phase catalytic processes	process where selectivity is limited by mass transfer resistances
Microreactors	feature extremely small dimensions in a sandwich-like structure consisting of a number of layers; advantages: integration of tasks (e. g., catalytic reaction, heat exchange and separation), high heat transfer rates, very low reaction volume/surface area ratios; disadvantages: economic feasibility	processes involving toxic or explosive reactants
Spinning disk reactors	very thin layer of liquid moves on the surface of a disk spinning at high RPM; advantages: high heat transfer coefficients	(very) fast liquid/liquid reactions (e. g., nitrations, sulfonations and polymerizations)
HIGEE devices	mass transfer operations carried out in rotating packed beds in which high centrifugal forces occur; advantages: enhancement of heat and momentum transfer; disadvantages: limited to specific applications	separation processes (e. g., absorption, extraction, distillation), G-L systems

faces various challenges, such as the minimization of chemicals inventory and the simplification of equipment complexity. Disasters such as the release of 40 metric tons of methyl isocyanate in Bhopal would never have happened if a conscious safety-oriented process had been designed.

---

i **Reactive distillation** – an example of the PI approach

Reactive distillation (RD) is a hybrid operation that combines two of the key tasks in chemical engineering, reaction and separation. The first patents for this processing route appeared in the 1920s, *cf.* Backhaus (1921a,b,c), but little was done with it before the 1980s Malone and Doherty (2000), Agreda and Partin (1984), when RD gained increasing attention as an alternative process that could be used instead of the conventional sequence chemical RD.

Table 2.13: Selected methods of process intensification (Stankiewicz and Moulijn, 2000).

Method	Scope
Multi-functional reactors	enhancement of chemical conversion by combining at least one more function (the definition of Task in Section 6.1) within the unit. Advantages: optimal utilization of heat of reaction, higher yields by continuous (and *in-situ*) removal of reaction products and equilibrium shift, controlled feed distribution. Examples: reverse-flow reactors, rotating monolith reactors, reactive distillation units (explanatory note *Reactive distillation – an example of the PI approach*), reactive crystallizers, reactive extruders, chromatographic reactors, membrane reactors and fuel cells.
Hybrid separations	integration of two or more separation mechanism. Advantages: compact equipment, limited occurrence of entrainment, flooding, channeling, or foaming, low operation pressure across unit and limited membrane fouling. Examples: membrane-assisted separations (absorption, distillation) and adsorptive distillation.
Alternative use of energy	use of centrifugal forces, ultrasound, solar energy, microwave energy, electric fields and plasma. Advantages: high local energy release, controlled droplet breakup.

The RD synthesis of methyl acetate by Eastman Chemicals is considered to be the textbook example of a task integration-based process synthesis (Stankiewicz and Moulijn, 2002; Stankiewicz, 2003, 2001; Li and Kraslawski, 2004; Siirola, 1996) (Fig. 2.9).

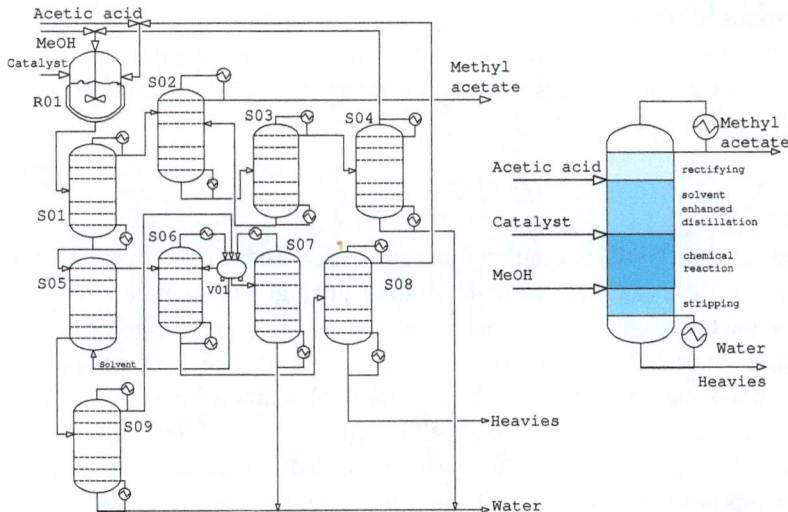

Figure 2.9: Schematic representation of the conventional process for the synthesis of methyl acetate (left) and the highly task-integrated RD unit (right). Legend: R01, reactor; S01, splitter; S02, extractive distillation; S03, solvent recovery; S04, MeOH recovery; S05, extractor; S06, azeotropic column; S07, S09, flash columns; S08, color column; V01, decanter.

Using this example one can qualitatively assess the inherent value of this processing strategy. The process costs are substantially reduced (~80 %) by the elimination of units and the possibility of heat integration. Using task integration-based synthesis the conventional process, consisting of 11 different steps and involving 28 major pieces of equipment, is effectively replaced by *one* highly task-integrated RD unit.

2.1.4.4 Sustainability

As a result of recent awareness of the human intervention on the environment, society is increasingly demanding sustainable processes and products. It is no longer innovative to say that the chemical industry needs to take into account sustainability of the biosphere. In current times, organizations are expected to actively embrace and apply sustainable manufacturing strategies. Since the 1980s, the global natural resource use has accelerated remarkably, while economic growth and population growth have been slowing (UNEP, 2016).

i   While there are countless definitions of sustainability, the one introduced by Brundtland (1987) is regarded as the most widely accepted. It is defined as the development that meets the needs of the present generations without compromising the ability of the future generations to meet their own needs.

In fact, overall the global economy has expanded more than threefold over the four decades since 1970, while the population almost doubled and global material extraction tripled. At the current unsustainable pace of use of natural resources, our generation will be responsible for the irreversible depletion of some resources and for critical damage to the planet. This accelerated growth in material use has been fueled by the growth in capital income and consumption, especially since 2000, and is conceptualized by the expression

$$I = P \times A \times T, \tag{2.1}$$

where I represents the material use (either domestic material consumption or material footprint), A the GDP per capita, T the material intensity and P the population.

The true impact of wealthy economies on global material use can be expressed by introducing the new indicator: material footprint of consumption. This metric reports the amounts of materials that are required for final demand (consumption and capital investment) in a country or region (UNEP, 2016) and is an effective proxy for the material standard of living (Fig. 2.10). By the end of 2010 the material footprint of Europe's consumption was estimated to be around 20 tonnes per capita and that of North America around 25 tonnes per capita.

As global economy is already surpassing some environmental thresholds or planetary boundaries at current levels of resource use, it is evident that in the foreseeable future there will be a resource deficit. In fact, in a recent analysis (Radjou and Prabhu,

2015) it has been predicted that at current rates of production and consumption, by 2030 we would require two planets to supply the resources humankind needs and to absorb the waste it generates. The current demand for energy, food, water and materials will drive up prices for natural resources, turning sustainability a decisive factor in the world's future economy. As embracing sustainable consumption and production patterns will enable an efficient resource use and the reduction of the impact of economic activities on the environment, it has been identified as one of the 17 goals of the 2030 Agenda for Sustainable Development (BSDC, 2017; UN, 2017).

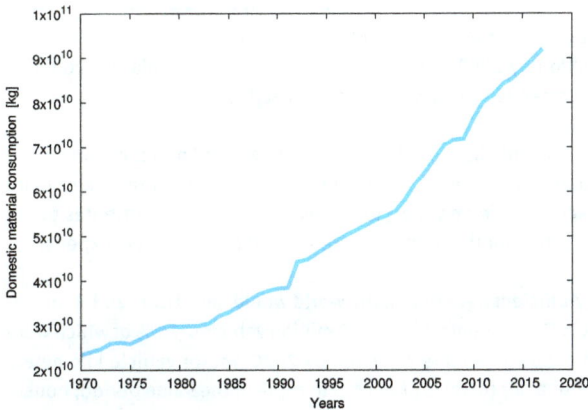

Figure 2.10: Effect of societal economy on global material use expressed in terms of material footprint of consumption (UNEP, 2016).

As highlighted in a recently published report by the United Nations (UN, 2017), there is a clear misalignment between current consumption and production patterns. The scientific community has agreed that we have a short window of opportunity to act aggressively to transition away from unsustainable approaches and engage in a drastic shift of society's consumption to lower levels supported by drastically different manufacture approaches.

From a manufacturing perspective, the road towards sustainable manufacturing within an organization not only involves the actual manufacture of a product or service, but spans over the design phase and the product life cycle management. Sustainable manufacturing can be regarded, in this context, as the economical creation of goods and services using technology and processes that are non-polluting, conserve energy and natural resources and are safe for employees, communities and consumers. The development of sustainable manufacturing approaches is fueled by drivers such as (i) the increasing costs of many highly demanded raw materials; (ii) the increasing waste emissions from manufacturing activities and their adverse impact on the environment; (iii) the constrained energy supply; and (iv) the social impact of manufacturing on workers' health and safety.

Despite these drivers, there are a set of internal and external barriers that affect the development of sustainable manufacturing practices within an organization. The internal barriers include the financial impact of implementing new methods, the level of preparedness of the organization to deal with new approaches, inherent failure risks associated with the introduction of new materials and/or processes and the gap in knowledge base of the personnel responsible for handling the changes in the organization's system. On the other hand, the external barriers include local/regional regulations, status of the infrastructure of the industry and legal framework concerning external competition, among others.

---

i **Sustainable consumption and production patterns (SDG-12): Facts and figures**

If the global population reaches 9.6 billion by 2050, the equivalent of almost three planets could be required to provide the natural resources needed to sustain current lifestyles.

Water

Less than 3 % of the world's water is fresh (drinkable), of which 2.5 % is frozen in Antarctica, Arctic and glaciers; man is polluting water faster than nature can recycle and purify water in rivers and lakes; more than 1 billion people still do not have access to fresh water; excessive use of water contributes to the global water stress; water from nature is free but the infrastructure needed to deliver it is expensive.

Energy

If people worldwide switched to energy-efficient light bulbs, the world would save USD 120 billion annually; in 2002 the motor vehicle stock in OECD countries was 550 million vehicles (75 % of which were personal cars); a 32 % increase in vehicle ownership is expected by 2020; motor vehicle kilometers are projected to increase by 40 % and global air travel is projected to triple in the same period; households consume 29 % of global energy and consequently contribute to 21 % of resultant CO_2 emissions; the share of renewable energy in final energy consumption has reached 17.5 % in 2015.

Food

Each year, an estimated one-third of all food produced – equivalent to 1.3 billion tons, worth around US$1 trillion – ends up rotting in the bins of consumers and retailers or spoiling due to poor transportation and harvesting practices; approximately 2 billion people globally are overweight or obese; land degradation, declining soil fertility, unsustainable water use, overfishing and marine environment degradation are all lessening the ability of the natural resource base to supply food; the food sector accounts for around 30 % of the world's total energy consumption and accounts for around 22 % of total GHG emissions.

Embracing sustainability within a manufacturing system is not a trivial task as it requires the organization to move from an operational space into a strategic level. At the early phases, sustainability-related aspects might be considered as add-on elements to processes. Analysis tools such as LCA are normally performed at this discovery phase. As the level of commitment increases, the organization moves into a opportunistic phase. At this phase sustainability-related aspects are gradually embedded within the organization's processes as significant benefits and improvements are identified. The last phase – the strategic phase – is characterized by the development of science behind sustainability in the organization's processes. Sustainability-related aspects are engrained in procedures, policies and instructions and in the culture of the organization.

Next to the embracement of sustainability within an organization's procedures, processes and culture, sustainable manufacture requires the definition of metrics. A set of metrics is expected to be valid over a long period of time, to be representative of the sustainability targets of the organization and to be properly understood and embraced by all interested parties within the organization. Commonly used metrics are related to economic, environmental and social factors (3P or Triple Bottom Line: People-Planet-Profit). In this framework the financial concept of bottom line – conventionally used by corporations – is expanded to include two other performance areas: social and environmental impact of the organization. As depicted in Figure 2.11, sustainability resides in the intersection of people, planet and profit areas.

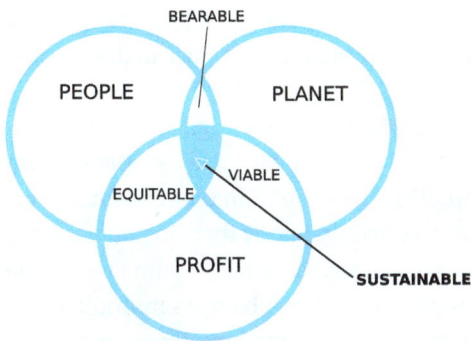

Figure 2.11: Sustainability in the context of Triple Bottom Line.

Examples of each of those factors are R&D investments in clean technologies and cost competitiveness, resource consumption and energy consumption, and worker health and safety and ethics.

The zero waste goal

The concept of waste has been widely used in the context of manufacturing systems as referring to any non-value-adding activity (NVA) (Gao and Low, 2014). Wasteful activities absorb resources while creating no value to the organization or production system. Waste – sometimes referred to as *muda* – can include activities involving excess of inventory, overproduction, unnecessary motion, movement and transportation of inventory, unnecessary processing or cycle times, unnecessary waiting, motion, movement and checking, defects, rework, rejects, scrap and – last but not least – unused employee creativity (Moore, 2007). Between 1948 and 1975 the Toyota Production System emerged as an integrated and systematic approach to improve the post-war operations of Toyota Motor Company. Being the precursor of Lean Management, TPS focuses on elimination of waste (muda) from the process while designing out overburden (muri) and inconsistency or unevenness (mura) (Eaton, 2013). By developing processes that are capable of delivering the required outputs as smoothly, flexibly and free of stress as possible, a minimum amount of resources is required. Seven categories of waste have been identified in the context of TPS (Ohno, 1988; Gao and Low, 2014): transportation, inventory, motion, waiting, overproduction, overprocessing and defect products (conventionally referred to as TIM WOOD).

A zero waste goal focuses on eliminating *all* the waste generated in any process within a manufacturing system. Approaches towards waste elimination in general, and material conservation in particular, include design efficiency and simplification.

2.1.4.5 Linear *versus* circular economy

In recent years the concept of circular economy has received increasing interest throughout all dimensions of human activities. Contrary to the linear architecture (steps: take ▷ make ▷ dispose), a circular economy (steps: reduce ▷ reuse ▷ recycle) is an economic system where products and services are traded in closed loops or cycles. In a circular economy, products do not have a life cycle with a beginning, middle and end, and, therefore, contribute less waste and add value to their ecosystem. This architecture is characterized as an economy which is restorative and regenerative by design and aims to keep products, components and materials at their highest utility and value at all times (MacCarthur-Foundation, 2015).

While the linear model is based on the principle that things (*e. g.*, materials, products, services) are disposed as waste when they are no longer needed, the circular economy considers that nothing useful should go to waste. Circular economy aims at decoupling economic growth from resource consumption by focusing on value retention. Value retention is achieved by keeping material streams as pure as possible during the complete value chain. Pure material streams can be used multiple times to provide a certain functionality or service, while incurring in investment only once. The principles and fundamental characteristics of circular economy drive four sources of value creation, as defined in Table 2.14.

Table 2.14: Sources of value creation in a circular economy (MacCarthur-Foundation, 2015).

Source	Scope
The power of the inner circle	the tighter the circle, the more valuable the strategy; inner circles preserve more of a product's integrity, complexity and embedded labor and energy
The power of circling longer	maximizing the number of consecutive cycles and/or the time in each cycle for products; each prolonged cycle avoids the material, energy and labor of creating a new product or component. For products requiring energy, though, the optimal serviceable life must take into account the improvement of energy performances over time
The power of cascaded use	diversifying reuse across the value chain
The power of pure inputs	contaminated material streams increase collection and redistribution efficiency while maintaining quality, particularly of technical materials, which in turn extends product longevity and thus increases material productivity

As depicted in Figure 2.12, the material cycles are closed, which allows for the use of all residual streams as valuable resources. Upon use all products are repaired or remanufactured to reuse them a second, third or fourth time.

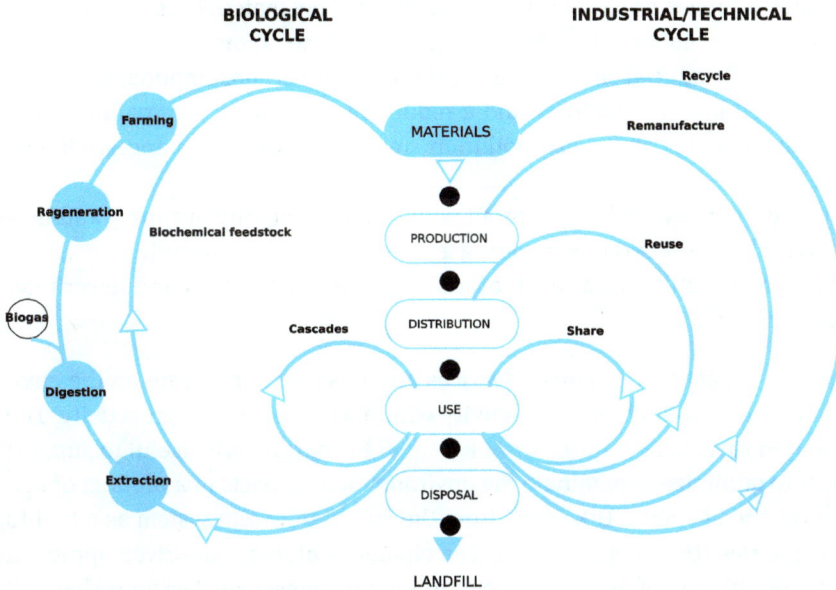

Figure 2.12: Schematic representation of circular economy. Modified from the Butterfly diagram by MacCarthur-Foundation (2015).

The model proposes two types of cycles, depending on the nature of the residual materials streams: biocycle and techno-cycle (MacCarthur-Foundation, 2015). As a clear differentiation is suggested between organic materials and synthetic and technical materials, the model relies on an efficient separation of biological and technical materials after use. Technical materials (*e. g.*, fossil fuels, plastics and metals) are finite and cannot be renewed. The techno-cycle involves the management of stocks of finite materials, where use replaces consumption and technical materials are recovered and mostly restored. Organic materials (*e. g.*, cotton, food and water) can be taken up in the ecosystem by means of biological processes. The biocycle encompasses the flows of renewable materials, which is enhanced by enabling the proper functionality of the ecosystem and biological processes. Consumption occurs within the cycle, provided that no contamination occurs with toxic substances and no ecosystem overload takes place. When the ecosystem is in proper balance, renewable (biological) materials are mostly regenerated.

The attraction that circular economy has gained in the last years resides on the significant opportunities it offers to industry and, by extension, to society. According to the study by MacCarthur-Foundation (2015), transitioning to a circular economy would represent a net benefit of *ca.* €2.8 trillion by 2030 in Europe only, which is *ca.* €0.9 trillion more than the current linear model. The opportunities and impact of circular economy expand over several aspects: economic, environmental, societal and entrepreneurial. These include the following specific opportunities:

- Economic: improved economic growth, substantial net material cost savings, creation of employment opportunities and increased innovation.
- Environmental: reduced emissions and primary material consumption, preserved and improved land productivity and a reduction in negative externalities (*e. g.*, land use, air, water and noise pollution, release of toxic substances, climate change).
- Entrepreneurial: new and bigger profit pools, greater security in supply and new demand for business services, building greater resilience as a result.
- Societal: greater utility as a result of more choice, lower prices and lower total cost of ownership.

Addressing the impact that products, services and businesses generate on the environment is a fundamentally relevant activity within a conscious design activity. The method referred to as Design for Environment (DfE) is particularly useful to support the attempts to minimize or eliminate the environmental impacts of a product or service over its entire life cycle. DfE stems from the cradle-to-cradle system as a tool to evaluate its progress (Rossi *et al.*, 2006). The challenge of this goal-driven approach involves the formulation of appropriate environmental-conscious decisions that will maintain or improve product quality with no negative cost impact, while reducing the environmental impact (*i. e.*, global warming, resource depletion, solid waste generation, air and water pollution and land degradation). Such decisions – as building blocks of DfE – should not be formulated separately, but embedded systematically within a product-process design approach. As a result of the increasing attention to environment-related characteristics of products and services, comprehensive lists of DfE guidelines have been generated. A selected sample of these guidelines is given in Table 2.15.

Table 2.15: Sample of DfE guidelines in the context of product and process design (Ulrich and Eppinger, 2012).

Area	Guideline	Impact
Materials	Use industrial materials that can be recycled continually	resource depletion, land degradation
Materials	Use natural materials that can be fully returned to Earth's natural cycles	resource depletion, solid waste, water and air pollution
Process	Avoid processes that produce unnatural, toxic materials	solid waste, water and air pollution
Energy	Use clean, renewable sources of energy	global warming, resource depletion, land degradation

Addressing all these process aspects, given the underlying aim of coping effectively with the dynamic environment of short process development and design times, has resulted in a wide set of technical responses. Examples of these responses include advanced process control strategies, real-time optimization and the synthesis of hybrid and intensified units. Focusing on the last case, by integrating several functions (or tasks as defined in Section 6.1) intensified units lead to substantial increases in process and plant efficiency (Grossman and Westerberg, 2000; Stankiewicz and Moulijn, 2002) and can be characterized by four terms: smaller, cheaper, safer and slicker (Stankiewicz and Moulijn, 2004). Embracing this four-dimensional strategy enables industry to operate in the broader context of sustainable technological development and safety, to facilitate short time-to-market and, last but not least, to minimize operational costs associated with scale-up activities.

2.2 Product-centered process design

It is now well established by industry and academia that the chemical industry focus has shifted from a process-centered orientation to a product-centered one (Hill, 2004; Bagajewicz, 2007). In recent decades the chemical industry has experienced a fundamental transformation in its structure and requirements (Edwards, 2006). In fact, during the last decades we have experienced how the commodity chemical business is gradually releasing its dominating role towards higher-added-value products, such as specialty chemicals, pharmaceuticals and consumer products. As shown in Figure 2.13, the footprint of revenues of the chemical industry by product type has evolved to respond to unprecedented market demands and constraints. The chemical industry has been gradually concentrating its activities and streamlining its assets towards value creation and growth by engaging in the manufacture of high-value chemical products.

Zooming exclusively in the production growth by sector in the EU (Figure 2.14), the specialty sector is clearly registering a steady increase in its growth year-over-year. Slightly less pronounced behavior has been reported in the consumer chemicals and polymer sectors. In all these cases, the performance by sector is exceeding the average of the entire category (*ca.* 0.8 % y-o-y). While still remaining an extremely relevant sector in terms of revenue, the petrochemical sector has registered a stagnant (or even no) growth during the last years in the EU. As a result of the relentless global competition with other markets, petrochemical sector growth in the EU – if any – is expected to be modest in the years to come and accentuated by the prominent role of other emerging economies. Figure 2.15 shows the average production growth of chemicals in key geographies. It is evident that the global landscape is becoming dominated by emerging economies outpacing industrial economies, and partly explained by the shift of manufacturing to Asia with its associated higher chemicals output growth.

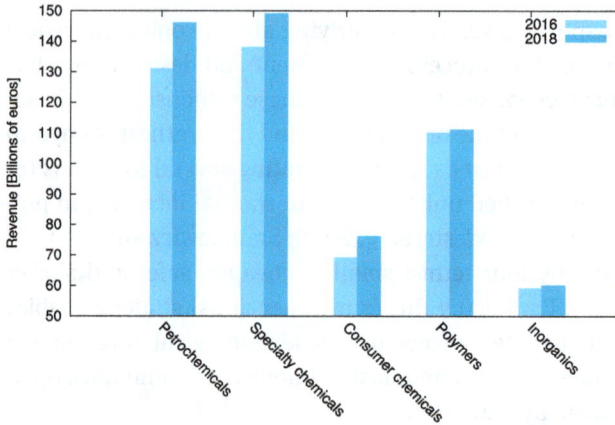

Figure 2.13: EU revenues of chemical products sales reported in the period 2016–2018 (CEFIC, 2016, 2018). Note: Pharmaceuticals are excluded.

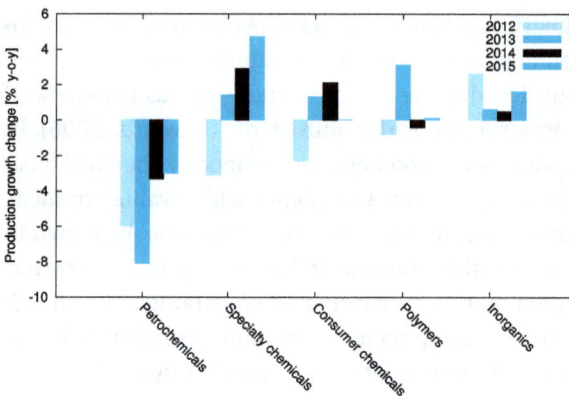

Figure 2.14: Production growth of chemicals by sector in the EU (CEFIC, 2016). Note: Pharmaceuticals are excluded.

Another proxy to stress the transformation in the structure and requirements of the chemical industry is given by the analysis of R&D expenditure by regions and sector. As depicted in Figure 2.16, Asia has experienced a strong and widespread growth in R&D at a substantial pace. With an estimate of *ca.* US$ 1.9 trillion in global R&D in 2015, the EU and US accounted for approximately half of the global expenditure. Within this regional expenditure in 2015, *ca.* US$ 620 billion was contributed by the top 1000 companies or corporations (PWC, 2017). Pharmaceutical industries reported a strong R&D expenditure, accounting for *ca.* US$ 120 billion (*ca.* 20 % of the global expenditure), followed by the foods and home/personal care sectors (US$ 27 billion, *ca.* 4 %) and energy sector (US$ 13 billion, *ca.* 2 %).

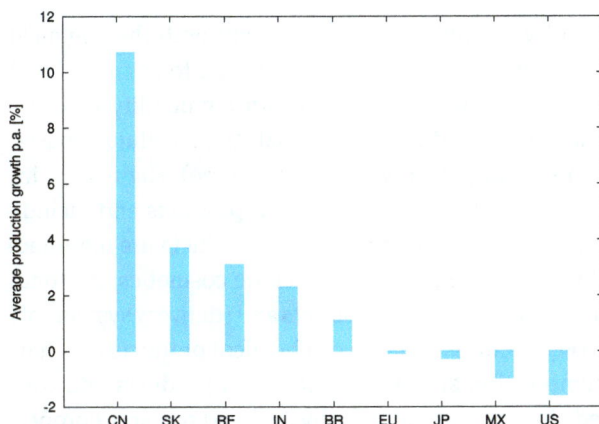

Figure 2.15: Average production growth per annum (2007–2017) (CEFIC, 2018). Legend: CN, China; SK, South Korea; RF, Russian Federation; IN, India; BR, Brazil; EU, European Union; JP, Japan; MX, Mexico; US, United States. Note: The pharmaceutical sector is excluded.

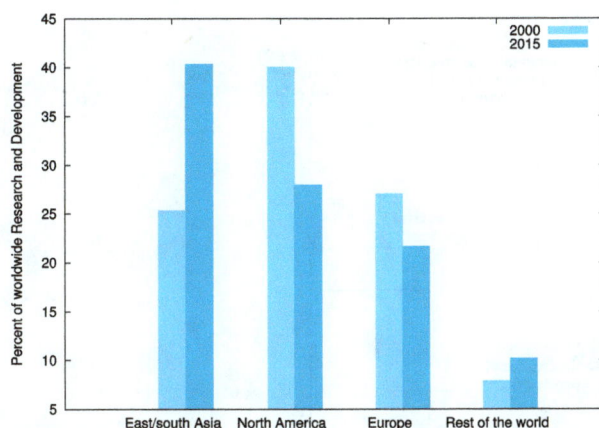

Figure 2.16: Regional share of worldwide R&D expenditures in the period 2000-2015 (NSB, 2018).

This shifting trend in the focus of the chemical industry is further reflected in the increasing number of scientific publications addressing product and process design (Kalakul *et al.*, 2018; Zhang *et al.*, 2017; Fung *et al.*, 2016; Mattei *et al.*, 2014; Smith and Ierapepritou, 2010; Seider *et al.*, 2009; Costa *et al.*, 2006; Edwards, 2006; Norton *et al.*, 2006; Ng, 2004; Schubert and Engel, 2004; Voncken *et al.*, 2004; Bröckel and Hahn, 2004; Gani, 2004a,b; Hill, 2004; Cussler and Wei, 2003; Wibowo and Ng, 2001, 2002), textbooks (Cussler and Moggridge, 2001; Seider *et al.*, 2004, 2010) and undergraduate/graduate courses in chemical process design (Moggridge and Cussler, 2000; Cussler and Wei, 2003; Shaw *et al.*, 2004; Saraiva and Costa, 2004).

From a manufacturing point of view and in close agreement with the chemical product tree of Figure 2.5, basic chemical products – also referred to as formulated products – are manufactured from natural resources and include commodity specialty chemicals (*e. g.*, ethylene, acetone, vinyl chloride), biomaterials (*e. g.*, cellulose-based materials) and polymeric materials (*e. g.*, polyethylene, polystyrene). High-value industrial products are manufactured from these basic chemical products and include fibers, paper, APIs, creams and pastes. Configured consumer products are manufactured from basic chemicals and industrial products and include cosmetics, personal care products, household products, composite food products and (delivery systems of) pharmaceuticals, among others (Figure 2.17). While basic chemical products are characterized exclusively by their composition and purity, industrial products and consumer products are characterized also by their thermo-physical and transport properties and by how the consumer perceives the product. As this consumer perception is directly connected to the expected performance or functionality of the product against the satisfaction of a need or desire of the consumer, both industrial products and consumer products are often termed performance or chemically structured products

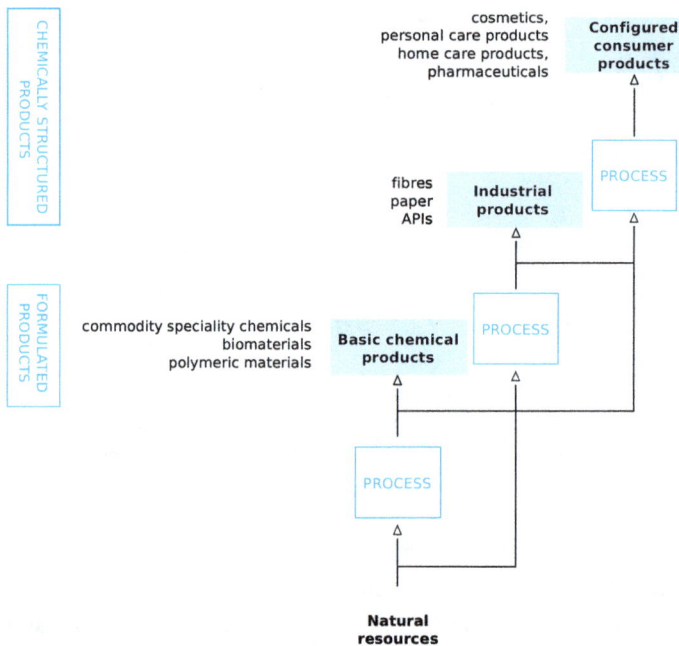

Figure 2.17: Relationship of formulated and chemically structured products (Bongers, 2009).

From a consumer-centered point of view, a product is a multi-dimensional concept and key entity for an organization. On the one hand, it has need satisfying capabilities to consumers/users and awakes interest of stakeholders, including potential buyers/consumers/users, trade intermediaries and members in the organization. On the other hand, a product represents a major opportunity for an organization to create value.

The design of products is of such a relevance to an organization's success, that several functions/groups within the organization are actively involved in activities associated with it. These cross-functional activities span from the initial concept for a product or service to its delivery/launch into the marketplace and its monitoring throughout its entire life. The design of a product is crucially important to determine its market acceptance, quality and reliability, producibility, availability (short time-to-market) and profit margins (low cost). Embracing effectively the design of products starts with the acknowledgment that the product has a life cycle, composed of four major stages, starting from its introduction into the market until its withdrawal from the marketplace (Smith, 2015). It is worth noting that before any product development activity is initiated, the marketing function of the organization is expected to have identified the consumer need(s) to be satisfied by the intended product or service. Next to this identification, marketing function should estimate the market opportunity (*e. g.*, projected volumes), target costs and selling prices.

2.2.1 Product life cycle

The evolution – or life cycle – of a product within the market where it is launched has been conventionally represented by a sequence of four stages (Figure 2.18). The length of each phase is dependent upon the product (*i. e.*, consumer need to be satisfied) and the industry in which the product competes (*i. e.*, speed of technological change).

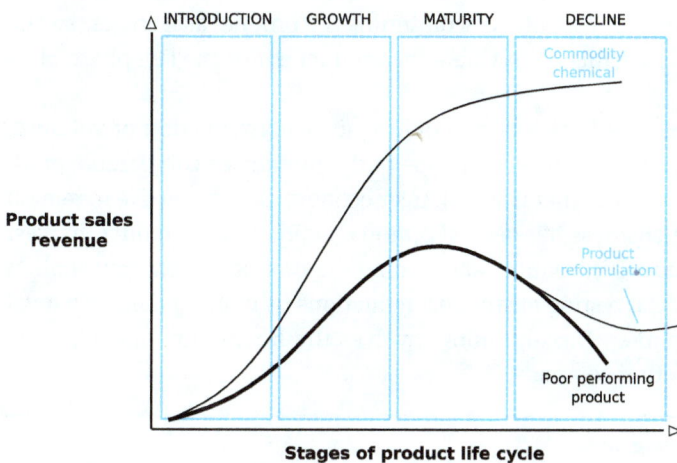

Figure 2.18: Life cycle representation of a product. Modified from Smith (2015), Brandt (1987).

Introduction. In this stage the business seeks to build product awareness and develop a market for the product. Product branding and quality level are established while intellectual property protection (*i. e.*, patents, trade marks) is obtained. The pricing strategy at this stage may be low penetration pricing or high skim pricing. Distribution is selective and promotion is aimed at early adapters. It is the most expensive and risky phase in the life cycle. If demand does not reach expectations, there is a potentially significant impact upon inventories planned to meet the anticipated demand. If demand is higher than expectations, there are possibly capacity constraints that prevent meeting actual demand. This phase is characterized by high profit margins and relatively low volumes until acceptance grows in the marketplace. Any error in the design of the product or service affecting the consumer acceptance will affect consumer acceptance. The error – if discovered – can be frequently corrected at a cost.

Growth. In this stage the business tries to build brand preference and increase market share. Product quality is maintained, and additional features or services are added. Distribution channels are added and promotion is aimed at a larger audience. During this phase, sales continue to increase at differing rates based upon market acceptance. As volume increases, product costs decrease. As the product or service gains high levels of acceptance, competitors enter the market, reducing prices and profit margins.

Maturity. In this stage the growth in sales diminishes. Competition may appear with me-too products (explanatory note *The Innovation Based Clustering Model*). The aim at this stage is to defend market share while maximizing profit. Product features may be enhanced, while pricing may be lowered. Distribution may be offered incentives and promotion emphasizes product differentiation. In this phase, total demand levels off and price competition is frequently quite severe. To stay ahead of the competition, product innovators should continually enhance their products or services. Upon reaching a point where profit contributions are no longer acceptable, the firm or corporation should strongly consider discontinuing the product as it enters the last phase of the life cycle. It is at this point where the product enters the last phase of its life cycle.

Decline. In this stage the business experiences a steady reduction of volumes. Competitors continue to offer improved versions of the product or comparable products at reduced prices to a point that it is no longer economically attractive to remain in the marketplace. The business has several options: maintain the product, harvest the product or discontinue the product. Maintaining the product needs rejuvenation by adding new features. Harvesting means cost reductions by offering them to a loyal niche segment in the market. Discontinuing involves the liquidation or selling the product or service to another company.

---

i   **The Innovation Based Clustering Model**
AC Nielsen and Ernst & Young carried out a research on new product launched in 1996–1997 in the European consumer goods sector. According to that research (Ernest-Young, 1999), manufactures spend

as much as 16 % of their net revenues on innovation, resulting in thousands of new products introduced to the market, most of them failing within 12 months. Based on the level of newness of the thousands of new products launched yearly, a model was proposed (IBC – Innovation Based Clustering) to classify them. While the term **new** might have a wide range of understandings or connotations, it is clear that a new product involves bringing a novel idea to practical use or an existing idea to new practical use. Because of the multi-dimensionality of innovation, where different actuators impose their own requirements and expectations, a product introduced to the market could be regarded new by some and a copy by others. The IBC introduces six product clusters, ranging from true new products to no-new products.

- **Classically innovative products.** These are breakthrough products that appear to the consumer to bring true innovation to a category, or that create new categories.
- **Equity-transfer products.** These are products that are new to the category but recognized by the consumer.
- **Line-extension products.** A line extension is a new version of a product within the same category.
- **Me-too products.** These are products that are substantially the same as existing ones.
- **Seasonal/temporary products.** These products have a short life cycle.
- **Conversion/substitution products.** These products replace ones already marketed without adding new value for the consumer.

Despite the relevance of innovation as growth fuel within an organization, new products introduced in the marketplace are by far not new. In fact, the research by AC Nielsen and Ernst & Young estimated that about 76 % of the products are me-too products, which hardly excite the consumers. True new products (aggregate of "classically innovative products" and "equity-transfer products" clusters) are rarely occurring, accounting for about 2 % of the total number of launched products, while "line extensions" represent about 6 % of all new items.

If the success of a product in the marketplace is evaluated on the basis of its weighted distribution after one year, Ernest-Young (1999) defined five groups, as listed below. About 6 % of the true new products, 3 % of the line extensions and as low as 1 % of the me-too products are considered **stars**.

Success group	Weighted distribution [%]
Dead	0 %
Almost or technically dead	0 %–5 %
Hanging in or in between	5 %–49 %
Successful	50 %–89 %
Very successful or stars	>90 %

Product range and variety is fundamentally relevant to an organization to remain competitive in the marketplace. By offering similar products within the same category, an organization is capable of maintaining (or even increasing) its market share, especially if a continuous inflow of new or enhanced products moving into the growth stage is achieved. These new products are expected to replace those products that are in either the maturity or decline stages of their life cycle. The range of product needs to be carefully selected, because if the range is too narrow, sales and customers may be lost to competitors, while if the range is too wide or products overlap, costs will increase and products will compete with each other rather than with competitors.

A key feature of an innovation-focused organization is such steady inflow of new or enhanced products, as this strategy allows the organization to stay ahead of competition, to improve product quality and reliability, to reduce costs, to improve profitability, to increase customer satisfaction and, ultimately, to increase market share. Organizations that focus on innovation and offer a wide product variety can achieve high levels of quality and profitability by three design initiatives: simplification, standardization and specialization.

– **Simplification** refers to the process of making something easier to accomplish, by reducing the number of items the manufacturer uses in its products, by taking advantage of volume pricing offered by suppliers if the same item can be used in multiple products, by reducing models, features and options, by embracing modular designs in multiple products and product families and by eliminating non-value adding steps in a process.

– **Standardization** involves the selection of the best-known method of accomplishing a given activity. Standardization can occur on different levels, and can range from standardization within a single organization to international standards that facilitate global commerce.

– **Specialization** refers to the concentration on a single activity, process, or subject. In product specialization, the manufacturer may focus on a narrow range of products with similar processes in high volume to satisfy low-cost requirements in price-sensitive markets. It enables the company to develop in-depth knowledge and technical expertise to facilitate fast-to-market quality products.

2.2.2 Structured products and fast-moving consumer goods

Within the performance products, those whose microstructure dictates directly their consumer preference (or dislike) play a predominant role. These structured products are ubiquitous in the agricultural, chemical, food, beverage, cosmetics and pharmaceutical industries (Bongers, 2009) and purchased, consumed and disposed on a daily basis worldwide (Table 2.16).

Table 2.16: Examples of structured products in the consumer sector. Legend: O/W, oil in water; W/O, water in oil; W/O/W, water in oil in water.

Micro-structure	Consumer product
Structured emulsion	dressings (O/W), margarine (W/O and W/O/W)
Structured liquids	hair conditioners, shower gels
Nanoemulsion	milk, APIs
Multi-phase system	ice cream
Granules	detergents, sweeteners, pesticides
Suspension	paint

Moreover, they are subject to intensive activities related to their development, manufacture, commerce and improvement by students and practitioners in the field of product and process design. Structured products are characterized by their complex microstructure, *i. e.*, the 3D arrangement of multiple components or set of components in different phases. While all components play a specific role within the structured system, their concentration is not of much importance, as long as the component functionality is delivered. Chemically structured products are characterized, in addition, by featuring complex rheology (*e. g.*, shear thinning or thickening, thixotropy behavior) and a metastable state (*i. e.*, microstructure is often at non-equilibrium), where the stability of the systems is strongly dependent on the state variables and the manufacturing route.

As depicted in Figure 2.19, the number of possible structures is relatively high and strongly determined by the manufacturing technology used to make the product (McClements, 2005; McClements *et al.*, 2007). The challenge to face comprises the development of products with the *right* structure, which in turn is governed by its manufacturing process. This intertwining relationship between product and process justifies and supports the need for an integrated and simultaneous design approach for product and process (Section 2.6).

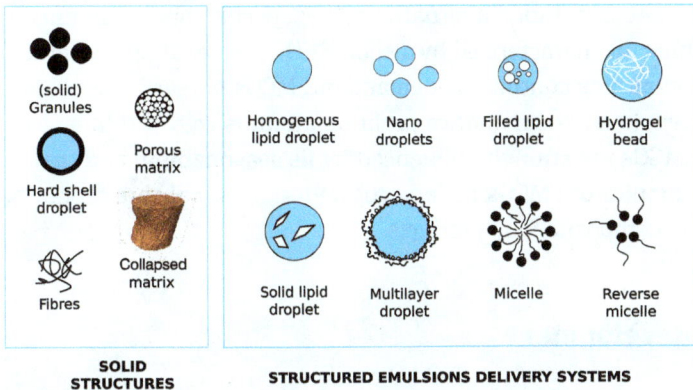

Figure 2.19: Overview of potential structures of consumer products and delivery systems of structured emulsions.

The technologies currently available in the manufacture of structured products are unconventional – as termed by Zhang *et al.* (2017) – and might consist of mixers, spraying and coating equipment, dryers, rollers, granulators, high-pressure homogenizers, emulsifying units and dry/wet etching units, among others. Despite being widely spread across multiple industries, structuring techniques (*e. g.*, emulsification, coating and agglomeration) are less well understood than manufacturing technologies of the bulk chemical industry. The applicability of structuring techniques to new systems and design spaces is continuously hampered by the intrinsic particularities of the system under consideration, including complex rheology and system stability. While the manufacturing processes define the micro- and nanostructure of the system, it is its in-use process that is essential to the assessment of the product performance. In fact, the performance of the products is determined by the behavior of the product when it is used, which is characterized by specific length/time scales and force fields. The in-use process might imply the breakdown or change of structure (*e. g.*, shear thickening behavior) and is governed by the same chemical principles of the manufacturing process. In designing the microstructure, the design practitioner conveys sound knowledge of transport phenomena and the physico-chemical or chemical changes that control the in-use process. Any product optimization is often possible by focusing on the in-use process of the product, including the force fields of the device/tool/subsystem used to deliver the product functionality (*e. g.*, shear stress during the mastication of a rehydrated food matrix).

There is a subset of structured products, which are characterized by their dynamics in the marketplace and user interaction. A fast-moving consumer good (FMCG) is a consumer product which is sold quickly at relatively low cost and low enterprise margin, sold in large quantities and generally replaced or fully used up over a short period of time. Moreover, if not available during the time frame of consumer demand, FMCGs are quickly substituted by competitors' alternatives. This specific type of product is intrinsically non-durable and characterized by a short SKU time, leading to a continuous cycle of consumption. The continuous demand of FMCGs trigger large volume sales that help to create a healthy cumulative profit on all units sold within a given period. Additionally, FMCGs are strongly influenced by its seasonability and that of their raw materials. Examples of FMCGs include cosmetics, personal care products, household products and composite food products.

2.3 New product development

Assessing whether an innovation is a true innovation and whether it is meant to be a successful innovation is a critical point to address during the development of new products. Current market innovations are the result of confronting market trends with the translation of consumer behavior and, on the other hand, the technological possibilities that have been incorporated in manufacturing them. It is a well-known fact that consumers are becoming more informed, critical and demanding when it comes to select a product or service (Deloitte, 2014). Meeting consumer expectations, which in turn reflect the dynamics of today's society, is reflected in continuously changing market trends. In the FMCG sector, for instance, market trends can include convenience, taste, ethnic cuisine, health and small packaging, among others. The challenge that arises from the need of fulfilling consumer expectations involves the translation of the relatively abstract consumer wishes into a concrete product through a product development process.

The development of new products (or better known as new product development, NPD) can be regarded as a series of steps that includes the conceptualization, design, development and marketing of newly created or newly rebranded goods or services. While product development aims at cultivating, maintaining and increasing a company's market share by satisfying a consumer demand, such satisfaction is not always achieved. In fact, as not every product will appeal to every customer, a target market for a product or service needs to be identified, researched and tested at the early phases of the product development process.

There is a wide spectrum of models or strategies in the development and execution of NPD. The most commonly adopted model is the Stage-Gate Innovation Process model, sometimes referred to as Staged Development Process, Robert Cooper Process or Milestone Review Process. An alternative type of NPD strategies is the Spiral Development Process, described below.

2.3.1 Stage-gate innovation process

All leading FMCG companies are rapidly transforming from general manufacturing hubs of loosely connected products to companies delivering health, wellness and nutrition. Manufacturing in a responsible and sustainable way products within strategic areas important for the company imposes technical challenges to work on and requires the development of R&D capabilities. Finding and creating business opportunities to be brought successfully to the market is the response of leading companies to the rapidly changing environment, characterized by slim profit margins and fierce competitiveness.

The Stage-Gate model is a project management tool to effectively manage such initiatives or projects (*e. g.*, new product development, process improvement and business change) throughout companies or corporations. In this context, each project is divided into stages separated by gates. At each gate, the continuation of the project is decided upon the performance against predefined and agreed Key Performance Indicators (KPIs). Typically the go/kill/hold/recycle decision is the responsibility of a steering committee. The process comprises a structured sequence of activities and flow of information (Ulrich and Eppinger, 2012). The **go** decision implies that the project is good enough to further pursue in the next stage of the innovation process; the **kill** decision implies that the project is not good enough and all activities associated with it need to be terminated immediately; the **hold** decision implies that while there is not sufficient merit in further pursuing the project at this moment, it should be kept on-hold and resumed at a later date; the **recycle** decision implies that the project is good enough to further develop, provided that specific changes are implemented.

While there is a wide range of variations of the Stage-Gate process, all of them share common characteristics aiming at delivering the right product with all associated benefits at a short time-to-market and at a reduced manufacturing expenditure. All variations involve a set of gates where go/kill/hold/recycle decisions are made, comprise well-defined stages and reviews, allow the iteration within each stage and do not permit cross-stage iteration.

Cooper (1990, 1996) was the first to formalize the Stage-Gate Innovation Process and distinguished five stages (Figure 2.20): (i) scoping, (ii) building business case, (iii) development, (iv) testing and validation and (v) launch and implementation. An initial preparatory stage (discovery) has been suggested as a way to identify and agree upon the initiative or project. A detailed description of the stages within the Stage-Gate Innovation Process is given in Table 2.17.

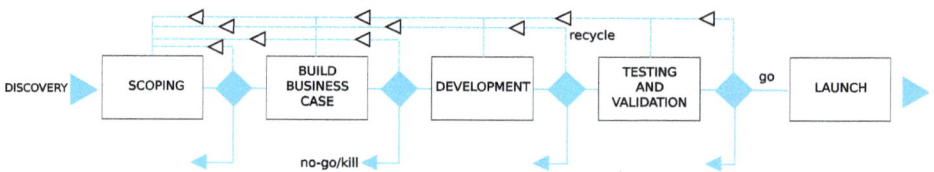

Figure 2.20: Stage-Gate Innovation Project.

Table 2.17: Stages within the Stage-Gate Innovation Process (Cooper, 1996, 1990).

Stage	Description
Scoping	evaluation of project and associated market; analysis of strengths and weaknesses and identification of added value to consumer/user; awareness of competitors' threats defines the decision to make
Building business case	a business plan is drawn up; last phase of concept development before the development of the actual product/service; substages to be considered: product definition and analysis, creation of business plan, creation of project plan and feasibility review
(Concept) development	key stage within the process; simple tests are carried out, involving consumer/user feedback; a timeline is created with specific milestones; a multi-disciplinary team is expected to provide expert advice; the output is a (series of) **prototype(s)** (explanatory note: *Prototyping*) to be extensively tested during the next phase
Testing and validation	manufacturing process and consumer acceptance are considered; substages to be considered: near testing, field testing and market testing
Launch and implementation	marketing strategy is key; volume estimation conducted and marketing campaign are developed.

Prototyping can be defined as any approximation of a product and service and is a fundamental activity throughout the development stage of the innovation process. Depending on the stage within the product development process or innovation process, prototyping can be crude or quick to test the feasibility of the concept (at the concept stage) or detailed or integrated to focus on specific details of one aspect (Ulrich and Eppinger, 2012).

In the development stage, prototypes are found to be extremely useful to learn the performance, usability or feasibility of the product, to communicate the performance to users or other interested party, to test the system's performance comprising a set of integrating subsystems and, finally, to support the milestones setting throughout the project planning. Depending on the scope of the prototypes, they can be classified in four types: analytical, physical, focused and comprehensive. Analytical prototypes are commonly mathematical models of the product, and as such can only exhibit behavior arising from the explicit modeled phenomena (Section 7.1.2.1). Physical prototypes, on the other hand, may exhibit unmodeled behavior, while being tangible approximations of the product. Focused prototypes implement one (or a few) of the attributes of the product, while comprehensive prototypes embrace and integrate multiple instances of focused prototypes.

Current computational capabilities to support the development of analytical prototypes are extremely strong and powerful. Examples of applications of modeling methods (3D CAD, CFD, FEA, among others) are abundant and spread over all sectors (*e. g.*, CPI, biotechnology, consumer products, product design). However, despite its relative universality, we can identify a set of limitations: (i) it requires specialized professionals to translate the situation (or design problem) into a mathematical model that predicts with accuracy and within a confidence level the behavior of the system; (ii) it is a computationally expensive exercise, which involves powerful hardware and specialized software; and (iii) the level of predictability of the behavior of the system is strongly dependent on the level of understanding of all physical/chemical/biological processes involved in the description of the system. As knowledge is sometimes unavailable, modeling relies on assumptions which can lead into unexpected behavior during the confirmation phase. In any case, process modeling supports effectively the designer to understand scenarios, predict behavior and carry out *in-silico* testing of the system. Both analytical and physical prototyping are resource-intensive activities, whose undertaking needs to be sufficiently justified. Physical prototypes are absolutely necessary, for instance, when they are compulsory due to regulation. This is the case, for example, in systems (*i. e.*, products, services, etc.) that directly involve consumers and their safety/health. Consider, for instance, the legal framework associated with FDA regulations, which require the physical testing of products in consumers before their actual launch. Pharmaceuticals need to be physically tested several times in long-lasting loops. It is no surprise that the launch of a new drug is characterized by an innovation loop of 5–10 years (Radjou and Prabhu, 2015). Moreover, and in alignment with the modeling limitations, physical testing/prototyping is required when the level of accuracy of simulation modeling will be too high to introduce risky uncertainties in the behavior of the system.

The development of prototypes requires a clear strategy. They need to be developed as a tool to reduce (a set of) uncertainties. As each prototype might involve the resolution of one specific uncertainty, the strategy requires the timing of each of those prototypes to be long enough to allow for a proper prototype–lesson learned cycle. The type of prototype to be developed to reduce uncertainty can be linked to the identification of the potential route of failure. This activity involves the identification and prioritization of failure modes.

The starting point of this model is the early product definition in planning and concept development stages. As we move along the later stages of the model, the uncertainties of the design problem decrease while the overall cost involved in the development increases.

A successful implementation of the model relies on the gradual elimination of uncertainties and assumptions. From an organizational perspective, it requires the continuous input of the steering committee and a clear set of communication rules and roles and responsibilities of each of the involved functions. For each function involved in the process a set of activities to be carried out at each phase of the project development process is identified. The strength of the process resides in its structured nature. Improving a staged-development process can be achieved by involving all parties during each decision making gate. A go/kill/hold/recycle decision should be based on the input of each player and should consider the overall product delivery as a team effort. In several instances, the decision is based on financial aspects, rather than strategic or social responsibility domains.

Figure 2.21 depicts a sample of the wide range of variations of the Stage-Gate Innovation Process as initially formalized by Cooper (Cooper, 1996, 1990). Next to all the characteristics of the innovation process mentioned above (*e. g.*, flow of information and decision and overall structure), the models can feature different numbers of stages depending on the nature of the project. Some of the models feature a converging section (*i. e.*, funnel) which represents the reduction of ideas as the process progresses stage after stage. Such funnel concept is commonly found in project progression tools and, as mentioned by Verloop (2004), conveys that during the process the number of innovation ideas/projects reduces as the poor ones are weeded out and the best ones get priority treatment.

Contrary to the Stage-Gate Innovation Process, the Spiral Development Model allows for all activities to be done in a repetitive and spiral fashion; see Figure 2.22). Because of its structure, this model has been widely used in the development projects in the software industry. Each activity is executed multiple times as the practitioners move from planning to testing phases. The various cycles that are performed could be compared to the several releases that software experiences throughout its life cycle (*e. g.*, beta version review, beta version release).

The Spiral Development Model is characterized by the planned iterations across all stages, comprising several cycles of concept ▷ integration ▷ testing. In most of its applications, the product definition is flexible, with no requirement of defining some features up front. As the iterations (or spirals) are executed, some of the features are detailed and/or fine-tuned, and the risk (or rate of failure) is subsequently decreased.

The development of new products does not always result in successful innovations. As mentioned by Kleef (2006), it is rather difficult and adventurous to estimate the rate of success in innovations. Firstly, the figures strongly depend on the type of product and market under review. Secondly, there is wide variation among different criteria that define success. In any case, it is widely accepted by and concurred within several studies that a great majority of new products never makes it to the market and those new products that enter the marketplace face very high failure rates. Failure rates have been reported to be in the range of 25 % to 65 %

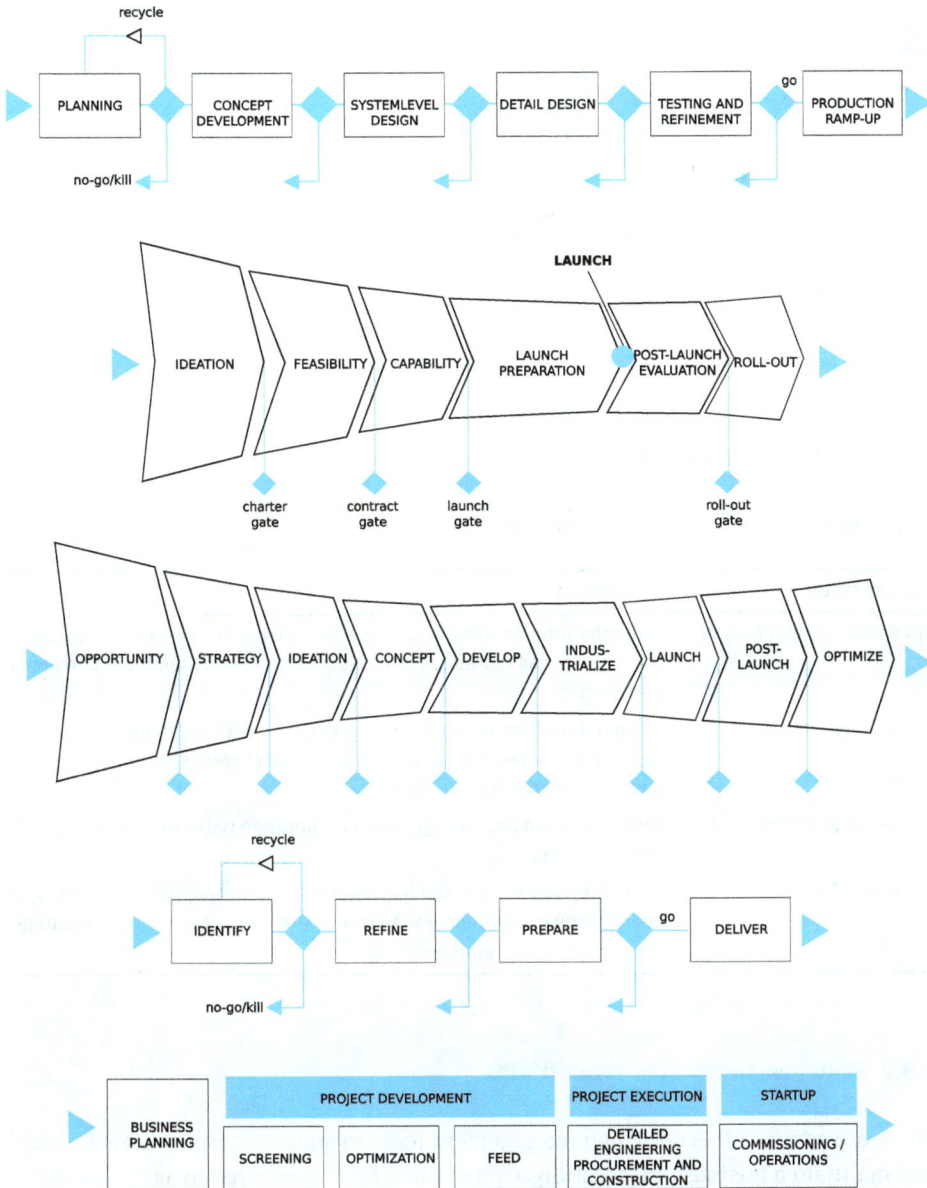

Figure 2.21: Examples of Stage-Gate Innovation Processes: product development process (Ulrich and Eppinger, 2012), innovation project management tool for project progression, NPD for innovation, idea to launch process, capital asset development process.

for specific market and product types and success definitions. Several factors, different in nature and dynamics, are responsible for NPD failure, as presented in Table 2.18.

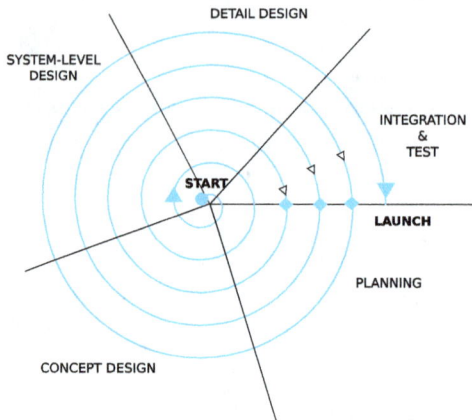

Figure 2.22: Spiral Development Process.

Table 2.18: Factors responsible for failure in NPD.

Class of factor	Description
Product-consumer factors	when the product or service is not new, distinct or valuable; moreover, the product has not responded effectively to the dynamics in consumer preferences
Environmental factors	when the product proposition is not aligned with the dynamics in technology; when the capacity of response of competition has not been considered thoroughly
Organizational factors	when there is a poor integration/co-operation between marketing and R&D functions
Marketing factors	when the market potential has been overestimated; the product has been poorly positioned in the market; the product was not adequately supported by the retail business

2.3.2 Management tools to support NPD

There are some well-established management tools to support corporations to build and maintain a business successfully. One of these tools is referred to as Ansoff's matrix and is regarded as one of the most well-known frameworks for deciding upon strategies for growth (Figure 2.23-left). It comprises four finite and proven basic strategies that are based around the fundamental question to whom is the corporation selling what (Brandt, 1987). The starting point for building the company of tomorrow is considered to be today's company. The first strategy – market penetration – involves delivering existing products/services to an existing set or type of customers. This strategy implies increasing the company's revenue by promoting the product, repositioning the brand, etc. The product is not altered – in any ways – and the corporation does not seek any new customers. If the commercial and financial environment suggest a change in the business strategy, potential ways to go involve market or product development. Market development is considered as a result of geographical expansion beyond the original, current market. The existing product range is offered to a new audience and requires the development of certain capabilities that are not necessarily available in the inventory of the new market (*e. g.*, management system). Exporting the product or marketing it in a new region are examples of market development. Product development is adopted as business building strategy when a new product is marketed to an existing set or type of customers. When consciously and systematically pursued, product development requires a unique set of skills/competences/functions, including product research and/or development. Product development might also imply that the corporation develops and innovates new product offerings to replace existing ones and markets them to existing customers. The fourth business strategy is referred to as diversification, which is executed when the corporation markets completely new products to new customers. Moving to new – and sometimes unknown – areas (products and markets) is a very big and tough challenge. There are two types of diversification: **related** and **unrelated** diversification. Related diversification means that the corporation remains in a market or industry with which it is familiar (*e. g.*, a soup manufacturer diversifies into cake manufacture). Unrelated diversification implies moving to domains where the corporation has no previous industry nor market experience (*e. g.*, a soup manufacturer invests in the rail business).

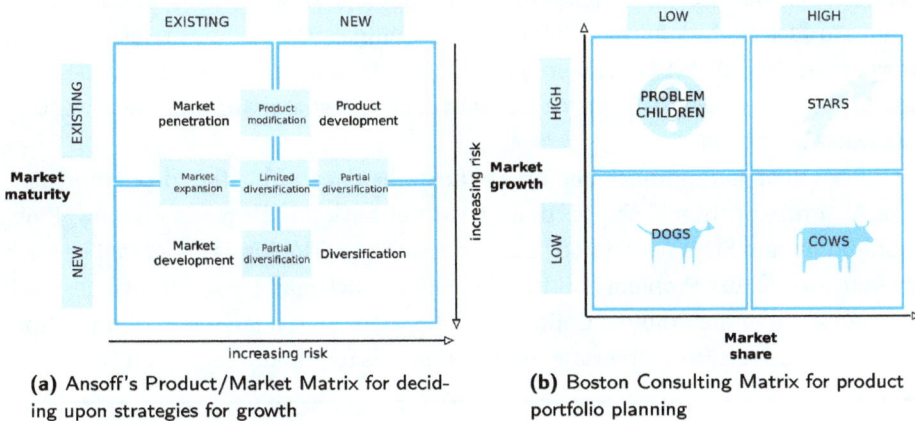

(a) Ansoff's Product/Market Matrix for deciding upon strategies for growth

(b) Boston Consulting Matrix for product portfolio planning

Figure 2.23: Strategic tools to support the product portfolio planning and priority setting.

Like Ansoff's Matrix, the Boston Consulting Matrix (Brandt, 1987) is a well-known business approach to product portfolio planning (Figure 2.23-right), while the learning curve theory is the cornerstone that explains the corporate growth associated with dominant share of a market. Market share analysis (in size and relative to competition) in combination with market potential (in growth rate) provides a powerful tool for planning hierarchy within a corporation. The [market share]–[market growth rate] matrix comprises four quadrants, represented by:

– Dogs: these are products with a low share in a slowly growing market; they do not generate cash for the company but tend to absorb it. Strategy: get rid of them as they are worthless or cash traps.

– Cash Cows: these are former Stars whose market growth rates have declined; these are products with a high share in a slowly growing market; they generate net cash for use elsewhere in the corporation and much more than is invested in them. Strategy: keep them temporarily in portfolio of products.

– Problem Children: these are products with a low share in a highly growing market; they consume resources and generate little in return; they absorb most money as it is attempted to increase market share; they can require years of heavy cash investment; if they do not occupy a leading market position before the growth slows, they become Dogs.

– Stars: these are the cash generators; these are products that are in highly growing markets with a relatively high share of that market; they tend to generate high amounts of income; they become eventually Cash Cows if they hold their market share; if they fail they become Dogs. Strategy: keep them and build upon them.

While this tool is simple to understand and easily implemented, it is worth mentioning that there are some intrinsic limitations. The main limitation revolves around the difficulties in accurately defining either the relative market share or the projected market growth rates (Brandt, 1987). The commonly adopted assumption considers that higher rates of profit are directly related to high rates of market share. Unfortunately, this may not always be the case.

The merit of this approach implies a strategic balance within the corporation portfolio. As a rule of thumb, the portfolio must not have any Dogs. Cash Cows, Problem Children and Stars need to be kept in equilibrium. The funds generated by Cash Cows are used to turn Problem Children into Stars, which may eventually become Cash Cows. As some of the Problem Children will become Dogs, a large contribution from successful products to compensate for the failures is required.

---

i The Learning Curve theory (Brandt, 1987): unit cost is a function of accumulated experience; that experience is a function of sales volume; and sales volume is a function of market share. Therefore, the company with the highest market share *vis-à-vis* its competitors should have relatively lower costs and, therefore, should generate more net cash. In short, cash flow is a direct function of market share, and gaining or holding a large market share relative to competitors is a strategically desirable thing to do.

The innovation paradox

The need to be consumer-focused in the development of innovations is justified by the fact that innovation is not only risky but also cost-intensive. According to Radjou and Prabhu (2015), over 80 % of new FMCGs fail at launch. Table 2.19 shows a sample of the financial losses incurred by FMCG and non-FMCG companies as a result of suboptimal innovations in past decades. On the other hand, innovation is essential for growth within companies, especially in saturated markets. Innovation is not only a key driver for profitability if properly conceived and rolled-out, but also serves as a critical vehicle of corporate image. Innovative companies often have a special place in the mind of the consumer, whose products services and associated brand names are becoming synonymous (*e. g.*, Gillette razors, iPad tablets, BVD shirts, sneaker shoes).

Table 2.19: Examples of financial losses due to suboptimal innovations brought to market. Source: CBInsights (2019).

Company – product	Situation
DuPont – CORfam	1963: leather alternative expected to cover over one-quarter of the entire US shoes market. Rigidness of the material made the shoes unbearable. Approximated loss: US$100 million.
Frito-Lay – WOW! Chips	1998: chips made with artificial fat Olestra, and supposed to pass harmlessly and inadvertently through the digestive tract. Financial loss: product recall, discontinued from the market and several legal suits.
Pepsi – Crystal Pepsi	1992: clear cola, whose flavor confused consumers. Financial loss: product recall and discontinued from the market
Atari – E.T. video game	1982: considered to be one of the contributors to the crash in the North American video gaming industry. Financial loss: US$20 million movie rights and 4 million cartridges produced.
Unilever – Persil Power	1990s: laundry detergent with extreme bleaching power, up to the extent of de-coloring and destroying cloth's integrity. Financial loss: product recall and several lawsuits.

The drive for innovation opposes the fear for innovation in the so-called innovation paradox. On the one hand, the drive for innovation aims for spectacular new concepts or services at high risk, while, on the other hand, the fear of innovation supports the low-risk delivery of marginally different products (me-too products; see explanatory note *The Innovation Based Clustering Model* in Section 2.2.1). Effective consumer research can support the decision making process within NPD.

It has been widely accepted that successful product innovations in the marketplace have a common set of characteristics, which include the following:

- **They are better**: in the eye of the consumer, these products are superior with respect to quality, cost-benefit ratio or functionality relative to competitors; failure rates are minimized by successfully fine-tuning product attributes with what the consumer really wants; these products satisfy consumer needs and wishes better than competition at a given time and place.
- **They are different**: they are true innovations and deliver real and unique advantages to consumers. According to Henard and Szymanski (2001), different products tend to be far more successful than me-too products.
- **They are formalized**: they are the result of structured processes or methodologies, rather than based on gut-feel approach. Although this characteristic is not fully inherent to the product but to the process associated with NPD execution, it has been agreed that products increase their likelihood of success if they are derived from a disciplined approach. By adopting a formal approach in NPD, information is more effectively shared, enhancing the decision making process (Cooper, 1999).
- **They are integrated**: they are an integral part of the business process, with emphasis on strategy rather than operation. Although this characteristic is not fully inherent to the product but more to the company strategy, it is fundamental that sufficient resources are allocated to the NPD initiative, that market entry is properly timed and that marketing and technological synergies are capitalized (Henard and Szymanski, 2001).

2.3.3 Competitive advantage

For many years the FMCG sector has been very successful in the use of new technologies with which quality, effectiveness and efficiency of production could be enhanced. In this scenario, referred to as the Technology-Push model, an enterprise competitive advantage is achieved by having a strong position in technology. Mastering technology becomes, in fact, the competitive advantage, while the reaction of the consumer is primarily of a supportive nature. As newly developed products are presented to the consumer for evaluation, they become the natural **gatekeepers** of the market. In the course of decades, this model gradually lost adapters despite the substantial and great progress that this approach brought to the sector. Under current market conditions several questions have arisen on the suitability of the model. Specially, the suitability is debatable on the basis of the following factors: (i) consumers are more critical in their choices; (ii) markets are further away from home; and (iii) technology is developing quickly. The combination of these factors has triggered the acknowledgment that a new market-relevant criterion needs to be defined, which accounts for the consumer behavior. In this model, referred to as the Market-Pull model, the competitive advantage arises from the superior consumer knowledge, while technology plays a supportive role. While these two approaches (Technology-Push and Market-Pull) sound incompatible at first sight, it is precisely their coherence that will lead to a synergic effect and, ultimately, turn an enterprise successful.

Understanding consumer needs is fundamentally relevant from a strategic NPD standpoint. Embracing consumer research at the early phases of NPD does not only allow bridging the gap between marketing and R&D functions (Kleef, 2006), but provides a competitive advantage in NPD. Consumer research is capable of generating more and better data than competition, provided a superior methodology is adopted. By creating a seamless marketing-R&D interface, data can be better integrated into the understanding of consumers beyond isolated facts. Moreover, by sharing the understanding derived from consumer research, a broad and actionable dissemination of data throughout the organization is possible.

Consumer behavior can be schematically represented in the wide context of effective consumer research (Figure 2.24). The consumer features a core role in the NPD landscape, where he/she requests the satisfaction of a need or wish. If we, as practitioners, know *what* the consumer wants, we can propose a product/service that will allow the satisfaction of the need. However, if we understand *why* the consumer has such a need, we are capable of anticipating market opportunities and consumer behavior. Understanding consumer behavior at the early phases of NPD facilitates not only a response before competitors do, but also helps expand the time horizon of innovation (Kleef, 2006). These two opportunities are accounted for in two building blocks: (i) the specific characteristics that the consumers want to find in the product and (ii) the development in the market (*e. g.*, socio-demographic, cultural, economic and psychological factors). Effectively intertwining both approaches within NPD will ensure the delivery of a product or service that is optimally fit-for-purpose. Hence, it features exactly the characteristics that the consumer is looking for. The definition of this new product or service corresponds to the *what*-component of the NPD activity, and is allocated conventionally to the marketing function of the company. The *how*-component corresponds to the R&D function, which is ultimately responsible for delivering the product or service based on the consumer requirements and aligned with the company strategy.

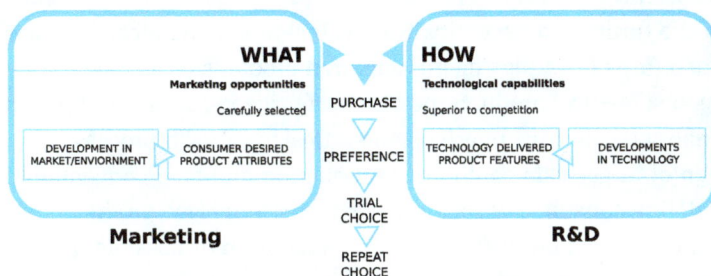

Figure 2.24: Schematic representation of consumer behavior in the context of NPD.

Consumer understanding, *i. e.*, research and conceptual frameworks, plays an important role not only in clarifying what the consumer wants, but also in bridging how those consumer requirements could take shape in a product or service. Both areas – marketing and R&D – are seamlessly connected by an interface.

Effective consumer research within NPD helps to identify new product ideas anticipating consumers' future needs and desires by bringing the voice of the customer (VOC) up-front. As mentioned by Kleef (2006), to develop a superior new product, consumer research needs to identify consumers' product attribute perceptions, including the personal benefits and values that provide the underlying basis for interpreting and choosing products (Figure 2.25). Moreover, NPD focuses on customer value delivery rather than product quality and translates consumer needs throughout the supply chain. Actionability is ensured through early co-operation of marketing and R&D functions. The delivery of customer values is meant to be effective (do the right thing), efficient (at minimum expenditure of resources), creative (in a distinctive way) and continuous (as a business process). The effective embracement of consumer-centered NPD will result in the generation of better ideas, fewer drop-outs and, ultimately, less incurred costs.

Figure 2.25: Relation between product characteristics and consumer values. Modified from Kleef (2006).

In this context, supply chain is defined as the mental picture of the flow of matter and important events happening to raw materials, intermediates and end-products until their end-use by consumers.

Due to profound changes in the structure of the chemical industry over the last 15 years, chemical engineering needs to be refocused. Specialty products have grown in importance, the biosciences offer novel products and processes and a sustainable chemical industry in the future is a necessity. High-value-added chemical products often imply the manufacture of complex microstructured products (*e. g.*, foods, coatings, catalysts, detergents, medical products, pesticides, fertilizers, dyes, paints) with very specific functionality to the consumer/customer. Design of such processes is a far cry from the design of large-scale continuous-production plants of conventional bulk chemicals. Translating consumer wishes and consumer-perceived product properties into physical and chemical product properties, generation of possible process routes to achieve those desired product properties is the core of product-led process engineering, or consumer-centered product-process engineering.

2.4 Design is not a trivial task

On a daily basis we encounter decision points. Each of these points requires a conscious choice based on a set of desires and expectations. The specific choice for a given product is embedded in a convoluted network of relationships, where the design of the product features centrally. In the wider context, this network involves multiple players, each of them having its own set of restrictions, boundaries, strategies and end-goals. Let us have a detailed look at the following situation (Figure 2.26). A given product or service is chosen by the consumer or end-user because it is preferred against a set of alternative products or services. This preference is directly related to the attributes this *product* or *service* has and how well it is performing and/or delivering its intended benefit.

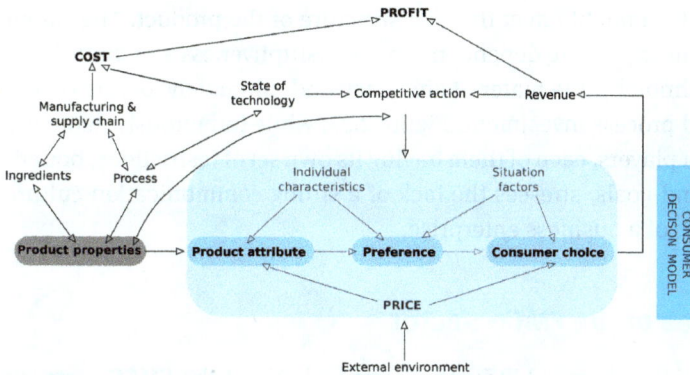

Figure 2.26: Convoluted network of players, attributes, properties and other factors in the decision making process.

From this point onwards, and to improve the readability of the document, "product" and "service" are considered interchangeable terms. The differences between a product and a service are listed in Table 2.5. **!**

While the consumer choice is strongly affected by situational factors under which the choice is made, individual product characteristics account for the product performance. The attributes of the product involve those characteristics close to the consumer heart, close to its perception. Examples of the attributes are creaminess in an ice cream, moist feeling in a skin cream and hiding power in a paint. The attributes can be translated into a set of quantifiable properties, which embrace physical, chemical and biological characteristics of the product. The selection of ingredients (or raw materials) and the selected production route strongly determine the properties of the product, which is translated into attributes. Both ingredients and processes are embedded in manufacturing and supply chain activities within a business enterprise.

These activities aim at delivering a product at a competitive production cost, which will assure the growth (or survival) of the enterprise. Beyond the consumer decision remit – consumer decision model – the product choice generates revenue (and top line profit) of the business.

The difference between the generated revenue and the cost of production represents the profit (or bottom line) of the enterprise. While the ultimate goal of a production process within a business enterprise would be the generation of profit, the role of external actuators needs to be borne in mind. As a result of a competitive environment and of dynamic consumer trends, the product properties are required to be revised and frequently optimized. This optimization will ultimately lead to improved product attributes, which will result in consumer preference against comparable products. Adaptation of formulations (or recipes) or functionalities, which might involve new raw materials and process improvement, might imply the uplift of manufacturing capabilities, which in turn might affect the cost structure of the product. The extent and directionality of the impact are dependent on the disruptiveness and maturity of the manufacturing technology, the material efficiency and, ultimately, on the cost of involved materials and process investment. Figure 2.27, while humorously depicting the intricate network of players, each of them having its own set of restrictions, boundaries, strategies and end-goals, stresses the lack of a strong communication culture among functional areas of a business enterprise.

2.5 The challenges of the FMCG sector

The change of focus of the chemical industry in general and of the FMCG sector in particular is the response to unprecedented market demands and constraints. This change of focus is evident from the continuous launches of novel products with exciting propositions, including weight management control (Foodbev, 2018e, 2019c,g), brain health (Foodbev, 2017b, 2018f), heart health (Foodbev, 2018i, 2019d), strength and immunity (Foodbev, 2019a,e) and beauty foods (Foodbev, 2018a,c), among others. Manufacturing in a responsible and sustainable way products within strategic areas, such as nutrition (Foodbev, 2018b,h, 2019f,i), health (Foodbev, 2019h), wellness (Foodbev, 2018d), convenience (Foodbev, 2017a, 2018j) and resources efficiency (Foodbev, 2017c,d, 2018g,k, 2019b), among others, imposes technical challenges to work on and requires the development of R&D capabilities to address them effectively. Building up technical capabilities should be the joint effort of academia and industry, with the clear aim of delivering a product with all associated benefits to the consumer at a short time-to-market and at a reduced manufacturing expenditure.

---

i Within the FMCG products, foods, beverages, personal care and home care products (Figure 2.28) are of significant relevance due to their market dynamics and rapidly changing consumer expectation and requirements.

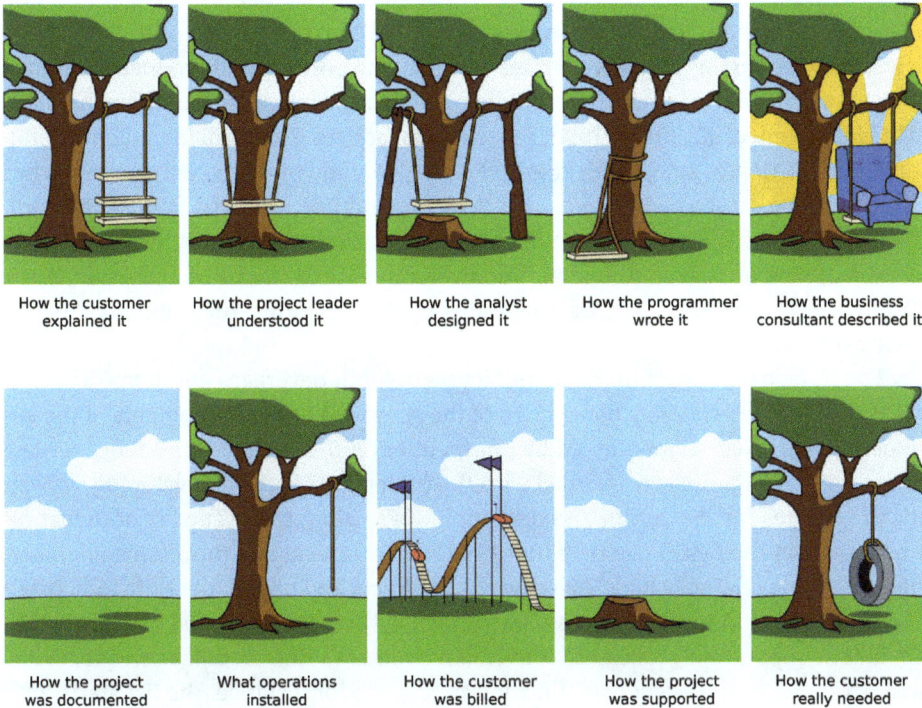

Figure 2.27: The non-trivial nature of product design.

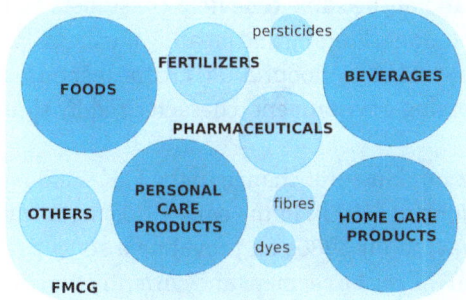

Figure 2.28: FMCGs in the context of consumer products.

Challenges that shall be faced by chemical and process engineering disciplines of the twenty-first century can been grouped in – but are not restricted to – four key areas: (i) optimization of product taste, enjoyment and nutritional profile; (ii) delivery of natural goodness; (iii) sustainability awareness and development of sustainable manufacturing practices; and (iv) on-shelf availability by supply chain optimization. The development of a Product-driven Process Synthesis approach is regarded to be the vehicle to achieve these objectives. Challenges are also occurring in the fields of fat technology, ice structuring, powders and particulated products manufacturing, structuring with natural ingredients and novel manufacturing mechanisms (Almeida-Rivera *et al.*, 2010a, 2011), separation of actives in natural ingredients (Monsanto *et al.*, 2014a,b, 2015, 2016; Zderic *et al.*, 2013a), advanced modeling approaches of multi-phase structured systems and active containing matrices (Monsanto *et al.*, 2012, 2013, 2014c; Zderic *et al.*, 2013b; Dubbelboer *et al.*, 2012, 2013, 2014a,b, 2018), novel methods and tools of operations involving FMCGs (van Elzakker *et al.*, 2011, 2012, 2013a,b, 2014a,b, 2017; Galanopoulos *et al.*, 2017) and frugal manufacturing, among others.

The industry engaged in the manufacture of food, beverages and home and personal care products is a key beneficiary of the advances and developments of the engineering activities. As mentioned by Charpentier (2007) chemical and process engineering are continuously concerned with the understanding and developing systematic procedures for the design and operation of FMCG systems. The list of industrial innovations in the FMCG business that stem from chemical and processing engineering activities is relatively long, covering nanosystems and microsystems to industrial-scale continuous and batch processes. The involved areas include food microstructure formation, separation processes, conversion processes and stabilization processes, among others. In the area of stabilization processes, for instance, the list starts with the idea of preserving foods in airtight containers in the early 1800s (Brul *et al.*, 2003) and spans until today's implementation of novel mild processing technologies (Datta and Ananthesarawan, 2001). All these innovations have changed and are changing the way we actually manufacture FMCGs, and at the same time help consumers feel good while satisfying their needs and expectations. Finding and creating business opportunities to be brought successfully to the market (Verloop, 2004) are the responses of food process industry to the rapidly changing environment, characterized by slim profit margins and fierce competitiveness.

This dynamic environment has led to a fundamental change in the FMCG industry emphasis. As mentioned by Bruin and Jongen (2003), in the early 1970s the focus was on making a product in a profitable way; in the 1980s on providing a service to the consumer based on products procured in the **make** segment and currently the emphasis is on delivering additional benefits to the consumer by combining several **services**. Clearly, the scope has moved from a make mode through a service mode into a **care** mode. The emphasis of the make segment was on improving manufacturing efficiencies (*e. g.*, batch *versus* continuous processing, improving process reliability, reducing waste products, effective energy utilization). While in the service segment consumer trends were strongly considered, leading to a noticeable market presence and change in consumer, the care segment is characterized by delivering a total package of care to the consumer, consisting of various products and services that are made by co-packers or other third parties. Coping effectively with this changing scope implies a performance level which requires a close collaboration among academic and industrial partners in the engineering community. The FMCG industry has reached a maturity mindset, where it recognizes that innovations are needed to deliver high-quality consumer products and those breakthrough innovations can only be developed quickly if industry taps onto external resources and ideas from academia.

2.5.1 Challenge #1: product profile improvement by microstructure creation

According to the World Health Organization (WHO, 2019), around the globe more than 1.9 billion adults – 18 years and older – are overweight and of these, 650 million are obese. These figures represent *ca.* 9 % and 13 % of the adult world population, respectively. Even more shocking is the fact that inadequate dietary practices are affecting young population groups. A recent report by the Centers for Disease Control and Prevention (CDC, 2007) revealed that the proportion of overweight US children aged 6–11 has doubled in 25 years, while the number of overweight adolescents has tripled. Globally, it has been reported (WHO, 2018) that the world has witnessed a more than tenfold increase in the number of obese children and adolescents aged 5–19 years in four decades, from just 11 million in 1975 to 124 million in 2016. An additional 213 million were overweight in 2016 but fell below the threshold for obesity.

Overweight and obesity (WHO, 2019)

Overweight and obesity are defined as abnormal or excessive fat accumulation that may impair health. Body mass index – BMI – is a simple weight-for-height index that is commonly used to classify overweight and obesity in adults. It is defined as a person's weight in kilograms divided by the square of his height in meters (kg/m^2).

For adults, (WHO, 2019) defines overweight and obesity as follows:

- overweight is a BMI greater than or equal to 25; and
- obesity is a BMI greater than or equal to 30.

BMI provides the most useful population-level measure of overweight and obesity as it is the same for both sexes and for all ages of adults. However, it should be considered a rough guide because it may not correspond to the same degree of fatness in different individuals. For children, age needs to be considered when defining overweight and obesity.

Global increases in overweight are attributable to a number of factors, including a global shift in diet towards increased intake of energy-dense foods that are high in fat and sugars but low in vitamins, minerals and other micronutrients (WHO, 2019). These trends in eating habits stretch all around the world and triggered the formulation of a key challenge within the food and beverage industry. This challenge involved the optimization of the nutritional profile of food products, but without compromising taste delivery and consumer enjoyment. Examples of such products include the effective use of natural structurants in the manufacture of low-dispersed phase structured emulsions. The consumer perception (*e. g.*, mouthfeel) should not be compromised while maintaining the calories intake (*i. e.*, nutritional value) at minimal values. Implementing natural structuring technologies requires a sound understanding – at different time and length scales – of the science underpinning the fundamentals of emulsification, product design, product-process integration and process synthesis. Let us take as example a full-fat mayonnaise in the dressings market. As most structured food products, the mayonnaise microstructure is composed of a dispersed phase of liquid and a continuous phase, with a length scale of the dispersed phase of about 10 µm (Figure 2.29), whereas the continuous phase may be of complex rheology and further structured by polymers or particulate networks (Norton *et al.*, 2006).

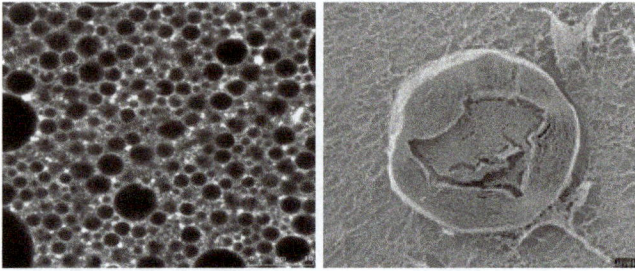

Figure 2.29: Confocal micrograph showing the microstructure of (a) an O/W emulsion and (b) a fiber containing structured system.

The dispersion of two or more nearly immiscible fluids is one of the core activities within the manufacture process of liquid structured products. Mixing plays a controlling role in the liquid–liquid system, being responsible for phenomena such as breakup of droplets, combination of droplets and the suspension of droplets within the system (Leng and Calabrese, 2004; Schubert *et al.*, 2006). As the end-use attributes of a liquid structured product or emulsion (*e.g.*, taste, smell, mouthfeel, color) depend primarily not on the overall product composition but rather on its microstructure (Schubert *et al.*, 2006), significant effort has been invested since the 1990s in the understanding and control of product structure and in the discovery of new ways of manufacturing emulsions (Bruin and Jongen, 2003). A fine dispersion of two immiscible liquids can be produced in many different ways, mechanical processes being the most frequently used (Schubert *et al.*, 2006). Within these processes four types of categories based on the mechanism of droplet disruption can be identified (Perrier-Cornet *et al.*, 2005; Schubert *et al.*, 2006): (i) rotor-stator systems, (ii) high-pressure systems (Valle *et al.*, 2000; Lander *et al.*, 2000; Perrier-Cornet *et al.*, 2005), (iii) ultrasound; and (iv) micropores (Sugiura *et al.*, 2004) (Figure 2.30).

Although each of these types features different strengths and scientific opportunities, a straightforward comparison is possible by adopting the approach suggested by Davies (1985) and further extended by Schubert and coworkers (Karbstein and Schubert, 1995). For different emulsifying devices this approach involves the estimation of the local power dissipation in the region of droplet breakup and the estimation

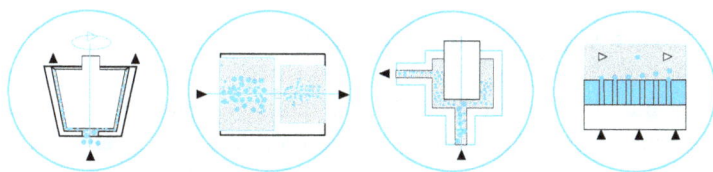

Figure 2.30: Types of mechanical process for droplet disruption: rotor-stator systems, high-pressure systems, ultrasound and micropores.

of its effect on the maximum drop sizes in the dispersion of two non-miscible fluids. Regarding the modeling of the governing phenomena in emulsion formation a level of maturity has been reached in the last decades. The current state of knowledge involves the availability of detailed phenomenological descriptive models of the breakup and coalescence of droplets in various flow regimes and emulsifying devices and for various ranges of interfacial properties. Thus, qualitative relations involving droplet breakup of high-viscosity fluids in laminar and turbulent flows were first introduced by Grace (1982) and later extended by Walstra (1993), among others. Coalescence was brought into the scene in the late 1980s and early 1990s through several contributions involving research on (non-)Newtonian turbulent multi-phase flow (Saboni *et al.*, 2002; Tjaberinga *et al.*, 1993). The formulation of droplet population balances involving breakup and coalescence mechanisms have been frequently addressed, usually resulting in new and more robust solving approaches (Attarakih *et al.*, 2003; Kostoglou, 2006; Nere and Ramkrishna, 2005) and feasibility of rigorous and short-cut modeling approaches (Agterof *et al.*, 2003; Steiner *et al.*, 2006). A sound understanding of breakage and coalescence phenomena and especially on how they interplay has allowed the mechanistic study on phase inversion (Groeneweg *et al.*, 1998; Klahn *et al.*, 2002; Nienow, 2004), duplex emulsion formation and the entrapment and release of active compounds (Benichou *et al.*, 2004). Finally, the role of the emulsifier properties in the emulsion formation has been often addressed with emphasis on adsorption kinetics (Perrier-Cornet *et al.*, 2005; Stang *et al.*, 1994) and on droplet breakup (Janssen *et al.*, 1994a,b).

Chemical and process engineers face, moreover, additional challenges involving the fundamental understanding of the interactions between the ingredients, especially for emulsions that contain a continuous phase structured by a (bio)polymer or particulate networks. Such scientific understanding should be extended to the synthesis of the production system, the design of the equipment units involved in the process and the selection of the most appropriate process conditions and control structure and strategy. In addition to the structure formation, nutritional and taste benefits can be attained by encapsulation technology and nanotechnology. A synergetic enhancement of the nutritional and taste profiles might be attained by using nutritional ingredients as encapsulators of taste components. This approach has been recently reported by Chiou and Langrish (2007), using natural fruit fibers to encapsulate bioactive compounds and at the same time removing the need for maltodextrin during spray-drying.

The applicability of nanotechnology in the FMCG industry spans over various domains, involving processing, materials, products and product safety (Weiss *et al.*, 2006). Examples of these applications are the protection of functional ingredients against environmental factors, development of new edible and functional films and synthesis of nanosensors and nanotracers (Tarver, 2006). Although the developments in nanotechnology assure an increasing number of applications in the FMCG industry, a challenge in this area for the years to come is the development of cost-effective and scalable technologies.

2.5.2 Challenge #2: product naturalness by mild processing

In view of the increasing demand on high quality and healthy products, considerable effort has been invested in the development and industrialization of goods that retain the naturalness of key ingredients. Within the FMCG sector, a number of ingredients are standing upfront as source of health benefits: tea, soy, fruits, vegetables, good oils, natural structurants, etc.

The benefits of tea are acknowledged by the ancient history of drinking tea leaf infusions in Asia, in particular in China. Tea is rich in flavonoids (Kivits *et al.*, 1997; Zderic *et al.*, 2015; Zderic and Zondervan, 2016) and contains polyphenols with antioxidant functionality.

The benefits of soy are numerous and stem from the richness of vitamins, minerals and low saturated fat level (Zderic *et al.*, 2014, 2016). The health benefits associated with the consumption of soy have made soy-based foods and drinks one of the fastest-growing markets in the world.

According to the Food and Agriculture Organization (FAO, 2007), some of the world's most common and debilitating nutritional disorders, including birth defects and mental and physical retardation, are caused by diets lacking vitamins and minerals. Low fruit and vegetable intake is a major contributing factor to these deficiencies.

Keeping the natural goodness of these key ingredients while ensuring hard-claimed benefits and product shelf-life sufficient for distribution and safe consumption requires specific skills, capabilities and technologies. These skills, capabilities and technologies involve areas as agronomy, (micro)encapsulation, mild processing and preservation. It is a well-accepted fact that most steps in a manufacturing process are responsible, to a certain extent, for the loss and reduced availability of vitamins, minerals and nutrients in foods and beverages. The loss of nutritional value is associated with the loss of quality, and *vice versa*. In this context, quality does not embrace only levels of vitamins, minerals and nutrients, but also microstructure, texture at the microscale, piece integrity, color, taste and microbiological stability. If the manufacturing process imbeds the complete supply chain, the quality loss of natural ingredients might begin at the (post-)harvest step and stretches until the consumption of the finished product by the end-consumer (Figure 2.31).

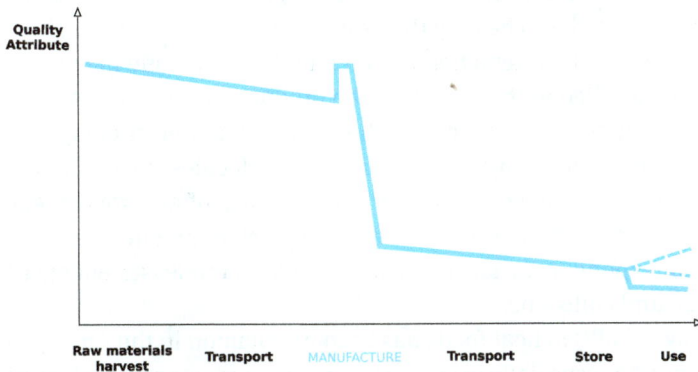

Figure 2.31: Simplified representation of quality loss of natural ingredients across the manufacturing supply chain.

Quality loss of foods and beverages can be the result of exposing the product over a large period of time to abusive temperatures, pressures, acidic or basic conditions, ultraviolet light or oxygen. Chemical and process engineers addressed during the last century the challenge of creating novel thermal processes to overcome the quality loss of foods while producing microbiologically safe products (Holdsworth and Simpson, 2007; Jones, 1897). These technologies were conceived to deliver – in most of the cases – minimal processing conditions (*e. g.*, time, temperature, energy input) to products and hence improve its overall quality.

It is worth reiterating that the quest of novel preservation technologies arises from the demand for convenient products with fresh-like characteristics closely resembling those of fresh products, coupled with shelf-life extension and added value during the conversion of raw materials (Tapia *et al.*, 2005). In addition to mildly processed, minimally processed can be non-sterile, as is the case for ready-to-eat products. Consumer safety is of paramount relevance in the processing of consumer products and non-negotiable for the food and beverage industry. Meeting consumer-sensorial requirements/wishes must never be at the expense of compromising consumer safety. Reaching a full satisfaction of all consumer expectations (sensory, safety, finance-related) is where the sweet spot lies. Examples of these novel processing technologies include volumetric heating processes (*e. g.*, Ohmic and microwave-assisted heating) (Coronel *et al.*, 2003, 2005; Coronel, 2005) and processes assisted by ionizing radiation, pulsed X-ray, ultraviolet light, pulsed electric fields (Zderic and Zondervan, 2016; Zderic *et al.*, 2013a,b), inductive heating, intense light pulses, combined ultraviolet light and low-concentration hydrogen peroxide, submegahertz ultrasound, filtration and oscillating magnetic fields, among others (Tapia *et al.*, 2005).

Conventional heating of consumer products relies on convective and conductive heat transfer to the product from a hot or cold medium, either directly or indirectly. Volumetric heating comprises a range of thermal technologies in which electromagnetic energy is directly converted into heat by the product. With these methods, heat is generated inside the product by interaction between molecules, while in conventional heating heat is first applied to the surface of the product and then transferred to its center by either conduction or convection. The conversion of electromagnetic energy into heat is achieved by the polarization of ionizable molecules, such as water, salts in solution and other polar molecules. Under the influence of electromagnetic energy these molecules try to align themselves to the rapidly changing field. As a result of not being able to respond as fast as the field, the material releases energy as heat, generating friction and collisions.

The use of microwaves (MW) to heat foods has become common in the American household and in the industry, especially in tempering frozen products and cooking of solid foods. The heating using microwaves is based on the generation of energy inside the product by the absorption of power transmitted by the microwaves. This method has several characteristics that differentiate it from conventional heating, including: (i) instant power turn off and on, (ii) speed, (iii) non-invasive as it does not rely on contact with hot surfaces or a hot medium, (iv) selectivity, and (v) more uniform heating. Microwave heating systems consist of a resonant cavity, either multi-modal or single-modal (Figure 2.32).

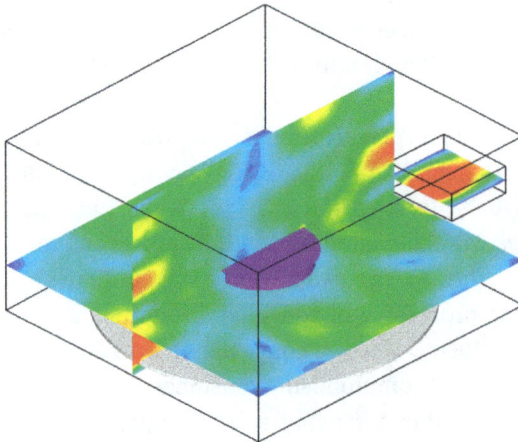

Figure 2.32: Thermal image of a microwave cavity.

Industrial applications of MW are normally built at 915 or 430 MHz. These frequencies have the advantage of providing penetration depths in the order of 50 to 100 mm, which are suitable for thermal processing (Coronel, 2005). Thus, microwave heating is a promising technology for heating materials that are difficult to heat conventionally, such as viscous liquids and heterogeneous products. Table 2.20 lists some of the numerous applications that have initially succeeded in being adopted widely.

Table 2.20: Current applications of microwave heating in the food industry (Coronel, 2005).

Product	Unit operation
Pasta, onions, herbs	Drying
Bacon, meat patties, chicken, fish sticks	Cooking
Fruit juices	Vacuum drying
Frozen foods	Tempering
Surimi	Coagulation

Microwave heating is not restricted to solid materials, which are conveyed by a belt. Applications for continuous processing of fluid food materials have been reported initially by Drozd and Joines (1999) and more recently by Coronel *et al.* (2005), Coronel (2005).

Ohmic heating is achieved by having electrical current (alternating current) passing through a column of product. The current is applied by electrodes located at both ends of a tubular heating chamber (Figure 2.33). The resistance of the product to the current will determine the dissipation of energy, and thus heating of the product (Coronel, 2005). Ohmic heating is very efficient, transforming 90 %–95 % of the energy into heat. Moreover, the product is heated uniformly when the resistance to electrical current is relatively uniform. However, if large differences exist between parts of the product or if the product is heterogeneous there may be zones that heat at different rates.

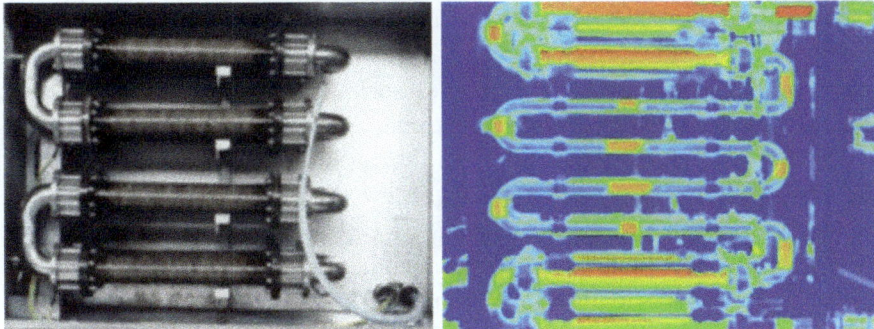

Figure 2.33: Ohmic heating system (right: thermal image during operation).

To achieve good heating the electrodes must be in contact with the column of product, which could be a potential source of fouling or contamination (Coronel, 2005). Continuous development of novel alloys (*e. g.*, titanium alloys) reduces the danger of contamination. Several industrial installations exist that use Ohmic heaters for sterilization of foods, especially fruits and vegetables (Table 2.21), with the tomato industry being the most active sector.

Table 2.21: Current applications of Ohmic heating in the food industry (Coronel, 2005).

Product	Unit operation
Fruits and vegetables	Pasteurization and sterilization
Tomato juice, paste, chunks	Sterilization
Low acid products and homogeneous products	Sterilization

Ohmic and MW heating are examples of two well-established mild processing technologies. Although they have been successfully implemented for many years in various processing applications (tempering, drying, baking and cooking of solids), they still impose technological challenges to be addressed by the chemical and process engineering. These challenges involve the development of systems that can handle materials and heterogeneous matrices containing chunks. Moreover, effort should be given to the development of complete aseptic systems based on Ohmic and MW technologies, among others.

2.5.3 Challenge #3: benign products by improved sustainability practices

Today's changing environment is characterized by the demands of society for new, high-quality, safe, clean and environmentally benign products. Moreover, consumers are increasingly interested in how effectively companies are addressing climate change. Measuring the extent of sustainability embodiment within the industrial sector is possible by estimating the Down Jones Sustainability Index (DJSI). The DJSI is an objective external indicator of companies' social, environmental and economic performance, using measures such as brand management, raw material sourcing, strategy for emerging markets and impact on climate strategy. As reported by Radjou and Prabhu (2015), over 80 % of Europeans believe that a product's environmental impact is a critical element in their purchasing decisions, with *ca.* 90 % of millennials willing to switch to more socially and environmentally responsible brands.

Environmentally conscious companies in the FMCG sector should always aim for a responsible approach to *farming practices*. These practices are the result of the sustained research on ways of farming that protect the environment and maximize social and economic benefits. Sustainable agriculture guidelines need to be defined to track progress against sustainable agriculture indicators, including water, energy, pesticide use, biodiversity and social capital. Despite the long time span of produce growing cycles and the colossal task of engaging with all suppliers, farmers and factories, sustainable agriculture initiatives have resulted in 100 % sustainable sourcing of key crops (*e. g.*, tomato) and a substantial (>90 %) decrease in the use of pesticide in agriculture-intensive markets (*e. g.*, India), among others.

Water management is also a key area to address within a sustainability-driven vision. According to the WHO (2019), around a billion people on the planet do not have access to safe drinking water. This water scarcity is predominantly occurring in developing markets and largely responsible for diseases such as cholera, diarrhoea and typhoid. Water management involves the usage of less water for multiple purposes. In the manufacturing process of FMCGs water is extensively used, predominantly in operations such as cooling, heating, washing and rinsing. It has been estimated that most FMCG companies use between one and seven liters of water per kilogram of product (van Ede, 2006). On the global scale, the consumption of water in the industry reaches *ca.* 665 billion cubic meters per year, which is approximately twice the amount used for domestic purposes (Worldmapper, 2016). Although water might be considered a non-expensive utility, its saving can be cost-effective if water consumption is linked to energy savings, recovery of raw materials and pollutants. A good housekeeping policy (Figure 2.34) is the first step towards more sustainable water consumption. From a processing perspective, more efficient operation policies should be defined to minimize water consumption.

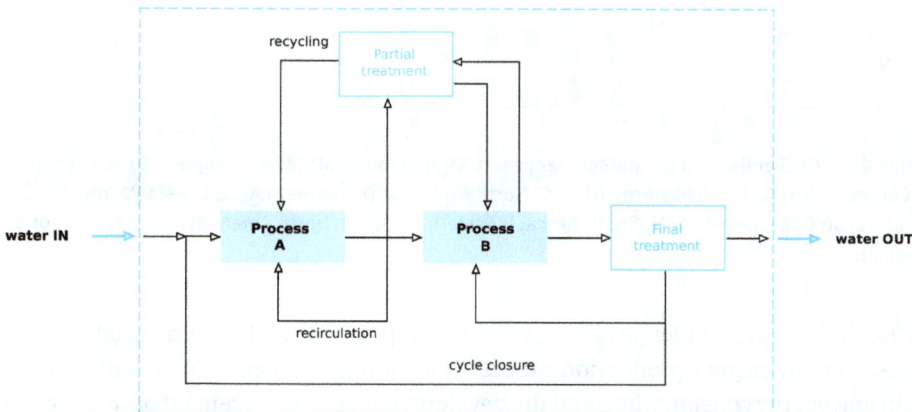

Figure 2.34: Water use, reuse, treatment and recycling (van Ede, 2006).

It is well accepted that global warming is already happening and that it is the result of human influence rather than a natural occurrence. The increasing and alarming level of (GHG) emissions worldwide is contributing to a large extent (*ca.* 98 %) to global warming. GHG emissions vary hugely between places and human groups due to differences in lifestyle and ways of producing energy (Worldmapper, 2016). As the level of industrialization can be directly related to the GHG emission per person per year (Figure 2.35), society regards the industrial sector as accountable for most of the enhanced greenhouse effect. Industry is facing, therefore, a mayor technological challenge with the ultimate goal of improving its societal image. Since the 2010s sustainability-conscious corporations have embarked upon aggressively reducing their GHG emissions as part of their corporate responsibility strategy. Promising results that have been achieved include orders of magnitude reduction of CO_2 emissions, by replacing non-renewable sources of energy with alternative sources, and the development of more climate-friendly refrigerants, among other initiatives. The challenges in this area aim at the reduction the impact of industry on the environment, with an emphasis on global warming, resource depletion, solid waste generation, air and water pollution and land degradation (Table 2.15).

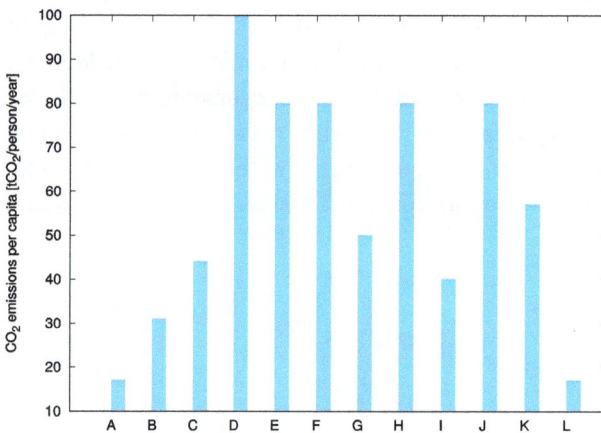

Figure 2.35: GHG emissions per person per year (tCO_2/person/year) (Worldmapper, 2016). Legend: A, Central Africa; B, South Eastern Africa; C, Northern Africa; D, Southern Asia; E, Asia Pacific; F, Middle East; G, Eastern Asia; H, South Africa; I, Eastern Europe; J, North America; K, Western Europe; L, Japan.

Particularly, those challenges include the development of tools and methods for the assessment of current production processes regarding GHG emissions, carbon footprint and energy consumption, and the development and embracement of sustainable manufacturing practices leading to new processes with low carbon footprint and low utilities consumption. Other challenges derived from the dynamics of current societal, technological and environmental spheres include areas as the development of more concentrated products to reduce packaging and transport costs and the development and implementation of sustainable agriculture policies.

Next to the challenges derived from the manufacture of products and services, a relevant source of environmental impact originates from the use of the product and consumer habits. As mentioned by Radjou and Prabhu (2015), two-thirds of GHG from a global FMCG corporation and half of the water consumption does not come from manufacture activities or transportation, but from their use. The ability of achieving more with less – referred to as frugal innovation – does not concern the industrial sector exclusively. Consumers need to shape our behavior to achieve more frugal outcomes.

2.5.4 Challenge #4: on-shelf availability by supply chain optimization

Current conditions of the chemical industry, and in the FMCG sector in particular, require a change in the way materials or goods flow through the value creation chain SOURCE ▷ MAKE ▷ DELIVER ▷ SELL. As mentioned by Radjou and Prabhu (2015), today's supply chains lack efficiency and flexibility because of the vast distance between the manufacturing location and the market where the product is consumed. As a result of this geographical gap, distribution costs are inflated and the level of responsiveness to changes in consumer needs is limited. In order to accommodate for potential and unexpected market changes, FMCG organizations hold large inventories, which contribute to extra costs.

FMCG processing activities and logistic operations are dynamic, stochastic and affected by uncertainties. Under these conditions, conventional operational/strategic policies are (locally) expertise-driven and most likely suboptimal. A challenge to address in this area is given by the need of coping effectively with the dynamics of the FMCG sector while creating a frugal supply chain. In this new framework – where more is achieved with less – manufacturers are required to have their products or services available when the consumer wants to buy them (on-shelf availability) and to have product propositions affordable for the consumers (value for money), while maintaining an innovative product portfolio (Bongers and Almeida-Rivera, 2011).

On-shelf availability requires a large inventory of products, which is penalized by substantial costs. At large product inventories, novel product innovations are costly due to a write-off of obsolete products, or due to the long lead-time before the old inventory is sold. On the other hand, manufacturing sites need to be efficient, which is achieved by manufacturing the products continuously for a longer time (*i. e.*, long run length) and keeping the number of changeovers from one product to the next minimal. Solving these conflicting objectives supports the delivery of a truly optimized demand-led supply chain.

Several approaches have been implemented to create a more responsive and frugal supply chain, including reshoring, local sourcing, sharing resources, distributing to the last mile (*i. e.*, access to users at remote locations), integrating manufacturing and logistics, using real-time demand signals and sharing data with partners (Radjou and Prabhu, 2015). Making the most of the existing assets supports the creation of a responsive and cost-effective supply chain. Flexing manufacturing assets have been successfully developed in several industrial cases, including the development of modular designs of manufacturing processes, continuous operation of traditional batch processes (*e. g.*, drug manufacture), mass customization at lower costs and decentralized production.

Challenge #4 comprises the development of realistic and complete network models to optimize on-shelf availability at lowest achievable cost in the presence of uncertainties. Novel modeling and optimization techniques for currently conflicting distributed length and time scales are foreseen to be developed to evaluate, optimize and synthesize such supply chains.

2.6 A novel approach for product design and process synthesis

Addressing the dynamic challenges of the FMCG sector requires the development of an appropriate design methodology. Such methodology should be equipped with the right tools and framework to solve the multi-dimensionality of the design problems, where multiple stakeholders, objectives and types of resources characterize the design space through various time and length scales. As brilliantly cast by A. Einstein, we cannot solve our problems with the same thinking we used when we created them. Aiming at a more structured approach toward the synthesis of consumer-centered product and their manufacturing processes, in the next chapters we propose a methodology that exploits the synergy of combining product and process synthesis workstreams. This methodology, Product-driven Process Synthesis (PDPS), is supported by decomposing the problem into a hierarchy of design levels of increasing refinement, where complex and emerging decisions are made to proceed from one level to another. The PDPS approach is one of the vehicles to embrace most of the design challenges in the FMCG sector, where key players are coping effectively with the dynamic environment of short process development and design times. While being novel and refreshing, this design methodology faces comparable challenges as any tool and method stemming from the chemical and process engineering community.

These challenges, which are accentuated during the development and deployment phases of the methodology, include the further embracement of a multi-scale modeling approach (from molecules to enterprise, as depicted in Figure 2.36), its casting into software tools for industrial use (Mattei *et al.*, 2014; Zondervan *et al.*, 2011; Kalakul *et al.*, 2018; Fung *et al.*, 2016; Gani, 2004b; Eden *et al.*, 2004; Zhang *et al.*, 2017; Ng, 2004) and the further exploration of a wider range of industrially relevant sectors and applications (Bröckel and Hahn, 2004; Wibowo and Ng, 2001, 2002). The following chapters address all levels of the PDPS approach, with emphasis on the scope of each design level and on a comprehensive description of the activities at each level.

The opportunities and benefits of the approach are stressed by addressing several case studies, which have resulted in either relatively important financial improvements in FMCG applications or in the birth of novel and differentiating alternatives to current manufacture practices.

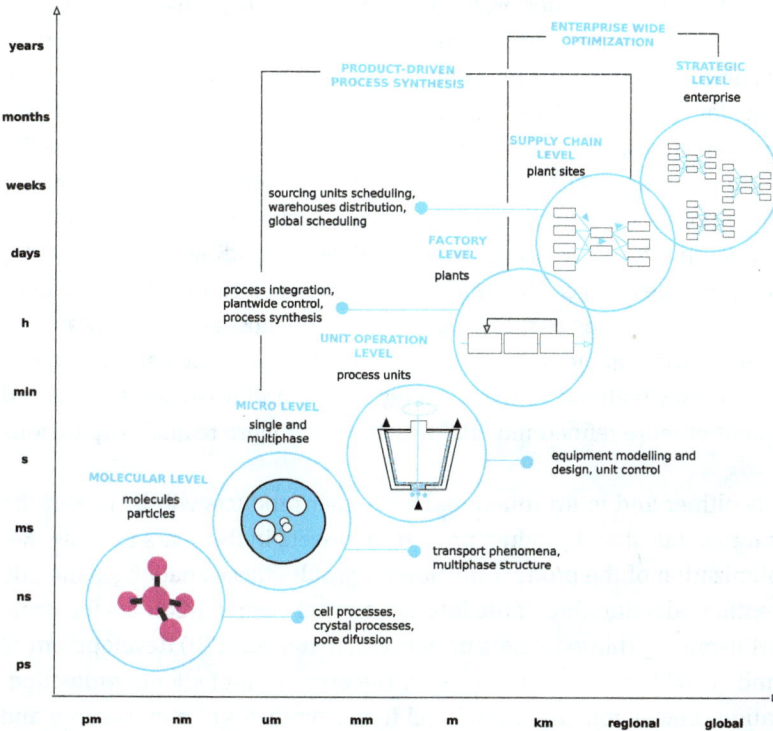

Figure 2.36: Multi-scale modeling spaces in the chemical industry. Modified from Marquardt *et al.* (2000).

2.7 Take-away message

The challenges to be faced by the process and chemical engineering disciplines in the FMCG sector are driven by a continuously societal demand for more efficient, controllable, sustainable, environmentally benign and safer processes that deliver healthier and multi-functional products. From a process perspective, the scope of consumer-product manufacturing activities is gradually broadening towards the optimization of multiple resources regarding spatial and temporal aspects. Time aspects and loss prevention are of paramount importance if the performance of a production plant is to be optimized over its manufacturing life span. Further challenges include the minimization of losses of mass, energy, run-time availability and, subsequently, profit. Poor process controllability and lack of plant responsiveness to market demands are just two issues that need to be considered by chemical and process engineers as causes of profit loss. It is a well-accepted fact that classical chemical engineering is shifting its focus towards process miniaturization. The recent advances at the micro- and nanoscales are empowering chemical and process engineers to pursue the quest of more efficient, controllable and safer processes at those length scales. In this area, a future challenge in the consumer products sector would be adopting a task-oriented design paradigm in place of the historical equipment-inspired design paradigm. This refocus will allow for the possibility of task integration and the design of novel unit operations or microsystems, which integrate several functions/tasks and reduce the cost and complexity of process systems. As mentioned by Zhelev (2005), PI via miniaturization might impose a set of additional challenges, including subsystem integration, flexibility, operability, safety issues, layout problems, building superstructures, converting batch processes into continuous microprocesses, formulation of the optimization problem and managing required resources in the most acceptable way. The drive for PI and more sustainable plant operations might lead in the near future to a different footprint of raw material sources, energy supply and waste generation and to the development of more refined models, which couple more realistically various phenomena.

Delivering healthier and multi-functional consumer products will be one of the key challenges for chemical and product-process engineers in the years to come. Regarding the optimization of the product nutritional profile, these challenges include (i) the fundamental understanding of the interactions between ingredients for structured emulsions involving (bio)polymers or particulate network; (ii) development of cost-effective and scalable nanotechnologies in the areas of ingredients protection, nanoencapsulation, new edible and functional films, synthesis of nanosensors and nanotracers and controlled rate of release of macronutrients; (iii) self-assembly of structures inside people; and (iv) encapsulation and targeted release of functional ingredients.

The challenges involving mild processing technologies include the development of technologies that can handle liquid products and heterogeneous products that contain chunks; the synthesis of aseptic systems based on Ohmic and microwave technologies; the synthesis of active packaging for microwave; better control strategy for Ohmic heating operations; the development of tools and methodologies to successfully handle the unpredictable nature of growing crops; and the development of tools and methods towards a superior nutrient/color retention and flavor release of processed fruits and vegetables. This last challenge might involve the use of natural actives to protect tissue structures during freezing or drying, the use of dehydro-freezing and infusions of fibers to improve rehydration, etc.

Regarding sustainability, future efforts should be focused on the development and implementation of manufacturing excellence practice with a stress on the paramount importance of better resource utilization. As society is increasingly demanding sustainable, environmentally benign and safer processes, the FMCG industry should take more actively into account sustainability of the biosphere. The challenges in this area include a global commitment of key players towards the reduction of GHG emissions and water/energy consumption; assessment of current production processes regarding GHG emissions and carbon footprint; establishment of sustainable manufacturing practices in all manufacturing sites; design of new processes with low carbon footprint and low utilities consumption; development of more concentrated products to reduce packaging and transport; and development of sustainable agriculture policies and engagement of all suppliers, farmers and factories worldwide.

Bibliography

Agreda, V. and Partin, L. (1984). Reactive distillation process for the production of methyl acetate (US4435595). Patent.

Agterof, W. G. M., Vaessen, G. E. J., Haagh, G. A. A. V., Klahn, J. K. and Janssen, J. J. M. (2003). Prediction of emulsion particle sizes using a computational fluid dynamics approach. *Colloids and Surfaces B: Biointerfaces*, 31, pp. 141–148.

Almeida-Rivera, C., Bongers, P., Egan, M., Irving, N. and Kowalski, A. (2011). Mixing apparatus and method for mixing fluids. US Patent App. 13/884,635.

Almeida-Rivera, C., Nijsse, J. and Regismond, S. (2010a). Method for preparing a fibre containing emulsion. US Patent App. 12/720,106.

Almeida-Rivera, C. P., Khalloufi, S. and Bongers, P. (2010b). Prediction of Supercritical Carbon Dioxide Drying of Food Products in Packed Beds. *Drying Technology*, 28(10), pp. 1157–1163.

Attarakih, M. M., Bart, H. J. and Faqir, N. M. (2003). Optimal Moving and Fixed Grids for the Solution of Discretized Population Balances in Batch and Continuous Systems: Droplet Breakage. *Chemical Engineering Science*, 58, pp. 1251–1269.

Backhaus, A. A. (1921a). Apparatus for the manufacture of esters (US1.400.849.20.12). Patent.

Backhaus, A. A. (1921b). Apparatus for the manufacture of esters (US1.400.850.20.12). Patent.

Backhaus, A. A. (1921c). Apparatus for the manufacture of esters (US1.400.851.20.12). Patent.

Bagajewicz, M. (2007). On the Role of Microeconomics, Planning, and Finances in Product Design. *AIChE Journal*, 53(12), pp. 3155–3170.

Becht, S., Franke, R., Geißmann, A. and Hahn, A. (2009). An industrial view of process intensification. *Chemical Engineering Progress*, pp. 329–332.

Benichou, A., Aserin, A. and Garti, N. (2004). Double Emulsions Stabilized with Hybrids of Natural Polymers for Entrapment and Slow Release of Active Matters. *Advances in Colloid and Interface Science* 108–109, 29–41.

BLS (2016). National Census of Fatal Occupational Injuries in 2016. Tech. rep., Bureau of Labor Statistics.

Bongers, P. (2009). Intertwine product and process design. Inaugural lecture/Eindhoven University of Technology, ISBN: 978-90-386-2124-1.

Bongers, P. and Almeida-Rivera, C. (2011). Optimal run length in factory operations to reduce overall costs. *Computer-Aided Chemical Engineering*, 29, pp. 900–904.

Brandt, S. (1987). *Strategic Planning in Emerging Companies*. Addison-Wesley.

Bröckel, U. and Hahn, C. (2004). Product Design of Solid Fertilizers. *Trans. IChemE – Part A*, 82, pp. 1453–1457.

Broughton, E. (2005). The Bhopal disaster and its aftermath: a review. *Environ. Health*, 4, pp. 1–6.

Brown, T. (2008). Design Thinking. *Harvard Business Review*, pp. 1–9.

Brown, T. (2019). *Change by Design*. Harper Business.

Bruin, S. and Jongen, T. (2003). Food Process Engineering: The Last 25 Years and Challenges Ahead. *Comprehensive Reviews in Food Science and Food Safety*, 2, pp. 42–81.

Brul, S., Klis, F. M., Knorr, D., Abee, T. and Notermans, S. (2003). Food Preservation and the Development of Microbial Resistance. In: *Food Preservation Techniques*. CRC Press, pp. 524–551. Chapter 23.

Brundtland, M. (1987). *Our common future: The world commission on environment and development*. Oxford University Press.

BSDC (2017). Better Business – Better World. Tech. rep., Business and Sustainable Development Commission.

BSI (2015). BS EN ISO 9000:2015 – Quality management systems Fundamentals and vocabulary. Tech. rep., British Standards Institution.

CBInsights (2019). When Corporate Innovation Goes Bad (https://www.cbinsights.com/). Electronic Citation.

CDC (2007). Organization website (http://www.cdc.gov). Electronic Citation.

CEFIC (2016). Facts and Figures 2016 of the Erupoean chemical industry. Tech. rep., European Chemical Industry Council.

CEFIC (2018). Facts and Figures 2018 of the European Chemical Industry. Tech. rep., European Chemical Industry Council.

Charpentier, J. C. (2007). Modern Chemical Engineering in the Framework of Globalization, Sustainability and Technical Innovation. *Industrial and Engineering Chemistry Research*, 46, pp. 3465–3485.

Chiou, D. and Langrish, T. A. G. (2007). Development and Characterisation of Novel Nutraceuticals with Spray Drying Technology. *Journal of Food Engineering*, 82, pp. 84–91.

Cooper, R. (1990). Stage-gate systems: a new tool for managing new products. *Business Horizons*, 33(3), pp. 44–54.

Cooper, R. (1996). Overhauling the new product process. *Industrial Marketing Management*, 25, pp. 465–482.

Cooper, R. (1999). The invisible success factors in product innovation. *Journal of Product Innovation Management*, 16(2), pp. 115–133.

Coronel, P. (2005). *Continuous flow processing of foods using cylindrical applicator microwave systems operating at 915 MHz*. Thesis/dissertation, North Carolina State University.

Coronel, P., Simunovic, J. and Sandeep, K. P. (2003). Thermal profile of milk after heating in a continuous flow tubular microwave system operating at 915 MHz. *International Journal of Food Science*, 68(6), pp. 1976–1981.

Coronel, P., Truong, V. D., Simunovic, J., Sandeep, K. P. and Cartwright, G. (2005). Aseptic processing of sweetpotato purees using a continuous flow microwave system. *Journal of Food Science*, 70(9), pp. E531–E536.

Costa, R., Gabriel, R. G., Saraiva, P. M., Cussler, E. and Moggridge, G. D. (2014). *Chemical Product Design and Engineering*. John Wiley and Sons, pp. 1–35.

Costa, R., Moggridge, G. D. and Saraiva, P. M. (2006). Chemical Product Engineering: An Emerging Paradigm Within Chemical Engineering. *AIChE Journal*, 52(6), pp. 1976–1986.

Cross, W. and Ramshaw, C. (2000). Process Intensification – laminar-flow heat-transfer. *Chemical Engineering Research and Design*, 64, pp. 293–301.

Crowl, D. and Louvar, J. (2011). *Chemical Process Safety – Fundamentals with Application*, 3rd ed. Prentice Hall.

Cussler, E. L. and Moggridge, G. D. (2001). *Chemical Product Design*. Cambridge Series in Chemical

Engineering, New York.

Cussler, E. L. and Wei, J. (2003). Chemical Product Engineering. *AIChE Journal*, 49(5), pp. 1072–1075.

Datta, A. K. and Ananthesarawan, R. C. (2001). *Handbook of Microwave Technology for Food Applications*. New York: Marcel Dekker.

Davies, J. T. (1985). Drop Sizes in Emulsion Related to Turbulent Energy Dissipation Rates. *Chemical Engineering Science*, 40(5), pp. 839–842.

Deloitte (2014). The Deloitte Consumer Review: The Growing Power of Consumers. Tech. rep., Deloitte.

Drozd, M. and Joines, W. T. (1999). Electromagnetic exposure chamber for improved heating. Patent.

Dubbelboer, A., Hoogland, H., Zondervan, E. and Meuldijk, J. (2014a). Model selection for the optimization of the sensory attributes of mayonnaise. In: *Proceedings of OPT-i*, pp. 1455–1461.

Dubbelboer, A., Janssen, J., Hoogland, H., Mudaliar, A., Maindarkar, S., Zondervan, E. and Meuldijk, J. (2014b). Population balances combined with Computational Fluid Dynamics: A modeling approach for dispersive mixing in a high pressure homogenizer. *Chemical Engineering Science*, 117, pp. 376–388.

Dubbelboer, A., Janssen, J., Hoogland, H., Mudaliar, A., Zondervan, E., Bongers, P. and Meuldijk, J. (2013). A modeling approach for dispersive mixing of oil in water emulsions. *Computer-Aided Chemical Engineering*, 32, pp. 841–846.

Dubbelboer, A., Janssen, J., Zondervan, E. and Meuldijk, J. (2018). Steady state analysis of structured liquids in a penetrometer. *Journal of Food Engineering*, 218, pp. 50–60.

Dubbelboer, A., Zondervan, E., Meuldijk, J., Hoogland, H. and Bongers, P. (2012). A neural network application in the design of emulsion-based products. *Computer-Aided Chemical Engineering*, 30, pp. 692–696.

Eaton, M. (2013). *The Lean Practitioner's Handbook*. KoganPage.

Eden, M., Jorgensen, S., Gani, R. and El-Halwagi, M. (2004). A novel framework for simultaneous separation process and product design. *Chemical Engineering and Processing*, 43, pp. 595–608.

Edwards, M. F. (2006). Product engineering: Some challenges for Chemical Engineers. *Transactions of the Institute of Chemical Engineers – Part A*, 84(A4), pp. 255–260.

Ernest-Young (1999). *Efficient product introductions – the development of value-creating relationships*. ECR Europe.

FAO (2007). Organization website (http://www.fao.org). Electronic Citation.

Foodbev (2017a). Breakfast snacks: health and convenience. *FoodBev*, 1(11), pp. 38–39.

Foodbev (2017b). Functional foods. *FoodBev*, 1(12), pp. 44–46.

Foodbev (2017c). Meaty snacks: the quest for top spot. *FoodBev*, 1(10), pp. 50–51.

Foodbev (2017d). Nanotechnology: fine-tune packaging. *FoodBev*, 1(13), pp. 18–19.

Foodbev (2018a). Anti-ageing drinks. *FoodBev*, 1(23), pp. 32–33.

Foodbev (2018b). Calcium: still an important claim? *FoodBev*, 1(21), pp. 40–41.

Foodbev (2018c). Collagen: beauty benefits. *FoodBev*, 1(18), pp. 20–21.

Foodbev (2018d). Extending shelf life: making food last longer. *FoodBev*, 1(14), pp. 42–43.

Foodbev (2018e). Functional tea: brewing up benefits. *FoodBev*, 1(16), pp. 16.

Foodbev (2018f). Healthier chocolate. *FoodBev*, 1(22), pp. 40–43.

Foodbev (2018g). Lightweight packaging: reducing weight, reducing cost. *FoodBev*, 1(18), pp. 42–43.

Foodbev (2018h). Meat alternatives. *FoodBev*, 1(18), pp. 31–33.

Foodbev (2018i). Probiotics: added functionality. *FoodBev*, 1(19), pp. 26–27.

Foodbev (2018j). The rise of meal kits: driven by convenience. *FoodBev*, 1(15), pp. 42–47.

Foodbev (2018k). Water reduction. *FoodBev*, 1(22), pp. 50–51.

Foodbev (2019a). Cannabis beverages. *FoodBev*, 1(27), pp. 38–41.

Foodbev (2019b). Circular economy. *FoodBev*, 1(26), pp. 44–45.

Foodbev (2019c). Gut health applications. *FoodBev*, 1(25), pp. 22–25.

Foodbev (2019d). Heart health. *FoodBev*, 1(24), pp. 12–18.

Foodbev (2019e). Immune boosting dairy. *FoodBev*, 1(28), pp. 34–35.

Foodbev (2019f). Infant nutrition. *FoodBev*, 1(28), pp. 14–17.

Foodbev (2019g). Low- and no-alcohol beer. *FoodBev*, 1(24), pp. 20–26.

Foodbev (2019h). Non-nutritive sweeteners. *FoodBev*, 1(24), pp. 14–15.

Foodbev (2019i). Superfoods. *FoodBev*, 1(25), pp. 42–43.

Fung, K., Ng, K., Zhang, L. and Gani, R. (2016). A grand model for chemical product design. *Computers and Chemical Engineering*, 91, pp. 15–27.

Galanopoulos, C., Odiema, A., Barletta, D. and Zondervan, E. (2017). Design of a wheat straw supply chain network in Lower Saxony, Germany through optimization. *Computer-Aided Chemical Engineering*, 40, pp. 871–876.

Gani, R. (2004a). Chemical product design: challenges and opportunities. *Computers and Chemical Engineering*, 28, pp. 2441–2457.

Gani, R. (2004b). Computer-aided methods and tools for chemcial product design. *Trans. IChemE – Part A*, 82(A11), pp. 1494–1504.

Gao, S. and Low, S. (2014). *Lean Construction Management: The Toyota Way*. Springer.

Gerven-van, T. and Stankiewicz, A. (2009). Structure, energy, synergy, time – the fundamentals of process intensification. *Ind. Eng. Chem. Res.*, 48, pp. 2465–2474.

Grace, H. (1982). Dispersion Phenomena in High Viscosity Immiscible Fluid Systems and Application of Static Mixers as Dispersion Devices in Such Systems. *Chemical Engineering Communications*, 14, pp. 225–277.

Groeneweg, F., Agterof, W. G. M., Jaeger, P., Janssen, J. J. M., Wieringa, J. A. and Klahn, J. K. (1998). On the mechanism of the inversion of emulsions. *Transactions of the Institute of Chemical Engineers – Part A*, 76, pp. 55–63.

Grossman, I. and Westerberg, A. (2000). Research challenges in Process Systems Engineering. *AIChE Journal*, 46(9), pp. 1700–1703.

Henard, D. and Szymanski, D. (2001). Why some new products are more succesful than others. *Journal of Marketing Research*, 38, pp. 362–375.

Herder, P. (1999). *Process Design in a changing environment*. Thesis/dissertation, Delft University of Technology.

Hill, M. (2004). Product and Process Design for Structured Products. *AIChE Journal*, 50(8), pp. 1656–1661.

Hill, M. (2009). Chemical product engineering – the third paradigm. *Computers and Chemical Engineering*, 33, pp. 947–953.

Holdsworth, D. and Simpson, R. (2007). *Thermal Processing of Packaged Foods*. Springer.

HSE (2017). Fatal Injuries Arising From Accidents at Work in Great Britain in 2017. Tech. rep., Health and Safety Executive.

Hughes, P. and Ferret, E. (2010). *International Health and Safety at work*. Taylor and Francis Group.

Janssen, J. J. M., Boon, A. and Agterof, W. G. M. (1994a). Influence of Dynamic Interfacial Properties in Droplet Breakup in Simple Plane Hyperbolic Flow. *AIChE Journal*, 43(9), pp. 1436–1447.

Janssen, J. J. M., Boon, A. and Agterof, W. G. M. (1994b). Influence of Dynamic Interfacial Properties in Droplet Breakup in Simple Shear Flow. *AIChE Journal*, 40(12), pp. 1929–1939.

Jones, F. (1897). Apparatus for electrically treating liquids. Patent.

Kalakul, S., Zhang, L., Fang, Z., Choudhury, H., Intikhab, S., Elbashir, N., Eden, M. and Gani, R. (2018). Computer aided chemical product design – ProCAPD and tailor-made blended products. *Computers and Chemical Engineering*, 116, pp. 37–55.

Karbstein, H. and Schubert, H. (1995). Developments in the continuous mechanical production of oil-in-water macro-emulsions. *Chemical Engineering and Processing*, 34, pp. 205–211.

Khalloufi, S., Almeida-Rivera, C. P. and Bongers, P. (2010). Supercritical-CO_2 drying of foodstuffs in packed beds: Experimental validation of a mathematical model and sensitive analysis. *Journal of Food Engineering*, 96(1), pp. 141–150.

Kiran, D. (2017). *Total Quality Management: Key Concepts and Case Studies*. BS Publications.

Kivits, G. A. A., Sman, F. P. P. and Tiburg, L. B. M. (1997). Analysis of catechins from green and black tea in humans: a specific and sensitive colormetric assay of total catechins in biological fluids. *International Journal of Food Science and Nutrition*, 48, pp. 387–392.

Klahn, J. K., Janssen, J. J. M., Vaessen, G. E. J., de Swart, R. and Agterof, W. G. M. (2002). On the escape process during phase inversion of an emulsion. *Colloids and Surfaces A: Physicochemical and Engineering Aspects*, 210, pp. 167–181.

Kleef, E. (2006). *Consumer research in the early stages of new product development*. Ph. D. thesis, Wageningen Universiteit.

Kletz, T. (1998). *Process Plants: a handbook for inherently safer design*. Taylor and Francis Group.

Kostoglou, M. (2006). On the Evolution of Particle Size Distribution in Pipe Flow of Dispersion Undergoing Breakage. *Industrial and Engineering Chemistry Research*, 45, pp. 2143–2145.

Lander, R., Manger, W., Scouloudis, M., Ku, A., Davis, C. and Lee, A. (2000). Gaulin Homogenization: A Mechanistic Study. *Biotechnology Progress*, 16, pp. 80–85.

Leng, D. and Calabrese, R. (2004). Immiscible liquid–liquid systems. In: E. Paul, V. Atiemo-Obeng and S. Kresta, eds., *Encyclopedia of Industrial Mixing*. Wiley Interscience. Chapter 12.

Lerner, K. and Lerner, B. (eds.) (2004). *The Gale Encyclopedia of Science*, 3rd ed. Thomson Gale.

Li, X. and Kraslawski, A. (2004). Conceptual design synthesis: past and current trends. *Chemical Engineering and Processing*, 43, pp. 589–600.

Lutze, P. and Górak, A. (2016). *Reactive and membrane-assisted separations*. De Gruyter.

MacCarthur-Foundation, E. (2015). Towars a Circular Economy: Business Rationale for an Accelerated Transition. Tech. rep., Ellen MacCarthur-Foundation.

Malone, M. F. and Doherty, M. F. (2000). Reactive Distillation. *Industrial and Engineering Chemistry Research*, 39, pp. 3953–3957.

Marquardt, W., Wedel, L. and Bayer, B. (2000). Perspectives on lifecycle process modeling. In: M. F. Malone, J. A. Trainham and B. Carnahan, eds., *Proceedings of AIChE Symposium Series*, vol. 96, pp. 192–214.

MARSH (2016). The 100 Largest Losses: 1974–2017. Tech. rep., Marsh and McLennan.

Mattei, M., Kontogeorgis, G. and Gani, R. (2014). A comprehensive framework for surfactant selection and design and for emulsion based chemical product design. *Fluid Phase Equilibria*, 362, pp. 288–299.

McClements, J. (2005). *Food Emulsions: Principles, Practices and Techniques*, 2nd ed. Boca Raton: CRC Pres.

McClements, J., Decker, E. and Weiss, J. (2007). Emulsion-Based Delivery Systems for Lipophilic Bioactive Components. *Journal of Food Science*, 72(8), pp. 109–124.

Moggridge, G. D. and Cussler, E. L. (2000). An Introduction to Chemical Product Design. *Transactions of the Institute of Chemical Engineers – Part A*, 78, pp. 5–11.

Monsanto, M., Hooshyar, N., Meuldijk, J. and Zondervan, E. (2014a). Modeling and optimization of green tea precipitation for the recovery of catechins. *Separation and Purification Technology*, 129, pp. 129–136.

Monsanto, M., Mestrom, R., Zondervan, E., Bongers, P. and Meuldijk, J. (2015). Solvent Swing Adsorption for the Recovery of Polyphenols from Black Tea. *Industrial and Engineering Chemistry Research*, 54(1), pp. 434–442.

Monsanto, M., Radhakrishnan, A., Sevillano, D., Hooshyar, N., Meuldijk, J. and Zondervan, E. (2014b). Modeling and optimization of an adsorption process for the recovery of catechins from green tea. *Computer-Aided Chemical Engineering*, 33, pp. 1441–1446.

Monsanto, M., Radhakrishnan, A. and Zondervan, E. (2016). Modelling of packed bed adsorption columns for the separation of green tea catechins. *Separation Science and Technology*, 51(14), pp. 2339–2347.

Monsanto, M., Trifunovic, O., Bongers, P., Meuldijk, J. and Zondervan, E. (2014c). Black tea cream effect on polyphenols optimization using statistical analysis. *Computers and Chemical Engineering*, 66, pp. 12–21.

Monsanto, M., Zondervan, E. and Meuldijk, J. (2013). Integrated optimization of black tea cream effect using statistical analysis. *Computer-Aided Chemical Engineering*, 32, pp. 31–36.

Monsanto, M., Zondervan, E., Trifunovic, O. and Bongers, P. (2012). Integrated optimization of the adsorption of theaflavins from black tea on macroporous resins. *Computer-Aided Chemical Engineering*, 31, pp. 725–729.

Moore, R. (2007). *Selecting the Right Manufacturing Improvement Tool*. Butterworth-Heinemann.

Nere, N. and Ramkrishna, D. (2005). Evolution of drop size distributions in fully developed turbulent pipe flow of a liquid-liquid dispersion by breakage. *Industrial and Engineering Chemistry Research*, 44, pp. 1187–1193.

Ng, K. (2004). MOPSD: a framework linking business decision-making to product and process design. *Computers and Chemical Engineering*, 29, pp. 51–56.

Nienow, A. (2004). Break-up, coalescence and catastrophic phase inversion in turbulent contactors. *Advances in Colloid and Interface Science*, 108, pp. 95–103.

NIOSH (2012). NIOSH Pocket Guide to Chemical Hazards.

Norton, I., Fryer, P. and Moore, S. (2006). Product/Process Integration in food Manufacture: Engineering Sustained Health. *AIChE Journal*, 52(5), pp. 1632–1640.

NSB (2018). Overview of the State of the U. S. S&E Enterprise in a Global Contect. Tech. rep., National Science Board.

Ohno, T. (1988). *Toyota production system: Beyond large-scale production*. MA: Productivity Press.

Perrier-Cornet, J. M., Marie, P. and Gervais, P. (2005). Comparison of Emulsification Efficiency of Protein-Stabilized Oil-in-Water Emulsion Using Jet, High Pressure and Colloid Mill Homogenization. *Journal of Food Engineering*, 66, pp. 211–217.

PWC (2017). The 2017 Global Innovation 1000 Study. Tech. rep., PwC.

Radjou, N. and Prabhu, J. (2015). *Frugal Innovation: how to do better with less*. The Economist.

Ramshaw, C. (1983). HIGEE distillation – An example of Process Intensification. *Chemical Engineering*, pp. 13–14.

Rossi, M., Charon, S., Wing, G. and Ewell, J. (2006). Design for the next generation: incorporation cradle-to-cradle design into Herman Miller products. *Journal of Industrial Ecology*, 10(4), pp. 193–210.

Saboni, A., Alexandrova, S., Gourdon, C. and Cheesters, A. K. (2002). Interdrop coalescence with mass transfer: comparison of the approximate drainage models with numerical results. *Chemical Engineering Journal*, 88, pp. 127–139.

Saraiva, P. M. and Costa, R. (2004). A Chemical Product Design Course with a Quality Focus. *Chemical Engineering Research and Design: Part A*, 82(A11), pp. 1474–1484.

Schubert, H. and Engel, R. (2004). Product and Formulation Engineering of Emulsions. *Chemical Engineering Research and Design*, 82(9), pp. 1137–1143.

Schubert, H., Engel, R. and Kempa, L. (2006). Principles of Structured Food Emulsions: Novel formulations and trends. *IUFoST*, 13(43), pp. 1–15.

Seider, W. D., Seader, J. D. and Lewin, D. R. (2004). *Product and Process Design Principles*. John Wiley.

Seider, W. D., Seader, J. D., Lewin, D. R. and Widagdo, S. (2010). *Product and Process Design Principles*. John Wiley and Sons, Inc.

Seider, W. D., Widagdo, S., Seader, J. D. and Lewin, D. R. (2009). Perspectives on chemical product and process design. *Computers and Chemical Engineering*, 33, pp. 930–935.

Shaw, A., Yow, H. N., Pitt, M. J., Salman, A. D. and Hayati, I. (2004). Experience of Product Engineering in a Group Design Project. *Chemical Engineering Research and Design: Part A*, 82(A11), pp. 1467–1473.

Siirola, J. J. (1996). Industrial applications of chemical process synthesis. In: J. Anderson, ed., *Advances in Chemical Engineering. Process Synthesis*. Academic Press, pp. 1–61. Chapter 23.

Silla, H. (2003). *Chemical Process Engineering – Design and Economics*. Marcel Dekker Inc.

Smith, B. V. and Ierapepritou, M. G. (2010). Integrative chemical product design strategies: Reflecting industry trends and challenges. *Computers and Chemical Engineering*, 34, pp. 857–865.

Smith, R. (2015). *Chemical Process Design* (1–29). Kirk–Othmer Encyclopedia of Chemical Technology, John Wiley and Sons.

Stang, M., Karbstein, H. and Schubert, H. (1994). Adsorption Kinetics of Emulsifiers at Oil-Water Interfaces and Their Effect on Mechanical Emulsification. *Chemical Engineering and Processing*, 33, pp. 307–311.

Stankiewicz, A. (2001). Between the chip and the blast furnace. *NPT Processtechnologie*, 8(2), pp. 18–22.

Stankiewicz, A. (2003). Reactive separations for process intensification: an industrial perspective. *Chemical Engineering and Processing*, 42, pp. 137–144.

Stankiewicz, A. and Moulijn, J. (2000). Process intensification: transforming chemical engineering. *Chemical Engineering Progress*, 96, pp. 22–34.

Stankiewicz, A. and Moulijn, J. (2002). Process intensification. *Industrial and Engineering Chemistry Research*, 41(8), pp. 1920–1924.

Stankiewicz, A. and Moulijn, J. (2004). *Re-engineering the Chemical Processing Plant – Process Intensification*. Marcel Dekker.

Steiner, H., Teppner, R., Brenn, G., Vankova, N., Tcholakova, S. and Denkov, N. (2006). Numerical Simulation and Experimental Study of Emulsification in a Narrow-Gap Homogenizer. *Chemical Engineering Science*, 61, pp. 5841–5855.

Stern, N. H. (2007). *The economics of climate change: the Stern review*. Cambridge, UK: Cambridge University Press.

Sugiura, S., Nakajima, M. and Seki, M. (2004). Prediction of droplet diameter for microchannel emulsification: prediction model for complicated microchannel geometries. *Industrial and Engineering Chemistry Research*, 43, pp. 8233–8238.

Tapia, M., Arispe, I. and Martinez, A. (2005). Safety and Quality in the Food Industry. In: G. Barbosa-Cánovas, M. Tapia and M. Cano, eds., *Novel Food Processing Technologies*. CRC Press, pp. 669–679. Chapter 33.

Tarver, T. (2006). Food Nanotechnology. *Food Technology*, 60(11), pp. 22–26.

Tjaberinga, W. J., Boon, A. and Cheesters, A. K. (1993). Model Experiments and Numerical Simualtions on Emulsification Under Turbulent Conditions. *Chemical Engineering Science*, 48(2), pp. 285–293.

Tsaoulidis, D. (2015). *Studies of intensified small-scale processes for liquid-liquid separations in spent nuclear fuel reprocessing*. Ph. D. thesis, University College London.

Tsouris, C. and Porcelli, J. (2003). Process Intensification – Has its time finally come? *Chemical Engineering Progress*, 99(10), pp. 50–55.

Ulrich, K. and Eppinger, S. (2012). *Product Design and Development*. New York: McGraw-Hill.

UN (2017). The Sustainable Development Goals Report 2017. Tech. rep., United Nations.

UNEP (2016). Global Material Flows and Resource Productivity. Tech. rep., United Nations Environment Programme.

Valle, D. D., Tanguy, P. and Carreau, P. (2000). Characterization of the Extensional Properties of Complex Fluids Using an Orifice Flowmeter. *Journal of Non-Newtonian Fluids Mechanics*, 94, pp. 1–13.

van Ede, J. (2006). Making Money Like Water? *Food Engineering and Ingredients*, 1(3), pp. 52–55.

van Elzakker, M., La-Marie, B., Zondervan, E. and Raikar, N. (2013a). Tactical planning for the fast moving consumer goods industry under demand uncertainty. In: *Proceedings of ECCE9*.

van Elzakker, M., Maia, L., Grossmann, I. and Zondervan, E. (2017). Optimizing environmental and economic impacts in supply chains in the FMCG industry. *Sustainable Production and Consumption*, 11, pp. 68–79.

van Elzakker, M., Zondervan, E., Almeida-Rivera, C., Grossmann, I. and Bongers, P. (2011). Ice Cream Scheduling: Modeling the Intermediate Storage. *Computer-Aided Chemical Engineering*, 29, pp. 915–919.

van Elzakker, M., Zondervan, E., Raikar, N., Grossmann, I. and Bongers, P. (2012). Tactical planning in the fast moving consumer goods industries: an SKU decomposition algorithm. In: *Proceedings of FOCAPO 2012*.

van Elzakker, M., Zondervan, E., Raikar, N., Hoogland, H. and Grossmann, I. (2013b). Tactical Planning with Shelf-Life Constraints in the FMCG Industry. *Computer-Aided Chemical Engineering*, 32, pp. 517–522.

van Elzakker, M., Zondervan, E., Raikar, N., Hoogland, H. and Grossmann, I. (2014a). Optimizing the tactical planning in the Fast Moving Consumer Goods industry considering shelf-life restrictions. *Computers and Chemical Engineering*, 66(4), pp. 98–109.

van Elzakker, M., Zondervan, E., Raikar, N., Hoogland, H. and Grossmann, I. (2014b). An SKU decomposition algorithm for the tactical planning in the FMCG industry. *Computers and Chemical Engineering*, 62(5), pp. 80–95.

Varma, R. and Varma, D. R. (2005). The Bhopal Disaster of 1984. *Bulletin of Science, Technology & Society*, 25(1), pp. 37–45.

Verloop, J. (2004). *Insight in Innovation – Managing Innovation by Understanding the Laws of Innovation*. Amsterdam: Elsevier.

Voncken, R. M., Broekhuis, A. A., Heeres, H. J. and Jonker, G. H. (2004). The Many Facets of Product Technology. *Chemical Engineering Research and Design: Part A*, 82(A11), pp. 1411–1424.

Walstra, P. (1993). Principles of Emulsion Formation. *Chemical Engineering Science*, 48(2), pp. 333–349.

Weiss, J., Takhistov, P. and McClements, J. (2006). Functional Materials in Food Nanotechnology. *Journal of Food Science*, 71(9), pp. R107–R116.

WHO (2018). World Health Statistics 2018 – Monitoring health for the SDGs. Tech. rep., World Health Organization.

WHO (2019). Organization website (http://www.who.org). Electronic Citation.

Wibowo, C. and Ng, K. M. (2001). Product-Oriented Process Synthesis and Development: Creams and Pastes. *AIChE Journal*, 47(12), pp. 2746–2767.

Wibowo, C. and Ng, K. M. (2002). Product-Centered Processing: Manufacture of Chemical-Based Consumer Products. *AIChE Journal*, 48(6), pp. 1212–1230.

Widen, S. (2019). How To Use Design Thinking And Agility To Reach Product Development Breakthroughs (http://www.forbes.com).

Worldmapper (2016). Official website (http://www.worldmapper.org). Electronic Citation.

Worldwatch-Institute-Staff (2014). *Vital Signs 2012 : The Trends That Are Shaping Our Future*. Island Press, Washington, United States.

Zderic, A., Almeida-Rivera, C., Bongers, P. and Zondervan, E. (2016). Product-driven process synthesis for the extraction of oil bodies from soybeans. *Journal of Food Engineering*, 185, pp. 26–34.

Zderic, A., Tarakci, T., Hosshyar, N., Zondervan, E. and Meuldijk, J. (2014). Process Design for Extraction of Soybean Oil Bodies by Applying the Product Driven Process Synthesis Methodology. *Computer-Aided Chemical Engineering*, 33, pp. 193–198.

Zderic, A. and Zondervan, E. (2016). Polyphenol extraction from fresh tea leaves by pulsed electric field: A study of mechanisms. *Chemical Engineering Research and Design*, 109, pp. 586–592.

Zderic, A., Zondervan, E. and Meuldijk, J. (2013a). Breakage of Cellular Tissue by Pulsed Electric Field: Extraction of Polyphenols from Fresh Tea Leaves. *Chemical Engineering Transactions*, 32, pp. 1795–1800.

Zderic, A., Zondervan, E. and Meuldijk, J. (2013b). Statistical analysis of data from pulsed electric field tests to extract polyphenols. In: *Proceedings of ECCE9*.

Zderic, A., Zondervan, E. and Meuldijk, J. (2015). Product-driven process synthesis for the extraction of polyphenols from fresh tea leaves. *Chemical Engineering Transactions*, 43, pp. 157–162.

Zhang, L., Fung, K., Wibowo, C. and Gani, R. (2017). Advances in chemical product design. *Reviews in Chemical Engineering*, pp. 1–2.

Zhelev, T. (2005). The Conceptual Design Approach – a Process Integration Approach on the Move. *Chemical Engineering Transactions*, 7, pp. 453–458.

Zondervan, E., Nawaz, M., Hahn, A., Woodley, J. and Gani, R. (2011). Optimal design of a multi-product biorefinery system. *Computers and Chemical Engineering*, 35, pp. 1752–1766.

3 A structured approach for product-driven process design of consumer products

3.1 Introduction

All leading FMCG companies are rapidly transforming from general manufacturing hubs of loosely connected products to companies delivering health, wellness and nutrition. Manufacturing in a responsible and sustainable way products within those strategic areas imposes technical challenges to work on and requires the development of research and development capabilities. These challenges and capabilities have the clear aim of delivering a product with all associated benefits (financial performance, environmental and societal impacts) at a short time-to-market and at a reduced manufacturing expenditure. Finding and creating business opportunities to be brought successfully to the market (Verloop, 2004) is the response of leading companies to the rapidly changing environment, characterized by slim profit margins and fierce competitiveness. A key activity in this innovation process is the actual creation of the conversion or manufacturing system.

Since its introduction, Process Systems Engineering (PSE) has been used effectively by chemical and process engineers to assist the development of chemical and process engineering. In tying science to engineering, PSE provides engineers with the systematic design and operation methods, tools that they require to successfully face the challenges of today's industry (Grossmann and Westerberg, 2000). Process Synthesis (PS) is one such method and focuses on the creation of the best conversion system that allows for an economical, safe and environmentally responsible conversion of specific feed stream(s) into specific product(s). According to Johns (2001), the aim of PS is the optimization of the logical structure of a chemical process, and specifically, PS focuses on the choice of chemicals (or feedstocks) employed including extracting agents, on the source and destination of recycle streams and on the sequence of processing steps (*e. g.*, chemical reaction, separation, extraction).

Originally conceived in the petrochemical sector at the end of the 1980s (Douglas, 1988; Siirola, 1996a), PS accounts for roughly 2 % of the total design costs and allows for fixing 80 % of the combined capital and operational costs (Siirola, 1996b). PS is not one step in the innovation process (Figure 3.1), but takes place in all conceptual process engineering steps within the process. In contrast to its relevance and impact on the expenditure of the entire plant, PS is normally carried out by copying existing processes or scaling up laboratory-scale non-optimized protocols.

Although the definition of PS might suggest a straightforward and viable action, process synthesis is complicated by the non-trivial tasks of (i) identifying and sequencing the physical and chemical tasks to achieve specific transformations; (ii) selecting feasible types of unit operations to perform these tasks; (iii) finding ranges of operating conditions per unit operation; (iv) establishing connectivity between units with

https://doi.org/10.1515/9783110570137-003

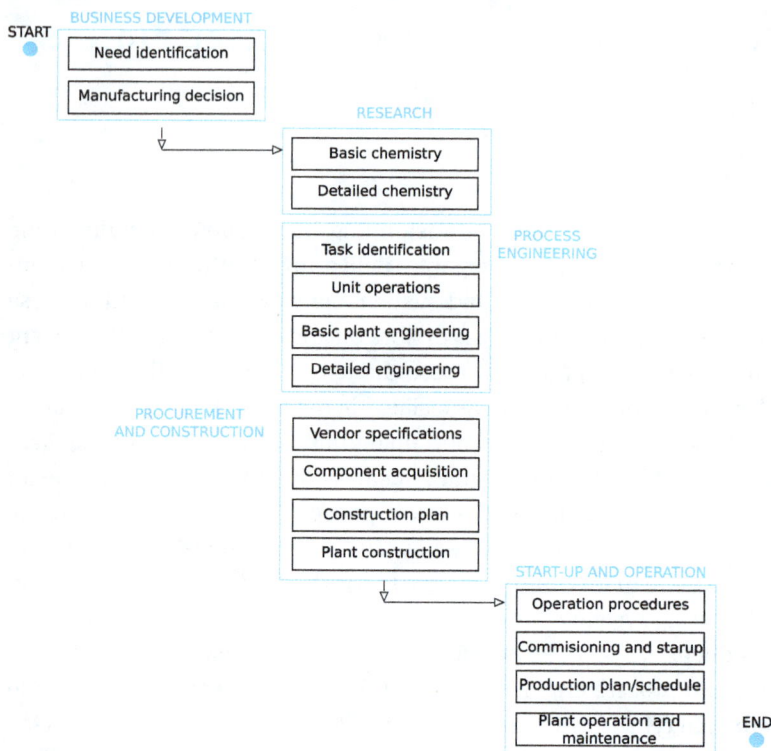

Figure 3.1: The sequence of the innovation process and the link to an organizational structure (Sirola, 1996c).

respect to mass and energy streams; and (v) selecting suitable equipment options and dimensioning; among others. Moreover, the synthesis activity increases in complexity due to the combinatorial explosion of potential options. This combination of many degrees of freedom and the constraints of the design space have their origin in one or more of the following (Almeida-Rivera, 2005; Almeida-Rivera *et al.*, 2017):

– there are many ways to select implementations of physical, chemical, biological and information processing tasks in unit operations/controllers;

– there are many topological options available to connect the unit operations, *i. e.*, flowsheet structure, but every logically conceivable connection is physically feasible;

– there is the freedom to pick the operating conditions over a physical range, while still remaining within the domain in which the tasks can be effectively carried out;

– there is a range of conceivable operational policies; and

– there is a range of geometric equipment design parameters.

As the number of possible combinations can easily run into many thousands (10^4 to 10^9, as estimated by Douglas, 1988) a structured methodology is required to beat com-

Table 3.1: Overview of process synthesis techniques (Johns, 2001).

N°	Technique	Scope
1	Systematic generation (hierarchical decomposition) (Douglas, 1988)	initial definition of a coarse structure that then is refined
2	Evolutionary modification (Peters and Timmerhaus, 1991)	incremental improvements to a base design
3	Superstructure optimization	generation of a large number of process alternatives embedded in a superstructure, with the best revealed by optimization
4	Expert panel	critical examination borrowed from work study
5	Superstructure optimization	generation of a large number of process alternatives embedded in a superstructure, with the best revealed by optimization
6	Graph-theoretic superstructure generation	use of graph theory to generate an optimizable superstructure
7	Thermodynamic pinch	sets thermodynamic minimum energy (cooling water, etc.) use target
8	Implicit enumeration	integer optimization after discretizing flow rates, etc.
10	Artificial intelligence	rule-based systems create or improve flowsheets
11	Design rationale	human reasoning being recorded and structured

plexity and, therefore, reduce the huge number of alternative processes into a much reduced set of potential feasible options.

In spite of its inherent complexity, the development of novel PS methodologies has lately gained increasing interest from academia and industry. This phenomenon is reflected in the number of scientific publications focusing on PS research issues and its applicability in industrial practice (Li and Kraslawski, 2004). For instance, the effective application of PS in industry has led to large cost savings, up to 60 % as reported by Harmsen *et al.* (2000), and the development of intensified and multi-functional units, *e.g.*, the well-documented methyl acetate reactive distillation unit (Siirola, 1996a; Harmsen and Chewter, 1999; Stankiewicz and Moulijn, 2002; Stankiewicz, 2003).

As outlined in Table 3.1 there are several fundamental approaches that have been envisaged for the synthesis of chemical process (Johns, 2001), three of them being the most widely used.

Systematic generation involves the synthesis of a flowsheet from scratch, building it up from smaller, basic components (Douglas, 1988; Siirola, 1996a). Initially intended to support hand design, it is considered to be the most widely used (semi-)formal design methodology. The design problem is decomposed into several design echelons where coarse-grained decisions need to be made. Decisions such as dividing a flowsheet into reaction and separation stages and identifying major recycles are taken at the early stages of the design process. These broad divisions are then refined to give more detail until individual unit operations are specified. In recent years, hierarchical decomposition rules have been successfully incorporated into AI programs (Johns, 2001).

Evolutionary modification starts with an existing flowsheet for the same or similar product(s) and involves modifying it as necessary to meet the desired objectives. Evolutionary rules generate a succession of potentially better flowsheets until no further improvement can be achieved. As mentioned by Johns (2001), recent variants of this technique apply simulated annealing and genetic algorithms. In simulated annealing, a range of adjacent processes is produced by random changes in the base process and the best from the generated set is identified. To find as many promising regions as possible, large random variations are allowed at the beginning of the design problem. As the annealing generation slows down smaller steps are employed. Genetic algorithms generate a **population** of alternative flowsheets, where new flowsheets are created by taking pairs of existing flowsheets and generating **children** by recombining features of their **parents** (heredity) and by random variation (mutation). The best processes are selected from the resulting large population and the iteration is repeated until a range of sufficiently good processes is attained.

Superstructure optimization views the synthesis problem as a mathematical optimization of a larger superflowsheet that contains embedded within it many redundant alternatives and interconnections (Ismail *et al.*, 2001; Papalexandri and Pistikopoulos, 1996; Proios and Pistikopoulos, 2005; Biegler *et al.*, 1997; Westerberg, 1989, 2004; Westerberg *et al.*, 1997; Westerberg and Subrahmanian, 2000). For both systematic generation and evolutionary modification the practitioner needs to derive algorithmic and heuristic methods. Optimization techniques (such as MINLP and MIDO) are normally used for the superstructure optimization approach (Karuppiah and Grossmann, 2006; Bansal *et al.*, 2000, 2003; Bansal, 2000; Bagajewicz, 2000). A flowsheet superstructure is generated that includes far more streams and units than would be needed in any practical process. By selectively removing streams and units, this superstructure can be reduced to a realistic process. The goal is to develop an automatic procedure to remove superfluous features and hence reveal the best process. Figure 3.2 illustrates such a superstructure for a simple process, involving separation, mixing and reaction activities in a multi-step process with several alternatives per step (Zondervan *et al.*, 2011).

The optimization problem includes integer variables that are zero if a stream or unit is removed. Although the solution of the optimization problem can be computationally expensive, it is supported by the steady improvement in computer speed and the major algorithmic advances of the last decades (Ye *et al.*, 2018; Biegler and Grossmann, 2004; Bansal *et al.*, 2000, 2002, 2003; Costa and Oliveira, 2001; Karuppiah and Grossmann, 2006; Yi and Reklaitis, 2002, 2003, 2004, 2006a,b, 2007; Pintarik and Kravanja, 2000; Sahinidis, 2004; Wolsey, 2003; Subrahmanyam *et al.*, 1994). The solution of the problem reveals the best process. As mentioned by Johns (2001) in a reference to the pioneering work of King and coworkers in the early 1970s, for

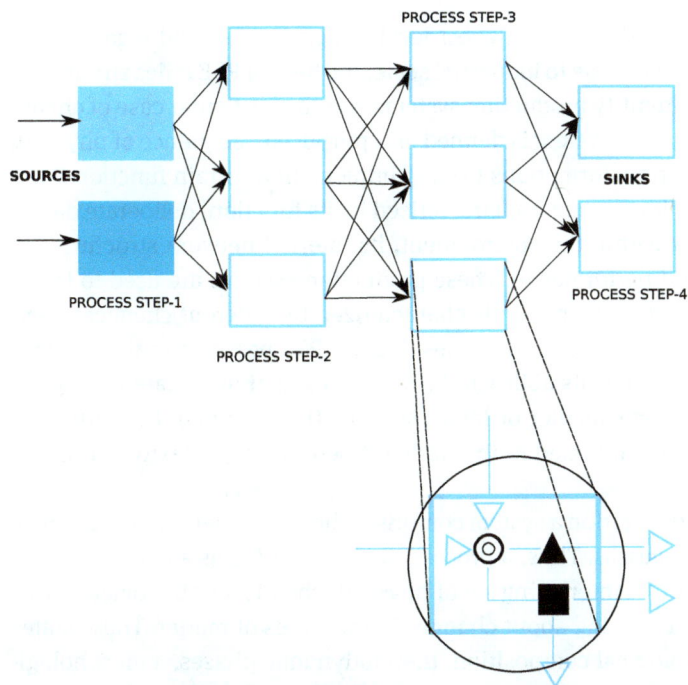

Figure 3.2: Superstructure of a simple reaction-separation-mixing system. Legend: ▲, side reaction; ■, reaction; ◎, mixing (Zondervan *et al.*, 2011; Zderic *et al.*, 2019).

superstructure optimization the search space is limited by the imagination of the engineer who defines the superstructure. To aid superstructure generation, engineers may use other techniques such as AI, thermodynamic pinch, implicit enumeration and graph-theoretic methods.

Despite their intrinsic advantages, each approach features drawbacks. For instance, systematic generation without exhausting generation of alternatives will probably result in already known and suboptimal flowsheets; the optimality of the alternative generated by evolutionary modification is strongly dependent on the starting flowsheet and on the methods to modify it; defining and formulating the superstructure for optimization involves extensive computational capabilities and an accurate phenomenological model that represents the systems.

3.2 Conceptual process design in the context of process systems engineering

Since its introduction, PSE has been used effectively by chemical engineers to assist the development of chemical engineering. In tying science to engineering PSE provides engineers with the systematic design and operation methods, tools that they require to successfully face the challenges of today's chemical-oriented industry (Grossmann and Westerberg, 2000). At the highest level of aggregation and regardless of length scale, *i. e.*, from microscale to industrial scale, the field of PSE relies strongly on engineers being able to identify production systems. For the particular case of chemical engineering, a production system is defined as a purposeful sequence of physical, chemical and biological transformations used to implement a certain function (Marquardt, 2004). A production system is characterized by its function, deliberate delimitation of its boundaries within the environment, its internal network structure and its physical behavior and performance. These production systems are used to transform raw materials into product materials characterized by different chemical identities, compositions, morphologies and shapes. From a PSE perspective the most remarkable feature of a system is its ability to be decomposed or aggregated in a goal-oriented manner to generate smaller or larger systems (Frass, 2005). Evidently, the level of scrutiny is very much linked to the trade-off between complexity and transparency.

At a lower level of aggregation a system comprises the above mentioned sequence of transformations or processes. Thus, a process can be regarded as a realization of a system and is made up of an interacting set of physical, chemical or biological transformations that are used to bring about changes in the states of matter. These states can be chemical and biological composition, thermodynamic phases, a morphological structure and electrical and magnetic properties.

Going one level down in the aggregation scale gives us the chemical plant. This is no more than the physical chemical process system. It is a man-made system, a chemical plant, in which processes are conducted and controlled to produce valuable products in a sustainable and profitable way. Conceptual Process Design (CPD) is carried out at the following level of reduced aggregation.

Since its introduction, CPD has been defined in a wide variety of ways. CPD and PSE activities are rooted in the concept of unit operations and the various definitions of CPD are basically process unit-inspired. The definition of CPD given by Douglas (1988) is regarded as the one which extracts the essence of this activity. Thus, CPD is defined as the task of finding the *best* process flowsheet, in terms of selecting the process units and interconnections among these units and estimating the optimum design conditions (Goel, 2004). The *best* process is regarded as the one that allows for an economical, safe and environmentally responsible conversion of specific feed stream(s) into specific product(s).

The CPD task is carried out by specifying the state of the feeds and the targets on the output streams of a system (Doherty and Buzad, 1992; Buzad and Doherty, 1995) and by making complex and emerging decisions.

CPD plays an important role under the umbrella of process development and engineering. As stated by Moulijn *et al.* (2001), process development features a continuous interaction between experimental and design programs, together with carefully monitored cost and planning studies. The conventional course of process development involves several sequential stages: an exploratory stage, a conceptual process design, a preliminary plant flowsheet, miniplant(s) trials, trials at a pilot plant level and design of the production plant on an industrial scale. CPD is used to provide the first and most influential decision making scenario and it is at this stage that approximately 80 % of the combined capital and operational costs of the final production plant are fixed (Meeuse, 2003). Performing individual economic evaluations for all design alternatives is commonly hindered by the large number of possible designs. Therefore, systematic methods, based on process knowledge, expertise and creativity, are required to determine which will be the best design given a pool of thousands of alternatives.

3.2.1 Developments in new processes and retrofits

From its introduction, the development of CPD trends has been responding to the harmonic satisfaction of specific requirements. In the early stages of CPD development economic considerations were by far the most predominant issue to be taken into account. Seventy years on, the issues surrounding CPD methodologies have been extended to encompass a wide range of issues involving economics, sustainability and process responsiveness (Almeida-Rivera *et al.*, 2004b; Harmsen *et al.*, 2000). Spatial and temporal aspects must be taken into account when designing a process plant. Additionally, the time dimension and loss prevention are of paramount importance if the performance of a chemical plant is to be optimized over its manufacturing life span. This broad perspective accounts for the use of multiple resources, *e. g.*, capital, raw materials and labor, during the design phase and the manufacturing stages. In this context and in view of the need to support the sustainability of the biosphere and human society, the design of sustainable, environmentally benign and highly efficient processes becomes a major challenge for the PSE community. Identifying and monitoring potential losses in a process, together with their causes, are key tasks to be embraced by a CPD approach. Means need to be put in place to minimize losses of mass, energy, run-time availability and, subsequently, profit. Poor process controllability and lack of plant responsiveness to market demands are just two issues that need to be considered by CPD engineers as causes of profit loss.

Li and Kraslawski (2004) have presented a comprehensive overview of the developments in CPD, in which they show that the levels of aggregation in CPD, *i. e.*, micro-, meso- and macroscales, have gradually been added to the application domain of CPD methodologies. This refocus on the design problem has led to a wide variety of success stories at all three scale levels, and a coming to scientific maturity of the current methodologies.

At the mesolevel the research interests have tended towards the synthesis of heat exchange networks, reaction path kinetics and sequencing of multi-component separation trains (Li and Kraslawski, 2004). A harmonic compromise between economic, environmental and societal issues is the driving force at the CPD **macrolevel**. Under the framework of multi-objective optimization (Clark and Westerberg, 1983), several approaches have been derived to better balance the trade-off between profitability and environmental concerns (Almeida-Rivera *et al.*, 2004b; Kim and Smith, 2004; Lim *et al.*, 1999). Complementary to this activity, an extensive list of environmental indicators, *e. g.*, environmental performance indicators, has been produced in the last decades (Lim *et al.*, 1999; Kim and Smith, 2004; Li and Kraslawski, 2004). At the CPD microlevel the motivating force has been the demand for more efficient processes with respect to equipment volume, energy consumption and waste formation (Stankiewicz and Moulijn, 2002). In this context a breakthrough strategy has emerged: abstraction from the historically equipment-inspired design paradigm to a task-oriented process synthesis. This refocus allows for the possibility of task integration and the design of novel unit operations or microsystems, which integrate several functions/tasks (Section 2.1.4.3) and reduce the cost and complexity of process systems (Grossman and Westerberg, 2000). An intensified unit is normally characterized by drastic improvements, sometimes of an order of magnitude, in production cost, process safety, controllability, time-to-market and societal acceptance (Stankiewicz, 2003; Stankiewicz and Moulijn, 2002).

3.2.2 Interactions between process development and process design

The aim of process design has, conventionally, been considered as the finding of equipment sizes, configurations and operating conditions that will allow for the economical, safe and environmentally responsible conversion of (a) specific feed stream(s) into (a) specific product(s). This task is in principle carried out only by specifying the state of the feeds and the targets on the output streams of a system (Doherty and Buzad, 1992; Buzad and Doherty, 1995). Frequently, however, the design activity is found to be severely complicated by the difficulty of identifying feasible equipment configurations and ranges of suitable operating conditions. This situation can be explained by the fact that the design of chemical process units is only a constitutive part of a larger-scale and, therefore, more complex problem: the joint consideration of process unit development and design. At this higher level of aggregation, the objective of the design problem includes sizing the unit and establishing the requirements of the design, screening the process feasibility and, at the top, identifying the process opportunities (Figure 3.3).

From this point of view, two parallel and interrelated activities can be identified: the design program and the development program. For a given level of design detail all the design tasks should be considered from top (*i. e.*, higher aggregation/process level) to bottom (*i. e.*, low aggregation/unit level). The information leaving the design program is compared with the desired outputs. If an acceptable agreement exists the level of detail is increased and the same strategy is adopted. For non-acceptable outputs the loop is performed at the same level of detail after redefining/reformulating appropriate assumptions/models/generalities in the design program. This is attained

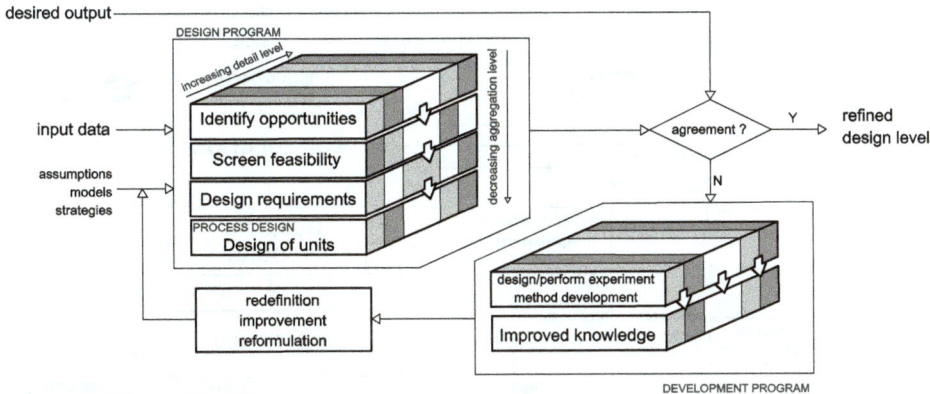

Figure 3.3: The design problem regarded as the combination of a design program and a development program (Almeida-Rivera, 2005).

by designing and performing an appropriate development program, which allows the designer to extend/improve/correct the existing knowledge and methodology, *e. g.*, removing the uncertainties introduced by estimation of parameters.

3.2.3 Structure of the design activity

If the general design paradigm (Siirola, 1996a,c; Biegler *et al.*, 1997) is used as the framework for the design task (*i. e.*, the bottom block of the **Design Program** in Figure 3.3), the overall design problem can be represented as a multi-stage goal-directed process, with dynamic flow of information between the design tasks, method development and knowledge development (Figure 3.4).

As depicted in Figure 3.4, synthesis can be regarded as a fundamental building block of the design task. It embraces a set of tools and methods to assist the designer in the creation of the *best* system (*i. e.*, product or service), which allows for an economic, safe and environmentally responsible conversion of specific feed(s) (*i. e.*, materials or information) into specific product(s) or services while satisfying – and hopefully exceeding – consumer expectations.

It can be noticed that the desired outputs from the design program might be grouped into two main categories: those related to the process structure and performance (*i. e.* minimum economic performance) and those related to temporal features of the unit(s) (*i. e.* operability and availability). For the sake of clarity, the definitions adopted for each category are included in Table 3.2.

Due to the size and conceptual/computational complexity of the design problem, a single comprehensive design strategy may not be available for years to come. As a first step towards such an integrated design approach, Almeida-Rivera and Grievink (2001), Almeida-Rivera *et al.* (2004a) propose a multi-echelon design approach, which allows the designer to address the multi-dimensionality of the product-process design

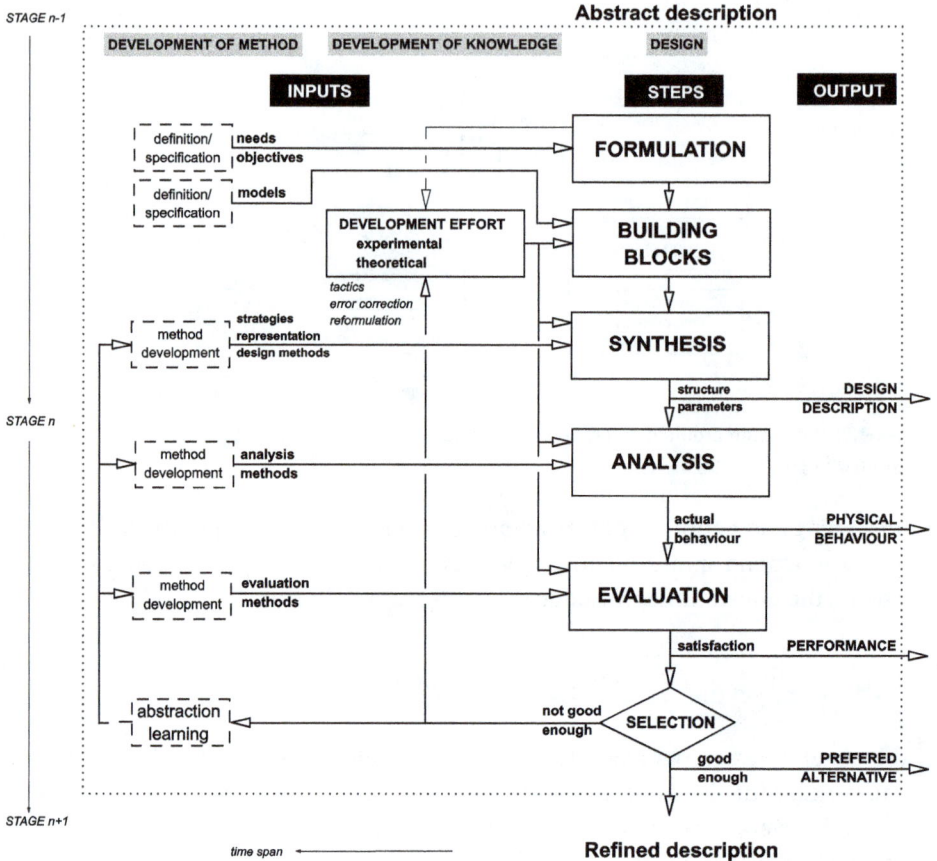

Figure 3.4: Overall design problem (Almeida-Rivera, 2005). Some ideas have been borrowed from Siirola (1996c), Biegler *et al.* (1997) and Bermingham (2003).

Table 3.2: Categories of information resulting from the design process.

Category of criteria	Scope of information
Structure	includes the complete spatial structure of the process, together with the control loops and operating conditions, *i. e.*, ranges of temperature, pressure, pH.
Performance	involves the estimation of criteria related to consumer acceptance, environment, safety & health, environment and economics. Technological aspects are addressed in the next category.
Operability	involves the ability to operate the unit(s) at preferred conditions in spite of disturbances and changes in operational policies.
Availability	is rarely considered and is a measure of the degree to which the unit(s) is/are in an operable state at any time.

problem. In general terms, the methodology is supported by a decomposition in a hierarchy of imbedded design spaces of increasing refinement. As a design progresses the level of design resolution can be increased, while constraints on the physical feasibility of structures and operating conditions derived from first principles analysis can be propagated to limit the searches in the expanded design space.

The general design paradigm of Siirola (Siirola, 1996a,c) is taken as the framework for the approach. This approach is intended to reduce the complexity of the design problem, and it should **not** be considered as a mechanistic recipe.

3.2.4 Life span performance criteria

Aiming at a design methodology that addresses the problem from a life span perspective, Grievink (2005) suggests a set of performance criteria. The motivation behind this perspective relies on the relevance of bearing in mind economic performance and potential losses over the process life span. Specifically, exergy efficiency and responsiveness aspects in the operational phase need to be considered. Moreover, in the context of consumer-centered design, the satisfaction of consumer needs is of paramount relevance and must be included in the scope of the performance criteria. Although safety and health aspects are of utmost importance to the chemical industry, we consider them to be implicitly outside the scope of our proposed methodology. As safety and health issues are key drivers for the innovation and technological breakthroughs that characterize manufacturing processes, they must be embedded within the inherent definition of the design problem. Moreover, we **only** consider sustainability aspects that are exclusively under the control of the process design. In this regard, we aim at operational efficiency, which basically means an effective utilization of resources in the process operation. Similar to life cycle assessment, we quantify the materials, energy and exergy used and assess the environmental impact of those flows, via loss minimization. Moreover, we address sustainability issues related to feedstock selection, (re)use of solvents, use of novel fields and sources of energy, among others.

The issues under consideration define a novel and extended performance space, which embraces aspects related to consumer satisfaction, environment, technology and economics, namely, the CETE performance set. Consumer satisfaction comprises the identification, quantification and satisfaction of consumer needs, consumer acceptance, level of acceptance and desirability, among others. Environmental considerations are directly linked to the minimization of mass, energy and exergy losses. Economic performance is translated quantitatively to the capital and operational costs of the design (*e. g.*, total annualized cost). Technical considerations are grouped into two subcategories: availability and operability. Availability embraces both reliability and maintainability, whereas operability includes process design attributes such as flexibility (steady-state), stability (multiplicity), controllability (dynamic) and switchability (start-up/shut-down). Controllability and switchability are lumped conveniently into a responsiveness criterion, which accounts for the dynamic response of the design in the presence of disturbances.

As these CETE-proposed performance criteria are of different natures, their formulation into an objective function involves the assignment of appropriate weighting factors. Assigning these values, however, is not a trivial task as it is an ultimate decision of the designer and it is done according to the requirements of the design problem (Clark and Westerberg, 1983). Pareto curve mapping is a convenient way to depict the effects of the weighting factors in a multi-objective optimization problem.

3.3 Process synthesis in the industry of consumer products

It is now well established by industry and academia that the focus of the chemical industry has shifted from a process-centered orientation to a product-centered one (Hill, 2004). During the last decades we have experienced how the commodity chemical business is gradually releasing its dominating role toward higher-added-value products, such as specialty chemicals and consumer products. This trend is further reflected in the increasing number of scientific publications addressing product and process design (Kalakul *et al.*, 2018; Zhang *et al.*, 2017; Fung *et al.*, 2016; Mattei *et al.*, 2014; Smith and Ierapepritou, 2010; Seider *et al.*, 2009; Costa *et al.*, 2006; Edwards, 2006; Norton *et al.*, 2006; Ng, 2004; Schubert and Engel, 2004; Voncken *et al.*, 2004; Bröckel and Hahn, 2004; Gani, 2004a,b; Hill, 2004; Cussler and Wei, 2003; Wibowo and Ng, 2001, 2002), textbooks (Cussler and Moggridge, 2001; Seider *et al.*, 2004, 2010) and undergraduate/graduate courses in chemical process design (Moggridge and Cussler, 2000; Cussler and Wei, 2003; Shaw *et al.*, 2004; Saraiva and Costa, 2004).

Stretching the boundaries of the synthesis activity toward products has brought challenges for the chemical and process engineers. Those refreshing problems need the development of novel methods and tools, involving areas like the fundamental understanding of the product-process interactions, multi-level modeling of consumer products, property models for products with internal microstructure and prediction of consumer liking and its dependence on ingredients and processes, etc. Whether product design is embedded in the process design activity still remains a topic of debate. As mentioned elsewhere (Cussler and Moggridge, 2001; Cussler and Wei, 2003; Moggridge and Cussler, 2000), if the emphasis is on product design, current methodologies of process design (*e. g.*, the hierarchy of decisions by Douglas, 1988) are not capturing the product design space. It is a need, therefore, to go beyond the process design hierarchy.

Table 3.3 shows the steps of process and product design as suggested by Douglas (1988) and Cussler and Moggridge (2001) and summarized by Bagajewicz (2007), respectively.

It is implied from this sequence of steps that process design is contained in the fourth step of the product design approach. Despite the maturity of most PS ap-

Table 3.3: Process design and product design steps (Douglas, 1988; Cussler and Moggridge, 2001).

Process design – hierarchy of decisions	Product design
Batch *versus* continuous	Identification of consumer needs
Input–output structure of the flowsheet	Generation of ideas to meet needs
Recycle structure of the flowsheet	Selection among ideas
General structure of the separation system	Manufacturing of product
Heat exchange network	

proaches for chemical products, they fall short when it comes to extending its scope and applicability to consumer products. This drawback of current approaches is derived from the intrinsic differences between bulk chemicals and consumer-centered or performance products, and include (Meeuse *et al.*, 2000) the following:
- performance products are typically structured products where the performance is determined by the internal microstructure of the product;
- unit operations are quite different, involving fewer reaction and separation tasks and more mixing, preservation and structuring;
- in the case of food or personal care products, the processes are generally multiproduct processes, where the same production line can accommodate the manufacturing of different products with different properties; and
- cleaning is an essential and non-negotiable task within the operational policy.

Thus, in contrast to bulk chemicals, structured products (Figure 3.5) are characterized not only by the level of each ingredient (*i. e.*, composition, purity, physical state, temperature, pressure, etc.), but also by the relative spatial arrangement of each ingredient and performance behavior. All these features are responsible for the exclusive attributes of structured products (*e. g.*, creaminess of an ice cream, spoonability of a mayonnaise, spreadability of a margarine, skin absorption of a body cream, oiliness of hand cream).

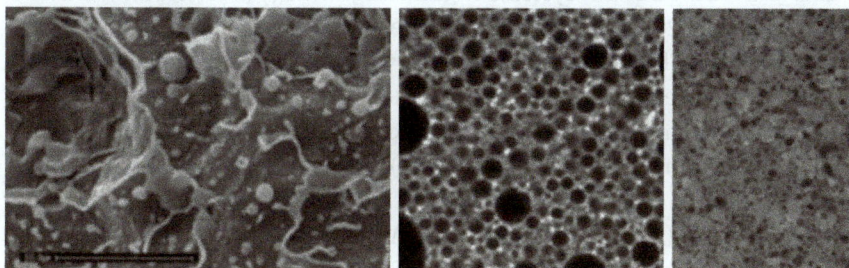

Figure 3.5: Left: lamellar-structured hair conditioner (Hill, 2004). Center: SEM micrograph of a high internal phase-structured emulsion. Right: SEM micrograph of an ice cream matrix.

The first attempt to widen the scope of PS to consumer products with internal structure was carried out by van-der-Stappen and coworkers (Meeuse *et al.*, 2000). More recent publications on product and process design in consumer products are those by Hill (2004), Meeuse (Meeuse, 2005, 2007; Meeuse *et al.*, 1999), Stappen-vander (2005), Linke *et al.* (2000) and Norton *et al.* (2006), among others. These and alike contributions might be grouped into three categories: those that involve sequential product and process design; those that involve simultaneous product and process design and those that involve anticipating sequential product and process design.

Each approach (simultaneous, sequential or anticipating sequential) has its own strengths and drawbacks. The sequential approach, for instance, is the most intuitive and, therefore, most adopted strategy. The generation of process alternatives is restricted by the *a priori* specification of the product, resulting in the need of multiple iterations. The simultaneous approach exploits the synergy of both design activities, leading to the optimal solution. However, it implies the formulation of the design problem as an optimization problem and, therefore, requires the complete phenomenological description of the underlying physical and chemical processes behind the product manufacturing. The anticipating sequential approach is a hybrid of those previous approaches and, therefore, borrows the best of both worlds, but also shares their limitations.

3.4 A Product-driven Process Synthesis approach

3.4.1 Generalities

Aiming at a more structured approach toward the synthesis of product and processes in the consumer-centered sector, in this book we propose a refreshed and integrated methodology: Product-driven Process Synthesis (PDPS). The concept of PDPS has the same scope and goal as the concepts of **Product and Process Design** coined by Hill (2004) and **Product-Oriented Process Synthesis** by Wibowo and co-workers (Wibowo and Ng, 2001).

The approach exploits the synergy of combining product and process synthesis workstreams and is based on the systems engineering strategy (Bongers and Almeida-Rivera, 2009; Bongers, 2009). Thus, it involves decomposing the problem into a hierarchy of design levels of increasing refinement, where complex and emerging decisions are made to proceed from one level to another. Moreover, each level in the PDPS methodology features the same, uniform sequence of activities, which have been derived from the pioneering work of Douglas (1988) and Siirola (1996a) and further extended by Bermingham (2003), Almeida-Rivera *et al.* (2004a, 2007), Stappen-vander (2005), Almeida-Rivera (2005) and Bongers and Almeida-Rivera (2009).

Current efforts in the development of PDPS methodology have been focusing on broadening the design scope to consumer preferences, product attributes, process variables (Almeida-Rivera *et al.*, 2007) and supply chain considerations and financial factors (de Ridder *et al.*, 2008). The industrial (and academic) relevance of the PDPS approach has been validated by the successful application to several studies involving recovery of high-value compounds from waste streams (Jankowiak *et al.*, 2015, 2012), synthesis of structured matrices with specific consumer-related attributes (Gupta and Bongers, 2011), generation of new-generation manufacturing technologies (Almeida-Rivera *et al.*, 2017) and extraction of high-value compounds from natural matrices (Zderic *et al.*, 2014, 2015, 2016; Zderic and Zondervan, 2016, 2017; Zondervan *et al.*, 2015), among other non-disclosed cases. The applicability of the approach to enterprise-wide scales has been demonstrated by studying the synthesis of biorefinery networks (Zondervan *et al.*, 2011; Zderic *et al.*, 2019; Kiskini *et al.*, 2016).

As a result to this quest, this methodology is our refined approach toward the synthesis of consumer products and is foreseen to define guidelines and trends for the synthesis of processes that are more efficient, controllable and safer and that deliver products or services that are meeting consumer expectations. In such a context the performance of a manufacturing process for consumer products needs to be optimized over its manufacturing life span, while accounting for the use of multiple resources in the design and manufacturing stages. Since such resources are of a different nature (*e. g.*, capital, raw materials and labor) with different degrees of depletion and rates of replenishment the performance of the process is characterized by multiple objectives with trade-offs.

The input of the PDPS methodology is a complete and comprehensive specification of the desired product(s), raw materials along with any other requirements the process needs to fulfill (*e. g.*, the capacity requirements, cost requirements, hygiene and safety standards and regulations). The output of a process synthesis exercise is a flowsheet structure, along with unit interconnections, operating conditions and key dimensions of key equipment units. Within PDPS the following concepts and terms are of interest. A product attribute is the relatively soft translation of consumer-related requirements or needs. When product attributes are mapped and quantified onto a physico-chemical-biological space, these attributes result in product properties (or specifications). These properties are regarded as features that differ between initial (ingredients) and final (final product) states and are fundamentally relevant to the perception of the end-consumer. Properties can involve the aspects, characteristics and facets of a product (*e. g.*, color, texture, droplet mean size, microbiological stability). The difference between the attributes in the initial and final states should be systematically eliminated via one or more fundamental tasks. A fundamental task – in this regard – can well be a resolution method for more than one property. Each task is characterized by a dominating mechanism in a specific operational window. Operational windows are defined in terms of specific driving forces, temperature, pressure, pH, electric/magnetic field strength and gradient, etc.

3.4.2 Structure of the methodology

Our proposed methodology consists of nine levels of increasing refinement, where complex and emerging decisions are made to proceed from one level to another (Figure 3.6). These levels address the following design spaces: framing level, consumer wants, product function, input–output level, task network, mechanism and operational window, multi-product integration, equipment selection and design and multi-product-equipment integration.

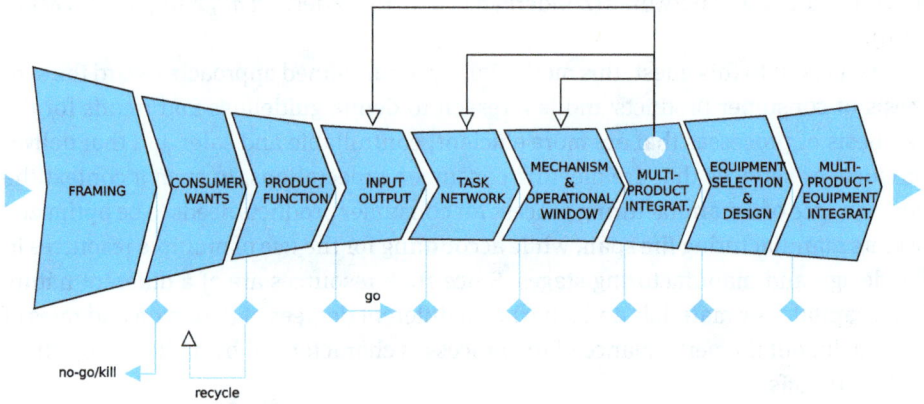

Figure 3.6: Product-driven Process Synthesis approach.

Each level of the methodology (summarized in Table 3.4) has been structured adopting the general design paradigm cycle (Siirola, 1996a): scope and knowledge ⟶ generate alternatives ⟶ analyze performance of alternatives ⟶ evaluate and select ⟶ report (Figure 3.7).

The CETE performance indicators at all levels include estimates of economic potential, turndown ratio, changeover time, sustainability, process complexity, scalability and hygiene design, among others, and all of them at various degrees of detail. As previously mentioned, the nature of a product-driven design problem involves the input of several actors with varying expectations and requirements. Under this construction, the design problem – and more specifically the **evaluate and select** activity – spreads over a multi-echelon domain and needs to be resolved by considering its multi-dimensionality. A multi-criterion decision making approach has been recently introduced to assist in the **evaluate and select** activity (de Ridder *et al.*, 2008). This approach, termed BROC (benefits, risks, opportunities and costs), has been validated in an industrial R&D scenario (Section 4.3.6) and is based on the combination of Quality Function Deployment (QFD) and Analytic Network Process (ANP).

QFD is a method for structured product planning and development that enables a development team to specify clearly the customer wants and needs and to evaluate

Table 3.4: Description of the levels of the PDPS approach.

Level	Generic description
0	**Framing level.** Embed the design into the overall project. Description of the background of the project and the business context, overall supply chain considerations (product portfolio, demand profiles, regional spread, etc.).
1	**Consumer wants.** Obtain the consumer wants (consumer likings, focus groups, interviews) in qualitative descriptions and translate them into quantifiable product attributes.
2	**Product function.** Map the quantifiable product attributes onto the product properties (property function), which are measurable; product concepts are generated (prototyping).
3	**Input–output level.** Make a complete specification of the output; choose inputs (ingredients) and the descriptors of the output (microstructure, flavor profile and microbiological status); determine performance parameters such as quality, economic potential, hygienic considerations, flexibility, pumpability and availability.
4	**Task network.** Define the fundamental tasks needed to go from input to output, taken from a cluster of tasks and its subgroups. Furthermore, tasks that require a certain sequence or that belong together without any doubt are grouped to reduce the number of sequencing possibilities. Then, a network is made from the selected tasks and clusters.
5	**Mechanism and operational window.** Select mechanism and principles that can be used to perform a task. This step includes the driving forces and kinetics. Furthermore, the operational window of the problem (time, P, pH, shear, T, etc.) is defined.
6	**Multi-product integration.** Compare the outcomes of Levels 3–5 for the different products to look for overlap and possibilities to combine the production.
7	**Equipment selection and design.** Select the operating units; consider the integration possibilities (e. g., by combining tasks with the same driving force that are close together in a task network) and controllability. The operational window from Level 5 is compared to the operating boundaries of the unit. The final design of the units (only of the key dimension) and final flowchart are made.
8	**Multi-product-equipment integration.** Optimize the interaction of the various unit operations in the flowsheet (plant-wide control); apply multi-stage scheduling of the multiple products, fed by the run-strategy based on the product demand and portfolio.

each proposed product or service capability systematically in terms of its impact on meeting those needs (Cohen, 1995). The backbone of QFD is the so-called House of Quality (HoQ) (Figure 3.8), which displays the customer wants and needs, and the development team technical response to those wants and needs. The concepts of QFD and HoQ are further described in Sections 4.3.4 and 4.3.5, respectively. ANP is a commonly used benchmarking technique to compare alternatives. It resembles a network, consisting of clusters of elements, which are the decision making criteria and the alternatives. The relations between elements depend on each decision making case. In the work by de Ridder *et al.* (2008) the concept of QFD was extended to cover consumer preferences, product and process characteristics, supply chain considerations and financial factors. It is evident that the BROC approach resembles the strategic planning tool SWOT. In the proposed methodology BROC is chosen over SWOT, as it embraces

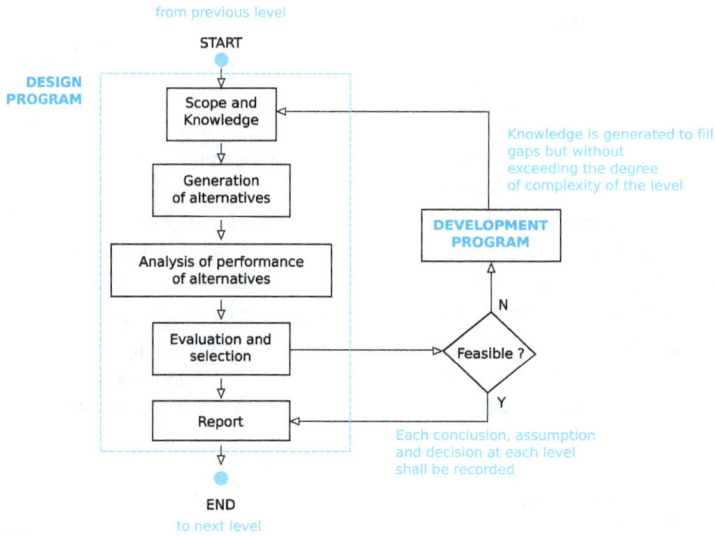

Figure 3.7: Sequence of activities at each level of the Product-driven Process Synthesis methodology.

ANP and is currently regarded as the most comprehensive framework for the analysis of societal, governmental and corporate decisions. Section 4.3.6 addresses in detail the development and applicability of the BROC approach to a consumer-centered product design problem.

At the framing level a full description of the background of the project and the business context are addressed. The required information includes the product portfolio, demand profile, category strategy, consumer liking, product history, supply chain strategy and – if available – current processing practice(s).

At the consumer wants level we obtain the consumer wants in qualitative descriptions (*e. g.*, great taste) and translate them into quantifiable product attributes (*e. g.*, creaminess). This translation can be represented by a consumer function (Bongers and Almeida-Rivera, 2008), as depicted in Figure 3.9. Individual and group interviews are the most commonly used tools to obtain the consumer wants, while QFD allows the formulation of the consumer function, and subsequently the mapping of those consumer liking attributes onto product attributes.

At the product function level we map the quantifiable product attributes onto a set of measurable product properties. This level is characterized by the translation of *soft* consumer-related attributes (*e. g.*, creaminess) to *hard* properties of the product (*e. g.*, size of air cells). This translation is articulated via the definition of a property function (Bongers, 2008). A parallel activity at this level includes the systematic generation of product/service concepts by prototyping (Section 2.3.1) and subsequent validation with consumers. As mentioned by Lockwood and Papke (2018), prototyping is one of the vehicles to accelerate learning within Design Thinking. Next to problem identifica-

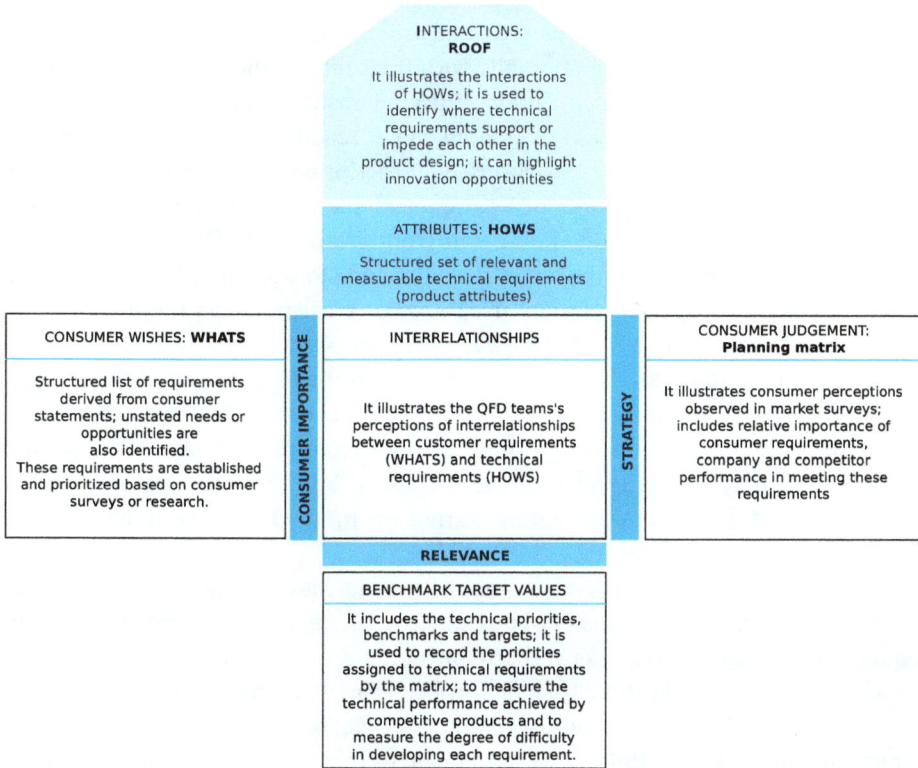

Figure 3.8: Quality Function Deployment as a tool to map consumer wants/needs onto the product attributes-space.

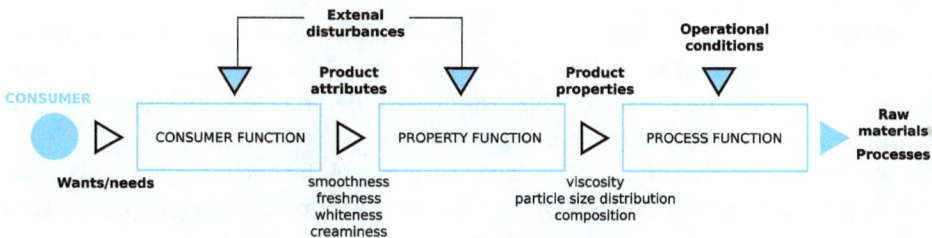

Figure 3.9: Consumer and property functions at the consumer wants and product function levels.

tion and emphatic collaboration, learning through hands-on experimentation is one of the key tenets to be addressed in this effective problem solving approach.

At the input–output level a complete specification of all exchange streams to the process (inputs/raw material(s) and target output/product(s)) is given. The exchange stream properties include physical, chemical, biological, sensorial and format attributes. These can be, for instance, flow, temperature, pressure, pH, droplet size distribution, microbiological status and color. The inputs include all ingredients with their amounts and, preferably, with their microstructure and aggregation state. The output includes a description of the microstructure, flavor structure, microbiological status and, in case of (bio)chemical reactions during the processing (*e.g.*, fermentation or Maillard reactions), a description of the outcome of these reactions. We determine performance parameters such as economic potential, hygienic considerations, flexibility, pumpability and availability. The generic product description contains aspects like target shelf-life and storage conditions, pack sizes, annual production capacity, product use, etc. For most products the definition of microbiological status is a straightforward activity. Namely, the product needs to be pasteurized or sterilized. This choice depends on the formulation of the product (pH, preservatives) and the required shelf-life and storage conditions (frozen, chilled, ambient). In addition to the specification of the exchange streams, the scope and knowledge activity at this level includes the list of boundary constraints and the relevant KPIs (*e.g.*, economic potential, turndown ratio, changeover time, sustainability).

At the fundamental tasks level we define the tasks that are required to go from input to output, taken from a cluster of tasks. Furthermore, tasks that require a certain sequence or that belong together without any doubt are grouped, to reduce the number of sequencing possibilities. Then, a network is made from the selected tasks and clusters. For chemical process, almost all processes can be described with only five different fundamental tasks (Siirola, 1996a): changes in molecular identity, changes in amount, changes in composition, changes in phase and changes in temperature/pressure. Similarly, Zhang *et al.* (2017) suggest a set of unconventional process clusters of relevance to consumer-centered product design: coating, agglomeration, synthesis, etching and breakage.

At the mechanism and operational window level we select mechanisms and principles that can be used to perform a task. This step includes the driving forces and kinetics. Furthermore, the operational window of the problem (time, pressure, pH, shear, temperature, etc.) is defined.

At the multi-product integration level we compare the outcomes of Levels 3 (input/output) to 5 (mechanism and operational window) for the different products, in order to look for overlap and possibilities to combine the production.

At the equipment selection and design level we select the operating unit. Integration possibilities should be considered (*e. g.*, by combining tasks with the same driving force that are close together in a task network). The operational window from Level 5 is compared to the operating boundaries of the unit. Then, the final design of the units (only of the key dimensions) and final flowchart are made.

At the multi-product-equipment integration level the interaction of the various unit operations in the flowsheet is optimized (plant-wide control). Multi-stage scheduling of the multiple products is applied, fed by the run-strategy based on the product demand and portfolio defined at Level 0.

A detailed description of the activities of each level is given in Table 3.5 and a list of fundamental tasks and mechanisms is given in Tables 6.3 and 6.6, respectively. It is relevant to acknowledge that both tables are provided as a guideline and need to be adapted and extended according to the specific requirements of the design problem.

A multi-echelon approach was introduced aimed at an integrated and generalized design methodology for consumer products (PDPS). Consumer acceptance and needs satisfaction (*i. e.*, desirability), environmental considerations (*i. e.*, sustainability), process responsiveness, safety and health considerations (*i. e.*, feasibility) and economic performance (*i. e.*, viability) are included as evaluation criteria under the umbrella of the CETE set. A decomposition of the design problem is proposed in this approach to give a hierarchy of design spaces of increasing refinement. The underlying foundation upon which this approach is built considers that form follows function. Thus, by adopting a task-oriented approach we postpone to later stages any unit operation design issue and exploit the opportunities of task-integration. This way of addressing the design/synthesis problems implies a paradigm shift toward a more powerful synthesis tool.

3.5 Take-away message

PDPS is foreseen to be the vehicle to embrace most of the challenges in the consumer products sector, where key players are coping effectively with the dynamic environment of short process development and design times. The challenges involving this synthesis approach are common to any tool and method stemming from the chemical and process engineering community. Those include the further embracement of the multi-scale modeling approach (from molecules to enterprise, Figure 2.36) and casting of the methodology into software tools for industrial use, among others. Various challenges are identified as next steps in the development of the approach. (i) Further developments of the methodology should focus on addressing multi-product processes. A challenge to face is stemming from the increasing complexity of a multi-product process, especially when different products involve different microstructures. For this case, multi-product-equipment integration would be also of key importance when it comes to the generation of the best production schedule. (ii) The development and validation of computational models (directional and/or deterministic) is a key activity for further research. The scope of the models should span various length and time scales with different degrees of complexity. They should aim at linking product microstructure, processing conditions and consumer liking based on a fundamental understanding of the governing phenomena. (iii) The quality of the final product is partly determined by the hygiene/cleaning systems in the factory. Therefore, aspects like modeling of microbiological growth and decontamination and soil removal become essential. (iv) We propose the BROC methodology to assist the designer in the decision of the best option between a pool of alternatives. A future endeavor involves the comprehensive collation of criteria, specific for each product category and stretching over the different areas (*i. e.*, consumer preferences, product and process characteris-

Table 3.5: Description of the activities of each level of the PDPS approach. Nomenclature: S&K, scope and knowledge; GA, generate alternatives; AP, analyze performance; E&S, evaluate and select; RE, report.

Activity	Description
Level 0: Framing level	
S&K	define background, objectives, constraints, assumptions, risks and deliverables, product portfolio, category strategy, product history, supply chain strategy, current processing practice(s)
Level 1: Consumer wants	
S&K	define consumer wants in qualitative descriptions; define quantifiable product attributes; translate them into quantifiable product attributes; list constraints
GA	establish relationships between consumer wants and quantifiable product attributes (*e. g.*, QFD and data mining techniques)
AP	apply heuristics, engineering judgment and rules of thumb to refine/simplify the consumer function; validate the consumer function with real data
E&S	evaluate alternatives based on BROC multi-criteria; select the most predictive consumer function
RE	compile relevant assumptions and decisions made
Level 2: Product function	
S&K	define set of measurable product properties; map the quantifiable product attributes onto the set of measurable product properties; list constraints
GA	establish relationships between quantifiable product attributes and measurable product properties (*e. g.*, QFD, data mining techniques and TRIZ); generate prototypes of products/services
AP	apply heuristics, engineering judgment and rules of thumb to refine/simplify the property function; validate the property function with real data
E&S	evaluate alternatives based on BROC multi-criteria; select the most predictive property function and concept
RE	compile relevant assumptions and decisions made
Level 3: Input–output	
S&K	define system boundaries; specify exchange streams (in and out) and properties (physical, chemical, biological, sensorial and format); list constraints; determine relevant CETE KPIs (*e. g.*, economic potential, turndown ratio, changeover time, sustainability)
GA	determine base case and alternative cases based on different options of inputs; create input–output structure for each alternative
AP	perform mass and energy balances
E&S	evaluate alternatives based on BROC multi-criteria; select input/output structure
RE	compile relevant assumptions and decisions made
Level 4: Task network	
S&K	define transformation pathways based on attribute changes; define fundamental tasks to achieve these transformations; list fundamental tasks; list constraints; determine relevant CETE KPIs (*e. g.*, sustainability, process complexity, economic potential)
GA	generate a network of fundamental tasks to achieve transformations; determine preliminary mass and energy balances
AP	apply heuristics, engineering judgment and rules of thumb to refine/simplify task network
E&S	evaluate alternatives based on BROC multi-criteria and select task network
RE	compile relevant assumptions and decisions made

Table 3.5: (continued)

Activity	Description
Level 5: Mechanism and operational window	
S&K	inventory mechanism for each fundamental task; define the operational window of the problem; determine relevant CETE KPIs (*e. g.*, economic potential, scalability, hygiene design, sustainability); list constraints
GA	link each fundamental task from the task network to the set of possible mechanisms
AP	determine for each mechanism the operational window (*e. g.*, shear regime, temperature, pressure, magnetic field, etc.) in combination with I/O level information and (short-cut) models
E&S	evaluate alternatives based on BROC multi-criteria; select a single mechanism alternative for each task in agreement with the problem operational window
RE	compile relevant assumptions and decisions made
Level 6: Multi-product integration	
▷ This level is applicable when multiple products are manufactured in the same process line.	
S & K	compare the outcomes of Levels 3–5 for each product; determine relevant CETE KPIs (*e. g.*, flexibility, product compatibility); list constraints
GA	create set of possible alternatives based on compatible product characteristics
AP	perform mass, energy, momentum and population balances for each integrated alternative
E &S	evaluate alternatives based on BROC multi-criteria; select the equipment unit
RE	compile relevant assumptions and decisions made
Level 7: Equipment selection and design	
S & K	map the operational window for each task; apply (short-cut) design methods for equipment; design rules of thumb; list constraints
GA	identify options for task integration in a single equipment unit; link each task to equipment unit
AP	perform mass, energy, momentum and population balances; determine operating condition of each equipment unit; size equipment unit
E&S	evaluate alternatives based on BROC multi-criteria; select the equipment unit; compare the operational window from Level 5 (mechanism and operational window) to operating boundaries of the unit
RE	compile relevant assumptions and decisions made
Level 8: Multi-product-equipment integration	
▷ This level is applicable when multiple products are manufactured in the same or different process lines within the plant site.	
S&K	define realistic disturbance scenario; define controllability criteria to assess the control scheme and policy; determine relevant performance indicators; list constraints
GA	design the control structure and control strategy; generate the multi-stage scheduling
AP	determine for each control structure and strategy the controllability performance index; validate the multi-stage scheduling
E&S	evaluate alternatives based on BROC multi-criteria; select a single control structure, control strategy and scheduling recipe
RE	compile relevant assumptions and decisions made

tics, supply chain considerations and financial criteria). (v) A comprehensive list of fundamental tasks, mechanisms and operational windows needs to be continuously updated, capturing state-of-the-art technologies and identifying blank spots in the knowledge domain. (vi) Product attributes need to be prioritized, as has been done for the chemical industry. (vii) And heuristics methods derived to simplify the task network (driven by expertise, common sense or a combination of both) should be continuously challenged and extended as new knowledge becomes available.

Bibliography

Almeida-Rivera, C. P. (2005). *Designing reactive distillation processes with improved efficiency.* Thesis/dissertation, Delft University of Technology.

Almeida-Rivera, C. P., Bongers, P. and Zondervan, E. (2017). A Structured Approach for Product-Driven Process Synthesis in Foods Manufacture. In: M. Martin, M. Eden and N. Chemmangattuvalappil, eds., *Tools For Chemical Product Design: From Consumer Products to Biomedicine*. Elsevier, pp. 417–441. Chapter 15.

Almeida-Rivera, C. P. and Grievink, J. (2001). Reactive distillation: On the development of an integrated design methodology. *Chemie Ingenieur Technik*, 73(6), p. 777.

Almeida-Rivera, C. P., Jain, P., Bruin, S. and Bongers, P. (2007). Integrated product and process design approach for rationalization of food products. *Computer-Aided Chemical Engineering*, 24, pp. 449–454.

Almeida-Rivera, C. P., Swinkels, P. L. J. and Grievink, J. (2004a). Designing reactive distillation processes: present and future. *Computers and Chemical Engineering*, 28(10), pp. 1997–2020.

Almeida-Rivera, C. P., Swinkels, P. L. J. and Grievink, J. (2004b). Economics and exergy efficiency in the conceptual design of reactive distillation processes. *AIChE Symposium Series*, 2004, pp. 237–240.

Bagajewicz, M. (2000). A review of recent design procedures for water networks in refineries and process plants. *Computers and Chemical Engineering*, 24, pp. 2093–2113.

Bagajewicz, M. (2007). On the Role of Microeconomics, Planning, and Finances in Product Design. *AIChE Journal*, 53(12), pp. 3155–3170.

Bansal, V. (2000). *Analysis, design and control optimization of process systems under uncertainty.* Thesis/dissertation, University of London. Imperial College of Science, Technology and Medicine.

Bansal, V., Perkins, J. and Pistikopoulos, E. (2002). Flexibility Analysis and Design Using a Parametric Programming Framework. *AIChE Journal*, 48(12), pp. 2851–2868.

Bansal, V., Perkins, J., Pistikopoulos, E., Ross, R. and van Schijndel, J. (2000). Simultaneous design and control optimisation under uncertainty. *Computers and Chemical Engineering*, 24, pp. 261–266.

Bansal, V., Sakizlis, V., Ross, R., Perkins, J. and Pistikopoulos, E. (2003). New algorithms for mixed-integer dynamic optimization. *Computers and Chemical Engineering*, 27, pp. 647–668.

Bermingham, S. (2003). *A design procedure and predictive models for solution crystallisation processes*. Thesis/dissertation, Delft University of Technology.

Biegler, L., Grossman, I. and Westerberg, A. (1997). *Systematic methods of chemical process design*. New York: Prentice Hall.

Biegler, L. and Grossmann, I. (2004). Retrospective on optimization. *Computers and Chemical Engineering*, 28, pp. 1169–1192.

Bongers, P. (2008). Model of the Product Properties for Process Synthesis. *Computer-Aided Chemical Engineering*, 25, pp. 1–6.

Bongers, P. (2009). Intertwine product and process design. Inaugural lecture/Eindhoven University of Technology, ISBN: 978-90-386-2124-1.

Bongers, P. and Almeida-Rivera, C. P. (2008). Product Driven Process Design Methodology. In: *AIChE Symposia Proceedings, Proceedings of the 100th Annual Meeting*. American Institute of Chemical Engineers, p. 623A.

Bongers, P. M. M. and Almeida-Rivera, C. (2009). Product Driven Process Synthesis Methodology. *Computer-Aided Chemical Engineering*, 26, pp. 231–236.

Bröckel, U. and Hahn, C. (2004). Product Design of Solid Fertilizers. *Trans. IChemE – Part A* , 82, pp. 1453–1457.

Buzad, G. and Doherty, M. (1995). New tools for the design of kinetically controlled reactive distillation columns for ternary mixtures. *Computers and Chemical Engineering*, 19(4), pp. 395–408.

Clark, P. and Westerberg, A. (1983). Optimization for design problems having more than one objective. *Computers and Chemical Engineering*, 7(4), pp. 259–278.

Cohen, L. (1995). *Quality Function Deployment: how to make QFD work for you*. Massachusetts: Addison-Wesley Publishing Company, Inc..

Costa, L. and Oliveira, P. (2001). Evolutionary algorithms apporach to the solution of mixed integer non-linear programming problems. *Computers and Chemical Engineering*, 25, pp. 257–266.

Costa, R., Moggridge, G. D. and Saraiva, P. M. (2006). Chemical Product Engineering: An Emerging Paradigm Within Chemical Engineering. *AIChE Journal*, 52(6), pp. 1976–1986.

Cussler, E. L. and Moggridge, G. D. (2001). *Chemical Product Design*. Cambridge Series in Chemical Engineering, New York.

Cussler, E. L. and Wei, J. (2003). Chemical Product Engineering. *AIChE Journal*, 49(5), pp. 1072–1075.

de Ridder, K., Almeida-Rivera, C. P., Bongers, P., Bruin, S. and Flapper, S. D. (2008). Multi-Criteria Decision Making in Product-driven Process Synthesis. *Computer-Aided Chemical Engineering*, 25, pp. 1021–1026.

Doherty, M. F. and Buzad, G. (1992). Reactive distillation by design. *Trans. Institution of Chemical Engineers – Part A*, 70, pp. 448–458.

Douglas, J. (1988). *Conceptual Design of Chemical Process*. New York: McGraw-Hill.

Edwards, M. F. (2006). Product engineering: Some challenges for Chemical Engineers. *Transactions of the Institute of Chemical Engineers – Part A*, 84(A4), pp. 255–260.

Frass, M. (2005). Begriffe der DIN 19226 – Regelung und Steuerung. Data File.

Fung, K., Ng, K., Zhang, L. and Gani, R. (2016). A grand model for chemical product design. *Computers and Chemical Engineering*, 91, pp. 15–27.

Gani, R. (2004a). Chemical product design: challenges and opportunities. *Computers and Chemical Engineering*, 28, pp. 2441–2457.

Gani, R. (2004b). Computer-aided methods and tools for chemcial product design. *Trans. IChemE – Part A*, 82(A11), pp. 1494–1504.

Goel, H. (2004). *Integrating Reliability, Availability and Maintainability (RAM) in conceptual process design*. Thesis/dissertation, Delft University of Technology.

Grievink, J. (2005). SHEET performance criteria. Personal Communication.

Grossman, I. and Westerberg, A. (2000). Research challenges in Process Systems Engineering. *AIChE Journal*, 46(9), pp. 1700–1703.

Grossmann, I. and Westerberg, A. W. (2000). Research Challenges in Process Systems Engineering. *AIChE Journal*, 46(9), pp. 1700–1703.

Gupta, S. and Bongers, P. (2011). Bouillon cube process design by applying product driven process synthesis. *Chemical Engineering and Processing*, 50, pp. 9–15.

Harmsen, G. J. and Chewter, L. A. (1999). Industrial applications of multi-functional, multi-phase reactors. *Chemical Engineering Science*, 54, pp. 1541–1545.

Harmsen, G. J., Hinderink, P., Sijben, J., Gottschalk, A. and Schembecker, G. (2000). Industrially applied process synthesis method creates synergy between economy and sustainability. *AIChE Symposium Series*, 96(323), pp. 364–366.

Hill, M. (2004). Product and Process Design for Structured Products. *AIChE Journal*, 50(8), pp. 1656–1661.

Ismail, S., Proios, P. and Pistikopoulos, E. (2001). Modular synthesis framework for combined separation/reaction systems. *AIChE Journal*, 47(3), pp. 629–649.

Jankowiak, L., Mendez, D., Boom, R., Ottens, M., Zondervan, E. and van-der Groot, A. (2015). A process synthesis approach for isolation of isoflavons from Okara. *Industrial and Engineering Chemistry Research*, 54(2), pp. 691–699.

Jankowiak, L., van-der Groot, A. J., Trifunovic, O., Bongers, P. and Boom, R. (2012). Applicability of product-driven process synthesis to separation processes in food. *Computer-Aided Chemical Engineering*, 31, pp. 210–214.

Johns, W. (2001). Process Synthesis: poised for a wider role. *CEP Magazine*, 1(4), pp. 59–65.

Kalakul, S., Zhang, L., Fang, Z., Choudhury, H., Intikhab, S., Elbashir, N., Eden, M. and Gani, R. (2018). Computer aided chemical product design – ProCAPD and tailor-made blended products. *Computers and Chemical Engineering*, 116, pp. 37–55.

Karuppiah, R. and Grossmann, I. (2006). Global optimization for the synthesis of integrated water systems in chemical processes. *Computers and Chemical Engineering*, 30, pp. 650–673.

Kim, K. J. and Smith, R. L. (2004). Parallel multiobjective evolutionary algorithms for waste solvent recycling. *Industrial and Engineering Chemistry Research*, 43(11), pp. 2669–2679.

Kiskini, A., Zondervan, E., Wierenga, P. A., Poiesz, E. and Gruppen, H. (2016). Using product driven process synthesis in the biorefinery. *Computers and Chemical Engineering*, 91, pp. 257–268.

Li, X. and Kraslawski, A. (2004). Conceptual design synthesis: past and current trends. *Chemical Engineering and Processing*, 43, pp. 589–600.

Lim, Y. I., Floquer, P., Joulia, X. and Kim, S. D. (1999). Multiobjective optimization in terms of economics and potential environment impact for process design and analysis in a chemical process simulator. *Industrial and Engineering Chemistry Research*, 38(12), pp. 4729–4741.

Linke, P., Mehta, V. and Kokossis, A. (2000). A novel superstructure and optimisation scheme for the synthesis of reaction-separation processes. *Computer-Aided Chemical Engineering*, 8, pp. 1165–1170.

Lockwood, T. and Papke, E. (2018). *Innovation by Design*. Career Press.

Marquardt, W. (2004). Lectures on "Modeling and Analysis of Chemical Process Systems". Delft University of Technology. Audiovisual Material.

Mattei, M., Kontogeorgis, G. and Gani, R. (2014). A comprehensive framework for surfactant selection and design and for emulsion based chemical product design. *Fluid Phase Equilibria*, 362, pp. 288–299.

Meeuse, F. M. (2005). Process Synthesis Applied to the Food Industry. *Computer-Aided Chemical Engineering*, 15(20), pp. 937–942.

Meeuse, F. M., Grievink, J., Verheijen, P. J. T. and Stappen-vander, M. L. M. (1999). Conceptual design of processes for structured products. In: M. F. Malone, ed., *Fifth conference on Foundations of Computer Aided Process Design*. AIChE.

Meeuse, M. (2003). *On the design of chemical processes with improved controllability characteristics*. Thesis/dissertation, Delft University of Technology.

Meeuse, M. (2007). Process Synthesis for structured food products. In: K. Ng, R. Gani and K. Dam-Johansen, eds., *Chemical Product Design: Towards a Perspective through Case Studies*, vol. 6. Elsevier, pp. 167–179. Chapter 1.

Meeuse, M., Grievink, J., Verheijen, P. J. T. and vander Stappen, M. L. M. (2000). Conceptual design of process for structured products. *AIChE Symposium Series*, 96(323), pp. 324–328.

Moggridge, G. D. and Cussler, E. L. (2000). An Introduction to Chemical Product Design. *Transactions of the Institute of Chemical Engineers – Part A* , 78, pp. 5–11.

Moulijn, J., Makkee, M. and van Diepen, A. (2001). *Chemical Process Technology*. London: John Wiley and Sons, Inc..

Ng, K. (2004). MOPSD: a framework linking business decision-making to product and process design. *Computers and Chemical Engineering*, 29, pp. 51–56.

Norton, I., Fryer, P. and Moore, S. (2006). Product/Process Integration in food Manufacture: Engineering Sustained Health. *AIChE Journal*, 52(5), pp. 1632–1640.

Papalexandri, K. and Pistikopoulos, E. (1996). Generalized modular representation framework for process synthesis. *AIChE Journal*, 42(4), pp. 1010–1032.

Peters, M. and Timmerhaus, K. (1991). *Plant design and economics for chemical engineers*. McGraw-Hill.

Pintarik, Z. and Kravanja, Z. (2000). The two-level strategy for MINLP synthesis of process flowsheets under uncertainty. *Computers and Chemical Engineering*, 24, pp. 195–201.

Proios, P. and Pistikopoulos, E. (2005). Generalized modular framework for the representation and synthesis of complex distillation column sequences. *Industrial and Engineering Chemistry Research*, 44(13), pp. 4656–4675.

Sahinidis, N. (2004). Optimization under uncertainty: state-of-the-art and opportunities. *Computers and Chemical Engineering*, 28, pp. 971–983.

Saraiva, P. M. and Costa, R. (2004). A Chemical Product Design Course with a Quality Focus. *Chemical Engineering Research and Design: Part A*, 82(A11), pp. 1474–1484.

Schubert, H. and Engel, R. (2004). Product and Formulation Engineering of Emulsions. *Chemical Engineering Research and Design*, 82(9), pp. 1137–1143.

Seider, W. D., Seader, J. D. and Lewin, D. R. (2004). *Product and Process Design Principles*. John Wiley.

Seider, W. D., Seader, J. D., Lewin, D. R. and Widagdo, S. (2010). *Product and Process Design Principles*. John Wiley and Sons, Inc.

Seider, W. D., Widagdo, S., Seader, J. D. and Lewin, D. R. (2009). Perspectives on chemical product and process design. *Computers and Chemical Engineering*, 33, pp. 930–935.

Shaw, A., Yow, H. N., Pitt, M. J., Salman, A. D. and Hayati, I. (2004). Experience of Product Engineering in a Group Design Project. *Chemical Engineering Research and Design: Part A*, 82(A11), pp. 1467–1473.

Siirola, J. J. (1996a). Industrial applications of chemical process synthesis. In: J. Anderson, ed., *Advances in Chemical Engineering. Process Synthesis*. Academic Press, pp. 1–61. Chapter 23.

Siirola, J. J. (1996b). Industrial Applications of Chemical Process Synthesis. In: J. L. Anderson, ed., *Advances in Chemical Engineering – Process Synthesis*. Academic Press. Chapter 23.

Siirola, J. J. (1996c). Strategic process synthesis: advances in the hierarchical approach. *Computers and Chemical Engineering*, 20(SS), pp. S1637–S1643.

Smith, B. V. and Ierapepritou, M. G. (2010). Integrative chemical product design strategies: Reflecting industry trends and challenges. *Computers and Chemical Engineering*, 34, pp. 857–865.

Stankiewicz, A. (2003). Reactive separations for process intensification: an industrial perspective. *Chemical Engineering and Processing*, 42, pp. 137–144.

Stankiewicz, A. and Moulijn, J. (2002). Process intensification. *Industrial and Engineering Chemistry Research*, 41(8), pp. 1920–1924.

Stappen-vander, M. L. M. (2005). Process Synthesis Methodology for Structured (Food) Products. *NPT Procestechnologie*, 6, pp. 22–24.

Subrahmanyam, S., Pekny, J. and Reklaitis, G. (1994). Design of Batch Chemical Plants under Market Uncertainty. *Ind. Eng. Chem. Res.*, 33, pp. 2688–2701.

Verloop, J. (2004). *Insight in Innovation – Managing Innovation by Understanding the Laws of Innovation*. Amsterdam: Elsevier.

Voncken, R. M., Broekhuis, A. A., Heeres, H. J. and Jonker, G. H. (2004). The Many Facets of Product Technology. *Chemical Engineering Research and Design: Part A*, 82(A11), pp. 1411–1424.

Westerberg, A. W. (1989). Synthesis in engineering design. *Computers and Chemical Engineering*, 13(4), pp. 365–376.

Westerberg, A. W. (2004). A retrospective on design and process synthesis. *Computers and Chemical Engineering*, 28, pp. 447–458.

Westerberg, A. W. and Subrahmanian, E. (2000). Product design. *Computers and Chemical Engineering*, 24, pp. 959–966.

Westerberg, A. W., Subrahmanian, E., Reich, Y. and Konda, S. (1997). Deigning the process design process. *Computers and Chemical Engineering*, 21, pp. S1–S9.

Wibowo, C. and Ng, K. M. (2001). Product-Oriented Process Synthesis and Development: Creams and Pastes. *AIChE Journal*, 47(12), pp. 2746–2767.

Wibowo, C. and Ng, K. M. (2002). Product-Centered Processing: Manufacture of Chemical-Based Consumer Products. *AIChE Journal*, 48(6), pp. 1212–1230.

Wolsey, L. (2003). Strong formulations for mixed integer programs: valid inequalities and extended formulations. *Math. Program. Ser B*, 97, pp. 423–447.

Ye, Y., Grossmann, I. and Pinto, J. (2018). Mixed-integer nonlinear programming models for optimal design of reliable chemical plants. *Computers and Chemical Engineering*, 116, pp. 3–16.

Yi, G. and Reklaitis, G. (2002). Optimal Design of Batch-Storage Network Using Periodic Square Wave Model. *AIChE Journal*, 48(8), pp. 1737–1773.

Yi, G. and Reklaitis, G. (2003). Optimal Design of Batch-Storage Network with Recycle Streams. *AIChE Journal*, 49(12), pp. 3084–3094.

Yi, G. and Reklaitis, G. (2004). Optimal Design of Batch-Storage Network With Financial Transactions and Cash Flows. *AIChE Journal*, 50(11), pp. 2849–2865.

Yi, G. and Reklaitis, G. (2006a). Optimal Design of Batch-Storage Network with Multitasking Semi-continuous Processes. *AIChE Journal*, 52(1), pp. 269–281.

Yi, G. and Reklaitis, G. (2006b). Optimal Design of Batch-Storage Network with Uncertainty and Waste Treatments. *AIChE Journal*, 52(10), pp. 3473–3490.

Yi, G. and Reklaitis, G. (2007). Optimal Design of Batch-Storage Network Considering Exchange Rates and Taxes. *AIChE Journal*, 53(5), pp. 1211–1231.

Zderic, A., Almeida-Rivera, C., Bongers, P. and Zondervan, E. (2016). Product-driven process synthesis for the extraction of oil bodies from soybeans. *Journal of Food Engineering*, 185, pp. 26–34.

Zderic, A., Kiskini, A., Tsakas, E., Almeida-Rivera, C. P. and Zondervan, E. (2019). Giving added value to products from biomass: the role of mathematical programming in the product- driven process synthesis framework. *Computer-Aided Chemical Engineering*, 46, pp. 1591–1596.

Zderic, A., Tarakci, T., Hosshyar, N., Zondervan, E. and Meuldijk, J. (2014). Process Design for Extraction of Soybean Oil Bodies by Applying the Product Driven Process Synthesis Methodology. *Computer-Aided Chemical Engineering*, 33, pp. 193–198.

Zderic, A. and Zondervan, E. (2016). Polyphenol extraction from fresh tea leaves by pulsed electric field: A study of mechanisms. *Chemical Engineering Research and Design*, 109, pp. 586–592.

Zderic, A. and Zondervan, E. (2017). Product-driven process synthesis: Extraction of polyphenols from tea. *Journal of Food Engineering*, 196, pp. 113–122.

Zderic, A., Zondervan, E. and Meuldijk, J. (2015). Product-driven process synthesis for the extraction of polyphenols from fresh tea leaves. *Chemical Engineering Transactions*, 43, pp. 157–162.

Zhang, L., Fung, K., Wibowo, C. and Gani, R. (2017). Advances in chemical product design. *Reviews in Chemical Engineering*, pp. 1–2.

Zondervan, E., Monsanto, M. and Meuldijk, J. (2015). Product driven process synthesis for the recovery of vitality ingredients from plant materials. *Chemical Engineering Transactions*, 43, pp. 61–66.

Zondervan, E., Nawaz, M., Hahn, A., Woodley, J. and Gani, R. (2011). Optimal design of a multi-product biorefinery system. *Computers and Chemical Engineering*, 35, pp. 1752–1766.

4 Formulation of design problems and identification of consumer wants

4.1 Introduction

The first level in the proposed PDPS strategy involves the framing of the design problem in the context of an overall business project. Description of the background of the project and the business context, overall supply chain and marketing considerations (product portfolio, demand profiles, regional spread, etc.) are evaluated at this stage. After the project is formulated, the team of designers is ready to embark upon the design activities by identifying, capturing and prioritizing the consumer needs and wants. The way in which needs are extracted from the customer statements, grouped, organized and ranked will depend on the product being considered. It will usually be a relatively simple task if the aim of the design problem is the improvement of an existing product. The more innovative the proposed product, the harder it may be to define the needs. The second level of the PDPS strategy focuses on obtaining the consumer wants and needs in qualitative descriptions and translate them into quantifiable product attributes.

4.2 Level 0: formulation of design problem

The starting point in any design activity is related to the formulation of the problem. Charles Kettering, the famed inventor (with over 186 patents) and engineer once said that a problem well-stated is half-solved. The scope, the background and the rationale behind a design problem are just a few of the key areas to address before embarking upon any development effort.

A prerequisite of this step is the availability of a functioning and motivated team. While there is a great number of definitions to what we refer to as teamwork, it is widely accepted that without teamwork, things do not get done (Lockwood and Papke, 2018). Based on its structure, a team can feature a functional, a project-based or matrix (or mixed) organization. In a functional organization a team focuses on activities that are done continuously in long-term assignments. This structure allows the team members to develop specialist/expert knowledge in a given field by learning from each other. Project organization addresses temporary and short-term activities. This way of working gives the team member the opportunity to develop new skills and knowledge while learning to be flexible. Matrix organizations aim to obtain the advantages of both functional and project organizations. One of the executions involves assigning projects to temporary teams recruited from the functions. Additionally, in a project-based organization expertise groups could be formed outside the projects.

An initial analysis of the purpose of the team is expected to be carried out, aiming at mapping the roles and strengths (skills and competencies) of each team member.

https://doi.org/10.1515/9783110570137-004

Moreover, a clear set of rules and code of conduct should be defined, agreed upon and communicated to all members. As a mechanism of control, performance measurements should be agreed upon and monitored throughout the whole project life cycle. Finally, key players outside the core team (stakeholders) should be identified and mapped and a communication strategy must be agreed upon accordingly. Depending on the interest in the project, the influence the stakeholder has and the attitude toward the project success, the stakeholder can be classified as (active or inactive) backer or (active or inactive) blocker. Additional terms for stakeholder include ambassador, saboteur, champion, sponsor, fence sitters and weak link. The communication and interaction requirements of each stakeholder range from minimal contact to thorough management, as depicted in Figure 4.1.

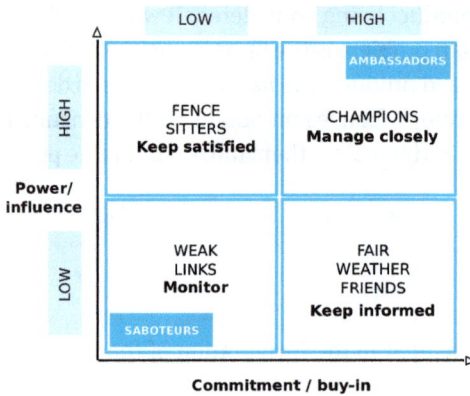

Figure 4.1: Stakeholder mapping and management strategy.

The framing level – also referred to as *mission statement* by Costa *et al.* (2014) – addresses the description of the background of the project and the business context. In fact, this level is characterized by focusing exclusively on the activities associated with the **scope and knowledge** aspect of the level. Areas to be covered include company and category strategy, current product portfolio, competitors analysis, manufacturing footprint and strategy, current manufacturing practices, market characteristics, consumer preferences (if available) and behavior, availability of resources and timing. At this level (Figure 4.2), the motivation of the project is provided with a strong emphasis not only on what is in the scope of the project, but also on what is *not* in the scope and on the resources (human, monetary, time) that are available. The majority of these aspects are commonly captured in a project brief template, like the BOSCARD template. As shown in Table 4.1, the BOSCARD template intends to include in a concise manner all relevant dimensions of the formulation of the design problem. The project brief needs to be communicated to, discussed with and signed-off by the project stakeholders.

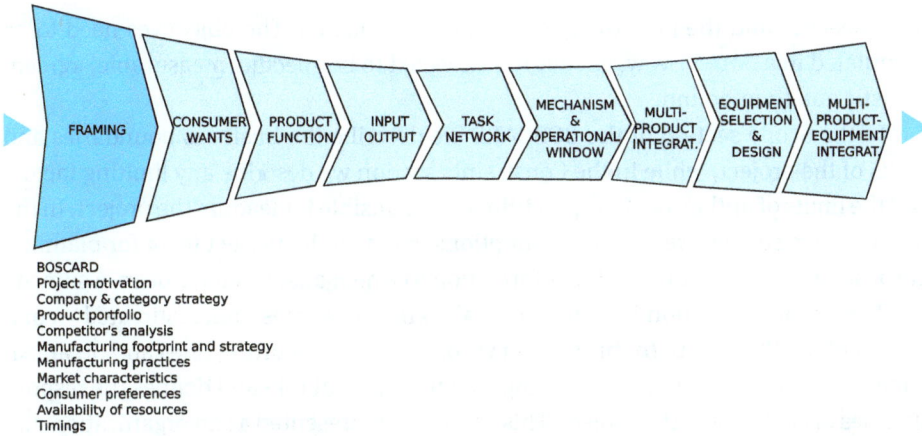

BOSCARD
Project motivation
Company & category strategy
Product portfolio
Competitor's analysis
Manufacturing footprint and strategy
Manufacturing practices
Market characteristics
Consumer preferences
Availability of resources
Timings

Figure 4.2: Framing level as part of the PDPS strategy.

Table 4.1: BOSCARD approach for the formulation of the design problem.

Activity	Description
Background	description of the context of the project, *e. g.*, strategic fit, starting point and opportunities
Objective	a SMART statement of business goals to be achieved and a statement of the result for the project team; a decision needs to be made about what is realistic, attainable and worthwhile
Scope	the team's own choice (empowered by the gatekeepers); statement: **we will...**
Constraints	restrictions from outside the project; a choice outside the team's direct influence; **we must...**
Assumptions	working assumptions to deal with uncertainty; should be checked and disappear over time, moving to scope or constraints
Risks	uncertainties that may make the project fail; need to be captured during risk analysis sessions; need to be identified, monitored and (if possible) mitigated via an action plan
Deliverables	SMART outputs, tangible and measurable, that will lead to the desired objective; they should cover the entire project

In the Background section we describe the circumstances that have led to this project being required. In particular, we explain the business rationale; why the project is being carried out and why now. Moreover, the relative priority of this project in relation to other projects and business activities is indicated. This section gives the *big picture* and puts the project into context.

In the Objective section we specify what needs to be achieved, including the overall business objectives and the specific project objectives. Moreover, emphasis is given on the definition of Key Performance Indicators (KPIs) to measure the fulfillment of the objectives and the periodicity of such measurements. The objectives need to be formulated in a SMART way, as they are expected to be specific, measurable, agreed, realistic and time-bound.

In the Scope section of the BOSCARD we describe the business boundaries and limits of the project, while in the Constraints section we describe any limiting factors and the limits of authority of the practitioner responsible for leading the project. In the Assumptions section we list the assumptions made by the project team for planning purposes and record any missing information to which answers have been assumed.

The Reporting section is a fundamental – and sometimes underestimated – component of the BOSCARD. In this section we describe the reporting lines and organizational structure of the project, detailing who fulfills what role and defines the responsibilities of each role in the project. This should be represented as an organization diagram with supporting text for responsibilities. In this section, the practitioner should set out how people will be kept informed about the progress of the project. More specifically, questions to be addresses in this section include when progress reporting will occur and at what levels, who is responsible for reporting what to whom, what format reporting will take and what the procedures are for exception reporting/problem escalation. Sometimes the R in BOSCARD refers to the *risks* associated with the execution of the project. These are given in terms of the likelihood of occurrence and their impact on the success of the project. Risks need to be identified, assessed, monitored and – if possible – mitigated by an effective action plan.

In the Deliverables section we describe the primary outcome at the end of the project. Moreover, it should explain what format these key deliverables should be in and how they will be verified and accepted (*i. e.*, signed-off by project stakeholders).

! **The six Ps of marketing**

The six Ps of marketing are designed to be an easy reminder of the key elements every marketing strategy should include to support the business strategy.

For each P, marketing should establish a goal and an outline of the plan to achieve it while maximizing or improving each P. **Product:** The goal is to thoroughly understand its features, and how it delivers benefits or value to customers at the right cost or price. It may include identifying new products and assisting in defining what features and characteristics in value dimension should be included in that product design. **People:** What are the organization's goals for acquiring new customers and retaining the existing ones? What are its goals for training salespeople? What goals does the organization have for building new strategic alliances, or improving its relationships with its promotional partners? **Price:** The goal is to thoroughly understand customers' perception of value; an effective evaluation of the value-cost or value-price ratio is crucial so that the selling price meets not only the company objectives, but also is appealing to the targeted market. **Promotion:** The goal is to promote product awareness and to stimulate demand. But it also includes creating or promoting awareness of the organization, its name, its image or its brand. **Place:** The goal includes plans for how the customers will get access to the product. Will the organization sell directly to its customers through stores, distributors, catalogs or through the Internet? Will it sell globally, or will the product only be sold regionally? **Performance:** The goal is to monitor how the corporation is performing and to ensure that the plans for improving the six Ps achieve performance; the goal is the establishment of appropriate performance measurements, monitoring performance against those measurements to make sure that they are driving the expected behaviors, and that those behaviors earn back, achieving the intended results, generating company profits and improving the return on investment.

Any plan for optimizing any of the six Ps must not negatively impact any other, and it must also be consistent with the higher-level corporate strategies and with all other functional business strategies. Cross-functional communication plays a paramount role in the identification of the six Ps and subsequently in the conduct of a marketing strategic plan.

There are other sets of Ps that have been defined, with the same rationale. For instance: proposition (the brand that consumers have in their heads; the brand values, benefits and reasons to believe); promotion (the message to the consumers); position (the place where the consumers buy the brand); packaging (the format that the consumer sees and buys); price (what the consumer pays, regular price or on promotion); and product (the product or service offered).

4.3 Level 1: identifying consumer wants

After the problem has been fully formulated, the second level of the PDPS approach is addressed. This level (Figure 4.3) focuses on customers, aiming at understanding their needs and wants and at defining the strategy towards meeting and exceeding their needs and expectations. Consumer wants are obtained in qualitative descriptions and translated into (relatively) quantifiable product attributes, represented by a consumer function (Figure 3.9).

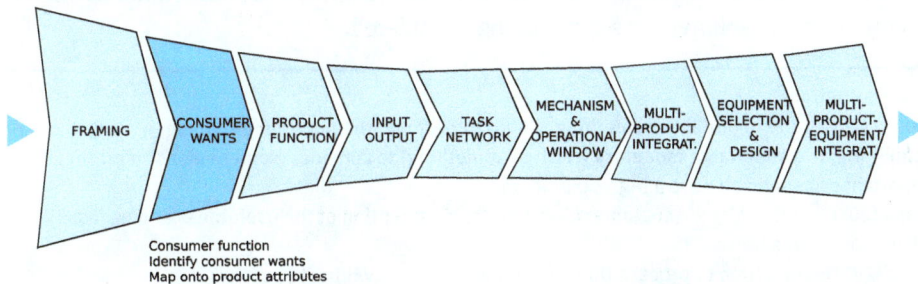

Consumer function
Identify consumer wants
Map onto product attributes

Figure 4.3: Consumer wants as part of the PDPS strategy.

Since the 1980s the marketplace has experienced a fundamental change towards conditions in which supply exceeds demand. The generalized reduction of disposable income has turned consumers more critical, demanding and conscious about their choices. Moreover, today's customers are becoming more connected, sceptical about the value-for-money proposition they receive and sophisticated in their demands with ever-increasing expectations of their product and service providers (Deloitte, 2014; Nielsen, 2018). From an organizational perspective, these marketplace conditions require organizations to respond quickly and, more importantly, to understand the consumer perceptions of value in the market they operate. Understanding the consumer's

perception of value is a key activity to undertake within any organization. Before engaging in any development or manufacturing activities, the organization needs to ask itself fundamental questions, such as:
- Why are consumers buying the products?
- How do consumers define value in products?
- Why are they buying specific products within an assortment?
- How can we offer products that the consumer will buy?
- How do we innovate the right products?

4.3.1 The perception and creation of value

The concepts of value, its perception and its creation are major concerns in many modern theories of production management. Although several definitions of value are available, all of them agree that value is a positive feature that should be targeted, pursued and encouraged. While the common perception of value is related to financial aspects (money, currency or cost), value is a multi-dimensional concept that might cover goals in quality and delivery too. In the eyes of the consumer, a product or service of value is perceived to bring a benefit to the consumer or to the organization. The difference in value perceived by the consumer between one product (or service) and another product (or service) will trigger the customer's preference. The perception of value is determined by a careful comparison of the features and characteristics offered or contained in each item or service being considered.

---

i Value **definitions**

Oxford (2018): the regard that something is held to deserve; the importance, worth, or usefulness of something; the material or monetary worth of something; to consider (someone or something) to be important or beneficial; have a high opinion of.

Ramu (2017): defined by the customer based on their perception of the usefulness and necessity of a given product or service.

Kiran (2017): any action or process that a customer would be willing to pay for.

A useful approach to understand the several components of value and their relationship is by expressing value as a mathematical formula:

$$\text{Value} = \frac{Q + D + I + U}{C}, \tag{4.1}$$

where Q denotes quality, including conformance to requirements, or the expectations or durability of a product; D is the delivery or quickness of response; I denotes innovation and is defined as the ability to offer something different or unique or leading-edge; U denotes utility and represents how well the product or service performs at its expected task, its ease of use or its effectiveness; C denotes the cost incurred in terms of time, labor and currency.

It can be inferred from this equation that value will increase if there is either an overall improvement in a product or service's quality, its speed, its innovation or its utility provided that the underlying costs remain constant. Alternatively, if Q, D, I and U remain constant while C decreases, an increase in value is expected. It goes without saying that any improvement in one or more of the dimensions valued by the consumer implies an additional cost.

What could be the impact of introducing a substitute product or service in the market on the perception of value of the original product? **?**

Value creation is concerned with generating high-quality new products, services or knowledge, and then converting them into innovations with a high-perceived (commercial) value and is a major concern in many modern theories of production management (Gao and Low, 2014). The creation of value can be regarded as a sequence of individual and linkable value adding activities. In fact, value stream is the series of activities that an organization performs (*e. g.*, procurement, design, production and delivery of products and services) to add value to the product or service in the eyes of the consumer (Ramu, 2017). Each link within the value chain adds (economic) value to the product or service, which contributes to the competitive advantage of the corporation.

A value stream often starts from a supplier and ends at the customer. A value stream comprises three main components (Figure 4.4): (i) the flow of materials from receipt of supplier material to delivery of finished goods and services to customers; (ii) the transformation of raw materials into finished goods, or inputs into outputs; and (iii) the flow of information required to support the flow of materials and transformation of goods and services.

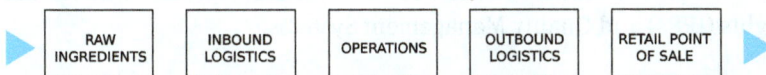

| ▶ | RAW INGREDIENTS | INBOUND LOGISTICS | OPERATIONS | OUTBOUND LOGISTICS | RETAIL POINT OF SALE | ▶ |

Figure 4.4: Classical value chain model (Porter, 1985).

Value Stream Mapping (VSM) is a commonly used tool to visualize the component of value creation throughout a process. In this simple yet powerful representation the flow of material, information, inventory, work-in-progress and operators, among others, is illustrated (Ramu, 2017). The analysis of VSM results in the identification of actions to improve the performance of a business. This activity is achieved by projecting the actual business activities on the generic value chain structure. Moreover, Value Stream Analysis can help uncover hidden wastes within the organization.

Value perception of a product of service can change over time and from market segment to market segment. In fact, regional markets may require different product designs, packaging, marketing displays or promotional methods. Regardless of the marketplace where it operates, a business can be seen as a chain of value creating activities (Porter, 1985), including five main activities and a set of supporting functions:

- *Inbound logistics:* includes reception, storage and inventory control of inputs (raw materials, process and production aids).
- *Operations:* is the core of value creation activities that transform inputs into the final product(s). Products are bound to delight the customer.
- *Outbound logistics:* includes all activities required to transport the finished product(s) to the customer, including warehousing, order fulfillment and transportation.
- *Marketing and sales:* are needed to get buyers purchasing the product(s). This includes channel selection, advertising and pricing strategies, among others. These activities are carried out to attract and retain customers.
- *Service:* is the maintenance and enhancement of the product value in the eyes of the customer through activities such as customer support and repair services. When done well these activities are fundamentally relevant to retain the customers.

As introduced by Porter (1985), next to these main or primary activities within the value creation chain, there are also support functions, including:

- *Procurement:* is the function of purchasing the raw materials and other inputs used in the value chain.
- *Technology management:* includes research and development, New Product Development (NPD), process optimization and automation, but also new technology in marketing, sales and service.
- *Human resources management:* includes recruitment, personal development and remuneration of personnel.
- *Business infrastructure:* includes financial control systems, legal and Intellectual Property Rights (IPRs) and Quality Management Systems.

---

i In this context, supply chain is defined as the mental picture of the flow of matter and important events happening to raw materials, intermediates and end-products until their end-use by consumers.

It is worth mentioning that every primary activity draws on the support activities (Figure 4.5).

4.3.2 Consumer focus: the consumer is king

As introduced in Section 2.1.4.1, one of the key principles of TQM is the focus on the consumer. Just meeting the consumer requirements may not be enough. Exceeding the requirements of the consumer may create a perception of value that encourages

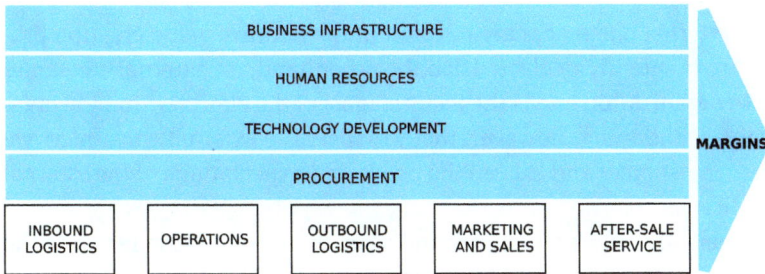

Figure 4.5: The value creation of a business and its supporting functions (modified from Hill, 2014; Porter, 1985).

the consumer to repeatedly come back for a product or service. In fact, a consumer is an entity characterized by a set of needs that need to be satisfied and expectations that need to be met (or exceeded). If a product or service that a business is offering is noted and preferred above the many alternatives a customer has to satisfy a particular need, the customer feels attracted to the offer. Moreover, the consumer feels delighted as the product or service offered either is the best value-for-money (cost reduction) or differentiates itself from similar offerings in a distinctive way worth a premium. As a result – provided that suitable business measures are in place – the consumer will come back to the offered product or service for further needs or repurchase of the product or service. These three elements – attract, delight and retain – are the functional tasks of a successful business. For instance, if the consumer has a bicycle helmet that offers sufficient protection against impact, the consumer is just satisfied but not excited. If the helmet is equipped with the right technology to inform the moves/actions to nearby riders, or to notify emergency medical services when an accident occurs, the customer might perceive a level of satisfaction beyond expectations. It might be likely that the consumer would adopt any service associated with the product (*e. g.*, subscription fee and membership). A constant focus on improvement of a product or service and meeting and exceeding consumer expectations is imperative to the success of TQM initiatives.

Customers can be categorized in numerous ways, depending on the role they play within the value creation supply chain (*e. g.*, suppliers, distributors, manufacturers, service providers and retailers), on their organizational affiliation (*i. e.*, internal and external) and on the specificity of their needs (*i. e.*, mainstream, extreme and lead users).

Regarding the role within the value creation supply chain, customers could be suppliers, manufacturers, retailers, distributors and end-users. From a strategic competitive advantage standpoint, each category of customer has its own set of requirements, which affect the marketing choices made by the producer as well as the production methodologies selected (Porter, 1985). An internal customer is that individual, department or subsection that receives the product or service internally within the organization. While internal customers are relevant in the iterative value creation process, their needs are frequently neglected. On the other hand, an external consumer resides outside the organization and is the focus of most of the efforts within the organization. It is worth mentioning, however, that to satisfy the external consumers we – as practitioners – must meet and exceed the needs and expectations of the internal consumers in the first place.

The fact that there are different markets and segments of users is reflected in the identification of three types of users: mainstream, extreme and lead users (Figure 4.6). Mainstream users represent the biggest market and are characterized by having a common set of needs. This group is extremely relevant to the financial performance of the organization. As the specificity of the set of needs increases, however, two additional types of users can be defined. Addressing and satisfying their needs might be crucial to market development and product development activities. Lead users are customers who are ahead of the game. They experience the need in advance of the mainstream market due to the intense and frequent use of the product or service. Additionally, lead users would benefit from improvements in the product for this product is extremely important for their life, business, etc. Any improvement in the product or service is highly valued by lead users. Interestingly, if such improvement is not readily available, lead consumers are more likely to have actually innovated the product or service to meet their specific expectations. This type of users are instrumental in identifying interesting (DIY) solutions and anticipate new needs or consumer requirements. Extreme users feature different (special) needs, which require them to utilize the product differently. Such special needs include limited mobility, strength or dexterity, visual or hearing impairment, etc. By observing extreme users potential solutions can be identified, which could be deployed to mainstream markets.

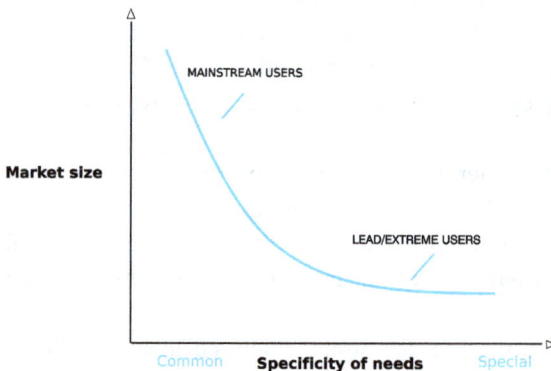

Figure 4.6: Types of users based on the specificity of their needs.

The starting point in the definition of the requirements for the product or service involves the end-customer dimension, where we evaluate whether customers' requirements are met in terms of quality, cost and delivery (Section 4.3.3.2). This process is conducted repeatedly as we move upstream – away from the end-consumer – along the other entities in Figure 4.4. The initial list of requirements collected from the end-customer entity is subsequently transmitted by the retailer to the next entity (*i.e.*, distributor) including its own set of needs. Although the end-consumer and retailer pursue the same set of basic quality characteristics from the manufacturer, they are valued differently (Hill, 2014). As we move upstream, we address the manufacturer entity, the most relevant value creation component. The manufacturer not only has an interest in meeting the end-consumer's requirements, but also has its own set of requirements that are necessary for its suppliers to provide. Key fundamental factors to consider in achieving both product conformance and reliability include the choice of materials and the manufacturer's production skills.

4.3.3 Consumer wants and needs, product specifications

A cornerstone of Level 1 of the PDPS approach consists of understanding what the consumer wants. Sometimes what consumers want is not always what they need. As mentioned by Costa *et al.* (2014), a need refers to real or imagined problem(s) a product or service solves and the functions it performs. On the other hand, a want is simply a product/service attribute that would be nice to have. This consumer want is related to a wish that is expected to be fulfilled by acquiring or using a product or service. Moreover, a consumer need is independent of the particular product concept to be developed. Often consumers are not fully aware of what they need, as they simply know what they want. Product specifications, on the other hand, are the technical features that a product or service should exhibit to satisfy the needs of the consumer. Product specifications are iteratively revised and updated during the product development process. As mentioned by Ulrich and Eppinger (2012), the first formulation of product specifications (**establish target specifications**) originates directly from the translation of consumer needs. These target specifications are aspirational and are fine-tuned after generating, developing and testing concepts. At the second instance (**set final specifications**), the specifications are locked and the design team commits to their delivery. The PDPS approach considers the translation of consumer needs to specifications (or properties) to take place in two steps. Level 2 comprises the initial translation of consumer needs to attributes, while Level 3 focuses on the translation of product attributes to properties. The generation, development and testing of concepts (or prototypes) is carried out at Level 3. The evaluation of the concepts by the consumer allows the designer to further refine the set of product specifications (or properties), resulting in a locked design prior to moving to Level 4. At the end of the design process, product specifications refer to the product concept selected, being based on technical and economic trade-offs and the characteristics of competing products, as well as customer needs (Costa *et al.*, 2014).

4.3.3.1 Identification of consumer needs

The first step in this activity is the identification of the actual consumer. In the context of product design, consumers shall not be understood as those who will buy the product or service, but those who will benefit from it (Costa *et al.*, 2014). Ideal consumers are those who are experts in a given type of product, will benefit the most by its improvement and are capable of accurately articulating their emerging needs.

There are several mechanisms to retrieve from the consumers information about their needs (Table 4.2), and all of them rely on active listening to consumers as a communication tool.

Table 4.2: Methods to gather information on consumer needs.

Mechanism	Description
Face-to-face meetings or interviews	direct retrieval of input, including the techniques of category appraisal and free elucidation (Kleef, 2006)
Focus groups	8–12 individuals: moderator focuses the attention on a predetermined set of topics in order to discuss views and opinions
Empathic design	multi-functional team is created to observe the actual behavior and environment of consumers

A key parameter to consider when it comes to selection of the method to gather consumer data is the size of the focal groups or the number of interviews to conduct or behavioral observations to carry out. According to the results of the study by Griffin and Hauser (1993), one 2-hour focus group reveals about the same number of needs as two 1-hour interviews. Next to the financial benefit of running interviews rather than focus group sessions (*i. e.*, requires a special trained moderator and recording infrastructure), interviews allows the design team to experience the use environment of the product or service. Empathic design (Section 2.1.2) is crucial for the understanding and identification of unarticulated needs, especially when extreme users are involved.

After information on consumer needs is gathered, it has to be interpreted, analyzed, assessed and organized. As a result of this analysis, a hierarchy of coherent needs is expected. By organizing the needs into a hierarchy, the practitioner eliminates redundant statements, consolidating those that are identical in meaning and then groups the needs based on similarity (Costa *et al.*, 2014). A further analysis of the list of consumer needs may require breaking them into secondary or tertiary needs or dropping carefully some of them due to practical, strategic or operational misalignment. Primary needs are the highest-level needs and embrace different domains within the need-space (*e. g.*, cost, operability and nutrition). Each primary need can be decomposed in a set of secondary needs. These needs provide specific information on the scope of the primary need. The grouping of primary and secondary needs is commonly carried out by the design team. While primary and secondary needs are normally identified and articulated by the consumer, the latent needs are exclusively identified by observation or immersion techniques (Radjou and Prabhu, 2015). These needs are hidden and, therefore, the customers may not even be aware of them. Moreover, latent needs are not easily expressed or articulated by consumers and hard to understand by interviewing consumers. Interestingly, as these needs are hidden to everybody, they are to competitors too. If these needs are effectively addressed and satisfied, products may delight and exceed consumer expectations.

There are several tools and methods to create the hierarchy of consumer needs. As mentioned by Costa *et al.* (2014), the hierarchy of consumer needs does not provide any information on the relative importance consumers give to each requirement. Any potential trade-off among consumer needs should be considered based on the relative importance of each of the needs. Three of the most commonly used methods to create the hierarchy of consumer needs are (i) quantification in terms of numerical weights and based on the experience and expertise of the design team; (ii) quantification in terms of numerical weights and based on the tailored-made consumer surveys; and (iii) application of Kano's method.

4.3.3.2 Kano diagram

The Kano diagram (Figure 4.7) is a quality-driven strategy to develop value stream and is useful in gaining a thorough understanding of the needs of customers (Kano *et al.*, 1984). While customers know exactly what they want, they may not be proficient at describing their needs or able to articulate their needs comprehensively. The Kano diagram is particularly important to close this knowledge gap and make available the information required to consistently design a winning product or service. It is a two-dimensional analysis tool of the value (quality, cost and delivery, QCD) of the product or service from the consumer's perspective. The vertical axis measures customer satisfaction, ranging from total dissatisfaction to total satisfaction, while the horizontal axis depicts the technical level of implementation of a given feature in a product or service.

The underlying feature of the Kano diagram is the classification of the consumer needs in basic, performance or WOW categories.

Basic features (sometimes referred to as **must-be** or **dis-satisfiers** or expected or **order qualifiers** or **order winners**) are those expected by the consumer, and are the minimum acceptable standard. Any technical failure causes customer dissatisfaction, and no level of technical excellence causes customer satisfaction. Customers are more dissatisfied if the product or service is less functional with respect to basic features, while customers' satisfaction never rises above neutral, no matter how functional the product or service is. Examples of basic features include reliability, safety, ease to use and nutritional value. Basic features, if fully implemented, allow a product or service to enter a market. *Performance* features (also known as **attractive quality** or **normal**)

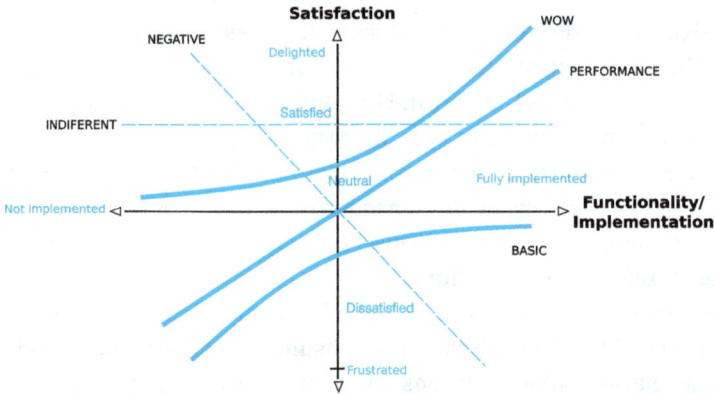

Figure 4.7: Kano diagram depicting the customer satisfaction or expectation and technical level of implementation in the product or service (Kano *et al.*, 1984; Szymczak and Kowal, 2016).

are those that, when offered, thrill or excite the consumer. Customer satisfaction is proportional to how far the feature is implemented in the product or service. In fact, these features can either satisfy or dissatisfy the customer depending on their presence or absence. Performance features allow a product or service to stay in the market. *WOW* features (sometimes referred to as **exciting** or **delighters**) are those unexpected by the consumer and that go beyond the consumer's current needs and expectations. Even if the implementation of the feature is imperfect, its presence causes excitement. WOW features – regardless of the level of implementation – allow a product or service to be the leader in the market. Examples of WOW features include heads-up displays in front windshields, GPS embedded in bike frames and molecular gastronomy dishes. The WOW features are commonly related to latent needs. When these needs are satisfied, the features generate excitement and consumer expectations are exceeded.

Improving the performance of a product or service in terms of a basic need that is already at a satisfactory level will be less valuable than improving the product performance in terms of a normal need. There is a fourth category of needs: the *non-issues*, which are features or characteristics that are not considered in the selection process. Examples include price of a product or service in a life threatening medical condition. Additional types of features also include indifferent and negative features. Indifferent features do not generate any effect on consumer satisfaction, regardless the level of implementation, while negative features are adversely impacting consumer satisfaction when implemented.

It is worth mentioning that over time, performance features become basic features. The exciting idea of today becomes the required product of tomorrow. A leading organization will constantly pulse its customers to identify the next WOWs. By identifying the best WOWs, a wide range of performance features and all the basic features at full satisfaction, the corporation will remain an industry leader.

4.3.4 Quality Function Deployment

Quality Function Deployment (QFD) is a Kaizen modeling tool extensively used to analyze consumer requirements (Imai, 1997). While QFD can be used as a planning tool for manufacturers engaged in TQM activities (Section 2.1.4.1), QFD is widely regarded as a methodology designed to ensure that all the major requirements of consumers are identified and subsequently met or exceeded. By capturing the voice of the customer (VOC) as input to the technical requirements for product design, QFD supports the practitioner to determine the product or service features customers may want or need in comparison with the technical requirements currently offered by an organization. According to Imai (1997), QFD enables management to identify the customer's needs, convert those needs into engineering and designing requirements and eventually deploy this information to develop components and processes, establish standards and train workers. QFD is structured in several interlinked design spaces and uses several management and planning tools to identify and prioritize customers' expectations (Figure 4.8). QFD is normally performed in four phases: product planning, design deployment, process planning and production planning. The main activities at each phase are given in Table 4.3.

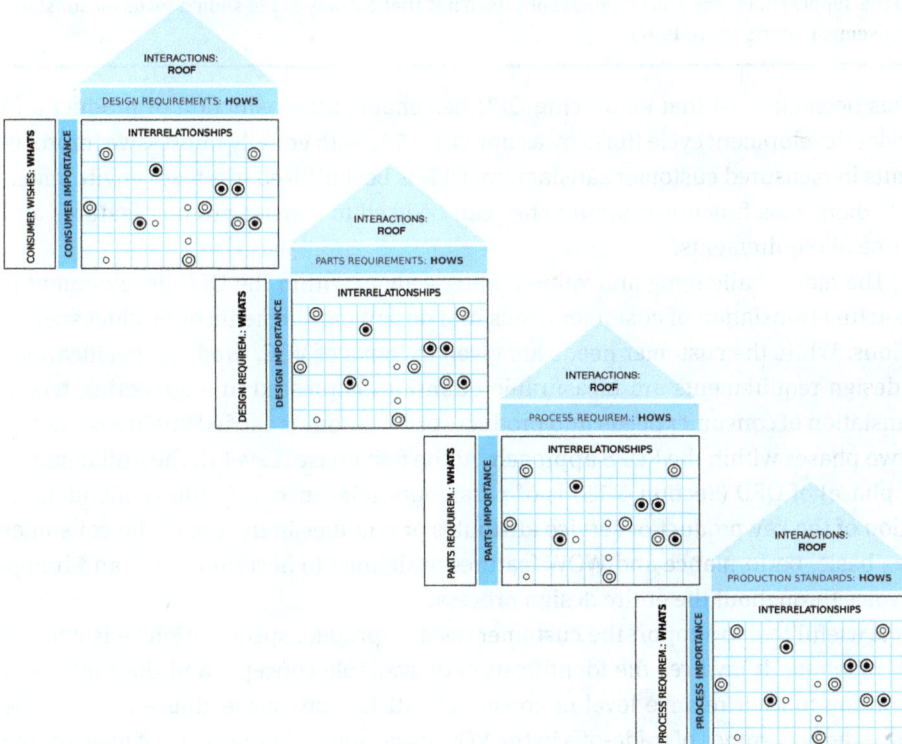

Figure 4.8: Quality Function Deployment in four matrices.

Table 4.3: Activities at each phase of QFD (Kiran, 2017).

Phase	Activities
Phase I: Product planning	determine customer requirements; translate customer requirements into design requirements
Phase II: Design deployment	select process concept; develop alternative processes and concepts; evaluate; analyze the relationship between the design requirements for each product feature; identify critical part characteristics
Phase III: Process planning	analyze and evaluate alternative designs for processes; compare relationship between process parameters and critical part characteristics; identify critical part characteristics; apply method study techniques to identify and eliminate non-value adding elements
Phase IV: Production planning	develop specific process controls; set up production planning and controls; prepare visuals of the critical process parameters for everyone to understand; train workers and ensure on-the-job guidance and supervision

ⓘ Kaizen

Kaizen is the Japanese word for improvement. In business, Kaizen refers to activities that continuously improve all functions and involve all employees, from the CEO to the assembly line workers. It also applies to processes, such as purchasing and logistics, that cross organizational boundaries into the supply chain. The Kaizen philosophy assumes that our way of life should focus on constant-improvement efforts (Imai, 1997).

It has been claimed that embracing QFD has enabled the reduction of product and service development cycle times by as much as 75 % with equally impressive improvements in measured customer satisfaction. QFD is best utilized in a team environment and when cross-functional approaches can be used to discuss product features and technical requirements.

The most challenging and value adding activity within the QFD development is the actual translation of customer needs into design requirements or product specifications. While the customer needs are given in terms of VOC, product specifications or design requirements are measurable quantities expressed in engineering terms. Translation of consumer needs into product specifications is carried out in a sequence of two phases within the PDPS approach. At the first phase (Level 1), the initial matrix (or phase) of QFD (Section 4.3.5) is of paramount relevance as it allows the identification of the key product or service attributes or qualities in the eye of the consumer (*i. e.*, basic, performance and WOW features as defined in Section 4.3.3.2) and brings its voice throughout the entire design process.

A useful tool to support the customer needs – product specifications – is perceptual mapping. It involves the identification of available concepts and their mapping according to their relative level of consumer satisfaction, suggesting – at the same time – the occurrence of trade-offs in the VOC space and/or in the space of measurable specifications/properties. As mentioned in Section 4.3.3, establishing product specifications/properties is a recursive activity within the PDPS approach. The iterative nature of the approach allows for the revision of the translation customer needs ▷ product attributes ▷ product properties/specifications, and at the same time the migration from an aspirational into a real and feasible space. Design practitioners should make sure that all the needs articulated by the consumers and identified by observations are adequately covered by one or more metrics. Specifications/properties can address multiple needs.

4.3.5 The QFD House of Quality

QFD uses a structured process called House of Quality (HOQ), which relates customer-defined requirements to the product's technical features needed to support and generate these requirements. The HoQ process involves five steps, as given in Table 4.4: (i) identifying customer requirements, (ii) identifying technical requirements, (iii) comparing customer requirements to design requirements, (iv) evaluating competitive efforts and (v) identifying technical features.

Table 4.4: Five steps in building the House of Quality.

Step	Description
Identification of consumer requirements	is generally accomplished through listening to the voice of the consumer. This becomes a structured list of the requirements of the consumers' wants and needs, involving the development of a more detailed view of the product in the arms of the consumer, especially focusing on the needs the consumer has and their translation into product or service requirements.
Identification of technical design requirements	involves the design requirements needed to support the consumer wants and needs. The design team identifies all the measurable characteristics of the product that are to meet the consumers' needs. Translation of the requirements as expressed by consumers into the technical characteristics of the product takes place at this step.
Comparison of consumer requirements with design requirements	involves also assigning ratings to signify the importance of the relationship. This is where the comparison is made to the needs of the consumers and the consumers' requirement priorities are quantified. Ratings are established to determine which of these priorities need to be worked on before others, or adjusted if there are issues with the consumer requirements that concern the design team. Improving one area might cause another area to deteriorate, and this is where these types of issues are discussed.
Evaluation of competitive efforts	includes doing competitive benchmarks and determining and measuring assessments of where the competition might offer different features.
Identification of technical features	is utilized in the final product design. As the final step in the House of Quality process, technical priorities and targets for integration into the final product design are established.

As shown in Figure 4.8, several HoQs can be used to translate consumer requirements into production planning. For a four-phase QFD system, the consumer's requirements are the input to the first HoQ, which translates the VOC into product characteristics or design requirements as seen by the manufacturer. The design requirements become the input for the second HoQ, where they are mapped onto parts requirements. The process continues and expands until production standards are identified, ultimately covering the complete value creation chain of the product or service: design ▷ engineering ▷ production ▷ sales and aftersale service ▷ support (Imai, 1997).

4.3.5.1 Building the HoQ: level 1

As input of the overall QFD system, the VOC is the most important element of the system. As illustrated in the first HoQ at Level 1 of the PDPS approach (Figure 4.9), populating the HoQ involves completing the relationship matrix (body of the house), the correlation matrix (roof), the competitive evaluation space and technical evaluation elements. Each requirement of the *VOC* (*i. e.*, the WHATs) is mapped onto *product attributes* (*i. e.*, the HOWs) with qualitative values for achieving the HOW (*i. e.*, maximize/minimize). The outcome design space of this HoQ involves concepts, which are qualitative in nature but more tangible than those extracted from the consumer requirements. In fact, we obtain the consumer wants (*e. g.*, via consumer likings, focus groups, interviews or observations) in *qualitative* descriptions and translate them into quantifiable product attributes.

The relationship between each consumer requirement and the set of product attributes is indicated by using symbolic representations (strong [⊙], medium [⊚], weak [∘] and non-existing). The Correlation Matrix (or roof) of the HoQ model provides a useful tool to examine the relationships among HOWs. Each HOW is compared with every other HOW to determine the strength of the paired relationship. Similar to the relationship matrix, symbolic representation is used to capture the relationship between each pair of HOWs, which range from strong positive [⊕], positive [+], negative [–] to strong negative [⊖]. Building the correlation matrix allows for not only the identification of conflicting and overlapping product attributes, but also the technical evaluation of competitors related to each product attribute and estimation of target values. The importance rating of each HOW can be estimated by aggregating all relationships per WHAT, including its relative relevance as identified by the consumers. The direction of improvement of each of the product attributes is also identified and symbolically represented using the following parameters: maximize [▲], minimize [▼] and hit target [▢].

Identifying product attributes is of utmost relevance. Not only does it allow for the understanding of the psychological nature and determinants of the attribute (*e. g.*, person, product, context, response task), but it also enables the development of improved

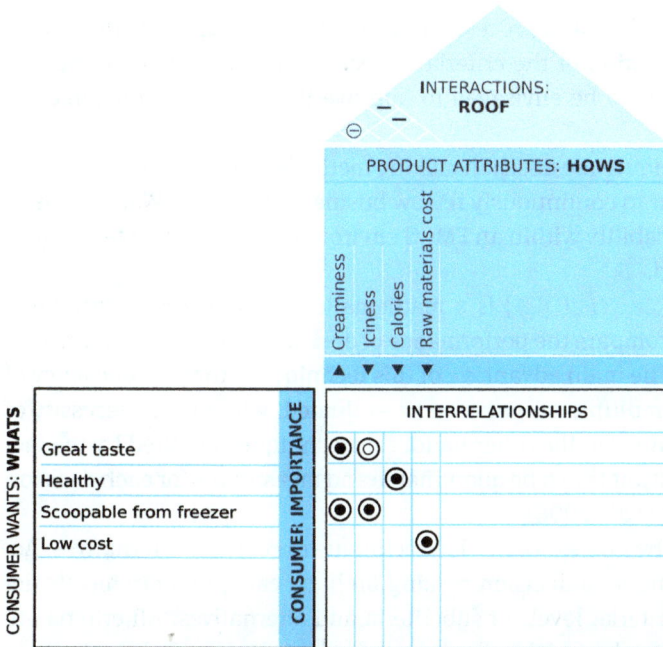

Figure 4.9: House of Quality – 1: consumer needs and product attributes in the development of an ice cream product.

measurement and further design and test intervention strategies. When multiple consumer functions are possible, each alternative is evaluated and assessed against a set of multi-dimensional criteria, referred to as the BROC decision making method (Section 4.3.6).

4.3.6 The BROC decision making method

De Ridder *et al.* (2008) developed a decision making methodology to assist the consumer-centered process design. This methodology allows for a broad design space accounting for criteria concerning consumer preferences, product attributes, process variables and supply chain considerations. Supply chain criteria include lead time, time-to-market, equipment transferability and seasonality, among others.

Improvement of decision making (DM) processes for the creation and operation of supply chains has been identified as one of the key challenges for the Process Systems Engineering community in the years to come (Grossmann and Westerberg, 2000). This process involves different aspects, characterized by different criteria with, sometimes, conflicting goals. To deal with the above, a multi-criterion decision making activity is suggested, including not only financial criteria but also criteria related to

consumer preferences, product and process characteristics and supply chain considerations. Next to the definition of the criteria to decide upon, a structured method
for decision making needs to be envisaged to improve the systematic implementation.

Benchmarking techniques are decision making methods, consist of decision making models and are meant to continuously review business processes (Watson, 1993).
Based on expected applicability within an FMCG environment, three such techniques
have been further studied.

– Data Envelopment Analysis (DEA) is a mathematical benchmarking technique
  normally applied to compare the performance of decision making units (*e. g.*, production processes). The main advantage of this technique is that the efficiency of
  multiple inputs and multiple outputs can be evaluated, without the necessity of
  using weighting factors. On the other hand, this technique is limited by the fact
  that all inputs and outputs must be quantifiable and measurable for each decision
  making unit (Cooper *et al.*, 2006).
– The Analytic Hierarchy Process (AHP) is a technique to compare alternatives. According to this technique, a decision making problem can be hierarchically decomposed in goal, criteria, levels of subcriteria and alternatives. All criteria are
  assumed to be independent and pair-wise comparisons are made by experts to
  determine weighting factors of the criteria. Pair-wise scores for all alternatives per
  criteria are given by the experts (Saaty, 1990, 2005).
– The Analytic Network Process (ANP) is a generalization of the AHP, where the assumption of a hierarchical structure is relaxed. It resembles a network, consisting
  of clusters of elements, which are the decision making criteria and the alternatives. The relations between elements depend on the decision making case (Saaty,
  1990, 2005).

The selection of the benchmarking technique is based on the applicability of the technique in the product and decision making process within an industrial setting and on
whether the criteria are quantifiable and interdependent. In view of the scope of the
PDPS approach, where the dependencies of product and process characteristics are of
relevance, the BROC approach relies on the ANP technique as benchmark.

As suggested by Partovi (2007), the BROC decision making method combines
the concepts of QFD and ANP. The concept of QFD was further extended to cover
consumer-driven product design, process design and supply chain design. Thus, the
extended QFD uses four HoQs to address and integrate consumer preferences, product and process characteristics and supply chain aspects within the decision making
activity. The network structure of this extended QFD concept and the financial factors
are implemented in ANP models of different levels of complexity (simple network,
small template and full template). For instance, the small template structure (Figure 4.10) is composed of two layers. In the first layer, the goal is divided into merits
(Benefits, Risks, Opportunities and Costs). In the second layer, subnets of these four

merits are obtained by using a QFD subdivision of the decision making criteria. According to this subdivision, clusters for decision making problems are divided into three groups: QFD factors, financial factors and alternatives. The clusters and nodes (*i. e.*, decision making criteria) in the subnets are pair-wise compared. The overall decision is reached provided that the BROC merits are *a priori* weighted. The decision making process involves going through various phases, as indicated by Saaty (2005). The following changes of and additions to Saaty's outline of steps are (i) inclusion of group decision, (ii) introduction of a step related to individual influence of decision makers, (iii) extension of the QFD concept covering consumer-driven product, process and supply chain, (iv) introduction of a structured way to arrive at the criteria, (v) specification of the timing of the different steps of the method and (vi) the reverse of the order of the steps related to the weighing of the merits and the cluster and node comparisons.

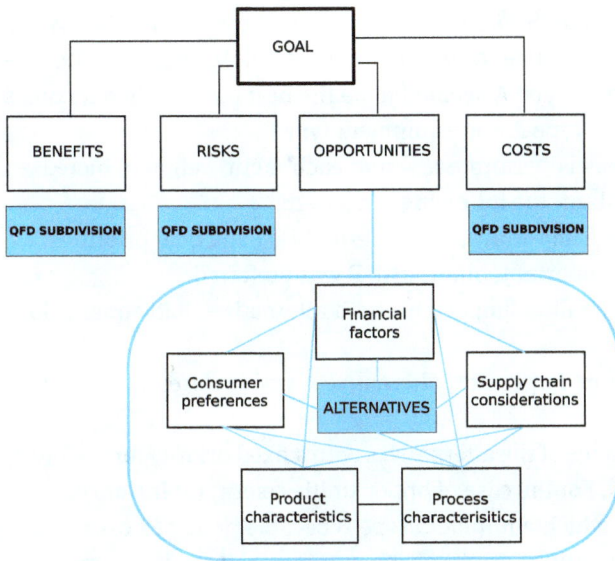

Figure 4.10: BROC: small template for Opportunities.

Example: which alternative is the most beneficial?
The applicability and scope of the BROC decision making method are demonstrated by means of a decision making case in an FMCG R&D project. The project involves different regions and parallel tasks involving design, development, implementation and launching of products and processes. In this project many decisions need to be made. An example of a decision is the choice between two types of production equipment, types A and B. The performance of type A is proven, whereas type B seems promising, but has a higher risk due to its non-fully proven performance. Equipment type B

is less expensive than type A. After having composed the appropriate decision making team, a long list of decision making criteria per merit has been compiled for the category products and processes. These criteria have been divided into consumer preferences, product characteristics, process characteristics, supply chain characteristics and financial factors. While this list does not include any non-negotiable criteria, in the event that alternatives conflict with one or more of these criteria the alternatives are not further considered for the decision making process. Next, a short list of criteria has been compiled for the particular case of deciding between equipment type A and type B. The short list (Table 4.5) has been compiled and agreed upon by key decision makers. Additionally, this short list of criteria has been divided into benefits, risks, opportunities and costs and further divided into consumer preferences, product characteristics, process characteristics, supply chain characteristics and financial factors. The small template was chosen as structure of the ANP model, as the number of clusters per subnet does not exceed 10. Per merit, the priority of the alternatives has been calculated using the software Super Decisions 1.6.0 and then multiplied by the weighting factors of the merits to come to the overall decision. According to the outcome of the software, equipment type A seemed to be the best option when it comes to deciding between equipment type A and equipment type B.

When a sensitivity analysis is performed, where each merit weight is increased while the others are kept constant, the following is noticed:
- If the benefits became more important, both alternatives had equal priorities.
- If the costs became more important, alternative B was preferred.
- If the opportunities became more important, both alternatives had equal priorities.
- When the risks became more important, alternative A was preferred.

The change of relative importance of the alternatives with a fixed priority for the BROC merits has been also analyzed. For the case of opportunities merit, for instance, there was a threshold score, above which equipment type B became preferred over equipment type A, keeping the weighting factors of the merits constant. This sensitivity analysis can be used effectively to establish the decision making criteria that should be changed and the extent of change to swap the decision in favor of equipment type A or type B.

! The role of competition

To develop strategies and business plans for competing in the marketplace, a firm needs to recognize all the different forms of its competition and understand the competition nearly as well as it understands its customers. There are two forms of competition: direct and indirect. While most organizations pay considerable attention to the direct competitors, they fall short in recognizing and scoping the impact of indirect competition. Reactive actions are conducted only after these competitors have eroded the market share significantly.

Table 4.5: BROC example: short list of decision making criteria for an FMCG R&D project.

Merits	QFD and Financial Cluster Groups	Cluster	Decision making criteria
Benefits	Consumer preferences	Mouthfeel; taste/flavor	Thickness; taste/flavor
	Product characteristics	Microstructure; storage stability; texture	Droplet size and distribution; processed fiber in continuous phase; firmness; shelf-life; critical stress; critical strain; viscosity continuous phase; viscosity emulsion
	Process characteristics	Characteristics operations; hygiene; safety; understanding of equipment	Maximum capacity equipment; scalability; cleanability; hygiene installation; production safety; experience with equipment; mechanisms; operating procedures; sustaining activities supplier; technical competence of local workforce
	Financial	Financial	Revenues
Costs	Process characteristics	Characteristics operations; hygiene	Labor intensity; plant impact; system complexity; time cleaning procedure
	Supply chain	Supply chain	Lead time; time-to-market
	Financial	Financial	Investment costs; maintenance costs; operating costs; own patents; pay-back time
Opportunities	Process characteristics	Product-process interaction; understanding of equipment	Current product opportunities; flexibility equipment; new product opportunities; competitive advantage; latest technology/knowledge/development in-house; learning for other projects
	Supply chain	Supply chain	transferability equipment to other sites
	Financial	Financial	Royalties from patents; potential savings
Risks	Process characteristics	Understanding of equipment	Risk of failure
	Financial	Financial	Business risk

4.4 Take-away message

The first level of the PDPS strategy (Level 0: Framing level) involves the formulation of the design problem. This level addresses the description of the background of the project and the business context. The BOSCARD template is introduced as a tool to capture the relevant formulation aspects of the problem. The next level of the methodology (Level 1: Consumer wants) is concerned with customers, with the understanding of needs and wants and with the strategy towards meeting and exceeding their needs and expectations. As customers have become more demanding, sceptical, critical and connected in recent decades, manufactures, suppliers, retailers, distributors, etc. are expected to improve their products and services. To deliver products and services that profitably delight customers, corporations need to understand the needs and wants of the customers in the market in which they operate. Level 1 of the PDPS approach addresses this challenge by considering that customers define value of a product or service in terms of the characteristics that they are willing to pay for. The Kano diagram is particularly useful in the identification of consumer requirements and their classification. Consumer requirements (or features) are classified in terms of the level of implementation and the consumer expectations and excitement. QFD supports the analysis of consumer requirements by capturing the VOC and translate it onto several interlinked design spaces. The first HoQ within QFD is of particular interest at this level, as it allows for the mapping of each requirement of the VOC (*i. e.*, the WHATs) onto product attributes (*i. e.*, the HOWs). The set of attributes – ranked according to the correlation matrix, direction of improvement and relevance of VOC requirements – are taken forward to the next level of the proposed methodology.

Exercises

- Prepare a BOSCARD form for the following products and services: a thickening agent for elderly consumers to be launched in central Europe and a management training for undergraduate students in Latin America.
- Prepare the HoQ-1 for the following consumer-centered products: a meatless hamburger, a container to transport beverages during commute and a long-lasting paint.
- There are *ca.* 23 million bicycles in the Netherlands. If you are expected to design an urban bicycle, could you list five basic, improvement and WOW features?
- Despite the massive risk reduction of serious head injury (*ca.* 60 %–70 %), the use of bike helmets is surprisingly low in many bike-oriented countries (like the Netherlands). Next to the lack of local mandatory use regulations, two-thirds of the bikers opt for not wearing helmets because current helmets are bulky and inconvenient to store or simply because bikers do not want to stand out among bike fellows. Draft a BOSCARD template to address the challenge of completely reimagining the urban bicycle helmet. Build a HoQ-1 by capturing the VOC of a representative population and mapping them onto a set of product attributes.
- Choose a product or service in the market and identify the primary, secondary and latent needs intended to be satisfied. Is this product or service successful in the market or is there any action that would increase its market share and consumer preference?

Bibliography

Cooper, W. W., Seiford, L. and Tone, K. (2006). *Introduction to Data Envelopment Analysis and its Uses*. Springer.

Costa, R., Gabriel, R. G., Saraiva, P. M., Cussler, E. and Moggridge, G. D. (2014). *Chemical Product Design and Engineering*. John Wiley and Sons, pp. 1–35.

de Ridder, K., Almeida-Rivera, C. P., Bongers, P., Bruin, S. and Flapper, S. D. (2008). Multi-Criteria Decision Making in Product-driven Process Synthesis. *Computer-Aided Chemical Engineering*, 25, pp. 1021–1026.

Deloitte (2014). The Deloitte Consumer Review: The Growing Power of Consumers. Tech. rep., Deloitte.

Gao, S. and Low, S. (2014). *Lean Construction Management: The Toyota Way*. Springer.

Griffin, A. and Hauser, R. (1993). The Voice of the Customer. *Marketing Science*, 12(1), pp. 1–27.

Grossmann, I. and Westerberg, A. W. (2000). Research Challenges in Process Systems Engineering. *AIChE Journal*, 46(9), pp. 1700–1703.

Hill, V. (2014). *A Kaizen Approach to Food Safety*. Springer.

Imai, M. (1997). *Gemba Kaizen: A Commonsense, Low-Cost Approach to Management*. McGraw-Hill.

Kano, N., Seraku, N., Takahashi, F. and Tsuji, S. (1984). Attractive quality and must-be quality. *Journal for Japanese Quality Control*, 14, pp. 39–48.

Kiran, D. (2017). *Total Quality Management: Key Concepts and Case Studies*. BS Publications.

Kleef, E. (2006). *Consumer research in the early stages of new product development*. Ph. D. thesis, Wageningen Universiteit.

Lockwood, T. and Papke, E. (2018). *Innovation by Design*. Career Press.

Nielsen (2018). What is next in emerging markets. Tech. rep., Nielsen Company.

Oxford (2018). Official website (https//en.oxforfdictionaries.com). Electronic Citation.

Partovi, F. (2007). An analytical model of process choice in the chemical industry. *International Journal of Production Economics*, 105(1), pp. 213–227.

Porter, M. E. (1985). *Competitive Advantage: creating and sustaining superior performance*. The Free Press.

Radjou, N. and Prabhu, J. (2015). *Frugal Innovation: how to do better with less*. The Economist.

Ramu, G. (2017). *The Certified Six Sigma Yellow Belt Handbook*. ASQ Quality Press.

Saaty, T. (1990). How to make a decision: The Analytic Hierarchy Process. *European Journal of Operational Research*, 48, pp. 9–26.

Saaty, T. (2005). *Theory and Applications of the Analytic Network Process: Decision Making with Benefits, Opportunities, Costs and Risks*. Pittsburgh: RWS Publications.

Szymczak, M. and Kowal, K. (2016). The Kano model: identification of handbook attributes to learn in practice. *Journal of Workplace Learning*, 28(5), pp. 280–293.

Ulrich, K. and Eppinger, S. (2012). *Product Design and Development*. New York: McGraw-Hill.

Watson, G. (1993). *Strategic benchmarking*. John Wiley and Sons, Inc.

5 Product function and generation of ideas

5.1 Level 2: product function

The next level of the methodology is characterized by the use of a second HoQ to move from one design-space to another. At this stage we take as input the (set of) product attributes obtained in the previous level, generate a set of (product) alternatives via prototyping and map those onto a (product) property design space (Figure 5.1). While product attributes are only qualitatively defined (*i. e.*, soft), product properties are fully quantified (*i. e.*, hard) and measurable. In the chain model, product attributes like creaminess, smoothness, freshness, softness and stickiness are translated into a set of quantifiable properties like composition, viscosity, particle size distribution, etc., through a property function (Figure 3.9). In alignment with the QFD strategy, the outcome of the first HoQ (the HOWs in Level 1) become the WHATs in Level 2. The HOWs in this level are consolidated based on the exploration of product ideas and the identification of measurable properties and technical requirements.

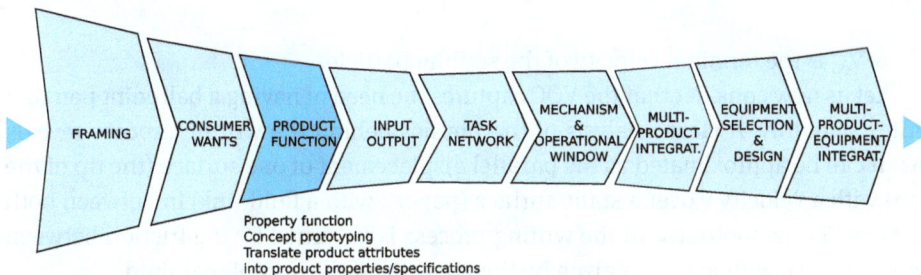

Property function
Concept prototyping
Translate product attributes
into product properties/specifications

Figure 5.1: Product function as part of the PDPS strategy.

Translating product attributes into properties can be compared to the **Set Target Specifications** step in the Product Development Process (Ulrich and Eppinger, 2012). In this step, each customer need – expressed in terms of the VOC – is translated into a set of measurable specifications. Until prototypes are developed and tested, the set of specifications remains as aspiration of the design team based on the needs articulated by the customer or inferred from observations. For each specification, there are two questions to answer: (i) how will it be measured? and (ii) what is the right value in terms of those measurable terms?

Translating product attributes into properties (or specifications) is by far not a trivial activity. It requires a sound understanding of the physical and (bio)chemical phenomena taking place within the boundaries of the system and of the nature of the attributes-properties relationships. The relevance of phenomenological understanding is addressed in the following examples.

https://doi.org/10.1515/9783110570137-005

Let us consider a multi-phase system comprising a liquid and a dilute dispersion of solids (*e. g.*, pigment in a paint). The VOC identified as a primary consumer want the stability of the system during a given period (*e. g.*, 2 years). This consumer need/want can be expressed as the attribute no noticeable settling of solids during that period of time. To translate this attribute into one or more properties, a phenomenological description of particle settling in a fluid is required. If we assume that the solid particles are spherical and that the fluid is Newtonian, the solids settling can be estimated to be governed by Stokes' (Griskey, 2002) law, resulting in the settling velocity

$$v_{solid} = \frac{1}{18} \frac{g \cdot \Delta\rho \cdot d^2}{\eta}, \tag{5.1}$$

where $\Delta\rho$ is the difference in density of solids and fluid, η is the viscosity of the fluid and d is the characteristic diameter of the solids.

For a given period of time and a maximum allowable displacement, viscosity is bound according to the expression

$$\eta > \frac{g \cdot \Delta\rho \cdot d^2}{18 \cdot v_\infty}, \tag{5.2}$$

where v_∞ is the terminal velocity of the settling particle.

Let us now consider that the VOC captures the need of having a ballpoint pen that writes smoothly. A close analysis of the physics behind the writing process reveals that it can be approximated as the parallel displacement of one surface (the tip of the pen) with a velocity v over a static surface (paper), with a fluid (ink) in between both surfaces. The smoothness of the writing process is governed by the friction between both surfaces, with in turn is given by the viscosity of the interplanar fluid.

---

i Viscosity controlling substances are normally included in the formulation of ink, including the blister-agent precursor thiodiglycol (2-(2-hydroxyethylsulfanyl)ethanol, CASRN: 111-48-8). Because of its dual use, thiodiglycol is a scheduled (*i. e.*, controlled) chemical under the Chemical Weapons Convention. As thiodiglycol is also a product of the hydrolysis of mustard gas, it can be detected in the urine of casualties.

Assuming a Newtonian behavior of the fluid, the force required to displace the pen on the substrate surface can be estimated according to the expression

$$F = \tau \cdot w \cdot b \tag{5.3}$$

$$= \eta \frac{v}{\delta} \cdot w \cdot b, \tag{5.4}$$

where τ is the applied stress, w and b are the width and length of the contact area, δ is the thickness of the film of fluid and v is the linear velocity of the pen on the substrate.

For a given pen geometry (*i. e.*, width and length) and thickness of the film, the viscosity of the fluid is the property that represents the consumer want and is given according to the expression

$$\eta < \frac{F_{max} \cdot \delta}{v \cdot w \cdot b}. \tag{5.5}$$

Let us now consider that the VOC captures the need of a pleasant mouthfeel of a high internal phase-structured emulsion (*e. g.*, mayonnaise). The sensorial attributes of structured emulsions are dictated by several physical properties of the system, including the droplet size distribution of the dispersed phase for high internal phase emulsions (HIPEs) and the viscosity of the continuous phase for low internal phase emulsions. For structured emulsions a reliable prediction of the rheological behavior is of key relevance. In the broad field covered by structured emulsions, the knowledge and prediction of the rheological properties of emulsions are vital for the design, selection and operation of the unit operations involved in the mixing, handling, storage and pumping of such systems (Pal, 2001a,b; Krynke and Sek, 2004). Moreover, in the FMCG sector reliable rheological models are continuously required to assist process engineers in activities related to quality control, interpretation of sensory data, determination of microstructure, functionality and compositional changes during processing (Holdsworth, 1993). In the open literature, a wide range of rheological models have been reported to predict the flow properties of structured products (Holdsworth, 1993). Among them, the generalized power law model of Herschel–Bulkley is commonly adopted as the most appropriate model for oil-in-water emulsions (Steffe, 1996). This three-term model requires a predetermined value of the yield stress τ_0 and involves the shear rate after yield $\dot{\gamma}$ of the emulsion. The model expression for the prediction of viscosity η can be written in the form given by Sherwood and Durban (1998), *i. e.*,

$$\eta_e = \begin{cases} \dfrac{\tau_0}{\dot{\gamma}} + K\dot{\gamma}^{(n-1)} & \text{if } \tau > \tau_0, \\ 0 & \text{if } \tau < \tau_0, \end{cases} \tag{5.6}$$

where n is the flow-behavior index, K is the consistency factor and the subscript e is related to the emulsion. The parameters K and τ_0 can be determined experimentally or can be found tabulated for a wide range of systems (*e. g.*, see Steffe (1996)). Alternatively, they can be estimated using empirical correlations like those reported by Princen and Kiss (1989) for HIPEs,

$$\tau_0 = \frac{2\sigma}{d_{32}} \Phi^{\frac{1}{3}} Y(\Phi) \tag{5.7}$$

and

$$K = C_K(\Phi - 0.73)\sqrt{\frac{2\sigma\eta_c}{d_{32}}}, \tag{5.8}$$

where Φ denotes the internal phase fraction of the emulsion, σ is the interfacial surface tension, η_c is the viscosity of the continuous phase, C_K is a model constant, d_{32} is the Sauter mean droplet diameter and $Y(\Phi)$ is a function given by the expression

$$Y(\Phi) = -0.08 - 0.114 \log(1 - \Phi). \tag{5.9}$$

The Sauter mean droplet diameter is defined as

$$d_{32} = \frac{\sum_{i=1}^{n_c} N_i d_i^3}{\sum_{i=1}^{n_c} N_i d_i^2}, \tag{5.10}$$

where N_i is the number of droplets in the size class i with mean diameter d_i and n_c is the number of size classes.

As the mouthfeel of HIPEs is characterized by the breakdown of the structure at a specific range of $\dot{\gamma}$ of *ca.* $50\,\text{s}^{-1}$ (Stokes *et al.*, 2013), we could identify viscosity as the property that quantifies the consumer need. More specifically, the particle size distribution of the dispersed phase – which in turn is characterized by d_{32} – defines the viscosity of the system and hence can be used, by extension, as the technical specification at this level.

5.1.1 Building the HoQ-2

As previously mentioned, the input at this level is the set of attributes which are derived from the articulation of consumer needs. As illustrated in the second HoQ of the PDPS approach (Figure 5.2), populating the HoQ involves completing the relationship matrix (body of the house), the correlation matrix (roof), the competitive evaluation space and technical evaluation elements. Each attribute (*i. e.*, the WHATs) is mapped onto *product properties* or *specifications* (*i. e.*, the HOWs). The outcome design space of this HoQ allows for the generation of concepts or prototypes, which are featuring the properties or meeting the specifications.

The relationship between each product attribute and the set of product properties is indicated by using similar symbolic representations as in HoQ-1: strong [⊙], medium [◉], weak [∘] and non-existing. The relationship matrix is populated by capturing the relationship between each pair of HOWs: strong positive [⊕], positive [+], negative [–] to strong negative [⊖]. Building the correlation matrix allows not only for the identification of conflicting and overlapping product properties, but also for the technical evaluation of competitors related to each product property. The importance rating of each HOW can be estimated by aggregating all relationships per WHAT. Moreover, the direction of improvement of each of the product properties is also identified and symbolically represented using the parameters maximize [▲], minimize [▼] and hit target [□].

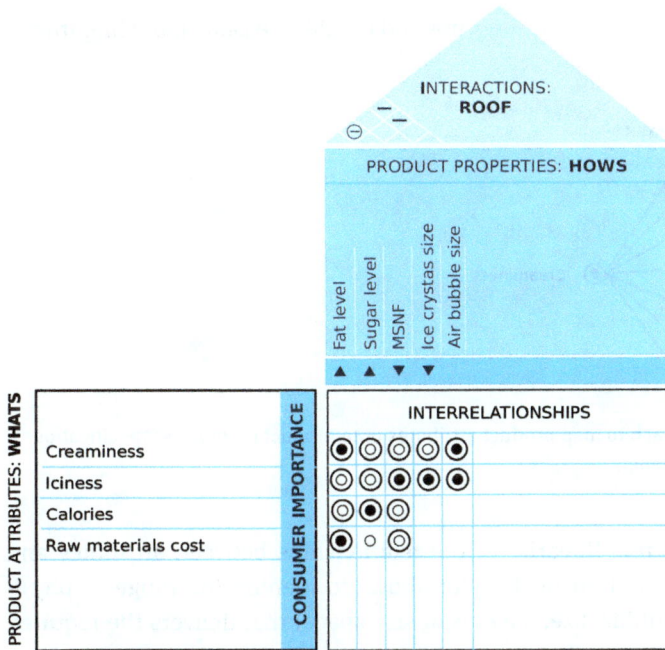

Figure 5.2: House of Quality – 2: product attributes and product properties (specifications) in the development of an ice cream product.

Product specification: is the translation in technical terms of one or more consumer needs. It is characterized by three elements: a metric, a set of values of the metric and the unit of the values. Examples of these elements are (i) [viscosity at $10s^{-1}$; (1.0–2.0); cps], (ii) [Sauter mean diameter; (10–15); μm], (iii) [mass; (5.0–7.0); kg] and (iv) [unit production cost; 3.14; €]. Product specifications and product properties are interchangeable concepts in the framework of PDPS.

An alternative method to translate the product attributes (or even the consumer wants) into product properties or specifications involves the application of data mining or AI techniques. When a set of reliable data points is available, a deterministic approach, like Convoluted Neural Networks (CNN), allows for the identification of correlation functions between inputs and outputs. The explanatory note *Convoluted Neural Network technique as a design tool* (Section 7.1.2.1) provides an overview of the method. Figure 5.3 depicts the structure of a CNN applied to the prediction of creaminess (product attribute) based on a set of product properties. The selection of the properties to be included in the model defines the accuracy of the prediction, together with the quality of the data. A similar approach was adopted by Almeida-Rivera *et al.* (2007) in the rationalization of raw materials for the manufacture of structured emulsions. Analytical properties and sensorial attributes (*i. e.*, consumer-liking attributes) were mapped onto each other and their relationships were established by integrating CNN and Partial Least Squares methods under the same modeling framework.

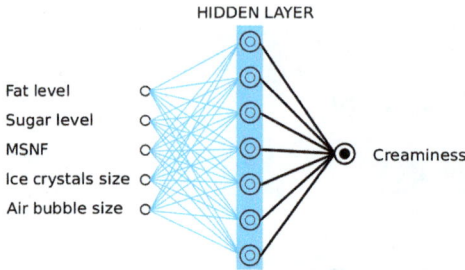

Figure 5.3: Data mining approach to map product attributes onto product properties (specifications) space: creaminess in an ice cream.

Figure 5.4 illustrates the non-linearity of the relationships between attributes and physical properties. The contour of the plot allows to identify the range of physical properties (*e.g.*, air bubble size, ice crystal size space) that delivers the required response (*e.g.*, the highest creaminess score).

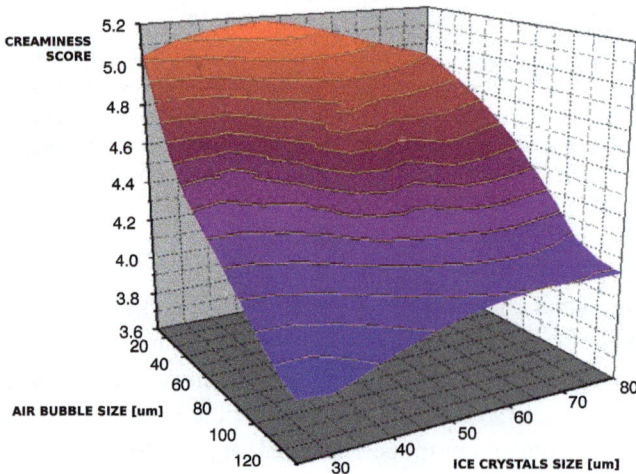

Figure 5.4: Dependency of physical properties (air bubble size and ice crystal size) with product attributes (creaminess) for the development of an ice cream product.

Next to the identification of the properties associated with each of the attributes (*i.e.*, property function), which are derived from bringing the VOC upfront, a paramount component of this stage is the generation of product ideas and the conduct of the iterative loop (prototype ▷ testing ▷ refining) (Section 2.3.1).

5.1.2 Generation of ideas

As defined by Ulrich and Eppinger (2012) a product concept (or idea) is an approximate description of the technology, working principles and form of the product. Moreover, the concept concisely addresses the description of how the product will satisfy the customer needs. The creative process of generating ideas or concepts can be regarded as solving a problem given by the gaps between product requirements and their fulfillment by systems (Gadd, 2011). As the generation of ideas or concepts spreads over different design spaces (*e. g.*, product, process, service), the solutions to be generated might embrace improvement both for operation and fulfillment of consumer requirements. The starting point of this activity involves a comprehensive formulation of the problem (Section 4.2), including understanding the gaps between requirements and current executions. As mentioned in Section 2.4, creativity-centered activities are rarely straightforward or trivial and require the application of a wide range of structured tools and methods, and sometimes a touch of luck. Failing in exploring the vast design space without the support of structured approaches will result in the incidence of dysfunctional behaviors, which translate into costly, unviable or unfeasible alternatives. Examples of these undesired situations include (i) a focus on a limited number of (biased) alternatives, (ii) a reduced level of commitment as a result of composing a non-multi-disciplinary team and (iii) failure to explore available solutions in-house, out-house or in other categories or industries.

Problem solving takes place – at various timelines and scales – throughout the entire product-process development sequence (design, manufacture, distribution, use and disposal). In each stage, problem solving requires the completion of four activities, which are reflected in each of the previous levels of the PDPS approach: (i) needs identification, (ii) system improvement or definition that allows for the satisfaction of those needs, (iii) resource allocation to deliver the system and (iv) the actual problem solving activity.

5.1.2.1 Brainstorming
Sometimes referred to as internal search, brainstorming is one of the commonly used techniques to generate ideas by randomly unlocking the tacit knowledge of the participants. While this technique is universally used and relatively easy to implement, brainstorming is extremely powerful provided a basic set of rules are followed.
1. Team composition: brainstorming is carried out by having a multi-disciplinary, non-hierarchical, constraint-free team; potential solutions might arise from focusing on the problem from a different point of view.
2. Judgement: the underlying rule is to defer judgment, for it affects the natural flow of ideas; no criticism must be allowed during the session; by encouraging the expression of any idea (including the wild and far-fetched), we stretch the boundaries of the solution space.

3. Number of ideas: one of the main goals is the generation of a large set of ideas to explore all corners of the solution space; as mentioned by Ulrich and Eppinger (2012), each idea acts as a stimulus for other ideas.
4. Building up ideas (snowballing): ideas that might be regarded as outlandish or unfeasible at first sight can often be upgraded, debugged or repaired during the session.

Variations of the conventional brainstorming session include brainwriting pool and 6-3-5 methods. In the brainwriting pool method, participants write their ideas (on a sheet of paper) once the problem has been explained. The sheet of ideas is placed in a pool, from which each participant pulls a sheet and add ideas to it. In the 6-3-5 method, six participants write three ideas on a sheet, which are passed five times among the participants. Each time, the participants add ideas to the sheet.

5.1.2.2 Morphological analysis

Morphological analysis – sometimes referred to as complete synthesis – is a relatively popular method for concept generation and based on problem decomposition. It involves splitting the problem into smaller subproblems and the systematic analysis or exploration of possible solutions of each of the subproblems. The strength of the method relies on a solid goal orientation in the search for solutions of each of the subproblems (Orloff, 2006). After a set of solutions are determined for each subproblem (or component) through internal (*e. g.*, brainstorming) and/or external (*e. g.*, patent review, competitor benchmarking) search, synthesis of possible combinations (or connections) is explored. Combinations with merit are included in a set of potential concepts, reviewed and further elaborated in the next phases or levels of the design process.

5.1.2.3 TRIZ

While working as a patent officer in 1945, Genrikh Saulovich Altshuller (1926–1998) observed that patent applications were ineffective and weak (Orloff, 2006). He recognized that bad solutions to problems ignored the key properties of problems that arose in the relevant systems. He concluded that all procedures were based on trial-and-error, on intuition and fantasy, and that none of these methods started with an investigation of the laws of the development of the systems and physical-technical conflicts that are part of the problem at hand. After having reviewed and analyzed over 10,000 inventions from 43 classes of patents (Section 5.1.2.5) and driven by his commitment to understanding creativity and innovation in others (Fox, 2008), Altshuller suggested that (Orloff, 2006) (i) the key to the solution to problems lies in the discovery and elimination of contradictions in the system; (ii) tactics and methods for solutions to problems can be created by analyzing important inventions; and (iii) the strategy of the solution of a problem must be supported by the laws of the development of technical systems.

Derived from the analysis of hundreds of thousands of patents, Altshuller determined that there are about 1,500 basic problems, which can be solved by applying one or more of the 40 universal answers or principles. In fact, Altshuller believed that inventive problems stem from contradictions between two or more elements.

Let us consider, for example, the following statement: *if we want more acceleration, we need a larger engine; but that will increase the cost of the car.* This simple example implies that more of something desirable also brings more of something less desirable, or less of something else also desirable.

While originally conceived to be an approach to solve engineering and design issues, currently TRIZ embraces a set of tools articulated around a model-based technology for generating innovative ideas and solutions (Fox, 2008). As mentioned by Gadd (2011), TRIZ is a unique, rigorous and powerful engineering-based toolkit which guides designers to understand and solve their problems by accessing the immense treasure of past engineering and scientific knowledge. Although there are several rigorous and powerful tools to support the designer after the problem solving activity has taken place, TRIZ is the only solution tool that helps practitioners beyond brainstorming at the actual concept's solution creation moment.

In the logic of TRIZ problem solving (Gadd, 2011), systems are created to provide functions which deliver benefits to fulfill needs. Whenever there is a mismatch between needs and the system's outputs, there are problems. To problem solve we have to understand the system, how it works and how well it fulfills the defined needs and also understand any requirements or benefits we are additionally seeking. TRIZ tools allow the designer to define the *system we have* and the *system we want* and to solve any problems by closing the gaps between them. The TRIZ toolkit comprises a set of rigorous, straightforward and strong tools and methods which support the generation of ideas and problem solving activity. An overview of those tools with a brief scope definition is given in Table 5.1.

The effectiveness of TRIZ relies on its systematic, auditable and repeatable nature. While plausible solutions to problems can be found using other solving tools or methods, TRIZ will ensure that no solutions are missed: the good ones but also the bad ones. As TRIZ is based on proven successful patents, there is limited (or even no) risk of failure of the solutions derived from the method. Moreover, TRIZ allows for a faster idea generation process, a broader range of ideas and a natural and self-reinforcing learning path. While the applicability of TRIZ hovers over a wide range of problems that require innovation and creativity components, this tool is not restricted to engineers or technical professionals, but also has been used successfully as a strategic management tool. TRIZ helps the practitioners to systematically unlock intrinsic knowledge and intelligently access the relevant world's knowledge (Gadd, 2011).

Table 5.1: Toolkit of TRIZ in the generation of ideas and problem solving (Gadd, 2011).

Tool	Scope
40 Principles	most important and useful TRIZ tool for solving contradictions accessed through contradiction matrix and separation principles
8 Trends of Evolution for perfecting systems	helps predict likely technological directions; useful when looking to improve an existing system; for future system development
Effects	engineering and scientific concepts arranged for easy use; a simple list of questions and answers to access all the relevant technical and scientific conceptual answers.
Thinking in Time and Scale	for problem context, understanding and solving
Ideal, Ideality, the Ideal Outcome, Ideal Solution, Ideal System and Ideal Resources	helps to overcome self-imposed constraints; for understanding requirements and visualizing solutions
Resources and Trimming	identifies all the potential resources available to solve a problem, even harmful/waste effects and products; the cheapest solution is likely to use an existing resource; for clever and low-cost solutions
Function Analysis and Substance Field Analysis	system analysis for understanding the interrelationship of functions; encourages functional thinking, and captures all the interactions within a system, even the harmful effects; particularly useful for cost reduction problems
76 Standard Solutions	for solving any system problems; creating and completing systems, simplifying systems, overcoming insufficiency, dealing with harms, future development and smart solutions for technical problems
Creativity Triggers	for overcoming psychological inertia and for understanding systems and visualizing solutions including Size-Time-Cost and Smart Little People
Smart Little People	engages the imagination, and helps break psychological inertia

5.1.2.4 Biomimicry

It is widely accepted that mother nature is not a muse exclusive to artists. In fact, the idea of looking to nature for inspiration has accompanied humankind since its early days. Even nowadays, where technology advances are steadily exceeding optimistic forecast, we are becoming increasingly aware that a large proportion of our inventions, innovations or creations can be found in nature. With *ca*. 3.8 billion years of R&D, nature is the most comprehensive, powerful and advanced laboratory.

Biomimicry, defined as the discipline engaged in the abstraction of good design from nature, has awaken increasing interest in the last decades. The list of examples of human innovations derived from observation to and emulation of natural forms and processes are countless and spread over all possible disciplines (Table 5.2).

The first step towards the application of biomimetic design principles and to create more efficient (and sustainable) systems is the understanding of the concept of

Table 5.2: Examples of nature-inspired designs and innovations.

Human design	Nature inspiration
Bullet train (shinkansen)	kingfisher peak shape: to overcome the intolerable noise caused by the air pressure changes every time the train emerged from a tunnel
Internal climate control in buildings	termite mounds: to facilitate gas exchange using internal air currents driven by solar heat.
Velcro strips	(French: *velours croché*) Hooked burrs of the burdock plant: to adhere two surfaces through loop-hook entanglement
Self-cleaning surfaces	lotus leaf (effect): to repel particles or droplets due to the presence of myriad crevices of its microscopically rough surface
Adhesive-free sticky surfaces	gecko feet: to adhere two surfaces by contact splitting

function. In the context of biomimicry, a function refers to the role played by an organism's adaptations or behaviors that enable it to survive (AskNature, 2019). Functions are related to specific needs of the organism, which are satisfied by implementing a biological strategy, mechanism or process. Similarly to the identification of consumer needs (Section 4.3.2), biomimicry requires the identification of functional needs and rephrases them using the formulation: HOW does nature do that? The biomimicry taxonomy is particular useful to this step and includes *ca.* 160 functions nested in groups and subgroups. The term context plays a key role in the identification of the strategy to accomplish a given task. Context can be defined as the surrounding environment and all other factors affecting the survival of the organism.

The biomimicry approach features several similarities with the PDPS strategy in terms of problem decomposition and systematic exploration. While biomimicry considers the sequence needs ▷ strategy ▷ function, the PDPS strategy follows the hierarchical sequence needs ▷ attributes ▷ properties ▷ fundamental tasks ▷ mechanism. Fundamental tasks and mechanisms (Chapter 6) are organized in a similar nested hierarchy as in the biomimicry taxonomy.

5.1.2.5 Patent review
Patents – together with other types of intellectual property rights – are a powerful source of information and inspiration for the design and development of products and services. The wider term intellectual property embraces not only the legally protectable ideas of products and processes, but also names, designs, breeds and artistic expressions (*e. g.*, literature, music and art). Considering the nature of the knowledge asset, intellectual property can fall into one of the following categories: patent, trademark, copyright, registered design and trade secret.

Intellectual property is not only an essential business asset in the knowledge economy, but also protects small innovative firms, helps guarantee standards for public benefit by means of licensed trademarks and enables the release of proprietary knowledge into the public domain under controlled conditions (EPO, 2019). From a societal standpoint the role of the patent system is to encourage technological innovation, promote competition and investment, provide information on the latest technical developments and promote technology transfer.

There is no positive definition of an invention under most patent laws, including the European Patent Convention (EPC). A possible definition in line with the majority of patent laws and that follows established case law (jurisprudence) establishes that an invention is the technical measures that are required to solve a problem. In accordance with this definition, an invention could include *products* showing certain (or a certain combination of) properties, *methods* to run a process at higher output or increased efficiency, *equipment adjustment* so that less maintenance is required and *new use* of a known product. Under the EPC, patents shall be granted for any inventions, in all fields of technology, provided they are new, involve an inventive step and are susceptible of industrial application.

The wealth and breadth of knowledge we can retrieve from patents – or any other form of intellectual property – allows for the possibility of an exhaustive screening of design spaces not considered otherwise. Moreover, patent review supports the identification of ideas that could trigger novel ideas, and – in the case of expired patents – allows for the direct right of use of the original invention. In fact, it has been estimated (EPO, 2019) that approximately 80 % of technical information is only available in patents in comparable detail and that only 10 % of such knowledge dwell is protected under active patents. Hence, about 90 % of the information in patents is freely available. Despite this encouraging statistic, it is estimated that globally up to 25 % of R&D resources are spent on ideas that are already available in patents. In the EU, for instance, this lost effort represents about 60 billion euros yearly.

ℹ Categories of intellectual property

Patent: gives the patent proprietor the exclusive right to benefit from the patented invention. Under Article 53 Dutch Patent Act, a patent shall confer on its owner the exclusive right (i) to make, use, put on the market or resell, hire out or deliver the patented product, or otherwise deal in it in or for his business, or to offer, import or stock it for any of those purposes and (ii) to use the patented process in or for his business or to use, put on the market or resell, hire out or deliver the product obtained directly as a result of the use of the patented process, or otherwise deal in it in or for his business, or to offer, import or stock it for any of those purposes [..].

Trademark: is any sign, capable of being represented graphically, which distinguishes the goods and services of one undertaking (company or organization) from those of another.

Copyright: protects any production of the human mind, such as literary and artistic works.

Registered design: is the outward appearance of the whole or parts of a product resulting from its features.

Trade secret: refers to the information that is not generally known or easily discovered, that has a business, commercial or economic value (actual or potential) because the information is not generally known and that is subject to reasonable efforts to maintain secrecy.

Plant variety right: is the exclusive exploitation right for new plant varieties.

There are other alternatives to patenting, including information disclosure, secrecy and a do-nothing approach. The underlying rationale behind the adoption of these alternatives is a cost-effective protection of exclusive right. Nevertheless, these strategies do not provide accountable protection against reverse-engineering or corporate data leaks.

5.2 Level 3: input–output

The next level of the methodology comprises the consolidation of all inputs and outputs of the design problem, as obtained from the previous levels (Figure 5.5). A complete specification of all exchange streams to the process is generated, as schematically shown in Figure 5.6-left for the case of the manufacture of an ice cream type product. These exchange streams comprise inputs/raw material(s) and target output/product(s), which are characterized by properties involving physical, chemical, biological, sensorial and format attributes. Examples of these properties include flow rate, temperature, pressure, pH, droplet size distribution, microbiological status, color and ergonomics, among others. The inputs include all (sources of) materials with their amounts, aggregation state and functionality within the product concept (Table 5.3). The output comprises a relatively complete description of the spatial distribution of the components within an intended microstructure (Figure 5.6-right), flavor profile and, in the case of (bio)chemical reactions during the manufacture (*e. g.*, fermentation or Maillard reactions), a description of the outcome of these reactions. Equally relevant, the output of this level includes the product or target specifications as obtained from translating consumer needs into measurable metrics.

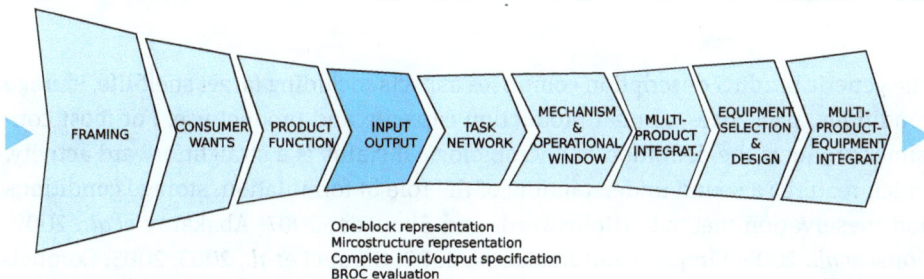

FRAMING · CONSUMER WANTS · PRODUCT FUNCTION · INPUT OUTPUT · TASK NETWORK · MECHANISM & OPERATIONAL WINDOW · MULTI-PRODUCT INTEGRAT. · EQUIPMENT SELECTION & DESIGN · MULTI-PRODUCT-EQUIPMENT INTEGRAT.

One-block representation
Mircostructure representation
Complete input/output specification
BROC evaluation

Figure 5.5: Input–output level as part of the PDPS strategy.

In addition to the specification of the exchange streams, the scope of this level requires the consolidation of a list of boundary constraints and relevant KPIs to assess and evaluate each potential alternative. Performance parameters include economic performance metrics (Section 5.2.1), turndown ratio, changeover time, microbiological safety and stability, flexibility, pumpability and availability, among others.

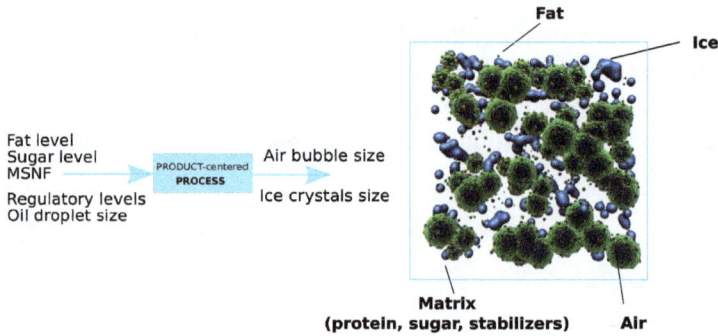

Figure 5.6: Input–output level of the PDPS strategy. Example: manufacture of an ice cream type product. Left: input–output representation. Right: expected microstructure.

Table 5.3: Overview of functionality of raw materials at Level 3 of the PDPS approach. Example: shower gel.

Raw material	Functionality
Alcohol	stabilization of system
Sodium chloride	thickener agent
Sodium hydroxide	pH balance
Salt of lauric acid	cleaning agent
Methylisothiazolinone	preservative
Colorants	coloring agent
Fragrances	scent agent
Styrene/acrylates copolymer	opacity agent
Water	continuous phase

The generic product description comprises aspects including target shelf-life, storage conditions, pack sizes, annual production capacity and product use. For most consumer products the definition of microbiological status is a straightforward activity, which requires a sound understanding of the role of formulation, storage conditions and preservation methods (Holdsworth and Simpson, 2007; Abakarov *et al.*, 2009; Donsi *et al.*, 2009; Simpson and Abakarov, 2009; Coronel *et al.*, 2003, 2005; Coronel, 2005).

Each potential alternative derived from the generation of product ideas is taken into account by executing individual input–output analyses and by evaluating appropriate KPIs for each case. Next to the requirement related to the non-negotiable satisfaction of mass and energy balances, the BROC multi-criteria decision making method (Section 4.3.6) could be adopted to support the identification of the most recommended alternative among a set of options characterized by (sometimes contradictory) trade-offs.

5.2.1 Economic evaluation

Next to the fulfillment of musts or non-negotiable requirements, at this level we are concerned about not spending any time on product alternatives that are not worth the effort. Examples of such musts include microbiological safety and stability, environmental impact, consumer acceptance or desirability and (expected) technical feasibility. From a business standpoint and in close agreement with the Design Thinking framework (Section 2.1.2), the alternatives to be further pursued should be viable. A rather simple metric to evaluate the viability of the alternative(s) is the economic potential (EP), which estimates the difference between the (expected) value of products plus by-products and the value of raw materials by the following formula:

$$EP = \sum_{i \in n_p} C_i + \sum_{j \in n_{bp}} C_j - \sum_{k \in n_{RM}} C_k, \tag{5.11}$$

where n_p represents the product streams, n_{bp} the by-products, n_{RM} the raw material streams and C_i the cost of the corresponding stream i.

A closely related metric to size the financial opportunity of each alternative is the Gross Margin (GM), which estimates the EP as a fraction of the cost of the feed streams as follows:

$$GM = \frac{EP}{\sum_{k \in n_{RM}} C_k} \cdot 100\,\%. \tag{5.12}$$

The criterion associated with the GM depends on the level of implementation complexity of the expected concept, maturity and risk of the target market and trade-off of affordability, among others. As a rule of thumb initial analysis should target a GM of at least 50 %. Next to EP and GM, there is an extensive list of metrics that could be used to assist the designer in the evaluation of the financial viability of the alternatives (Table 5.4).

When it comes to the evaluation of the financial performance of a project or business, it has been widely accepted that a set of principles should be observed and respected. Those principles – given in the explanatory note *Generally accepted accounting principles* – provide for a consistent, reliable and realistic strategy towards project evaluation. Next to these principles, there is a set of metrics that assist the quantification of the business performance, given in terms of statements on liquidity, performance and investment (Table 5.4). Each type of investment plays a specific role at given times within the business cycle. This cycle describes the flow of financial resources from investment until reinvestment or dividend payment to shareholders. To start a business, investment of capital is required. With the expectations of a return of their investment in the future, investors buy shares of the company (share money). With the invested cash assets for the company can be purchased and resources can be

Table 5.4: Metrics to assist financial evaluation of alternatives and financial statements to control the performance of a project/business.

Metric	Scope and calculation
Net Present Value	value today of future cash flows; earnings in the future discounted for inflation (or internal rate of return); the higher the number the better. It is calculated as $NPV = C_0 + \sum_{t=1}^{N} \frac{C_t}{(1+r)^t}$, where N represents the number of years, C_0 the capital invested, C_t the earnings in year t and r the discount rate.
Internal Rate of Return	rate of return made on the project, relating the cash generated to the cash invested (e. g., CAPEX, A&P); the higher the number the better.
Payback Time	period of time required to repay the original investment; time it takes to recover the initial amount invested, i. e., when you get your money back and start making a profit; the shorter time the better.
Balance Sheet	referred to as statement of financial position or condition; reports on a company's assets, liabilities and net equity as of a given point in time.
Income Statement	referred to as Profit and Loss statement (P&L); reports on a company's results of operations over a period of time.
Statement of Cash Flows	reports on a company's cash flow activities, particularly its operating, investing and financing activities.
Total Shareholder Return	value of a (public) company, determined by what the shareholders are prepared to pay for the shares. It is calculated as $TSR = \frac{P_e - P_b + \sum D}{P_b}$, where P represents the price of a share and e denotes end, b beginning and D the dividends paid to the shareholders.
Return on Capital Employed	allows to identify what is the *best* company/project to invest in.
Gross Profit	first-level estimation of difference between turnover and material costs.

employed to engage in production activities. These production activities result in products and/or services, which generate a profit. Upon acquisition of the products and/or services by customers or consumers cash is generated for the business. This cash is either reinvested to further increase production capacity and efficiency or (partly) used to pay back the shareholders.

---

[i] **Generally accepted accounting principles**
Principle of regularity: regularity can be defined as conformity to enforced rules and laws. This principle is also known as the principle of consistency.
Principle of sincerity: according to this principle, the accounting unit should reflect in good faith the reality of the company's financial status.
Principle of the permanence of methods: this principle aims at allowing the coherence and comparison of the financial information published by the company.
Principle of non-compensation: one should show the full details of the financial information and not seek to compensate a debt with an asset, a revenue with an expense, etc.
Principle of prudence: this principle aims at showing the reality *as is*: one should not try to make things look prettier than they are. Typically, a revenue should be recorded only when it is certain and a provision should be entered for an expense which is probable.

Principle of continuity: when stating financial information, one should assume that the business will not be interrupted. This principle mitigates the principle of prudence: assets do not have to be accounted at their disposable value, but it is accepted that they are at their historical value.
Principle of periodicity: each accounting entry should be allocated to a given period, and split accordingly if it covers several periods. If a client prepays a subscription (or lease, etc.), the given revenue should be split to the entire time span and not counted for entirely on the date of the transaction.

It is worth mentioning that EP (equation (5.11)) and gross profit (Table 5.4) can be used interchangeably at this level of the PDPS approach. As the level of design detail increases the gross profit incorporates not only the material costs, but also capital costs, energy costs, labor costs and supply chain costs. Table 5.5 lists the main components of each contribution to the production costs.

Table 5.5: Components of product cost per unit measure of final product costs in a simplified economic evaluation.

Element of product cost	Components
Production cost	production directs, direct labor, depreciation production, other production hall cost
Production indirects & other	production indirects, other production cost
Production services	utilities, repair & maintenance, warehousing/handling/transport, other production services
Material cost	raw materials, packaging materials

A key component of the estimation of production cost is the calculation of capital cost. The sequence to follow for the estimation of capital cost starts with the identification of all assumptions, including PFD, production rate, uncertainty factors and other related factors (*e. g.*, electrical, automation, etc.). Next, the appropriate equipment type is selected and the Equipment Flow Diagram or Process Flowsheets (Section 7.1.2.3) are built. Mass balance exercise is then carried out, resulting in the sizing of the selected equipment. Prices of equipment units are then obtained from specialized databases and/or recent quotations. A material balance and financial models are then populated and the final capital cost is estimated. The work by Marouli and Maroulis (2005) provides a systematic analysis of capital cost data to estimate the appropriate factor models for rapid cost estimation in food plant design.

5.3 Take-away message

At Level 2 of the PDPS approach (Product function) we map the product attributes obtained form Level 1 onto a product property design space. This translation is carried out by defining a *property function*. This function can be obtained through the population of a second HoQ which translates product attributes into product properties. A key challenge while consolidating the second HoQ is the complexity of understanding/identifying the physical and (bio)chemical phenomena that are taking place within the boundaries of the design problem. Resolving this challenge will allow the identification of the relevant properties to further pursue. An alternative way to define the property function involves the use of data mining or AI techniques (*e. g.*, CNN, PLS). Next to the definition of the property function, this level is characterized by the generation of product concepts. Brainstorming, morphological analysis, TRIZ, biomimicry and patent review, among others, are extremely relevant sources of knowledge to support the generation of concepts.

Level 3 of the approach focuses on the consolidation of *all* inputs and outputs of the design problem. Consolidation of a list of constraints and relevant KPIs is also carried out at this level. Product prototypes or concepts are included as output descriptors of the design problem.

Bibliography

Abakarov, A., Sushkov, Y., Almonaci, S. and Simpson, R. (2009). Multiobjective optimization approach: Thermal Food Processing. *Journal of Food Science*, 9(74), pp. 471–487.

Almeida-Rivera, C. P., Jain, P., Bruin, S. and Bongers, P. (2007). Integrated product and process design approach for rationalization of food products. *Computer-Aided Chemical Engineering*, 24, pp. 449–454.

AskNature (2019). Organization website (http://www.asknature.org).

Coronel, P. (2005). *Continuous flow processing of foods using cylindrical applicator microwave systems operating at 915 MHz*. Thesis/dissertation, North Carolina State University.

Coronel, P., Simunovic, J. and Sandeep, K. P. (2003). Thermal profile of milk after heating in a continuous flow tubular microwave system operating at 915 MHz. *International Journal of Food Science*, 68(6), pp. 1976–1981.

Coronel, P., Truong, V. D., Simunovic, J., Sandeep, K. P. and Cartwright, G. (2005). Aseptic processing of sweetpotato purees using a continuous flow microwave system. *Journal of Food Science*, 70(9), pp. E531–E536.

Donsi, F., Ferrari, G. and Maresca, P. (2009). High-pressure homogenization for food sanitization. In: *Global Issues of Food Science and Technology*. Elsevier.

EPO (2019). Corporate website (http://www.epo.org). Electronic Citation.

Fox, M. (2008). *Da Vinci and the 40 answers*. Canada: Wizard Academy Press.

Gadd, K. (2011). *TRIZ for Engineers: Enabling Inventive Problem Solving*. John Wiley and Sons, Inc.

Griskey, R. (2002). *Transport phenomena and unit operations: a combined approach*. Wiley Interscience.

Holdsworth, D. and Simpson, R. (2007). *Thermal Processing of Packaged Foods*. Springer.

Holdsworth, S. D. (1993). Rheological models used for the prediction of the flow properties of food products: a literature review. *Transactions of the Institute of Chemical Engineers – Part C*, 71, pp. 139–179.

Krynke, K. and Sek, J. P. (2004). Predicting viscosity of emulsions in the broad range of inner phase concentrations. *Colloids and Surfaces A: Physicochemical and Engineering Aspects*, 245, pp. 81–92.

Marouli, A. Z. and Maroulis, Z. B. (2005). Cost data analysis for the food industry. *Journal of Food Engineering*, 67, pp. 289–299.

Orloff, M. (2006). *Inventive Thinking through TRIZ*. Berlin: Springer.

Pal, R. (2001a). Evaluation of theoretical viscosity models for concentrated emulsions at low capillary numbers. *Chemical Engineering Journal*, 81, pp. 15–21.

Pal, R. (2001b). Single-parameter and two-parameter rheological equations of state for nondilute emulsions. *Industrial and Engineering Chemistry Research*, 40, pp. 5666–5674.

Princen, H. M. and Kiss, A. D. (1989). Rheology of foams and highly concentrated emulsions. *Journal of Colloid Interface Science*, 128(1), pp. 176–187.

Sherwood, J. D. and Durban, D. (1998). Squeeze-flow of a Herschel-Bulkey fluid. *Journal of Non-Newtonian Fluid Mechanics*, 77, pp. 115–121.

Simpson, R. and Abakarov, A. (2009). Optimal scheduling of canned food plants including simultaeous sterilization. *Journal of Food Engineering*, 90, pp. 53–59.

Steffe, J. F. (1996). *Rheological Methods in Food Process Engineering*, 2nd ed. Michigan: Freeman Press.

Stokes, J., Boehm, M. and Baier, S. (2013). Oral processing, texture and mouthfeel: From rheology to tribology and beyond. *Current Opinion in Colloid and Interface Science*, 18, pp. 349–359.

Ulrich, K. and Eppinger, S. (2012). *Product Design and Development*. New York: McGraw-Hill.

6 Network of tasks, mechanisms and operational window

6.1 Level 4: fundamental tasks and task network

The next level of the methodology is characterized by the identification of fundamental tasks (Figure 6.1). Being probably the most creativity demanding level, this stage of the PDPS approach defines the backbone structure of the entire process and strongly relies on process synthesis methods to exhaustively generate alternatives (Table 3.1 in Section 3.1). After the product specifications or product properties have been defined, a process function is required to systematically eliminate the differences in such properties or specifications between the initial and final states of the intended product or service,

$$\min_{\text{process function}} \Theta_{\text{input}} - \Theta_{\text{output}}, \tag{6.1}$$

where Θ denotes the set of properties identified as relevant in the previous level.

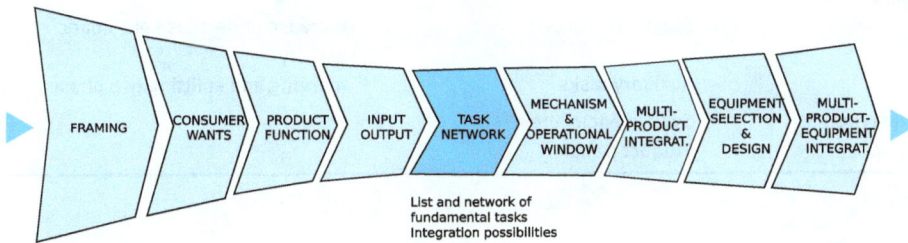

Figure 6.1: Task network as part of the PDPS strategy.

The concept of task has its motivation in the synthesis of intensified units. The synthesis of methyl acetate by reactive distillation is probably the most fascinating and clear textbook example of task-based process design. A comprehensive description of this study case is given in the explanatory note *Reactive distillation – an example of the PI approach* in Section 2.1.4.3. This process function can be regarded as the resolution method for (one or more) properties and comprises the fundamental transformation(s) – or task(s) – required to achieve the desired product specification(s). Defining the fundamental tasks needed to move from input to output spaces requires a sound understanding of the physico-chemico-biological processes involved in the change of product properties. Within the design of (bio)chemical processes, with only five generic types of fundamental tasks almost all processes can be described (Siirola, 1996): changes in molecular identity, changes in amount, changes in composition, changes in phase and changes in temperature/pressure. These types of tasks can be associated with more specific classes of tasks, as given in Table 6.1.

https://doi.org/10.1515/9783110570137-006

Table 6.1: Types and classes of fundamental tasks.

Type of tasks	Class of task	Scope/examples
Changes in molecular identity	(bio)catalyzed chemical reaction	creation of new molecular identity
Changes in phase	Change the number of thermodynamic phases	introduction of new thermodynamic phases; formation and/or removal of thermodynamic phases (e. g., aeration and crystal nucleation)
	Change the spatial distribution of the phases	distributive mixing
	Change the distribution properties of a phase	dispersive mixing, size reduction of solid particles, cell wall disruption, elongation/shape change/aspect ratio change
	Change the microbiological properties of a phase	
	Change the thermodynamic state of a phase	formation and/or removal of thermodynamic phases
Changes in composition	Separation of two phases	
	Enzymatic tasks	increase or decrease enzymatic activity
	Auxiliary tasks	merging and splitting two phases
Other type	System enclosure Product transfer	

It is worth mentioning that the changes in distribution properties relate to the changes in number, characteristic size and shape (when related to size changes) of the entities, while the changes in spatial distribution are associated with the shape changes when related to changes in spatial organization and not to changes in the distribution.

When each task is represented by a block and connected according to their requirement and occurrence within the product function, a network or superstructure of tasks is obtained (Figure 6.2). The complexity of the task network (in size and number) increases considerably due to the combinatorial explosion of options. This combination has its origin in the many topological options available to connect the tasks, which can result in the number of possible combinations easily being in the order of many thousands. As performing individual evaluations for each task network is commonly hindered by the large number of possible alternatives, tasks that require a certain sequence or that belong together without any doubt are grouped.

By combining or grouping tasks, integration possibilities could be considered, as has been widely and repeatedly proven by the current intensified processes (Section 2.1.4.3). Moreover, the combination of heuristics (sometimes referred to as rules of thumb), engineering judgement-based assumptions, simplified models and creativity

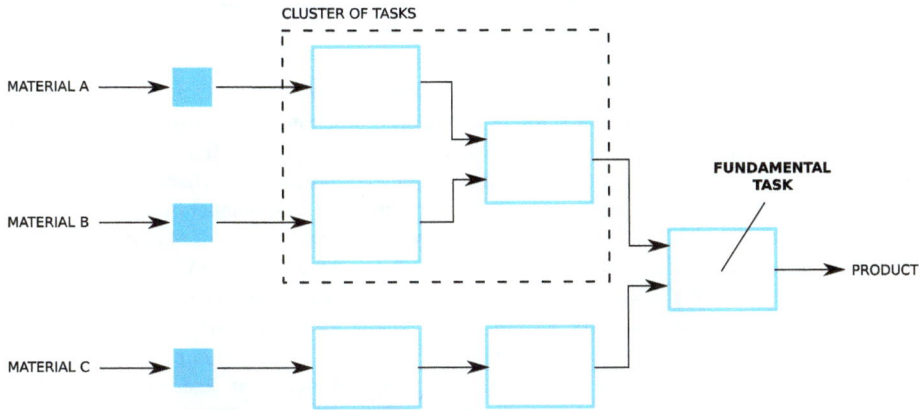

Figure 6.2: Fundamental task building blocks of a task network.

allowed for further pruning the number of potentially possible network alternatives to a manageable number. The nature of the heuristics responds to a need for operation simplification, logistics simplicity and material efficiency, among others. Examples of heuristics, which are strongly dependent on the type of product/service to be achieved, are included in Table 6.2.

Table 6.2: Examples of heuristics in the identification of fundamental tasks and consolidation of task networks.

Number	Heuristic
H1	powder materials present in a combined stream are first dissolved, then mixed
H2	merging of two systems is implicitly performed when two or more streams are combined
H3	changes in interfacial composition occur before any disruptive mixing step
H4	materials that require hydration are first hydrated before any pasteurization step
H5	materials that are shear-sensitive must not be intensively milled
H6	materials that require specific pretreatments are sourced already pretreated

Let us consider, for example, the synthesis of a high internal phase emulsion. From Level 3 of the PDPS approach, it has been suggested that the mean droplet size of the dispersed phase is the property to be controlled and resolved. A target value of this property has been specified (*e. g.*, δ_{out} = 10 µm), which ensures consumer satisfaction. If the initial state of this property is assumed to be given by bulk conditions (*e. g.*, δ_{in} = 250 µm), the fundamental task to resolve this difference in distribution property is dispersive mixing, *i. e.*,

$$\delta_{in} \xrightarrow{\text{dispersive mixing}} \delta_{out}, \tag{6.2}$$

where δ denotes the mean droplet size of the dispersed phase.

ℹ Intermezzo: introduction to (high internal phase) emulsions

Structured emulsions are a particularly important class of chemically formulated products widely common in the food and beverage, agricultural chemical, consumer products and pharmaceutical industries. Unlike commodity chemicals, which are characterized almost exclusively by their composition, structured emulsions have specific end-use properties that are intimately connected to their microstructure. Thus, in contrast to bulk chemicals, structured products are characterized not only by the level of each ingredient (*i. e.*, composition, purity, physical state, temperature and pressure), but also by the relative spatial arrangement of each ingredient and performance behavior. All these features are responsible for the exclusive attributes of structured products (*e. g.*, creaminess of an ice cream, spoonability of a mayonnaise and spreadability of a margarine). As the product and process design problems for such complex, multi-phase, microstructured materials are strongly coupled, the overall design problem involves specification of the chemical formulation and the processing conditions that produce the desired microstructure and physical properties. Reaching a desired microstructure involves not only the close marriage between product composition and processing, but also the development of a process synthesis methodology embracing both (*e. g.*, de Ridder *et al.*, 2008). A model-based approach is a fundamental building block of such synthesis methodology.

According to the process scheme (Figure 6.3), the ingredients are continuously discharged from buffer tanks to a static mixer at a given ratio where they are roughly mixed. The resulting premix is fed to a mixing vessel, which operates under vacuum conditions and is equipped with a given number of scrapers. From the bottom of the vessel the premix passes through a colloid mill, where most of the emulsification process takes place. The resulting stream is then recirculated to the mixing vessel. As soon as the product specifications are met, the product is withdrawn from the unit through a three-way valve. In addition to its emulsification purpose, the colloid mill in this configuration acts as a recirculation pump (Almeida-Rivera and Bongers, 2009, 2010a).

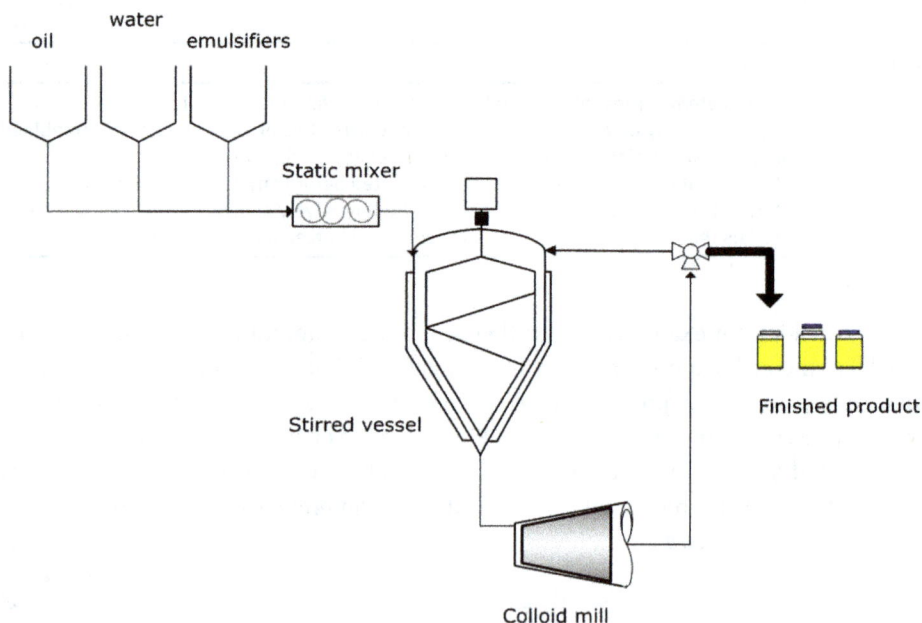

Figure 6.3: Schematic representation of the Fryma Delmix process in the manufacture of HIPEs.

The sensorial attributes of structured emulsions are dictated by several physical properties of the system, including the droplet size distribution of the dispersed phase for high internal phase emulsions (HIPEs) and the viscosity of the continuous phase for low internal phase emulsions (LIPEs). HIPEs are characterized by a high phase volume ($\Phi > 65\,\%$), where the desired product quality is mainly related to the Sauter mean droplet diameter, d_{32}, which is defined as

$$d_{32} = \frac{\sum_{i=1}^{n_c} N_i d_i^3}{\sum_{i=1}^{n_c} N_i d_i^2},$$ (6.3)

where N_i is the number of droplets in the size class i with mean diameter d_i and n_c is the number of size classes.

It is evident that the most challenging activity at this level is the identification of the fundamental tasks required to transition from the input to the output states of attributes, specifications or properties. Probably the most inspiring industrially relevant example of task identification is given by the synthesis of methyl acetate by reactive distillation (Section 2.1.4.3). A thorough analysis of all transformations occurring within the conventional process – which comprises 11 different steps and involves 28 major pieces of equipment – allowed Siirola (1996) to replace all units by a single highly task-integrated reactive distillation unit. In fact, the technological breakthrough was derived from systematically looking at tasks (or transformations) rather than process steps or unit operations only.

For structured products, a thorough understanding of the underlying target product microstructure, which plays a key role in the derivation of sequence-based design heuristics, is of utmost relevance. A comprehensive – but by no means complete – list of fundamental tasks is given in Table 6.3. Following the methyl acetate synthesis roadmap, fundamental tasks were identified via analysis of current processes (reengineering-based approach).

Upon identification of the fundamental tasks and consolidation of network of tasks, Level 4 of the PDPS approach requires conducting a preliminary numerical evaluation by performing mass and energy balances. Tasks that require a certain sequence or that belong together without any doubt are grouped and integrated, if possible. After heuristics, rules of thumb, engineering judgement-based assumptions and/or simplified models are applied, and the resulting network(s) of tasks are evaluated based on BROC multi-criteria (Section 4.3.6). The evaluation of the alternatives will support the identification of the most prominent option(s) to further elaborate in the next levels of the approach.

Let us consider the manufacture of a structured emulsion containing an artificial thickener as an example. The set of ingredients/raw materials comprises a continuous phase liquid (*e.g.*, water), fat-soluble liquids, a dispersed phase (*e.g.*, oil), water-soluble liquids, a slow-moving emulsifier, powders in minor quantities and the artificial thickener. Based on a consumer research, the attribute of the desired product is related exclusively to the thickness of the product, which is translated into the

Table 6.3: List of selected classes of tasks and fundamental tasks.

Fundamental task	Subtask
Class of task: A. Change the number of thermodynamic phases	
A1 Solvation (hydration)	A1a Dissolution
	A1b Solubility (dynamic equilibrium)
A2 Crystallization	A2a Primary nucleation
	A2b Secondary nucleation
	A2c Growth
	A2d Attrition
	A2e Dissolution (=A1a)
	A2f Selective removal
	A2g Precipitation
A3 (Partial) sublimation	
A4 (Partial) evaporation	
A5 (Partial) condensation	
A6 (Partial) deposition	
A7 (Partial) ionization	
A8 (Partial) deionization	
Class of task: B. Change the spatial distribution of the phases	
B1 Distributive (de)mixing	B1a Liquid–liquid
	B1b Liquid–solid
	B1c Solid–solid
	B1d Liquid–gas
	B1e Solid–gas
B2 Transfer of a component from one phase to another	
B3 Change interfacial composition	
Class of task: C. Change the distribution properties	
C1 Dispersive mixing	C1a Droplet breakup
	C1b Coalescence
	C1c Interface stabilization
	C1d Ostwald ripening
C2 Size reduction of solid particles	
C3 Cell wall disruption	
C4 Elongation/shape change/aspect ratio change	
Class of task: D. Change the microbiological properties of a phase	
D1 Pasteurization (vegetative)	
D2 Sterilization (spores)	
D3 Growth of microorganisms	
D4 Freezing/cooling	
Class of task: E. Change the thermodynamic state of a phase	
E1 Melting	
E2 Freezing	
E3 Evaporation	
E4 Condensation	
E5 Sublimation	
E6 Deposition	
E7 Ionization	
E8 Deionization	

Fundamental task	Subtask
Class of task: F. Enzymatic tasks	
F1 Increase enzyme activity	
F2 Decrease enzyme activity	
Class of task: G. Separation of two phases	
G1 Separation of a system into two systems with different composition	
Class of task: H. System enclosure	
H1 Filling	
H2 Packing	
H3 Shaping	
Class of task: I. Product transfer	
I1 Transfer from one position to another	
Class of task: J. Stabilization	
J1 Physical stabilization	
J2 Chemical stabilization	
Class of task: K. Auxiliary tasks	
K1 Merging two streams	
K2 Splitting two streams	

rheological variables: critical strain, critical stress and viscosity at $10\,s^{-1}$. The values of these specifications have been found to be 100 %; 2500 dyne cm^{-2}; 1800–2400 cps. The structure of the task network is obtained by identifying the tasks required by each stream separately. For instance, the artificial thickener needs to be hydrated and then pasteurized, before it is combined with an emulsified stream. The oil phase needs to be combined with the fat-soluble materials and then combined with a stream comprising water, the water-soluble materials and the emulsifier. The combined stream is then emulsified and the resulting stream is combined with the pasteurized thickener material. The obtained network of tasks is depicted in Figure 6.4.

6.2 Level 5: mechanism and operational window

As mentioned in the previous level of the methodology, the differences between the properties (or attributes) in the initial and final states are systematically eliminated via one or more fundamental tasks. Each task might be achieved by one or more dominating mechanisms in specific operational windows. In Level 5 of the PDPS approach (Figure 6.5), the operational window of a mechanism is defined in terms of the specific driving force that causes the transformation to occur and the operating conditions (*e.g.*, temperature, pressure, pH, electric/magnetic field strength and gradient) under which the transformation takes place. Moreover, the numerical framework of

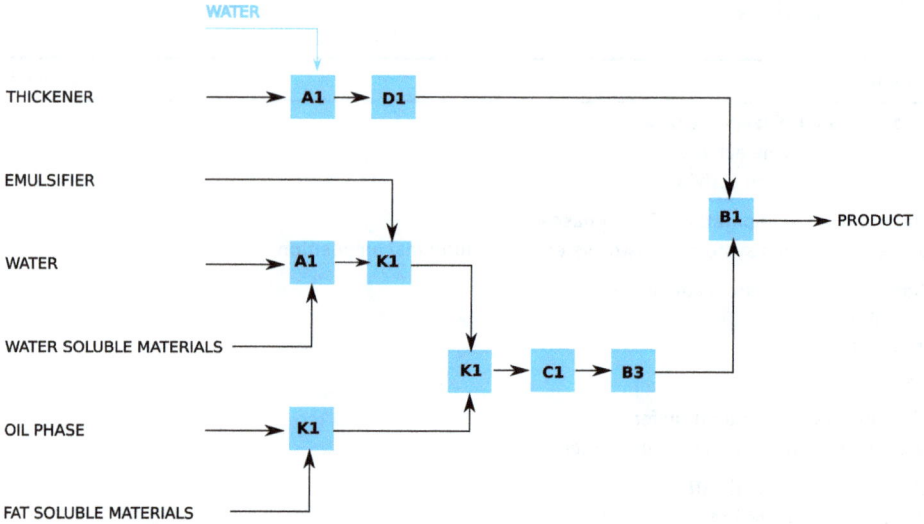

Figure 6.4: Network of fundamental tasks in the manufacture of a structured emulsion containing an artificial thickener. Figures denote the tasks listed in Table 6.3.

this level includes a complete description of the mechanism that causes the transformation to occur (**why**), the kinetic rate at which the transformation proceeds (**how fast**) and the estimation of the *threshold* that needs to be overcome before a change actually occurs.

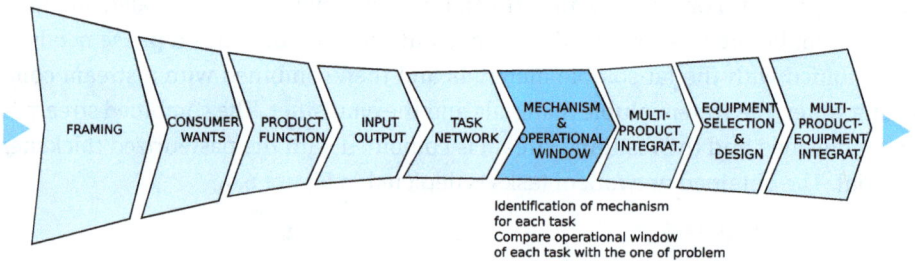

Figure 6.5: Mechanisms and operational window as part of the PDPS strategy.

The number of mechanisms that could potentially deliver a fundamental task is steadily increasing due to the high pace of technological advances and the emergence of novel manufacturing technologies. Similar to the identification of fundamental tasks, inventorying all mechanism for each fundamental task is certainly not a trivial activity. It requires a sound understanding of the underlying phenomena during the transformation, up-to-date knowledge of current technologies and a solid assessment of the feasibility of the technological implementation. This feasibility requirement is strongly aligned with the Design Thinking framework, as depicted in Figure 2.4 in Section 2.1.2.

Let us consider as tutorial example of the identification of mechanisms and operational window, the fundamental task associated with the reduction of mean droplet size of the dispersed phase in the manufacture of a structured emulsion. As mentioned by McClements (2005), a wide range of devices have been developed for the manufacture of structured emulsions. While the right choice of such units largely depends on the scale of the manufacturing step (*e. g.*, laboratory, pilot-plant or industrial scales), the desired product attributes and properties, the operating and capital expenditure, it is worth realizing that these units could be operating under different mechanisms and within specific operational windows. An overview of the most commonly used devices in the FMCG sector is given in Table 6.4.

Table 6.4: Overview of selected devices for the manufacture of structured emulsions (McClements, 2005). Legend: LV, laminar viscous; TV, turbulent viscous; TI, turbulent inertial; CI, cavitational.

Emulsification unit	Flow regime	E_v (J m^{-3})	d_{32}^{min} (µm)
Colloid mill	LV/TV	10^3-10^8	1
High-pressure homogenizer	TI/TV/CI/LV	10^6-10^8	0.1
Ultrasonic jet homogenizer	CI	10^6-10^8	1
Microfluidizer	TI/TV	10^6-10^8	< 0.1
Membrane emulsification	Injection	$< 10^3-10^8$	0.3

As proposed by Karbstein and Schubert (1995), the volumetric energy input (E_v) can be used as a reference variable to compare the effectiveness of droplet size reduction in the devices. The minimum attainable Sauter diameter is represented by the variable d_{32}^{min} and is considered the product specification or property to target (Almeida-Rivera *et al.*, 2010). Despite the inherent mechanistic difference among devices, the current industrial practice toward the selection of units is based on *ad hoc* experience-based knowledge, rather than on rigorous phenomena understanding. This approach, nevertheless, has resulted in the identification of proven technologies across the manufacturing of base chemicals and industrial and configured consumer products (Almeida-Rivera and Bongers, 2009, 2010a,b, 2012).

The volumetric energy input can be estimated as the product of the introduced mean power density (ϵ_v) and the mean residence time of the droplets in the high-shear zone of the device (t_r) (Schubert and Engel, 2004; Schubert *et al.*, 2003; Karbstein and Schubert, 1995), *i. e.*,

$$E_v = \epsilon_v \times t_r = \frac{P}{Q_v}, \tag{6.4}$$

where P is the power input drawn by the device and Q_v is the volumetric flow rate of the emulsion.

The applicability of the concept of E_v in the selection, design, scale-up and control of emulsification equipment is further broadened as its definition involves variables and parameters easily accessible. For instance, in the particular case of valve-based devices, the volumetric energy input is directly related to the pressure drop across the unit,

$$E_v = \Delta p, \tag{6.5}$$

where Δp is the pressure drop inside the emulsification zone.

Mapping target product attributes (e.g., analytical [mean droplet size] and sensorial [mouthfeel]) as a function of the volumetric energy input for each technology allows us to easily identify the operational window of each unit, given by the E_v range where the device is expected to deliver a finished product with desired attribute(s).

The relationship between the target product attributes and the volumetric energy input can be represented by the following generic expression:

$$\Psi = C_\Psi \times E_v^{-b_\Psi}, \tag{6.6}$$

where Ψ denotes the target product attribute and C and b are constants that depend on the emulsifying agents, properties of the continuous and disperse phase and the emulsification technology used (Schubert et al., 2006).

It is to acknowledge that equation (6.6) has been previously reported when Ψ denotes the maximum stable droplet diameter (Davies, 1985, 1987) and when Ψ represents the Sauter diameter of droplets (Karbstein and Schubert, 1995; Davies, 1987). Moreover, the validity of equation (6.6) is given by the fulfillment of two conditions: the external stress exerted on the droplet surface exceeds the form maintaining interfacial force locally for a sufficiently long time, and the residence time in the dispersion zone of the device is long enough to assess the extent of emulsification.

Emulsification technologies can be benchmarked based on their availability to deliver products with desired attributes. This benchmarking activity will lead to the identification of the *optimal* operating window – expressed in terms of E_v – for each mechanism. In this regard, benchmarking can be considered as a supporting approach for the identification of mechanisms and settling of operational windows. Aiming at representing the most commonly occurring phenomena responsible for droplet breakup, five commercial technologies have been included and analyzed (Table 6.5 and Figure 6.6).

As reported by Almeida-Rivera et al. (2010), the benchmarking study involves the analysis of the results of an exhaustive emulsion process design and operation benchmark study using an oil-in-water emulsion with a phase volume around the close packing density of spheres ($\phi_{disperse} > 0.74$). Using the volumetric energy input as independent variable, the mean droplet diameters – expressed as d_{43} – achieved by each unit

Table 6.5: Types of emulsification devices to achieve the reduction in mean droplet size of the dispersed phase in the manufacture of a structured emulsion (Almeida-Rivera *et al.*, 2010).

Type	Description
Type A	rotor-stator, single-pass configuration where the premix fed to the center of the device is drawn up into the high-speed rotor, where it is accelerated and expelled radially, at high velocity, through the openings in the stator
Type B	rotor-stator, multiple-pass configuration, where the disperse phase droplets are disrupted in a conical gap between the rotor and stator; the emulsion is recirculated through the high-shear zone until product specifications are met
Type C	valve configuration operated in a single-pass strategy, where the premix is pumped through a narrow orifice under high pressure
Type D	rotor-stator, comparable to type B but operated with a single-pass strategy, where the disperse phase droplets are disrupted in a conical gap between the rotor and stator
Type E	rotor-stator, single pass configuration, tight tolerance and multiple high-shear zones; this novel design involves a device capable of generating controllable, predictable and uniform deformation rates independently of flow rates and stream rheology, leading to processes more amenable to scale-up and flexible

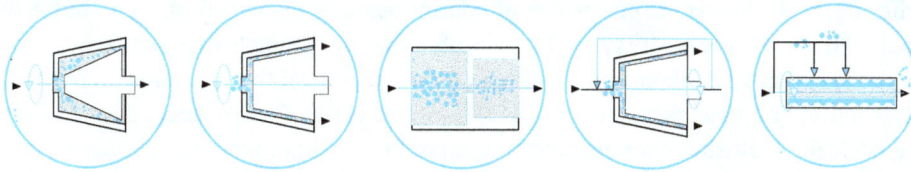

Figure 6.6: Schematic representation of types of emulsification units. Shaded areas represent moving/rotating parts. Type A: rotor-stator, single-pass; Type B: rotor-stator, multiple-pass; Type C: valve configuration, single-pass; Type D: rotor-stator, conical; Type E: rotor-stator, single pass, tight tolerance (Almeida-Rivera *et al.*, 2010).

were mapped in Figure 6.7 for units of type A, B and D. The target attribute was assumed to be within an arbitrary target range (5.5 μm < d_{43} < 10 μm). This generalized representation – plot E_v *versus* Ψ – offers a powerful and simply elegant method to compare the performance of emulsification units and establish the operational window to deliver a target specification. By defining a generic variable (E_v) that captures the essence of the phenomenon without being perturbed by particularities of each experimental design point, we effectively circumvent the combinatorial nature of the design problem. Moreover, when the product characterization variables (*e. g.*, mean droplet size, viscosity at a given shear rate) are plotted against E_v, all experimental data fall under an E_v master curve. With this elegant graphical representation, the practitioner can easily point out the specific E_v range that is expected to deliver a final product with specific analytical and sensorial attributes.

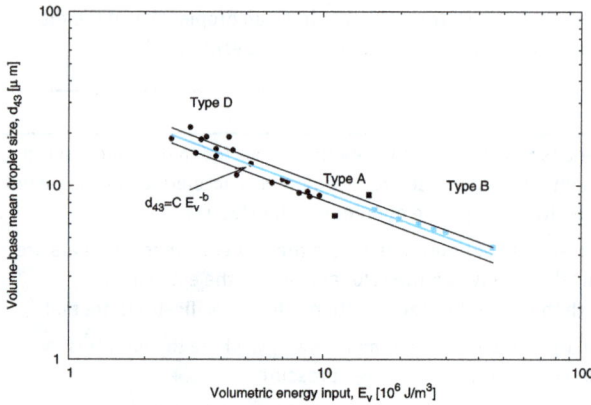

Figure 6.7: Mapping of mean droplet size as a function of volumetric energy input for unit types A, B and D. Parameters of equation (6.6): $C_\psi = e^{11}$; $b_\psi = 0.55$. Error lines of the fitting curve at $\pm 10\%$ are also included (Almeida-Rivera et al., 2010).

As implied from the benchmarking of devices, the governing mechanism behind the reduction of mean droplet size of the dispersed phase in the manufacture of structured emulsions (i. e., droplet breakup) is highly dependent not only on the physical properties of the system (e. g., viscosities of continuous and dispersed phases), but also on the flow conditions. The type of flow profile that the droplets experience depends on the mechanical configuration and operational regime of the unit and can be characterized by the flow parameter $\alpha \in [-1, 1]$ (Stork, 2005). For instance, stirred tanks are basically featuring simple shear in the laminar regime, whereas turbulent inertial and cavitation mechanisms are responsible for droplet breakup in high-pressure homogenizers. Despite the sound expertise generated over the last decades regarding droplet disruption and coalescence mechanisms (Nienow, 2004; Tjaberinga et al., 1993; Saboni et al., 2002), the effort has been exclusively channeled to those units where either single-shear flow ($\alpha = 0$) or rotational flow ($-1 < \alpha < 0$) dominates. Contrary to that research focus, both single-shear flow and rotational flow are rarely the dominant droplet breakup mechanisms in commercial emulsion technologies. The spectrum of current emulsification devices was broadened by exploring units based on extensional flow ($0 < \alpha \leq 1$) (Almeida-Rivera and Bongers, 2010b, 2012). In this type of flow regime, normally referred to as **shear-free** flow, a preferred molecular orientation occurs in the direction of the flow field. Moreover, it is characterized by the absence of competing forces to cause rotation, resulting in a maximum stretching of molecules and large resistance to deformation (Steffe, 1996). As turbulent flow is responsible for droplet breakup in high-stress emulsification units, we focus on this flow regime. Moreover, it is worth noting that from an industrial perspective, production of emulsions in the laminar regime ($Re_p < 10^{-2}–10^{-6}$) (Grace, 1982) is a highly energy demanding operation for systems where the ratio of viscosities of the disperse and continuous phases, η_d/η_c, is less than 0.1 and even impossible for systems where

$\eta_d/\eta_c > 4$ (Figure 6.8). Under single-shear conditions, the droplets are not able to deform as fast as the flow regime induces deformation (Walstra, 1993).

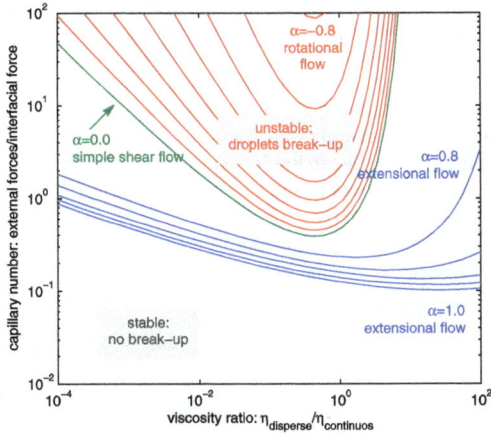

Figure 6.8: Critical capillary number for droplet breakup in single-shear, extensional flow and only rotational flow; α represents the flow type. Modeling parameters are those reported in Grace (1982) and de Bruijn (1989).

A similar exercise can be carried out for each fundamental task identified in Table 6.3, resulting in a comprehensive list of mechanisms and operational windows, shown in Table 6.6. It is worth acknowledging that the current list is indicative only and not expected to be either comprehensive or complete.

It is worth mentioning that as each fundamental task could be potentially carried out by one or more mechanisms, the number of potential alternatives is simply exploding to unmanageable levels. A rather simple approach towards the reduction of alternatives within the PDPS approach involves the validation of the operational window of the design problem against the individual windows of the potential mechanisms. The operational window of the problem can be retrieved from the formulation of the problem and expressed in terms of legal, consumer-relevant, safety or technological requirements.

For instance, let us consider that the fundamental task that was identified to transform a given attribute (*e. g.*, concentration of vegetative spores in the matrix of the product) from input to output states is D1: pasteurization. The operational window of the intended product specifies that the matrix is temperature-sensitive, as relatively low temperatures impact the organoleptic and nutritional profile of the material. Under this constraint, task D1 cannot be delivered by any mechanism involving high temperatures or where high-temperature operating conditions are required. Potential mechanisms include high-pressure, chemical, PEF, size exclusion and irradiation techniques.

Table 6.6: List of selected mechanisms and operational windows. Legend: d_{drop}, mean droplet diameter; d_{hole}, membrane hole diameter; ϵ_{m}, local power draw/dissipation per unit mass; $\dot{\gamma}$, shear rate; η, viscosity; ρ, density; E_v, volumetric energy input.

Fundamental task	Mechanism or driving force	Operational window
Class of task: A. Change the number of thermodynamic phases		
A1 Solvation (hydration)	Solubility	Solid to liquid
A2 Crystallization	Supersaturation	Liquid to solid
A3 (Partial) sublimation	Gradient of vapor pressure	Solid to gas
A4 (Partial) evaporation	Gradient of vapor pressure	Liquid to gas
A5 (Partial) condensation	Mean temperature difference	Gas to liquid
A6 (Partial) deposition	Chemical potential difference	Gas to solid
A7 (Partial) Ionization	Electric field	Gas to plasma
A8 (Partial) Deionization	Electric potential difference	Plasma to gas
Class of task: B. Change the spatial distribution of the phases		
B1 Distributive (de)mixing	B11 Mechanical	Macroscopic scale/agglomeration
	B12 Pneumatic	
	B13 Magnetic	
	B14 Ultrasonic	
B2 Transfer of a component from one phase to another	B21 Diffusion	Chemical potential gradient (charge, concentration)
	B22 Free convection	Density gradient
	B23 Forced convection	Pressure gradient, momentum
	B24 Partitioning	Activity gradient
B3 Change interfacial composition	B31 Diffusion	Chemical potential gradient (charge, concentration)
	B32 Free convection	Density gradient
	B33 Forced convection	Pressure gradient, momentum
	B34 Partitioning	Activity gradient
Class of task: C. Change the distribution properties		
C1 Dispersive mixing	C11 Elongational stress	E_v: 10^3–10^8 J m^{-3}; $d_{\mathrm{drop}} < 4 \times d_{\mathrm{hole}}$
	C12 Shear stress	ϵ_{m}: 10^5–10^9 W/kg; E_v: 10^3–10^6 J m^{-3}; $d_{\mathrm{drop}} = f(\dot{\gamma}, \eta, \text{formulation})$
	C13 Magnetic	droplet size $= f$(magnetic field intensity, magnetic susceptibility, ρ)
	C14 Ultrasonic	droplet size $= f$(field intensity, physical properties material)
	C15 Cavitation	E_v: 10^6–10^8 J m^{-3}
	C16 Turbulent flow	E_v: 10^3–10^8 J m^{-3}
	C17 Laminar flow (elongational + shear)	
C2 Size reduction of solid particles	C21 Attrition	
	C22 Impact	
	C23 Ultrasonic	
	C24 Cutting	
	C25 Explosion	Pressure decompression
	C26 Use of enzyme(s)	
	C27 Use of chemical(s)	

Table 6.6: (continued)

Fundamental task	Mechanism or driving force	Operational window
C3 Cell wall disruption	C31 Internal cell phase change	
	C32 Electro-magnetic fields (PEF, ultrasound)	Electric field strength, ultrasound frequency
	C33 Shear	
	C34 Use of enzyme(s)	
	C35 Use of chemical(s)	
C4 Elongation	C41 Elongational stress	Shape change/aspect ratio change
	C42 Shear stress	
	C43 Magnetic	
	C44 Ultrasonic	
	C45 Cavitation	
	C46 Turbulent flow	
	C47 Laminar flow	Elongational + shear stresses
	C48 Electrical field	
Class of task: D. Change the microbiological properties of a phase		
D1 Pasteurization (vegetative)	D11 Size exclusion	
	D12 Thermal: indirect	
	D13 Thermal: Ohmic	
	D14 Thermal: microwave	
	D15 Thermal: RF	
	D16 Chemical	
	D17 PEF	
	D18 Pressure	
	D19 Irradiation	
D2 Sterilization (spores)	D21 Thermal: indirect	
	D22 Thermal: Ohmic	
	D23 Thermal: microwave	
	D24 Thermal: RF	
	D25 Chemical	
	D26 PEF	
	D27 Pressure	
	D28 Irradiation	
Class of task: E. Change the thermodynamic state of a phase		
E1 Melting	E11 Free energy difference	Solid to liquid
E2 Freezing	E21 Free energy difference	Liquid to solid
E3 Evaporation	E31 Gradient of vapor pressure	Liquid to gas
E4 Condensation	E41 Mean temperature difference	Gas to liquid
E5 Sublimation	E51 Gradient of vapor pressure	Solid to gas
E6 Deposition	E61 Chemical potential difference	Gas to solid
E7 Ionization	E71 Electric potential difference	Gas to plasma
E8 Deionization	E81 Electric potential difference	Plasma to gas

Table 6.6: (continued)

Fundamental task	Mechanism or driving force	Operational window
Class of task: G. Separation of two phases		
G1 Separation of a system into two systems with different composition	G11 Molecular/particle size	
	G12 Electrical charge	
	G13 Solubility	Partition coefficient; concentration gradient
	G14 Chemical affinity	Partition coefficient
	G15 Chemical reaction	Rate constants
	G16 (Vapor) pressure	
	G17 Gravity	Density ratio of phases
	G18 Shear	
	G19 Combination of G11 and G12	
Class of task: J. Stabilization		
J1 Physical stabilization	J11 Freezing	
	J12 Cooling	
J2 Chemical stabilization	J21 Use of chemical(s)	
	J22 Use of enzyme(s)	

After inventorying all mechanisms for each fundamental task and excluding those violating the operational window of the problem, a subset of mechanism-based networks is identified and analyzed against relevant KPIs (*e. g.*, economic potential, scalability, hygiene design and sustainability). The BROC multi-criteria decision making approach is deployed to support the selection of a single mechanism alternative for each task in accordance with the problem operational window.

Let us consider as an example the extraction of polyphenols from tea leaves. As suggested by Zderic *et al.* (2019), Zderic and Zondervan (2017), this separation processes is represented by the network given in Figure 6.9.

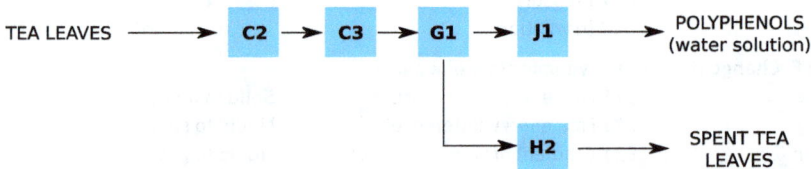

Figure 6.9: Network of mechanisms in the extraction of polyphenols from tea leaves (Zderic *et al.*, 2019; Zderic and Zondervan, 2017). Figures denote the tasks listed in Table 6.3.

According to Table 6.6, the potential mechanisms to deliver the proposed tasks could be the following:

- C2 (size reduction of solid particles): attrition, impact, ultrasonic, cutting, explosion;
- C3 (cell wall disruption): internal cell phase change, electro-magnetic fields, shear, use of enzymes, use of chemicals;
- G1 (separation of a system into two systems with different composition): molecular/particle size, electrical charge, solubility, chemical affinity, chemical reaction, vapor pressure, gravity, shear, combination of molecular size and electrical charge;
- J1 (physical stabilization): freezing, cooling;
- H2 (packing): packing.

By means of a comprehensive experimental program, Zderic *et al.* (2019), Zderic and Zondervan (2017) screened the complete design space of mechanisms' alternatives. The resulting processing route features a pulsed electric field as a mechanism for task C3, because of its non-thermal and non-invasive nature. The operational window for this mechanism was explored experimentally, resulting in a set of optimal operation parameters which achieve an extraction yield larger than 30 %.

6.3 Take-away message

Level 4 (Fundamental task network) of the PDPS approach addresses the resolution of the process function. It involves the identification of the fundamental tasks required to systematically eliminate the property differences between the initial and final states of the intended product or service. Upon identification of the tasks, a network is consolidated, resulting in the backbone of the (manufacturing) processes.

Level 5 (Mechanism and operational window) focuses on the selection of the *best* mechanism to deliver each task. Both tasks and mechanisms are selected from a list of alternatives. Next to the selection of mechanisms, al Level 5 the operational window of the intended mechanism is compared with the one articulated in the formulation of the design problem (Level 0: Framing level).

Exercises
- Identify at least two extra mechanisms for the fundamental tasks C1 (dispersive mixing), C2 (size reduction of solid particles) and C3 (cell wall disruption).
- Agglomeration techniques include layering and sintering. To which task do these mechanisms belong? Motivate your answer.

Bibliography

Almeida-Rivera, C. and Bongers, P. M. M. (2009). Modelling and experimental validation of emulsification process in continuous rotor-stator units. *Computer-Aided Chemical Engineering*, 26, pp. 111–116.

Almeida-Rivera, C. and Bongers, P. (2010a). Modelling and experimental validation of emulsification processes in continuous rotor stator units. *Computers and Chemical Engineering*, 34(5), pp. 592–597.

Almeida-Rivera, C. P. and Bongers, P. (2010b). Modelling and simulation of extensional-flow units. *Computer-Aided Chemical Engineering*, 28, pp. 793–798.

Almeida-Rivera, C. P. and Bongers, P. (2012). Modelling and simulation of extensional-flow units. *Computers and Chemical Engineering*, 37, pp. 33–39.

Almeida-Rivera, C. P., Bongers, P., Irving, N., Kowalski, A. and Egan, M. (2010). Towards the Selection of the Most Efficient Emulsification Unit: A Benchmark Study. In: *Proceedings of the World Congress on Emulsions 2010*.

Davies, J. T. (1985). Drop Sizes in Emulsion Related to Turbulent Energy Dissipation Rates. *Chemical Engineering Science*, 40(5), pp. 839–842.

Davies, J. T. (1987). A physical interpretation of drop sizes in homogenizers and agitated tanks, including the dispersion of viscous oils. *Chemical Engineering Science*, 40(7), pp. 1671–1676.

de Bruijn, R. (1989). *Deformation and breakup of drops in simple shear flow*. Ph. D. thesis, Delft University of Technology.

de Ridder, K., Almeida-Rivera, C. P., Bongers, P., Bruin, S. and Flapper, S. D. (2008). Multi-Criteria Decision Making in Product-driven Process Synthesis. *Computer-Aided Chemical Engineering*, 25, pp. 1021–1026.

Grace, H. (1982). Dispersion Phenomena in High Viscosity Immiscible Fluid Systems and Application of Static Mixers as Dispersion Devices in Such Systems. *Chemical Engineering Communications*, 14, pp. 225–277.

Karbstein, H. and Schubert, H. (1995). Developments in the continuous mechanical production of oil-in-water macro-emulsions. *Chemical Engineering and Processing*, 34, pp. 205–211.

McClements, J. (2005). *Food Emulsions: Principles, Practices and Techniques*, 2nd ed. Boca Raton: CRC Pres.

Nienow, A. (2004). Break-up, coalescence and catastrophic phase inversion in turbulent contactors. *Advances in Colloid and Interface Science*, 108, pp. 95–103.

Saboni, A., Alexandrova, S., Gourdon, C. and Cheesters, A. K. (2002). Interdrop coalescence with mass transfer: comparison of the approximate drainage models with numerical results. *Chemical Engineering Journal*, 88, pp. 127–139.

Schubert, H., Ax, K. and Behrend, O. (2003). Product engineering of dispersed systems. *Trends in Food Science and Technology*, 14, pp. 9–16.

Schubert, H. and Engel, R. (2004). Product and Formulation Engineering of Emulsions. *Chemical Engineering Research and Design*, 82(9), pp. 1137–1143.

Schubert, H., Engel, R. and Kempa, L. (2006). Principles of Structured Food Emulsions: Novel formulations and trends. *IUFoST*, 13(43), pp. 1–15.

Siirola, J. J. (1996). Industrial Applications of Chemical Process Synthesis. In: J. L. Anderson, ed., *Advances in Chemical Engineering – Process Synthesis*. Academic Press. Chapter 23.

Steffe, J. F. (1996). *Rheological Methods in Food Process Engineering*, 2nd ed. Michigan: Freeman Press.

Stork, M. (2005). *Model-based optimization of the operation procedure of emulsification*. Ph. D. thesis, Delft University of Technology.

Tjaberinga, W. J., Boon, A. and Cheesters, A. K. (1993). Model Experiments and Numerical Simualtions on Emulsification Under Turbulent Conditions. *Chemical Engineering Science*, 48(2), pp. 285–293.

Walstra, P. (1993). Principles of Emulsion Formation. *Chemical Engineering Science*, 48(2), pp. 333–349.

Zderic, A., Kiskini, A., Tsakas, E., Almeida-Rivera, C. P. and Zondervan, E. (2019). Giving added value to products from biomass: the role of mathematical programming in the product- driven process synthesis framework. *Computer-Aided Chemical Engineering*, 46, pp. 1591–1596.

Zderic, A. and Zondervan, E. (2017). Product-driven process synthesis: Extraction of polyphenols from tea. *Journal of Food Engineering*, 196, pp. 113–122.

7 Equipment selection and design

7.1 Level 7: equipment selection and design

This level of the methodology is characterized by the selection and design of operating units (Figure 7.1). These units are selected based on their capability of delivering the mechanism selected at Level 5. Moreover, the possibility of integrating the tasks identified in Level 4 (alternatively, the mechanisms at Level 5) is considered. This integration involves, for instance, combining tasks with the same driving force that are close together in the task network. The operation window from Level 5 is compared to the operation boundaries of the unit. The final design of the units (only of the key dimension) and final flowchart are considered at this level.

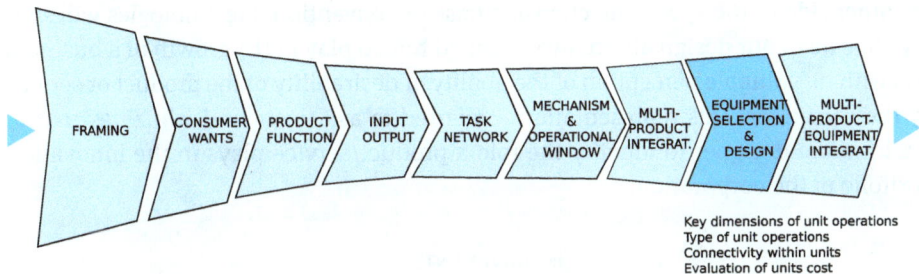

Key dimensions of unit operations
Type of unit operations
Connectivity within units
Evaluation of units cost

Figure 7.1: Unit selection and design as part of the PDPS strategy.

Throughout the evolution of chemical engineering the concept of paradigms has been recurrently addressed. As mentioned by Stankiewicz and Moulijn (2002), this discussion distinguished the first paradigm as early as 1920: the unit operations paradigm. This paradigm focused on the study of the major equipment for chemical manufacturing, including its construction and performance. One of the principles underlying this paradigm was the realization that any process could be regarded as the combination of unit operations and unit processes. Unit operations – in this context – consist of physical changes (*e. g.*, mass and heat transfer), whereas unit processes consist of chemical changes (*e. g.*, oxidation, esterification) (Silla, 2003). In the early 1960s, the unit operation paradigm was replaced by the transport phenomena paradigm. Due to its perceived limitations regarding process creativity, novelty and invention, this second paradigm is gradually losing its leading position. Although a consensus about the third paradigm has not been reached yet, this *should* embrace both the steadily increasing relevance of consumer-centered product engineering and the approaches to achieve such products (*e. g.*, process intensification, green chemistry, risk-based process safety).

A selection of tutorial examples of mathematical models of operating units with relevance to consumer-centered sectors are given in the next sections.

https://doi.org/10.1515/9783110570137-007

7.1.1 Selection of operating units

By establishing the sequence of fundamental tasks and mechanisms, process design variables like general arrangement of units and operation windows are also specified. The identification of the *best* unit to deliver a specific task based on a selected mechanism and within a given operating window is of paramount relevance. While this unit might be commercially available from specialized vendors, this ideal scenario is not always happening. In fact, breakthrough developments in CPI have been the result of bridging the gap between industry/market requirements and the current enabling technology base. Examples of such technology innovations are given in Table 2.12 in Section 2.1.4.3. By mapping the enabling technologies against the value perception of a product or service (Figure 7.2), which could be consumer- or corporate-based, it is evident that true innovations require radical technologies to succeed in the market. At the other side of the spectrum, choosing base or conventional technologies will simply allocate to our design alternative a limited role to play in the growth of a business and with no valuable perception of the quality of desirability of the product or service by the consumer. This representation – referred to as Consumer-Technology matrix – allows the designer to identify the role a product/service plays in the innovation portfolio of the corporation.

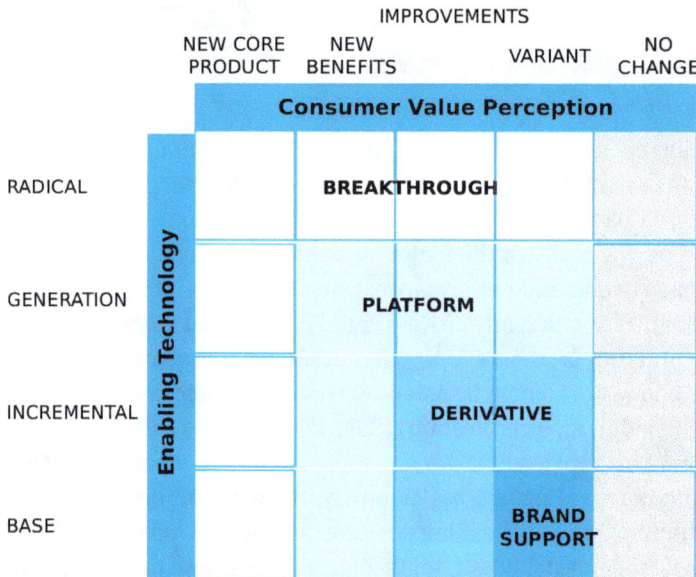

Figure 7.2: Consumer-Technology matrix as a tool to depict the relevance of technology selection.

Contrary to this factual market situation, current unit selection approaches strongly rely on applying the best proven practice and engaging in an intense – and sometimes financially exhausting – journey of laboratory ▷ pilot-plant ▷ semi-industrial ▷ full-scale trials. It is worth restating that this approach does not apply any systematic exploration or design of innovative solutions.

7.1.2 Design of operating units

The design of units is a fundamentally relevant activity within the PSE field. Despite the availability of countless rules of thumb to support the design of processing units, the sound description of the phenomena occurring within the physical boundaries of the device is still regarded as the most reliable design approach. Rules of thumb are governing or play a key role in everyone's daily activities, activities that require us to make decisions when we are not capable of finding the best course of action. Design or engineering rules of thumb allow for a preliminary or conceptual statement regarding suitable sizes or performance of the units, without the need of extended or detailed calculations (Walas, 1990). Moreover, if correctly applied, rules of thumb support preliminary cost estimations and the merit or motivation of engaging in detailed design activities. While the development of engineering rules of thumb is mostly derived from the expertise of practitioners, these design statements need to be individually evaluated regarding their level of reliability, range of applicability and extent of extra/intrapolation. It is worth mentioning that a considerable segment of existing rules of thumb reside as corporate in-house know-how as they support the competitive advantage of enterprises. A summary of relevant engineering rules of thumb for processing unit operations is outlined in Table 7.1.

7.1.2.1 Modeling in the design of operating units
The development and use of models to assist the process designer reside in the heart of PSE. As mentioned by Grossmann and Westerberg (2000), PSE provides engineers with the systematic design and operation methods, tools that they require to successfully face the challenges of today's industry. The role of modeling in design is described by Harmsen *et al.* (2018) by defining the terms designing and a design. A design is the representation of an artifact resulting from a designing activity. This activity is motivated by the satisfaction of a need, which can reside in organizational, financial and/or consumer-centered spaces. This artifact is represented by a design, which can be expressed in the form of one or more mutually consistent models. Each model is characterized by its specific scope (*e. g.*, physical, financial and operational), the level of deployability and communication within an organization. While mathematical models are the most commonly employed formulations in design, they are not the only ones. As given by Harmsen *et al.* (2018), there are several types of models and types of formulation (Table 7.2). In fact, the development of models – or modeling – provides high-quality information to underpin decision support in process and product innovation and process design and operation. Moreover, it is commonly regarded

Table 7.1: Summary of relevant engineering rules of thumb (Walas, 1990).

Operation	Rules of thumb
Compression and pumping	Compression ratio in *multi-stage unit*: about the same in each stage (*n*), ratio = $(P_n/P_1)^{1/n}$. Efficiency of *rotary compressors*: 70 %, except liquid liner type (50 %). Steam requirements of a *three-stage ejector*: 100 lb steam/lb air to maintain a pressure of 1 Torr. Fans are used to raise the pressure about 3 %, blowers raise to less than 40 psig, and compressors to higher pressures.
Cooling	In commercial units, 90 % of saturation of the air is feasible. *Countercurrent-induced draft towers* are the most common in CPI; Efficiency: able to cool water within 2 °F of the wet bulb. Evaporation losses: 1 % of the circulation for every 10 °F of cooling range; windage or drift losses of mechanical draft towers: 0.1 %–0.3 %; blowdown: 2.5 %–3.0 % of the circulation necessary to prevent excessive salt buildup. Water in contact with air under adiabatic conditions eventually cools to the wet bulb temperature.
Drying	Drying times: few seconds in *spray dryers*, up to 1 hour or less in *rotary dryers* and up to several hours or even several days in *tunnel shelves* or *belt dryers*. Drying times: *continuous-tray and belt dryers* for granular material (of natural size or pelleted to 3–15 mm): 10–200 min. *Rotary cylindrical dryers*: superficial air velocities: 5–10 ft/sec; residence times: 5–90 min; holdup of solid: 7 %–8 %. An 85 % free cross-section taken for design purposes. In countercurrent flow, the exit gas: 10–20 °C above the solid; in parallel flow, temperature of the exit solid: 100 °C; rotation speeds: about 4 rpm; product of rpm and diameter [ft]: 15–25.
Liquid–liquid extraction	*Sieve tray towers*: holes: 3–8 mm; velocities through the holes: below 0.8 ft/sec; tray spacings: 6–24 in; tray efficiency: 20–30 %. Dispersed phase: is the one that has the higher volumetric rate except in equipment subject to backmixing. It should be the phase that wets the material of construction less well. Since the holdup of continuous phase usually is greater, that phase should be made up of the less expensive or less hazardous material.
Filtration	*Continuous filtration*: not be attempted if 1/8 in. cake thickness cannot be formed in less than 5 min. Rapid filtering accomplished with *belts*, *top feed drums* or *pusher-type centrifuges*.
Heat exchange	Standard tubes: 3/4 in. OD, 1 in. triangular spacing, 16 ft long; a shell 1 ft in diameter accommodates 100 sqft; 2 ft diameter, 400 sqft, 3 ft diameter, 1,100 sqft. Minimum temperature approaches 20 °F with normal coolants, 10 °F or less with refrigerants. Estimates of heat transfer coefficients (Btu/(hr)(sqft)(°F)): water to liquid, 150; condensers, 150; liquid to liquid, 50; liquid to gas, 5; gas to gas, 5; reboiler, 200. Max flux in reboilers, 10,000 Btu/(hr)(sqft).
Mixing	Mild agitation: achieved by circulating the liquid with an impeller at superficial velocities of 0.1–0.2 ft/sec, and intense agitation at 0.7–1.0 ft/sec. Proportions of a stirred tank relative to the diameter *D*: liquid level = *D*; turbine impeller diameter = *D*/3; impeller level above bottom = *D*/3; impeller blade width = *D*/15; four vertical baffles with width = *D*/10.

Table 7.2: Types of models in the context of product and process design (Harmsen *et al.*, 2018).

Type of model	Scope and features
Physical	look like the real system they represent (car, building, manufacturing plant, product prototypes, mock-ups); can be small-scale or full-scale
Schematic	are pictorial representations of conceptual relationships of the design. Examples: molecular structures drawings, (bio)chemical reaction pathways, product structure drawings (spatial distribution of phases with each size, shape, form), product structure drawing in different stages of manufacturing or use graphs showing relationships (suggested, measured) between product and process variables (reaction rate *versus* concentrations, heat transfer rate *versus* temperature differences, payback time *versus* capital investment), tables (containing symbols and data), pure component properties, stream specification, process stream summary, diagrams (*e. g.*, HoQ, process block diagram, process flow diagram)
Verbal	contain words (organized in lines of text, bullet lists, paragraphs, chapters or (text) tables) that represent systems or situations that (may) exist in reality. These words provide: design problem context, knowledge relevant to solve the problem and analyze and evaluate performance, assistance in construction of (parts) of the design (equipment CAD drawing), assistance in operating, maintaining and repairing (manuals) and justifications about design decisions and recommendations
Mathematical	most abstract and precise models in a structural language; these models are systematic and precise and often restricted with respect to their domain of valid application

as a tool to deploy corporate knowledge effectively. When properly executed, modeling results in accelerated innovation, better process designs, better operations, better product designs and effective risk management. In the design of units, modeling allows for the identification of knowledge gaps in the product-process interactions and, ultimately, enables the focus on these bottlenecks. Scale-up activities strongly benefit from modeling as it allows the relatively risk-free single-step translation from bench-scale equipment to factory-scale equipment. This translation contributes largely to significant reductions in time-to-market. From an operational perspective, modeling allows fault diagnosis by comparing the actual operation of the process with the desired performance.

7.1.2.2 Mathematical modeling as design tool

The development of mathematical models to assist the practitioner in the design of units is a powerful and continuously growing activity within PSE and gradually becoming a standard asset within design teams. While the level of detail and description of mathematical models spreads over a wide range, all mathematical models aim at the same targets: support the decision making process to find the *best* alternative that meets a set of objectives, support the designer to understand scenarios, predict behavior and carry out *in silico* testing of the system. Starting from the lowest level of granularity, mathematical models can be directional, data mining-based or phenomena-driven at varying levels of detail. As mentioned by Silla (2003), at the beginning of a process design, simple sizing procedures or models are sufficient to determine whether the alternatives under consideration are delivering the expectations (*e. g.*, preliminary production cost, financial viability, production throughput). Mathematical models of manufacturing processes are often of high complexity, covering unit operations for (bio)chemical reactions and mass, momentum and energy transfer, including thermodynamics, kinetics and transfer phenomena (Harmsen *et al.*, 2018). Such detailed mathematical models are extensively used in process simulators and mathematical optimization software in all innovation phases. Current computational capabilities play a crucial role in the development of such mathematical models. While examples of applications of computational methods (*e. g.*, 3D CAD modeling, Computational Fluid Dynamics [CFD] and Finite Element Analysis [FEA]) are abundant and spread over all industrial sectors (*e. g.*, chemical process industry, biotechnology and consumer products), their relative universality is limited by the following facts:

- Mathematical modeling requires specialized professionals to translate the situation (or design problem) into a mathematical model that predicts with accuracy and within a confidence level the behavior of the system.
- It is a computationally expensive exercise, which involves powerful hardware and specialized software.
- The level of predictability of the behavior of the system is strongly dependent on the level of understanding of all physical/chemical/biological processes involved in the description of the system. As knowledge is sometimes unavailable, modeling relies on assumptions which can lead to unexpected behavior during the confirmation phase.

Building a mathematical model requires the execution of a set of steps, as outlined in Table 7.3 (Harmsen *et al.*, 2018). One of the most commonly accepted methods to simplify the development of mathematical models involves the decomposition of the design problem into a set of small, mutually connected and potentially solvable subproblems. This decomposition approach can be recognized as one of the features of the PDPS methodology and can be considered one of the thrusts of process synthesis.

A sound modeling of a process or unit serves as starting point of its translation into tangible artifacts. This step – referred to as prototyping (Section 2.3.1) – is a common practice in most design environments and is even compulsory due to regulation. This is the case in systems (*i. e.*, products and services) that directly involve consumers and their safety/health. Take for example FDA regulations, which require the physical testing of products in consumers before their actual launch. Pharmaceuticals need

Table 7.3: Common steps in the development of mathematical models (Harmsen *et al.*, 2018).

Step	Activity
1	Select the object of modeling, and justify a modeling effort from anticipated benefits
2	Agree on goal(s) of model and intended application domain(s) for future model use
3	Find a suitable conceptual representation of the object to be modeled
4	Develop a mathematical model of the object for its structure/behavior/performance
5	Code the mathematical model in software with verification of correctness of coding
6	Select a numerical solver with enough numerical precision for intended applications
7	Validate the coded model by comparing computed answers with experimental data
8	Select case studies based on interaction scenarios between object and its surroundings
9	Apply model to case studies and analyze computed results on achieved accuracies
10	Evaluate computed results on significance for goals and new insights
11	Report model, case studies, numerical results and the evaluation of results

to be physically tested several times in long-lasting loops. It does not come as a surprise that the launch of a new drug is characterized by an innovation loop of 5–10 years. Moreover, and in alignment with the modeling limitations, physical testing is required when the level of accuracy of the simulation model will be high enough to introduce risky uncertainties in the behavior of the system. The development of a concept requires both analytical and physical prototyping. Analytical modeling is required to capture the mechanical resistance of the concept (via FEM methods), durability of moving parts (*e. g.*, ball rings in the wheels, handles and locks) and esthetics-related aspects. Physical models are certainly required to assess the consumer interaction (*e. g.*, ergonomics, touch-feel dimension). If the costs associated with the various physical concepts are not expected to be the limiting factor of the project business case, a series of physical prototypes should be developed. Risks are then effectively addressed and managed within the delivery time frame.

As suggested by Bongers and Almeida-Rivera (2011), the principle of describing a process mathematically is to use the most simple model that fulfills the purpose. The predictions of the model can be viewed as a chain of phenomena, in which a rough description of all phenomena provides better predictions than a detailed description of only one phenomenon. Within the phenomena that can be described, the following are commonly considered within the scope of mathematically modeling:

- material properties: incompressible fluid mixture having a non-Newtonian rheology;
- energy sources: heat transfer, scraping friction, crystallization and viscous dissipation;
- droplet breakup and coalescence: population balance;
- product formulation specifics: thermal conductivity, SFC curve, specific heat, rheology properties (viscosity as a function of shear rate, flow curves, thixotropy behavior, etc.).

Votator in the crystallization of fat blend (Bongers and Almeida-Rivera, 2011; Bongers, 2006)

Crystallization of the fat blend is performed in a scraped surface heat exchanger (votator, Figure 7.3), where rotating scraper blades continually remove fat crystals from the cooled surface and thus maintain a high heat transfer rate.

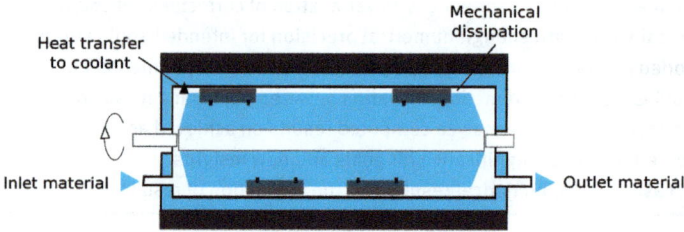

Figure 7.3: Schematic representation of a votator in the crystallization of fat blend.

It is within the votator barrel that much of the product structuring occurs. The quality of the final product depends to a large degree on how these structuring processes have been carried out. In order to optimize the process for a given product formulation or to maintain a desired product quality on scale-up, it is necessary to know the local conditions inside the heat exchanger and how these change with changing operating conditions. Since direct measurement of temperature and shear conditions in a freezer barrel is difficult to achieve, a mathematical modeling approach is required to predict these quantities. The mathematical model is developed by considering the votators as a series of stages, comprising continuously stirring tank reactors.

SPECIFIC HEAT MODEL: the specific heat of the oil/fat/water mixture (Cp) can be described by the following expression:

$$Cp = f(X \times C_p^f + (1 - X) \times C_p^0) + (1 - f)C_p^w(T), \tag{7.1}$$

where X is the fraction of dispersed solid fat crystals, f is the total fat content in the system and the superscripts f, o and w represent the fat, oil and water phases, respectively.

VISCOSITY MODEL: the viscosity of the structured material (η) is equal to the viscosity of the fat phase divided by the total fat content, $i.\,e.$,

$$\eta = \frac{\eta_f}{f}, \tag{7.2}$$

where the fat phase viscosity can be described by

$$\eta_f = b_1 \times \eta_0 \times e^{\frac{b_2 \times X}{1 - b_3 \times X}}, \tag{7.3}$$

with

$$\eta_0 = a_1 \times e^{\frac{a_2}{T}}, \tag{7.4}$$

where a_1, a_2 are constants and b_1, b_2, b_3 are shear-dependent constants.

CRYSTALLIZATION MODEL: experimental work has shown that growth of β' crystals only occurs on transformed α crystals. Nucleation of β' crystals is apparently a mechanism which is not attending in the forming of β' crystals. Transition of α to β' crystals can occur in two manners: directly through the solid phase or indirectly through the liquid phase. Furthermore, it is assumed that transition of α to β' crystals will not occur as long as the product is supersaturated to the saturation line, *i. e.*, crystallization of α is occurring. However, β' crystallization will occur as the product is also supersaturated to the β' saturation line. Therefore, three mass balances are used:

$$\frac{d\alpha}{dt} = k_\alpha^g(\alpha_s - X) - k_\delta \times \alpha(i) + \frac{w}{m}(\alpha_{\text{in}} - \alpha_{\text{out}}), \tag{7.5}$$

$$\frac{d\delta}{dt} = k_\delta \times \alpha + \frac{w}{m}(\delta_{\text{in}} - \delta_{\text{out}}), \tag{7.6}$$

$$\frac{d\beta'}{dt} = k_{\beta'}^g \times \delta(\beta'_s - X) + \frac{w}{m}(\beta'_{\text{in}} - \beta'_{\text{out}}), \tag{7.7}$$

where δ is the fraction of transformed α crystals, k^g is the growth rate, k_δ is the transition rate, m is the mass (or holdup) of the system and w is the mass flow.

PROCESS MODEL: mass, energy and impulse balances are formulated for each of the stages in the model.

Mass balance: for an incompressible liquid, $dm/dt = 0$.

Energy balance: the energy generating terms mechanical (viscous dissipation and scraping), cooling and crystallization are taken into account. The changes in solid fat phase volume and morphology determine the crystallization energy. We have

$$m\frac{dh}{dt} = w(h_{\text{in}} - h_{\text{out}}) + Q_{\text{viscous}} + Q_{\text{scraping}} + Q_{\text{crystallization}} + Q_{\text{cooling}} \tag{7.8}$$

and

$$dh = C_p \times dt, \tag{7.9}$$

where h is the enthalpy.

Viscous dissipation is calculated using laminar mixing theory, *i. e.*,

$$Q_{\text{viscous}} = \tau\dot{\gamma}V + \frac{w}{m}\Delta P = \eta_f\dot{\gamma}^2V + \frac{w}{m}\Delta p, \tag{7.10}$$

where τ is the shear stress, $\dot{\gamma}$ is the shear rate, V is the volume and Δp is the pressure drop along the volumetric element.

Scraping friction was estimated using an empirically derived equation (Trommelen, 1967) based on the thickness of the frozen layer at the wall,

$$Q_{\text{scrapping}} = c_1 X^{c_2} (T - T_{\text{coolant}})^{\frac{5}{3}} N_r^{\frac{1}{5}} L N_{\text{blades}}, \tag{7.11}$$

where c_1 and c_2 are parameters determined experimentally, N_r is the rotational speed of the unit, L is the length of the unit and N_{blades} is the number of blade rows.

Heat transfer to the coolant (through the barrel wall) is calculated by assuming that the product-side wall heat transfer coefficient (htc) is the limiting resistance and, therefore, the coolant-side htc can be ignored. The product-side htc is estimated using an empirical correlation (Trommelen, 1967) based on penetration theory:

$$Q_{\text{refrigeration}} = \bar{\alpha} \times (T - T_{\text{coolant}}) \pi \times D \times L, \tag{7.12}$$

with

$$\bar{\alpha} = 2 \times \Phi_{\text{corr}} \sqrt{\rho \times C_p \times N_{\text{blades}} \times N_r \times D \times \lambda}, \tag{7.13}$$

$$Q_{\text{crystallization}} = (H_\alpha \times k_\alpha^g \Delta_{S_\alpha} + H_{\beta'} k_\beta^g \delta \Delta S_{\beta'} + (H_{\beta'} - H_\alpha) k_\delta \bar{\alpha}) \times \rho_f \times V, \tag{7.14}$$

where Δ_{S_α} is the supersaturation of alpha crystals, $\Delta S_{\beta'}$ is the fraction of transformed β' crystals, H is the heat of fusion, $\bar{\alpha}$ is the mean htc, Φ_{corr} is a correction factor which accounts for experimental inaccuracies and λ is the thermal conductivity.

Impulse balance: the flow as a function of the pressure drop $|\Delta p|$ between two consecutive elements is described by (Fredrickson and Bird, 1958)

$$w = \frac{\rho \times \pi \times D^4 \times |\Delta P|}{96 L \eta_f} \left(1 - \frac{D_{\text{core}}}{D} \right), \tag{7.15}$$

where D_{core} is the diameter of the core.

MODEL VALIDATION: the model has been validated with measurements of outlet temperature on an industrial-scale votator and using a commercial formulation. As depicted in Figure 7.4, the model predicts the product outlet temperature accurately.

Rotor-stator unit in the synthesis of oil-in-water HIPE (Almeida-Rivera and Bongers, 2009, 2010a)

The production of an oil-in-water HIPE is achieved using a continuous rotor-stator unit, which is composed of three interconnected building blocks: a static mixer, a mixing vessel and a colloid mill (Figure 7.5). The mathematical model to be developed is intended to predict the dynamic behavior of a continuous rotor-stator unit for the production of oil-in-water emulsions. Each model building block is formulated accounting for the estimation of the governing breakup mechanism within the boundaries of each block, the population balance of droplets and, ultimately, the distribution of

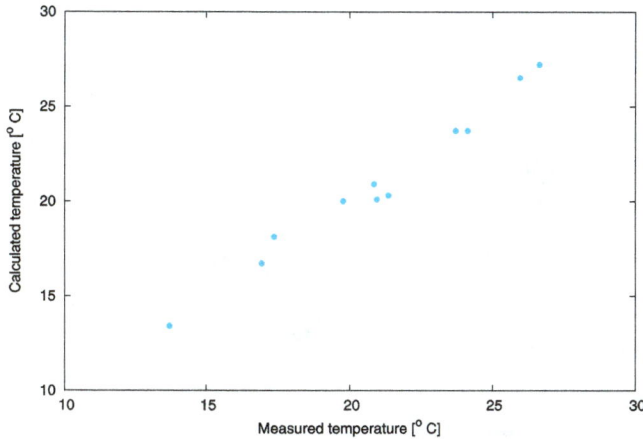

Figure 7.4: Experimental validation of the model: outlet temperatures for different cooling and throughputs (Bongers and Almeida-Rivera, 2011).

droplet sizes at the exit of each block. Input variables and parameters are required to solve the set of differential and algebraic expressions. Such parameters involve the physical properties of the materials and operational parameters. The complete system is consolidated by linking each building block according to the process topology. Several assumptions are introduced to the model, without compromising its reliability and the representation of the actual process. With the same goal in mind, very simplified models are used to describe the evolution of droplet size (*i. e.*, population balance) and physical properties of the system (*e. g.*, viscosity). The level of detail of each building block is chosen to be as simple as possible to fulfill the purpose, without being too simplistic. The main assumption behind each building block model is that the emulsion is water-continuous. A comprehensive list of assumptions is outlined by Almeida-Rivera and Bongers (2010a).

POPULATION BALANCES: are constructed to account for the disappearance and appearance of droplets/particles at each size class in a size distribution. A relevant aspect of these balances is the estimation of the prevailing mechanism (breakup or coalescence) at each size class as a function of droplet diameter, acting forces on the droplet and time domain. Knowing the governing mechanism at a given size class allows the calculation of the rate of production and breakage of droplets. Thus, depending on the mean droplet diameter, the stress exerted on it and the residence time at a given size class, a droplet is disrupted into a given number of daughter droplets, coalesces with a colliding droplet or remains unchanged in size. The population balance equations for a size class of mean volume v are given by the contributions of convective, birth and death terms (Nere and Ramkrishna, 2005; Raikar *et al.*, 2006), *i. e.*,

$$\frac{dn(v,t)}{dt} = \phi_{in}(v,t) - \phi_{out}(v,t) + \int_{v'>v}^{\infty} N(v')S(v')B(v,v')n(v',t)\,dv' - S(v)n(v,t), \quad (7.16)$$

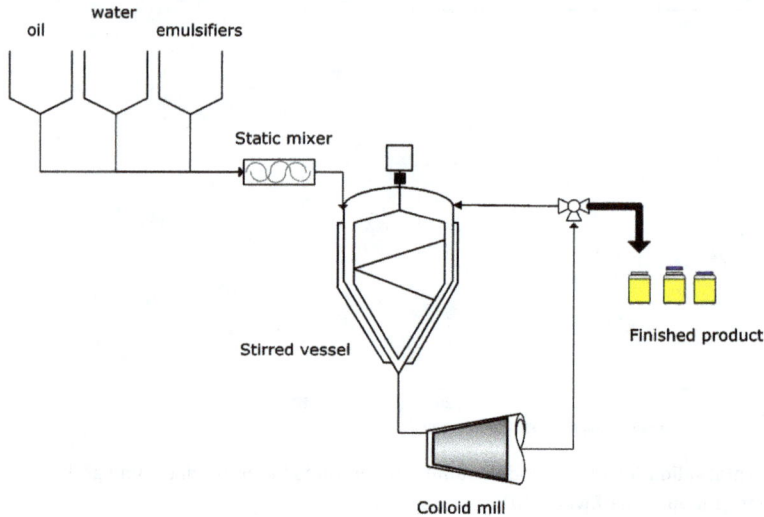

Figure 7.5: Schematic representation of a rotor-stator unit in the synthesis of oil-in-water HIPE.

where $n(v, t)dv$ represents the number of droplets per unit volume of the dispersion at time t with volumes between v and $v + dv$, $\phi(v, t)$ is the convective flow of droplets of volume v, $N(v', z)$ is the number of daughter droplets produced by breakage from parent droplets of volume v', $B(v, v')$ is the droplet size distribution of daughter droplets of volume v produced by breakage of parent droplets of volume v', $S(v)$ is the breakup frequency (or rate) of droplets of volume v and $n(v)$ is the number density of droplets of volume v. The initial condition of the differential system is $n(v, 0)$ and represents the number density of the coarse emulsion.

A simplified population balance is applied for each of the building blocks of the model and, furthermore, a discretized version of equation (7.16) is used. The discretization in volume classes is defined such that

$$v_i = 2v_{i+1}. \tag{7.17}$$

Under the assumption of droplet sphericity, the corresponding class diameter is given by

$$d_i = \left(\frac{6v_i}{\pi}\right)^{\frac{1}{3}}. \tag{7.18}$$

For a binary breakup mechanism, for instance, the droplets with diameter d_i are assumed to break into droplets with diameter d_{i-1} with a volume-related breakage function G, resulting in the following generic expression:

$$\frac{d}{dt}v(d_{i-1}) = Gv(d_i), \tag{7.19}$$

$$\frac{d}{dt}v(d_i) = -Gv(d_i),\qquad(7.20)$$

where the first ordinary differential equation accounts for the birth of small droplets and the second predicts the death of the same volume of large droplets.

The main advantage of this approach is the compliance with the law of mass conservation, while the main disadvantage relies on the limited accuracy due to the limited number of volume elements.

RHEOLOGICAL MODEL: for reliable prediction, the rheological behavior of structured emulsions is of key relevance. The development of improved and even more accurate models is justified by the wide range of industrial applications involving structured systems. In the broad field covered by structured emulsions, the knowledge and prediction of the rheological properties of emulsions are vital for the design, selection and operation of the unit operations involved in the mixing, handling, storage and pumping of such systems (Pal, 2001a,b; Krynke and Sek, 2004). Moreover, in the FMCG sector reliable rheological models are continuously required to assist process engineers in activities related to quality control, interpretation of sensory data, determination of food structure, functionality and compositional changes during food processing (Holdsworth, 1993). In the open literature, a wide range of rheological models have been reported to predict the flow properties of structured products (Holdsworth, 1993). Among them, the generalized power law model of Herschel–Bulkley is commonly adopted as the most appropriate model for oil-in-water emulsions (Steffe, 1996). This three-term model requires a predetermined value of the yield stress τ_0 and involves the shear rate after yield $\dot{y}$ of the emulsion. The model expression for the prediction of viscosity η can be written in the form given by Sherwood and Durban (1998), i. e.,

$$\eta_e = \begin{cases} \dfrac{\tau_0}{\dot{y}} + K\dot{y}^{(n-1)} & \text{if } \tau > \tau_0, \\[2mm] 0 & \text{if } \tau < \tau_0, \end{cases}\qquad(7.21)$$

where n is the flow-behavior index, K is the consistency factor and the subscript e is related to the emulsion. The parameters K and τ_0 can be determined experimentally or can be found tabulated for a wide range of systems (e. g., see Steffe (1996)). Alternatively, they can be estimated using empirical correlations like those reported by Princen and Kiss (1989) for HIPE,

$$\tau_0 = \frac{2\sigma}{d_{32}}\Phi^{\frac{1}{3}}Y(\Phi)\qquad(7.22)$$

and

$$K = C_K(\Phi - 0.73)\sqrt{\frac{2\sigma\eta_c}{d_{32}}},\qquad(7.23)$$

where Φ denotes the internal phase fraction of the emulsion, σ is the interfacial surface tension, η_c is the viscosity of the continuous phase, C_K is a model constant and $Y(\Phi)$ is a function given by the expression

$$Y(\Phi) = -0.08 - 0.114\log(1-\Phi). \tag{7.24}$$

STATIC MIXER MODEL: the droplet size distribution at the exit of the static mixer is described as a log-normal distribution with specified values of initial mean diameter and variance ($d_{32}^0 = 100\,\mu m$ and $\sigma_v = 0.18$, respectively), i. e.,

$$v(d_i) = \frac{1}{\sigma_v\sqrt{2\pi}}e^{-\frac{(\log(d_i)-\log(d_{32}^0))^2}{2\sigma_v^2}}. \tag{7.25}$$

COLLOID MILL MODEL: the flow through the colloid mill ϕ_{CM} is estimated accounting for the pressure generated by the rotational action of the mill, the pressure drop in the unit and the pressure drop in the recirculation loop. The pressure gain in the cylindrical geometry of the colloid mill $\Delta P_{g,CM}$ can be estimated according to a modified expression of the model developed by Bird *et al.* (1959),

$$\Delta P_{g,CM} = \rho_e\left(\frac{\omega_{CM}\pi D_{ri}}{1-\lambda_D^2}\right)^2\left(2\ln\lambda_D + \frac{1}{2}\lambda_D^{-2} - \frac{1}{2}\lambda_D^2\right), \tag{7.26}$$

where ρ_e is the emulsion density, ω_{CM} is the colloid mill rotational speed, D_{ri} is the inner diameter of the rotor and λ_D is defined as

$$\lambda_D = \frac{D_{ri}}{D_{ro}}, \tag{7.27}$$

where D_{ro} is the outer diameter of the rotor.

The pressure drop in the colloid mill can be predicted according to the following expression:

$$\Delta P_{CM} = \frac{12\eta_e\phi_{CM}}{\alpha\pi D_{ri}h^3}\ln\left(\frac{1}{1+\alpha l_r}\right), \tag{7.28}$$

where h is the mill gap, l_r is the length of the rotor and α is a geometry parameter defined as

$$\alpha = \frac{D_{ro}-D_{ri}}{D_{ri}l_r}. \tag{7.29}$$

The pressure drop along the recirculation loop is estimated as

$$\Delta P_p = \phi_{CM}\frac{128L\eta_{elp}}{\pi D_p^4}f_n(n), \tag{7.30}$$

where L is the length of the pipe, $\eta_{e|p}$ is the emulsion viscosity estimated at the flow conditions in the pipe, D_p is the pipe diameter and f_n is a function dependent on the rheological model (e. g., $f_n(n) = (\frac{3n+1}{n})^n$ for a power law fluid).

Combining equations (7.26)–(7.30) leads to an expression for the estimation of the flow through the colloid mill, i. e.,

$$\phi_{CM} = \frac{\Delta P_{CM}}{\eta_{e|p}\frac{128}{\pi D_p^4}f_n(n) - \eta_e\frac{12}{a\pi D_n h^3}\ln(\frac{1}{1+al_r})}. \tag{7.31}$$

A detailed analysis of solution of equation (7.31) reveals that the throughput of the colloid mill can be limited at large mill gaps. Thus, for units with geometries such that $\frac{l_r}{h} \leq 10$, the mill throughput can be restricted, as depicted in Figure 7.6. Preliminary calculations also confirm that the throughput of the colloid mill decreases with decreasing d_{32} values. This trend is expected as the consistency factor C_K increases as d_{32} decreases.

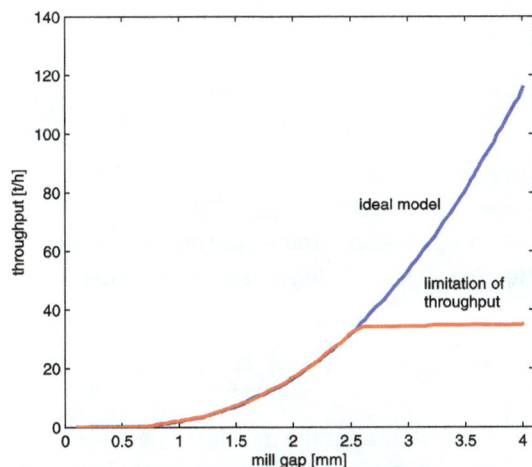

Figure 7.6: Limitation of the colloid mill throughput as a function of mill gap. The throughput limitation is noticed at $l_r/h \leq 10$.

The breakup mechanism of droplets within the colloid mill is divided into sections depending on the droplet diameter: (i) stable droplets, (ii) viscous or inertial breakup and (iii) breakup behind the gap between rotor and stator. Droplet breakup occurs when the external stress exerted on the droplet surface exceeds the form maintaining interfacial force locally for a sufficiently long time. Thus, breakup of droplets materializes when a set of conditions is fulfilled:

- for the inertial breakup mechanism: the size of a droplet is larger than the length of Kolmogorov energy dissipating eddies, l_k; the pressure fluctuations are much larger than the pressure due to interfacial forces;

- for the viscous breakup mechanism: the size of a droplet is smaller than l_k; the viscous forces are larger than the interfacial forces; and
- for both mechanisms: the residence time of a droplet in the volume element (where breakup is expected) is longer than the time required for breakup, t_b.

Thus, for the inertial breakup mechanism, the first condition is expressed as follows:

$$l_k < d_{32}(v), \tag{7.32}$$

where the length of energy dissipating Kolmogorov eddies, l_k, is given by

$$l_k = \left(\frac{(\frac{\eta_e}{\rho_e})^3}{\epsilon} \right)^{1/4}, \tag{7.33}$$

where ϵ is the total specific power dissipation and ρ_e is the density of the emulsion.

To predict when breakup occurs we introduce the Weber dimensionless number We, which relates inertial and interfacial forces. Breakup takes place when the Weber number exceeds a critical value We^{cr},

$$We(v) = \frac{\rho_e \bar{u}^2 d_{32}(v)}{\sigma} > We^{cr}, \tag{7.34}$$

where $We(v)$ is the Weber number of droplets in size class v which relates the inertial and interfacial forces and $\bar{u}$ is the averaged local fluid velocity.

Based on the Weber number and the critical Weber number, the maximum droplet size that will not be broken can be estimated (Davies, 1985; Steiner et al., 2006), i. e.,

$$d^{max} = 0.66 \left(\frac{\sigma^3}{\epsilon^2 \rho_e^3} \right)^{\frac{1}{5}}, \tag{7.35}$$

where d^{max} is the maximum unbreakable droplet size in the inertial regime.

The last necessary condition to guarantee breakup is the sufficiently long residence time of a droplet in the region with certain turbulent flow intensity. This condition is reflected in the following expression:

$$t_b = 0.4\pi \left(\frac{(3\rho_d + 2\rho_c)(d_{32}/2)^3}{24\sigma} \right)^{\frac{1}{2}}, \tag{7.36}$$

where ρ_d is the density of the disperse phase and ρ_c is the density of the continuous phase.

In the viscous mechanism, the droplets are smaller than the turbulent eddies and breakup occurs due to viscous forces in the eddies' bulk. Namely, the following condition holds:

$$l_k \geq d_{32}(v). \tag{7.37}$$

Additionally, the breakup mechanism requires the viscous forces exerted on the droplet to be larger than the interfacial forces. The ratio between both forces is termed capillary number Ω, whose critical value should be exceeded for the breakup to occur, $i.\,e.$,

$$\Omega(v) = \frac{\eta_e \dot{\gamma} d_{32}(v)}{\sigma} > \Omega^{\mathrm{cr}}. \tag{7.38}$$

The maximum stable droplet diameter, below which no viscous breakup takes place in the colloid mill, can be estimated as

$$d^{\mathrm{max}} = \frac{2\sigma \Omega^{\mathrm{cr}}}{\eta_e \dot{\gamma}}. \tag{7.39}$$

Similarly, the residence time of a droplet inside the viscous eddies should be long enough to guarantee breakup ($i.\,e.$, $t_b < t_r$) (Tjaberinga $et\,al.$, 1993),

$$t_b = \frac{\eta_c d_{32}(\lambda_\eta)^{0.315}}{2\sigma} t_b^*, \tag{7.40}$$

where λ_η is the ratio of viscosities ($\lambda_\eta = \eta_d/\eta_c$) and t_b^* is a model constant.

MIXING VESSEL MODEL: in the mixing vessel droplet breakup is confined to a region of approximately 10 % of the vessel volume, next to the propeller blades (Park and Blair, 1975). A simplified model is used to predict the behavior of the vessel, involving the decomposition of the unit into two CSTR models. Thus, mixing and holdup variation are described in one large CSTR ($V_I = 0.9V$), whereas mixing and breakup occur in a small CSTR ($V_{II} = 0.1V$). Moreover, both CSTRs are interconnected by the flow of material V_{I-II} (Figure 7.7).

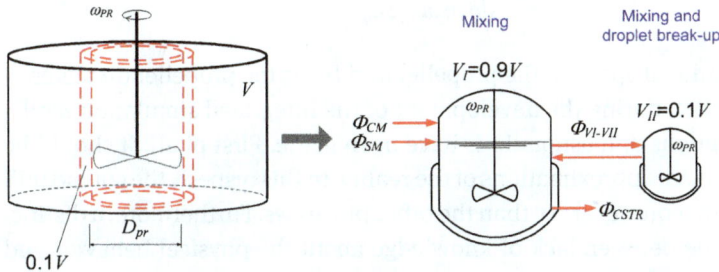

Figure 7.7: Simplified CSTR model for the mixing vessel. The vessel is decomposed into two CSTR models.

The variation of holdup in the mixing vessel is estimated by carrying out a mass balance in the large CSTR, assuming constant density of the emulsion,

$$d\frac{V_I}{dt} = \Phi_{\mathrm{CM}} + \Phi_{\mathrm{SM}} - \Phi_{\mathrm{CSTR}}, \tag{7.41}$$

where the subscript SM is related to the static mixer and CM to the colloid mill.

Under the assumption of only mixing in V_I and constant density, the time evolution of the droplet size distribution of the outer CSTR, DSD_I, can be written as

$$V_I \frac{d}{dt} DSD_I = (\Phi DSD)_{\text{CM}} + (\Phi DSD)_{\text{SM}} + \Phi_{VI-VII}(DSD_{II} - DSD_I) - \Phi_{\text{CSTR}} DSD_I. \quad (7.42)$$

The change in DSD in the small CSTR, DSD_{II}, is the result of the mixing of the corresponding streams and the breakup due to the propeller action, *i. e.,*

$$V_{II} \frac{d}{dt} DSD_{II} = \Phi_{VI-VII}(DSD_I - DSD_{II}) + G \cdot DSD_{II}, \quad (7.43)$$

where matrix G accounts for the droplet breakup mechanism in the CSTR.

For droplets larger than the critical diameter (equations (7.39) and (7.35)), binary breakup is assumed, resulting in an expression similar to equation (7.20),

$$\frac{d}{dt} v(d_i) = -g(d_i)v(d_i) + g(d_{i+1})v(d_{i+1}), \quad d_i \geq d^{\text{max}}, \quad (7.44)$$

where $g(d_i)$ is the droplet breakage rate, which can be approximated as

$$g(d_i) = \lambda_g (d_i - d^{\text{max}})^{n_g}, \quad (7.45)$$

where λ_g and n_g are model constants (Hsia and Tavlarides, 1983).

The critical droplet diameter can alternatively be estimated by considering the second CSTR as an annulus with a diameter equal to the propeller diameter, leading to the following expression:

$$d^{\text{max}} = \frac{\sigma}{\sqrt{\rho_e \eta_e \omega_{pr}^3 D_{pr}}}, \quad (7.46)$$

where ω_{pr} is the rotational speed of the propeller and D_{pr} is the propeller diameter.

MODEL VALIDATION: during the development of the integrated simulation model a number of assumptions and estimations have been made. First of all, it should be noted that the models are approximations of the reality. In this respect, the colloid mill can be described with more accuracy than the other processes. Furthermore, a distinction needs to be made between lack of knowledge about the physical behavior and uncertainty in the model parameters. The integrated execution of all building blocks is validated against experimental data at pilot-plant and industrial scales and using commercial units and ingredients. As depicted in Figure 7.8 for the case of pressure gain in the colloid mill and mean droplet diameter of the finished product, the degree of agreement is remarkable considering the simplification of the phenomenological descriptions. Thus, despite the limited knowledge in some key domain areas and the level of complexity of the PBE and physical properties descriptors, the simulation model is able to successfully predict trends observed during industrial and pilot plant trials.

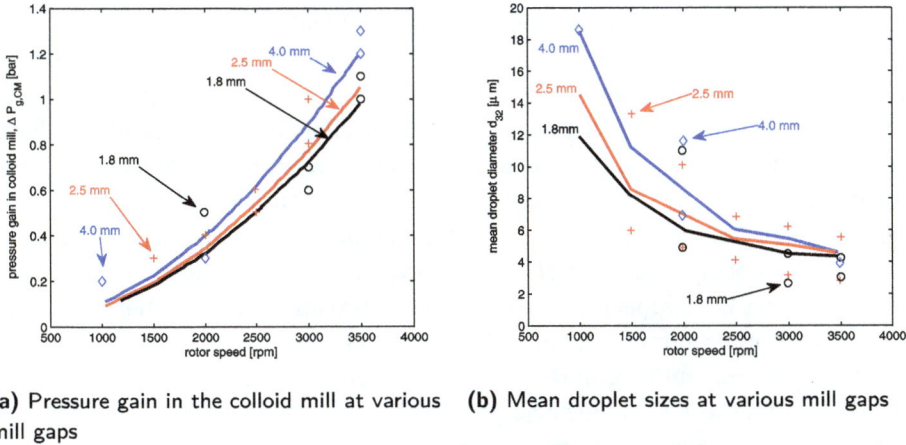

(a) Pressure gain in the colloid mill at various mill gaps   (b) Mean droplet sizes at various mill gaps

Figure 7.8: Comparison between experimental and simulated data. Continuous curves represent the simulation results; symbols represent experimental data (Almeida-Rivera and Bongers, 2010a).

The sensitivity of key model parameters is, additionally, determined by perturbing the nominal parameter values. The parameters under considerations are the initial mean diameter in the static mixer d_{32}^0 (equation (7.25)), the variance σ_v (equation (7.25)) and the rheology parameter C_K (equation (7.23)). The summary of this analysis is given in Table 7.4.

Table 7.4: Sensitivity of model parameters.

Uncertainty source	d_{32} [µm]
$\Delta d_{32}^0 = +50\%$	$\approx 10\%$
$\Delta\sigma_v = +50\%$	$\approx -1\%$
$\Delta C_K = -25\%$	$\approx -10\%$
$\Delta C_K = +25\%$	$\approx +20\%$

It can be extracted from this analysis that the rheology parameter C_K has the largest influence on the product attributes' predictions. A more refined modeling approach should then focus on the incidence of this parameter, while experimental designs should aim at a more accurate estimation of its value.

Roller forming unit for fat containing slurry

This unit is expected to deliver a solid slab of solids with the right mechanical strength and appropriate surface texture. The input to the roller former is a stream of loosely bound, fat containing wet powders coming from an upstream mixing step.

The roller forming unit comprises three rollers with an adjustable relative displacement (Figure 7.9). These displacements allow setting the gap between each pair of rollers at any desired value. The rollers are positioned in an inverted triangle arrangement. The upper two rollers are referred to as roller 1 and roller 2, while the lower roller is arbitrarily referred to as roller 3. The gap between roller 1 and roller 2 is denoted as gap 1, while the gap between roller 1 and roller 3 is denoted as gap 2. The gap between roller 2 and roller 3 is fixed and does not play any role in the compaction of the powders. Roller 1 and roller 2 counterrotate, while roller 3 rotates in the same direction as roller 1. A hopper is localized above the two upper rollers, where material is accumulated and continuously fed to the pair of rollers 1 and 2. The material entering the roller forming unit is compressed by the upper rollers, resulting in the creation of a slab with a theoretical thickness gap 1. This slab is then further compressed to a thickness gap 2, provided that gap 2 < gap 1.

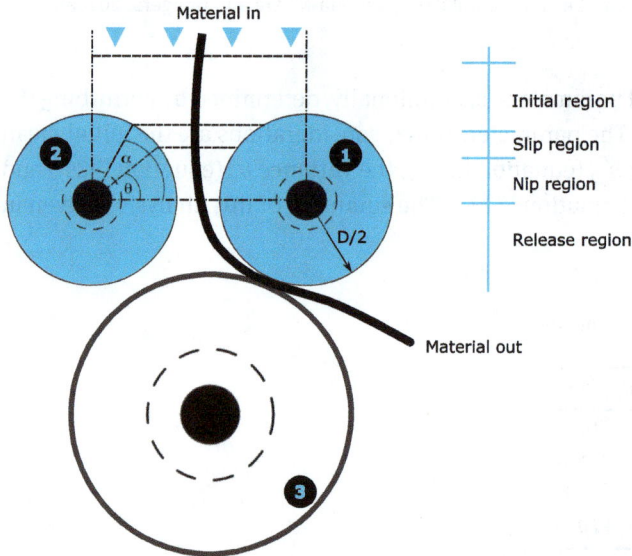

Figure 7.9: Schematic representation of an extrusion unit for fat containing slurry. Roller 1 is the one at the upper right-hand side, roller 2 is the one at the upper left-hand side and roller 3 is the lower roller.

DESCRIPTION OF PHYSICAL PHENOMENA: due to the compaction exerted on the wet powders, they get consolidated to form a continuous slab. Besides this mechanical compaction, the binding of the powders is enhanced by the partial crystallization of the fat present in the material. This phase change is accompanied by an increase in the mechanical strength and stability of the cube. These two properties might be of extreme relevance to maintain the product integrity during and after packing and handling. From a microstructure angle, the fat crystallization allows the completion of the binding process initiated in the mixing step. This phase transition phenomenon can be achieved by tempering the rollers accordingly. An additional benefit of the rotating rollers is given by the possibility of creating a porous structure on the surfaces of the slab. Thus, by having profiled rollers the resulting slab features a non-smooth texture at a lower density value. The extent of compaction – which is mainly given by the gap between rollers – is a key parameter responsible for the mechanical stability of the slab, while visual attributes (smooth and moist surface) are dependent on operating parameters like roller speed and surface roughness. Besides the parameters related to the mix formulation (fat content, internal friction coefficient, particle size distribution, etc.) the parameters associated to the operation of the roller forming unit can be summarized as follows: temperature of the rollers, speed of the rollers, gap between the rollers, rollers' material and feed pressure.

The modeling problem implies the prediction of the physical and mechanical properties of the slab leaving the roller forming unit. The integrity of the slab and the density/extent of compaction are regarded as the variables that define such properties. Thus, the problem involves the prediction of the density of the slab at the exit of the roller forming unit. Moreover, the model should be able to assess the impact of operating conditions (*e. g.*, roller speed, gap between rollers, feed temperature and rollers' material) on the density of the slab. To develop the roller forming model a multi-stage approach has been adopted. Thus, three processing stages have been considered to occur within the boundaries of the roller forming unit:

- Step 1: *Roller compaction*: at this stage the wet loose powders are compacted by the counterrotating rollers (roller 1 and roller 2) at gap 1. The resulting slab features a theoretical thickness given by such gap.
- Step 2: *Compression*: at this stage the incoming slab of thickness gap 1 is further compressed by the co-rotating rollers (roller 1 and roller 3) at gap 2. The resulting slab features a theoretical thickness given by gap 2, where gap 2 < gap 1.
- Step 3: *Expansion*: at this stage the compressed slab leaves the roller forming unit and experiences a relaxation of the structure. This is manifests as an increase in the thickness of the slab, as the pressure is released.

STEP 1: THE ROLLER COMPACTION MODEL of the proposed mechanism is based on the modeling of rolling compaction of powders. The Johanson model is one of the first to attempt to describe this powder compaction. Johanson (1984) developed an analytical model that enables one to predict the roller surface pressure, force and torque from the physical characteristics of the powder and the dimensions of the rolls. The main advantage of this model is that it requires only a limited number of experimental parameters for the powder: the angle of wall friction and the angle of internal friction (both measurable from shear testing experiments) and a compressibility factor (determined from uniaxial compression).

According to the Johanson model, the compaction process can be divided into three zones (Figure 7.9), defined by the behavior of the material (Kleinebudde, 2004;

Reynolds *et al.*, 2010; Bindhumadhavan *et al.*, 2005; Peter *et al.*, 2010; Guigon and Simon, 2003):

- **Feeding or entry zone**, where the densification is solely due to the rearrangement of particles under relatively small stresses created by the feeding method (*i. e.*, screw feeder). This zone is also referred to as slip zone.
- **Compaction zone**, where particles fracture and/or deform plastically under heavy stress provided by the rolls. This region is believed to begin at a point called the nip angle where the powder no longer slips and begins to stick to the wall. As a result the compaction zone is also referred to as the nip region.
- **Exit zone**, where a great decrease in pressure takes place as the compacted slab is ejected and can expand due to elasticity. As the material is pushed out of the process zone, it picks up speed and begins to move faster than the roller. This increase in speed causes slip in the opposite direction before the product finally loses contact with the roller. The beginning of the ejection region is sometimes referred to as a **neutral point** because it sets the boundary between the region where the material moves at the same speed as the wall surface.

The friction forces acting on the slab are greater in the region where the roller moves faster than the material. The difference between the frictional forces in the two regions produces a net frictional force that pulls the powder into the roller gap. Generally, if the wall friction coefficient is too low, the material cannot be drawn through the roller press. Based on a phenomenological description of the roller compaction step we could identify the following operating, geometrical and material parameters that play a key role in slab creation: roller force, roller torque, roller velocity, feed pressure, gravity, inertia, roller diameter, roller width, gap size, internal (effective) angle of friction, cohesion, admissible stress, compressibility, bulk density and friction between powder and roller surface.

The development of the Johanson model has been extensively covered in the open literature (Kleinebudde, 2004; Reynolds *et al.*, 2010; Johanson, 1984; Peter *et al.*, 2010; Johanson, 1965; Teng *et al.*, 2009; Guigon and Simon, 2003; Yusof *et al.*, 2005). It is based on the Jenike yield criteria for steady-state particle flow in silos and hoppers. Under the assumption of slip conditions along the roller surfaces in the feed region, Johanson demonstrated that the pressure gradient in the slip region is given by the following expression:

$$\left(\frac{d\sigma}{dx}\right)_{\text{slip}} = \frac{4\sigma(\pi/2 - \theta - v)\tan\delta_E}{(D/2)[1 + (S/D) - \cos\theta][\cot(A - \mu) - \cot(A + \mu)]}, \qquad (7.47)$$

where θ is the angular position at the roller surface, such that $\theta = 0$ corresponds to the minimum gap, σ is the normal stress, v is the acute angle given by the Jenike-Shield plot (Bindhumadhavan *et al.*, 2005), D is the roller diameter, S is the roller gap, δ_E is the effective angle of internal friction, μ is the wall friction coefficient and A is given by the following expression:

$$A = \frac{\theta + v + (\pi/2)}{2}. \tag{7.48}$$

In the nip region, a more simplified model is assumed, where no slip occurs along the roller surface and all material trapped between the two rollers at the nip angle must be compressed into a slab with a thickness equal to the roller gap. Moreover, the pressure at any angle $\theta < \alpha$ can be estimated according to the following relationship:

$$\sigma_\theta = \sigma_\alpha \left[\frac{\rho_\theta}{\rho_\alpha} \right], \tag{7.49}$$

where α denotes the nip angle as shown in Figure 7.9 and K is the compressibility factor that is determined from the slope of logarithmic plots of the density as a function of pressure in uniaxial compaction.

The pressure gradient for the nip condition is given by the expression

$$\left(\frac{d\sigma}{dx} \right)_{\text{nip}} = \frac{K\sigma_\theta (2\cos\theta - 1 - (S/D))\tan\theta}{(D/2)[(1 + (S/D) - \cos\theta)\cos\theta]}. \tag{7.50}$$

The cornerstone of the Johanson model relies on the assumption that the pressure gradients in the slip and nip regions at the nip angle α are equal. Thus,

$$\left(\frac{d\sigma}{dx} \right)_{\text{slip}} = \left(\frac{d\sigma}{dx} \right)_{\text{nip}}. \tag{7.51}$$

Solving this equality for the nip angle implies *a priori* knowledge of material properties (compressibility factor, internal friction angle and wall friction coefficient) and process parameters (roller diameter and gap). Compressibility factors can be estimated by uniaxial compressibility tests, while internal and wall friction coefficients can be measured in a ring shear unit. Based on this derivation, step 1 of the proposed mechanism involves the following material-related parameters: average particle size, size distribution, angle of internal friction, angle of wall friction, compressibility and bulk density. Additionally, the following process-related parameters are of interest: nip angle, roller gap, roller speed, roller surface friction and pressure profile. Despite its wide applicability, the Johanson approach falls short when it comes to the prediction of the roller velocity effect on the slab creation. Thus, as pointed out (Reynolds *et al.*, 2010), the Johanson model does not consider the roller velocity as a sensitive operational parameter at all. This shortcoming might be explained by the fact that scale-up of roller compaction processes can be achieved to a limited extent by increasing throughput for a given piece of equipment.

STEP 2: THE COMPRESSION MODEL addresses the compression of the existing slab of a thickness gap 1 to a thinner slab of thickness gap 2, where gap 2 < gap 1. The mathematical description of this step is represented by a simplified relationship between density and roller gap,

$$\rho_2 = \rho_1 \left(\frac{\text{gap 1}}{\text{gap 2}} \right)^n. \tag{7.52}$$

In this simple model, the density of the slab is considered to vary with the roller gap, in an inversely proportional form. The exponent n is meant to capture any deviation from the ideal behavior.

STEP 3: THE EXPANSION MODEL accounts for the expansion or relaxation of the existing slab into a thicker one. This relaxation has been experimentally reported, and is accountable for the final size of the slab and, subsequently, its density and mechanical strength. In this step, the effect of roller velocity on the slab attributes is addressed, particularly on the slab density. The modeling approach used in the step involved data mining techniques in general, and Convoluted Neural Networks (CNNs) in particular. The input variables for the model include the two roller gaps, velocities of the three rollers, velocity of the belt and the density of the slab before expansion, as given by the step 2 model.

ⓘ Convoluted Neural Network technique as a design tool

A neural network consists of a number of nodes, called neurons, connected to the inputs and outputs. The neurons weigh all inputs and provide an output via the activation function. The complexity of the neural networks used will be determined by the number of nodes in the hidden layer (2, 3, 5 or 7). The activation function used in this study is a hyperbolic tangent function. In mathematical terms, the output of neuron j is defined by

$$y_j = \tanh\left(\sum_{i=1}^{i=n} w_i \times u_i \right),$$ (7.53)

where y_j is the output of neuron j, $w_i \times u_i$ is the input from neuron i (or input i), weighted with w_i.

The weightings w_i in the neural network are determined by an optimization algorithm using the error between the measured outputs and the outputs predicted by the neural network. As reported by Bongers (2008), one hidden-layer neural network is sufficient to describe all input/output relations. More hidden layers can be introduced to reduce the number of neurons compared to the number of neurons in a single-layer neural network. The same argument holds for the type of activation function and the choice of the optimization algorithm. In addition to a multi-linear regression, four neural network models have been considered for each prediction with 2, 3, 5 and 7 nodes in the hidden layer. A large subset of the experimental data has been used as training set and the rest of the data to validate the model. To avoid coincidental results, the whole training-validation procedure is repeated for a given number of times with different training sets and validation sets. In each repetition, the validation set has been selected randomly. The prediction model type is finally selected on the basis of the root mean square of the prediction error,

$$\epsilon = \sqrt{\frac{1}{N-1} \sum_{i=1}^{i=N} (y_i - \underline{y_i})^2},$$ (7.54)

where y_i is the predicted output for case i, y_i is the predicted output for case i and N is the number of cases.

MODEL VALIDATION: Experimental trials are used to validate the accuracy of the mathematical model. Various data sets are collected at different operating conditions of the roller forming unit. Based on preliminary simulation results, these experimental trials are designed so that two operational constraints of the optimal region are met: $v_2/v_3 < v_{max}$ and gap $2 \times$ gap $1^{-1} \sim 1$.

The measured density of each experimental point is compared with the predicted data obtained by the model. The comparative plot is given in Figure 7.10. As a general observation we can point out that the model predicts – in most of the tested cases – a slab density with an acceptable accuracy and within experimental error expected at pilot-plant trials (*ca.* 10 %). Several process disturbances might be responsible for this slight deviation between experimental and predicted values, including non-constant temperature of the mix during the trial, relatively inaccurate protocol to measure density and transient operation of the unit from point to point. Despite these possible shortcomings – which might be easily resolved – the level of predictability of the model is satisfactory. This fact encourages the deployability of the model as a scale-up tool for roller forming units. As one of the submodels involves a data mining algorithm, the availability of new experimental data will increase the model accuracy and extend its window of applicability.

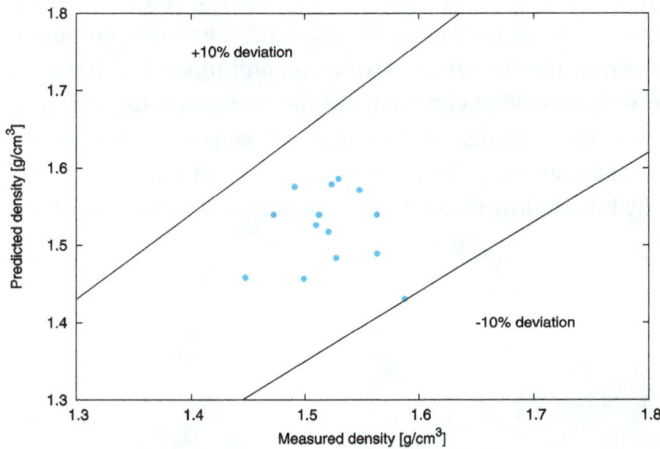

Figure 7.10: Measured and predicted density values for the roller forming unit. Dashed lines define a +10 % and −10 % deviation from the symmetry plane $x = y$.

Supercritical CO_2-assisted dryer for thermally sensitive materials (Almeida-Rivera *et al.*, 2010, 2011; Khalloufi *et al.*, 2009, 2010)

Dehydration assisted by supercritical fluids has recently awaken considerable interest as a promising processing alternative for products with acceptable quality at reduced expenditure. At this thermodynamic condition, the supercritical fluid cannot be condensed even with extreme compression. In the supercritical region, there is a continuous transition between the liquid and gas phases, with no possible distinction between these two. Beyond this point, the special combination of gas-like and liquid-like properties makes the supercritical fluid an excellent solvent for the extraction industry. Fluids at pressure and temperature above their critical points exhibit a solvent power (due to the density) closer to that of liquids and a viscosity and diffusivity comparable to those of gases. Liquids are well known to feature high solvating power and are, therefore, strong solvents. On the other hand, gases have high diffusivity and low viscosity, which results in enhanced transport properties. Supercritical fluids have attractive properties for extraction not only because they penetrate the sample faster than liquid solvents and transport the material extracted from the sample faster but also because they dissolve solutes from food matrices. Among the fluids used at supercritical conditions in extraction applications, carbon dioxide is currently the most desirable and widely used fluid in foods, personal care and home care products and pharmaceuticals. Despite the extensive list of benefits of SC-CO_2-assisted drying, the applicability and feasibility of this technology to the drying of solid matrices is still under investigation, with very limited available data.

The drying unit is mainly equipped with a drying chamber, a recirculation pump and a zeolite reactor, as schematically shown in Figure 7.11. The process is usually carried out in five steps: (1) the drying chamber is filled with the product to be dried; (2) the whole system is pressurized by feeding in SC-CO_2 until the target pressure is reached; (3) the SC-CO_2, which carries the released water, is continuously recirculated at a constant flow rate through the drying chamber and the zeolite reactor; (4) when the humidity of SC-CO_2 is low and remains constant, the system is smoothly depressurized and the product is taken out from the drying chamber; and (5) the zeolite in the reactor is regenerated by hot air flow-through. The temperature within the drying chamber is controlled.

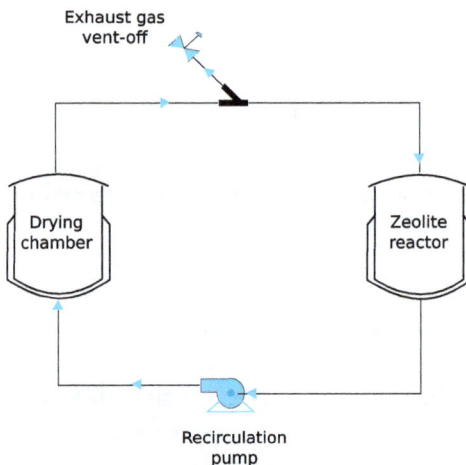

Figure 7.11: Schematic representation of a supercritical CO_2-assisted dryer of thermally sensitive materials. Water is extracted from the moist material by concentration gradient and carried by the SC-CO_2 stream to the zeolite material, where it will be absorbed.

MASS BALANCES: the stream of SC-CO$_2$ carries the water released by the moist material, and the zeolite adsorbs this water. In this example we focus on the phenomena occurring within the boundaries of the drying chamber. The variation of water concentration in the SC-CO$_2$ stream can be determined by a mass balance performed in a differential element (DE) of a particle bed (Figure 7.12). The solid particles and the void space present in the bed are segregated into distinct regions, resulting in the following mass balance:

$$\text{Flow}|_z - \text{Flow}|_{z+dz} = \text{Accumulation}|_{dz}^{\text{solid+fluid}}. \tag{7.55}$$

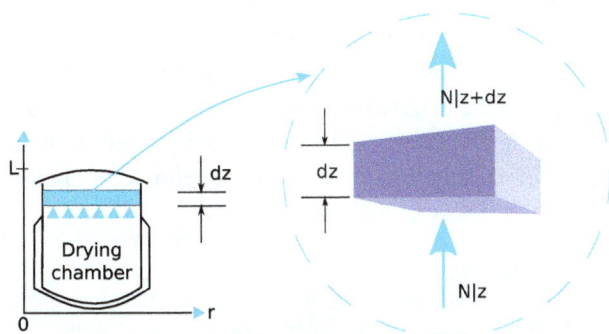

Figure 7.12: Schematic representation of the drying chamber and mass transfer model in the differential element (DE). The height of DE is represented by *dz*.

The mathematical expressions derived for each term of equation (7.55) are listed in Table 7.5.

Table 7.5: List of relevant terms of the mass balance expression. Nomenclature: *m* is the mass, *A* is the surface perpendicular to the SC-CO$_2$ flux, ε is the porosity of the bed, ρ is the density, *C* is the water concentration expressed as kg of water/kg solid matrix or kg of SC-CO$_2$, $\mathcal{D}$ is the axial diffusion coefficient and the subscripts f and S are related to fluid and solid matrix, respectively.

Term	Expression
Water accumulated within the solid matrix	$\frac{dm_S}{dt} = \frac{d}{dt}[dzA(1-\varepsilon)\rho_S C_S]$
Water accumulated in the fluid phase (SC-CO$_2$)	$\frac{dm_f}{dt} = \frac{d}{dt}[dzA\varepsilon\rho_f C_f]$
Water flux passing within the porosity through the differential element	$N = (Q\rho_f C_f) + (JA\varepsilon)$
Partial water flux due to the diffusion phenomenon	$J = -\mathcal{D}\rho_f \frac{dC_f}{dz}$
Flux of SC-CO$_2$ passing through the packed bed	$Q = \varepsilon AU$

As shown in Table 7.5, the water flux is the result of two mechanisms. The first is axial dispersion, which occurs because of the differences in water concentration in the flow direction. Therefore, the molecular mass transfer will take place in the opposite direction of the flow to decrease the mass gradient. The second mechanism is due the convection phenomenon, which takes into account the water mass that is transported with the SC-CO$_2$ stream. In the case of the homogeneous particle bed, the velocity of SC-CO$_2$ in the bed void space can be related to the flux, column cross-sectional area and porosity of the bed. Because the solution is usually diluted, the SC-CO$_2$ density is used instead of the solution density. The proposed model implies no accumulation of SC-C$_2$ in the solid particles. This imposed condition is used to assess the validity of the mass balance equations.

As a first approximation to solve the problem, the following assumptions have been introduced to the model that describes the SC-CO$_2$ drying process: (1) constant temperature, pressure, density, porosity, diffusion coefficient and flow rate of SC-CO$_2$; (2) negligible radial concentration gradients in the drying chamber; (3) water-free flow of SC-CO$_2$ at the entrance of the drying chamber; (4) homogeneously sized particle bed; and (5) interfacial mass transfer mainly governed by convection. Introducing these assumptions leads to the set of expressions listed in Table 7.6.

Table 7.6: List of relevant terms of the simplified mass balance expressions and rate of drying. Nomenclature: K is the external mass transfer coefficient, a is the specific solid surface, L is the total length of the packed bed, V is the total volume of the packed bed, r is the drying rate and the subscripts avg and rel denote average values and values relative to the water released during the drying process, respectively.

Expression

MASS BALANCE

$$\frac{dC_f}{dt} = D\frac{d^2 C_f}{dz^2} - U\frac{dC_f}{dz} - \left(\frac{1-\varepsilon}{\varepsilon} \cdot \frac{\rho_S}{\rho_f}\right)\frac{dC_S}{dt}$$

$$\frac{dC_S}{dt} = -\left(\frac{1-\varepsilon}{\varepsilon} \cdot \frac{\rho_S}{\rho_f}\right)Ka[C_f^* - C_f]$$

$z = 0 \wedge t > 0 \qquad : \qquad \frac{\partial C_f}{\partial z}|_{z=0} = \frac{U}{D_z}C_f$

$z = L \wedge t > 0 \qquad : \qquad \frac{\partial C_f}{\partial z}|_{z=L} = 0$

$0 \leq z \leq L \wedge t = 0 \quad : \quad C_f(z) = 0$

$\qquad\qquad\qquad\qquad\qquad C_S(z) = C_{S0}$

RATE OF DRYING

$$m_S(t) = \rho_S \int_{V_S} C_S \, dV_S$$

$$m_S(t) = \rho_S V_S C_{S,\text{avg}}(t)$$

$$C_{S,\text{avg}}(t) = \frac{1}{V_S} \int_{V_S} C_S \, dV_S$$

$$m_{\text{rel}}(\infty) = m_S(0) - m_S(\infty) = \rho_S V_S [C_S(0) - C_S(\infty)]$$

$$m_{\text{rel}}(t) = m_S(0) - m_S(t) = \rho_S V_S [C_S(0) - C_{S,\text{avg}}(t)]$$

$$r(t) = \frac{dw(t)}{dt} = \frac{1}{m_{\text{rel}}(\infty)}\frac{dm_{\text{rel}}(t)}{dt}$$

This set includes the corresponding differential and algebraic equations and the initial and boundary conditions. The first two equations on the upper section of Table 7.6 define the driving forces controlling the mass transfer phenomena and the equilibrium condition between the solid matrix and the SC-CO2 and are used to link the values of C_f and C_S.

RATE OF DRYING: the governing expressions to estimate the drying rate are outlined in Table 7.6. For a non-uniform drying distribution, the mass of the remaining water within the food matrix at time t, $m_s(t)$, is obtained by integration.

The model can be numerically solved for each differential element of height dz along the packed bed. The ordinary differential equations (ODEs) are discretized using an explicit finite difference method. Finally, the discretized model is implemented and solved using an explicit Runge–Kutta solver. Although some small perturbations can be observed in the water concentration within SC-CO$_2$ at the inlet of the drying chamber, the accuracy of the numerical solution does not depend on the value of the integration step (dz).

The data published by Brown *et al.* (2008) for SC-CO$_2$ drying of carrots are used to validate the accuracy of the model. This data set comparison indicates that globally the present model gives reasonably accurate predictions. A small deviation between the experimental and simulated data is noticed at the final stage of the drying process; more specifically, concerning only the last experimental point. This deviation is probably caused by the violation of one or more of the assumptions used to build this model. For example, some of the preliminary experimental observations suggest that in some cases, especially for carrot samples, the shrinkage phenomenon could be significant and, therefore, non-negligible. If this is the case, the specific interfacial area of the solid matrix would increase (Ratti, 1994; Krokida and Maroulis, 1997, 1999; Rahman, 2001) and the length/diameter of the packed bed would decrease. In such conditions, the simulations would underestimate the drying rate, which is in agreement with the results obtained in this example (Figure 7.13).

As reported by Khalloufi *et al.* (2009, 2010), Almeida-Rivera *et al.* (2010), there is a variation in water concentration within the solid matrix and SC-CO$_2$. Due to numerical solution, some perturbations of the water concentration within SC-CO$_2$ at the inlet of the drying chamber could be identified. Decreasing the integration space step (dz) allows one to smoothen and even to eliminate this perturbation. However, this minor improvement in the quality of the results comes at the expense of the computational time. According to the simulation results the model gives globally coherent results on the water concentration profiles in both the solid matrix and the fluid.

Extrusion unit in the synthesis of multi-phase frozen structures (Almeida-Rivera *et al.*, 2017; Bongers, 2006)

A schematic representation of a Single Screw Extruder (SSE) for the production of a multi-phase frozen food structure (*e. g.*, ice cream) is given in Figure 7.14. The unit can

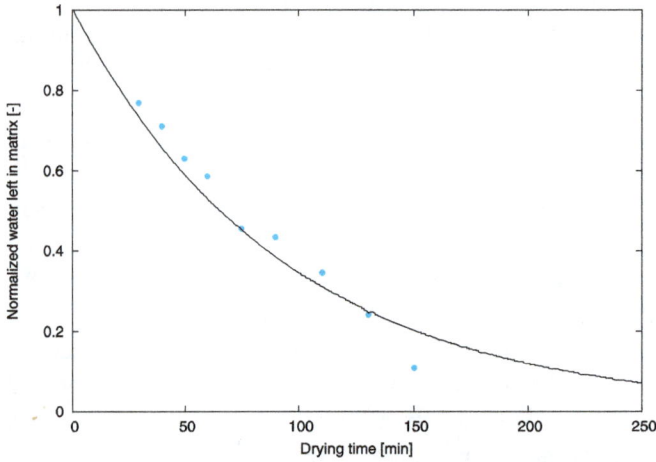

Figure 7.13: Comparison of the numerical solution of the present mathematical model (solid curve) with the experimental data (points) published by Brown *et al.* (2008).

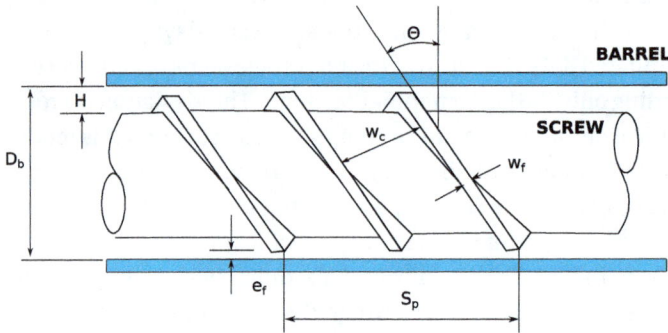

Figure 7.14: Schematic representation of an extrusion unit in the synthesis of multi-phase frozen structures.

be seen as a modified scraped surface heat exchanger, where the dasher of a scraped surface heat exchanger is replaced by a screw having a lower rotational speed. The SSE comprises the following components:

- **Barrel:** where the actual freezing of the product takes place. Instead of a dasher in scraped surface heat exchangers, a close fit screw is rotating at low speed inside the barrel.
- **Screw:** by rotation of the screw, product is transported from inlet to outlet, while product is scraped off the inner surface of the barrel and mixed into the bulk.
- **Cooling jacket:** the level of coolant inside the cooling jacket should be such that the outside of the barrel is flooded by the boiling coolant. Usually ammonia is applied as coolant.

- **Control system**: the temperature of the coolant is controlled via the evaporating pressure, while the level of coolant in the jacket is controlled via a restriction valve on the liquid coolant supply. A basic level of product control can be achieved by controlling the inlet pressure of the extruder via the rotational speed and controlling the outlet temperature by the evaporating pressure.

The model accounts for the following system characteristics: (i) a compressible fluid with a non-Newtonian rheology, (ii) heat transfer accounting for scraping friction, crystallization and viscous dissipation and (iii) detailed descriptors of the physical properties of the structured product (thermal conductivity, ice phase curve, rheology parameters, specific heat and density).

MODEL BOUNDARIES AND DEVELOPMENT: to achieve a reliable mathematical model the interactions between the process and its environment through the boundaries around the process must be specified. Actuators on the extruder include rotational speed, flow/temperature/composition of inlet stream and coolant temperature. On the other hand, the extruder responds by influencing the following variables: rotor shaft torque, temperature of outlet stream and pressure at inlet and outlet points. In order to develop a model that describes the relevant phenomena, a number of assumptions need to be made:

- Instantaneous crystallization of water into ice: the residence time of the product inside the extruder is much longer than the time needed for crystallization of the water. Therefore, the ice phase curve is used to determine the ice fraction as a function of temperature.
- The ideal gas law relates pressure, volume and temperature of the ice cream product.
- Physical properties are a function of temperature and air phase volume only. It is assumed that the absolute level of the pressure has no influence other than on the air phase volume.
- The physical properties of the individual ice cream components are independent of temperature.
- The ice cream flow can be described by a liquid flow. Although crystallization takes place inside the extruder, it is assumed that ice cream will still behave like a (very viscous) fluid.
- Power law viscosity model: ice cream is known to be a non-Newtonian fluid. The simplest extension from Newtonian rheology is a power law rheology.
- Convective heat transfer-dominated process: the Péclet number can be used to determine the relative importance of convective and conductive heat transport. For the ice cream extruder the Péclet number is in the order of 100, implying that the process is convection-dominated.
- Heat losses to the environment are negligible. It is assumed that the extruder is perfectly insulated.

- The cooling cylinder is always flooded with coolant. It is assumed that the liquid level in the ammonia jacket is well controlled.
- Heat resistance at the ammonia side is negligible. The ammonia coolant is assumed to be boiling, hence a very high heat transfer rate can be achieved compared to the wall and product side.
- No axial heat conduction takes place in the tube wall. The large radial heat flux from the ice cream to the ammonia dominates the axial heat flux.
- Heat capacities of the screw, wall and shell are negligible.
- Mass of the air phase is negligible compared to the mass of the solid and liquid phases. The ice cream mix density is more than two orders of magnitude larger than the air density, hence the mass of air can be neglected.
- An ice layer in the flight clearance forms an effective seal between the channels.

The extruder process model has been divided into a series of continuous stirred tank reactors, as schematically represented in Figure 7.15.

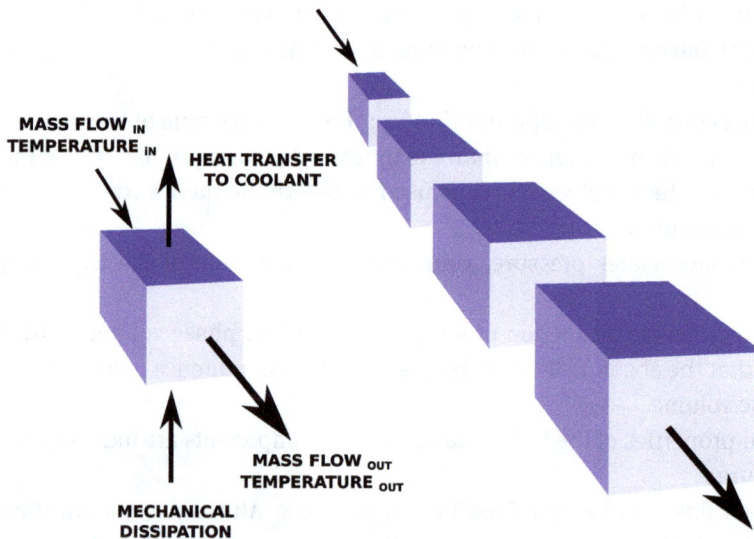

Figure 7.15: Schematic representation of the extrusion unit as a series of CSTRs models.

MASS BALANCE: the total mass balance for one section is given by the following expression:

$$\frac{dm}{dt} = w_{in} - w_{out},$$

(7.56)

where w is the mass flow, m is the mass of material and the subscripts in and out denote the input and output positions along the unit.

Using the ratio $r = \frac{m_{gas}}{m_{liquid} + m_{gas}}$, the mass balance can be written for each of the two components, *i. e.*,

$$\frac{dm_{gas}}{dt} = r_{in} w_{in} - r_{out} w_{out},$$

$$\frac{dm_{liquid}}{dt} = (1 - r_{in}) w_{in} - (1 - r_{out}) w_{out}. \tag{7.57}$$

ENERGY BALANCE: the energy balance for one section is given by the following expression:

$$\frac{dmh}{dt} = (w_{in} h_{in} - w_{out} w_{out}) + Q_{viscous} + Q_{scraping} + Q_{crystallization} + Q_{refrigeration} + E_{flow}, \tag{7.58}$$

where $\frac{dmh}{dt}$ is the total accumulation of energy, $(w_{in} h_{in} - w_{out} w_{out})$ is the change of energy by convective transport, $Q_{viscous}$ is the heat generation due to viscous dissipation, $Q_{scraping}$ is the heat generation due to scraping forces, $Q_{crystallization}$ is the heat generation due to crystallization, $Q_{refrigeration}$ is the heat removed by the refrigerant and E_{flow} is the heat generated due to pressure difference.

Viscous dissipation: the viscous power dissipation can be expressed by

$$Q_{viscous} = (\tau \dot{\gamma} V)_{channel} + (\tau \dot{\gamma} V)_{flight}, \tag{7.59}$$

where the shear rates ($\dot{\gamma}$) are defined as

$$\dot{\gamma}_{channel} = \frac{\pi D_c N_r}{H},$$

$$\dot{\gamma}_{flight} = \frac{\pi D_f N_r}{e_f}, \tag{7.60}$$

where D is the diameter, N_r is the rotational speed, H is the channel depth, e_f is the flight clearance, V is the volume, τ is the shear stress and the subscripts c and f are related to channel and flight, respectively.

Scraping energy: the scraping energy can be approximated according to the following empirical expression:

$$Q_{scraping} = c_1 s (T - T_{coolant})^{\frac{5}{3}} N_r^{1.5} L_{section}, \tag{7.61}$$

where c_1 is an experimental constant, s is the ice crystal content, $L_{section}$ is the length of section of the unit and T is the temperature.

Refrigeration heat: the heat transferred from the product into the coolant is determined in three steps. Firstly, it is assumed that the amount of heat transferred can be described by the temperature difference between the coolant and the bulk product times a heat transfer coefficient times the surface area, *i. e.*,

$$Q_{refrigeration} = \alpha_{total} A (T_{coolant} - T), \tag{7.62}$$

where α is the heat transfer coefficient and A is the contact surface.

The overall heat transfer coefficient can be then decomposed into a heat transfer coefficient between the product and the inner side of the wall, a heat transfer coefficient between the inner and outer side of the wall and a heat transfer coefficient between the outer side of the wall and the coolant, where

$$\frac{1}{\alpha_{\text{total}}} = \frac{1}{\alpha_{\text{product side}}} + \frac{1}{\alpha_{\text{coolant}}} + \frac{D_{\text{wall}}}{\lambda_{\text{wall}}}, \tag{7.63}$$

where λ is the thermal conductivity and wall is related to the wall.

The heat transfer coefficient between the coolant and the wall is very large, *e. g.*, for boiling NH_3 in an SSHE it is approximately 12,000 W/(m²K). As the thermal conductivity of the wall is constant, the focus is on the heat transfer coefficient between the inner side of the wall and the product. In an SSHE it is assumed that between two scrapings a layer of ice crystals may be formed. This layer is thin compared to the bulk, hence the Penetration Theory can be applied to calculate the heat transfer coefficient. In a single screw extruder the product is not scraped off the inner wall due to the clearance with the flights. To account for this effect we assume a stagnant layer of ice between the flights and the barrel wall. The resulting heat transfer coefficient is expressed as follows:

$$\frac{1}{\alpha_{\text{product side}}} = \frac{1}{\alpha_{\text{product}}} + \frac{e_f}{\lambda_{\text{product}}}, \tag{7.64}$$

where α_{product} can be estimated according to the expression

$$\alpha_{\text{product}} = 2\sqrt{\frac{\rho_{\text{product}} C_{p,\text{product}} N_{\text{start}} N_r D \lambda_{\text{product}}}{\pi(\frac{w_c}{\tan\theta})}}, \tag{7.65}$$

where $C_{p,\text{product}}$ is the specific heat of the product, N_{start} is the number of thread starts, w is the width and θ is the pitch angle.

Viscous flow: the generation of heat due to axial viscous flow is expressed by

$$E_{\text{flow}} = \frac{w}{\rho_m}\Delta p, \tag{7.66}$$

where ρ_m is the density and Δp is the pressure drop.

Crystallization heat: under the assumption of instantaneous crystallization, the amount of ice crystals is fully determined by the temperature inside a section. The amount of heat liberated by ice formation equals to

$$Q_{\text{crystallization}} = -H_{\text{ice}}(w_{\text{in}}\Gamma_{\text{in}} - w_{\text{out}}\Gamma_{\text{out}}) + Q_{\text{viscous}}, \tag{7.67}$$

where H_{ice} is the latent heat of the ice phase and Γ is the mechanical work.

Mechanical energy balance: comprises the energy dissipated by the product by rotating the screw and the energy used to scrape the product from the frozen layer. The rotor shaft torque is expressed by

$$2\pi N_r M_r = Q_{\text{scraping}} + Q_{\text{viscous}},$$

$$M_r = \frac{Q_{\text{scraping}} + Q_{\text{viscous}}}{2\pi N_r}, \tag{7.68}$$

where M_r is the rotor shaft torque.

The expressions developed for the mass flow and pressure drop for a non-Newtonian fluid ($n \neq 1$) are outlined in Table 7.7.

Table 7.7: Principal expressions for the flow of a non-Newtonian fluid in the extruder model. Nomenclature: K is the consistency factor and n is the power law constant.

Expression

MASS FLOW

$w_{\text{total}} = F_d \times w_{\text{drag}} + F_p \times w_{\text{pressure}}$

$w_{\text{drag}} = \frac{1}{4}\rho\pi D_c^2 Nr \cos\theta w_c \left(\frac{D_f^2 \ln(\frac{D_f}{D_c})}{D_f^2 - D_c^2} - \frac{1}{2} \right)$

$w_{\text{pressure}} = \rho \frac{nH^2 w_c}{2(2n+1)} \left(\frac{\Delta pH}{2KL} \right)^{\frac{1}{n}}$

$F_d = \frac{16w_c}{\pi^3 H} \sum_{i=1,3,5\ldots}^{\infty} \frac{1}{i^3} \tanh(\frac{i\pi H}{2w_c})$

$F_p = 1 - \frac{192H}{\pi^5 w_c} \sum_{i=1,3,5\ldots}^{\infty} \frac{1}{i^5} \tanh(\frac{i\pi H}{2H})$

PRESSURE DROP

$p_{\text{gas}} = p_{\text{normal}} \frac{\rho_{\text{gas}}(T_{\text{normal}} + T)}{p_{\text{normal}} T_{\text{normal}}}$

$\rho_{\text{gas}} = \frac{m_{\text{gas}}}{V_{\text{gas}}} = \frac{m_{\text{gas}}}{V_{\text{section}} - (\frac{m_{\text{liquid}}}{\rho_{\text{liquid}}})}$

The accuracy of the mathematical model was assessed by comparing the estimated outcome with measured data obtained from experimental trials at pilot-plant and semi-industrial scales. As depicted in Figure 7.16, the level of predictability of the model is remarkable. Not only the relative difference between measured and predicted data is within the experimental error of measurements, but the model is sufficiently robust to be used as a design and scale-up tool.

Extensional flow-based emulsification unit in the synthesis of oil-in-water HIPE (Almeida-Rivera and Bongers, 2010b, 2012)

In this example the modeling and simulation aspects of a unit characterized by the extensional flow of material are addressed. In this type of flow regime, normally referred to as shear-free flow, a preferred molecular orientation occurs in the direction of the

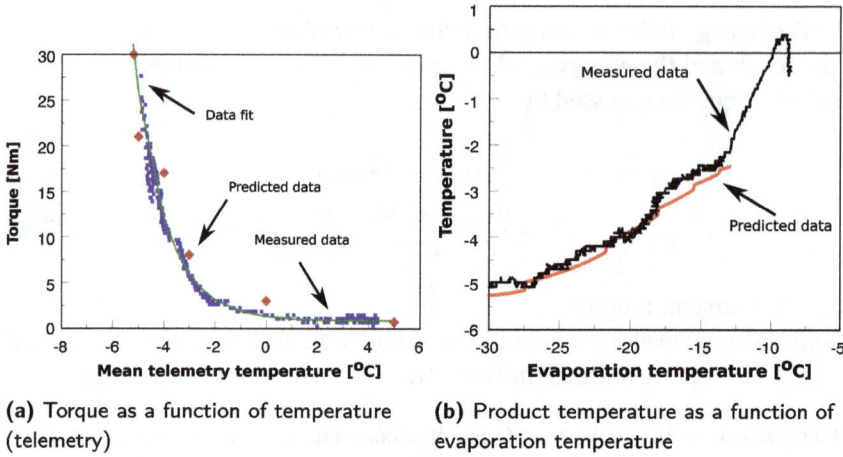

(a) Torque as a function of temperature (telemetry)

(b) Product temperature as a function of evaporation temperature

Figure 7.16: Comparison between experimental and simulated data of the extrusion model at a given rotational speed. No ice formed.

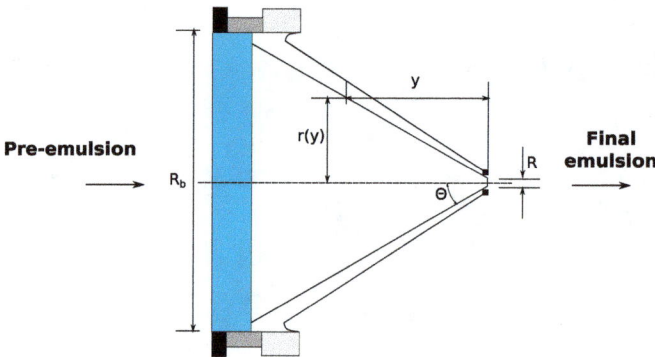

Figure 7.17: Schematic representation of a highly extensional element of the nozzle type.

flow field. From a mechanical point of view, an extensional-flow unit is materialized, for instance, in a converging element of the nozzle type (Figure 7.17).

Moreover, it is characterized by the absence of competing forces to cause rotation, resulting in a maximum stretching of molecules and large resistance to deformation (Steffe, 1996). As turbulent flow is responsible for droplet breakup in high-stress emulsification units, this flow regime is the focus. Note that from an industrial perspective, production of emulsions in the laminar regime ($\mathrm{Re}_p < 10^{-2}–10^{-6}$) (Grace, 1982) is a highly energy demanding operation for systems where the ratio of viscosities of the disperse and continuous phases, η_d/η_c, is less than 0.1 and even impossible for systems where $\eta_d/\eta_c > 4$ (Figure 7.18). Under single-shear conditions, the droplets are not able to deform as fast as the flow regime induces deformation (Walstra, 1993).

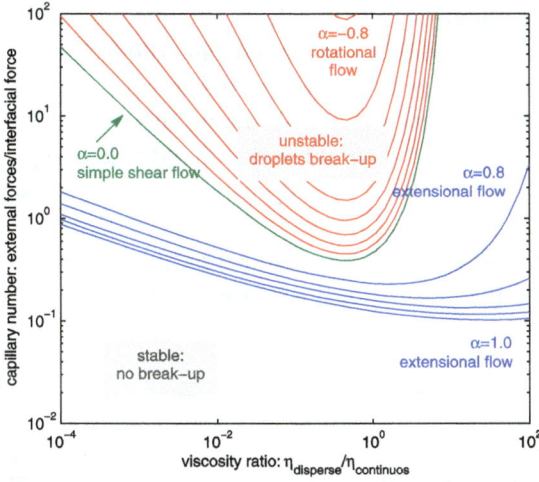

Figure 7.18: Critical capillary number for droplet breakup in single shear, extensional flow and only rotational flow; α represents the flow type. Modeling parameters are those reported in Grace (1982) and de Bruijn (1989).

The dynamic modeling of the production process of an oil-in-water structured emulsion involves the description of all first principle phenomena together with a reliable estimation of relevant physical properties.

POPULATION BALANCES: are constructed to account for the disappearance and appearance of droplets/particles at each size class in a size distribution. A relevant aspect of these balances is the estimation of the prevailing mechanism (breakup or coalescence) at each size class as a function of droplet diameter, acting forces on the droplet and time domain. Knowing the governing mechanism at a given size class allows us to calculate the rate of production and breakage of droplets. Thus, depending on the mean droplet diameter, stress exerted and residence time at a given size class, a droplet is disrupted into a given number of daughter droplets, coalesces with a colliding droplet or remains unchanged in size.

The population balance equations for a size class of mean volume v are given by the contributions of convective, birth and death terms at each location z within the unit (Nere and Ramkrishna, 2005; Raikar *et al.*, 2006; Almeida-Rivera and Bongers, 2010b, 2012), *i. e.*,

$$\frac{dn(v,z,t)}{dt} = \phi_{\text{in}}(v,z,t) - \phi_{\text{out}}(v,z,t) + \int\limits_{v'>v}^{\infty} N(v',z)S(v',z)B(v,v',z)n(v',z,t)\,dv'$$
$$- S(v,z)n(v,z,t). \tag{7.69}$$

The initial condition of the differential system is $n(v,0)$ and represents the number density of the coarse emulsion. The variable z represents all the spatial coordinates within the boundaries of the system. An analytical solution for the number-based population balance (equation (7.69)) is provided by Nere and Ramkrishna (2005). Their approach relies on simplifications of the breakage frequency and experimental correlations. In this contribution, the population balance is solved numerically by discretization of the size classes in the birth term (third term at the right-hand side of equation (7.69)). Further information on this modeling approach can be found in Raikar *et al.* (2006).

DROPLET BREAKUP: occurs when the external stress exerted on the droplet surface exceeds the form maintaining interfacial force locally for a sufficiently long time, *i. e.,*

$$\sigma(z) > P_L(z) \tag{7.70}$$

and

$$t_r(v, z) > t_b(v, z), \tag{7.71}$$

where σ is the deformation stress, P_L is the capillary pressure (or Laplace pressure), t_r is the residence time and t_b is the breakup time.

The ratio between the external forces exerted on each droplet and the counteracting interfacial force that is responsible for keeping the droplet in a spherical shape defines a key dimensionless variable in droplet breakup. This variable – termed capillary number for the viscous breakup mechanism – enables one to predict whether or not breakup will occur under the operating conditions (y-axis in Figure 7.18).

In the turbulent flow regime the breakup of droplets is caused by two size-dependent mechanisms:
-   for the inertial breakup mechanism: the size of a droplet is larger than the length of Kolmogorov energy dissipating eddies, l_k; the pressure fluctuations are much larger than the pressure due to interfacial forces; and
-   for the viscous breakup mechanism: the size of a droplet is smaller than l_k; the viscous forces are larger than the interfacial forces.

As stated in equation (7.71) for both mechanisms, the residence time of a droplet in the volume element where breakup occurs should be longer than the time required for breakup (t_b).

Thus, for the inertial breakup mechanism, the first condition is expressed as follows:

$$l_k < d_{32}(v), \tag{7.72}$$

where the radius of the droplet is expressed as Sauter mean droplet diameter d_{32},

$$d_{32} = \frac{\sum_{i=1}^{n_c} N_i d_i^3}{\sum_{i=1}^{n_c} N_i d_i^2}, \tag{7.73}$$

where N_i is the number of droplets in the size class i with mean diameter d_i.

The length of energy dissipating Kolmogorov eddies, l_k, is given by,

$$l_k = \left(\frac{\eta_e}{\rho_e}\right)^{3/4}\frac{1}{\epsilon^{1/4}},\qquad(7.74)$$

where η_e is the viscosity of the emulsion, ρ_e is the density of the emulsion and ϵ is the total specific power dissipation.

To predict when breakup occurs the Weber dimensionless number We is introduced, which relates inertial and interfacial forces. Breakup takes place when the Weber number exceeds a critical value We^{cr},

$$We = \frac{\rho_e \bar{u}^2 d_{32}}{\sigma_s} > We^{cr},\qquad(7.75)$$

where $\bar{u}$ is the averaged local fluid velocity and σ_s is the interfacial surface tension.

Based on the Weber number and the critical Weber number, the maximum droplet size that will not be broken can be estimated (Davies, 1985; Steiner *et al.*, 2006) as follows:

$$d^{max} = 0.66\left(\frac{\sigma_s^3}{\epsilon^2 \rho_e^3}\right)^{\frac{1}{5}}.\qquad(7.76)$$

The last necessary condition to guarantee breakup is the sufficiently long residence time of a droplet in the region with certain turbulent flow intensity. This condition is reflected in the following expression:

$$t_b = 0.4\pi\left(\frac{(3\rho_d + 2\rho_c)(d_{32}/2)^3}{24\sigma_s}\right)^{\frac{1}{2}},\qquad(7.77)$$

where ρ_d is the density of the disperse phase and ρ_c is the density of the continuous phase.

In the viscous mechanism, the droplets are smaller than the turbulent eddies and breakup occurs due to viscous forces in the eddies' bulk. Namely, the following condition holds,

$$l_k \geq d_{32}.\qquad(7.78)$$

Additionally, the breakup mechanism requires the viscous forces exerted on the droplet to be larger than the interfacial forces. Thus, the critical value of capillary number (Ω^{cr}) should be exceeded for the breakup to occur, *i. e.*,

$$\Omega = \frac{\eta_e \dot{\gamma} d_{32}}{\sigma_s} > \Omega^{cr}.\qquad(7.79)$$

The maximum stable droplet diameter, below which no viscous breakup takes place, in the unit can be estimated as

$$d^{max} = \frac{2\sigma_s \Omega^{cr}}{\eta_e \dot{\gamma}},\qquad(7.80)$$

where $\dot{\gamma}$ is the shear rate.

Similarly, the residence time of a droplet inside the viscous eddies should be long enough to guarantee breakup (*i. e.*, $t_b < t_r$) (Tjaberinga *et al.*, 1993),

$$t_b = \frac{\eta_c d_{32}(\lambda_\eta)^{0.315}}{2\sigma_s} t_b^*, \tag{7.81}$$

where λ_η is the ratio of viscosity of the disperse and continuous phases ($\lambda_\eta = \eta_d \times \eta_c^{-1}$) and t_b^* is the model constant.

DROPLET COALESCENCE: understanding droplet coalescence in emulsions has been a hot research topic for the last decades (Cheesters, 1991; Saboni *et al.*, 2002). Droplet coalescence occurs under the following conditions:
- droplet surfaces are not sufficiently stabilized by emulsifier agent and
- the contact time between adjacent droplets is long enough for the phase between them to be effectively drained.

According to Nienow (2004), the coalescence mechanism involves the following events: (i) two adjacent droplets collide with a frequency given by their size and energy input; (ii) according to the coalescence efficiency both droplets will either coalesce into a larger droplet or separate. The coalescence efficiency has been found to be a function of the contact time between droplets, the draining time of the phase between them and the interface characteristics between phases (Karbstein, 1994; Karbstein and Schubert, 1995). Combining the collision frequency (λ_{col}) and coalescence efficiency (φ_{col}) results in the estimation of the coalescence frequency (Γ). Thus,

$$\Gamma = \lambda_{col}\varphi_{col}. \tag{7.82}$$

The collision efficiency is a function of the interphase characteristics and might be computed, as suggested by Tjaberinga *et al.* (1993), as

$$\lambda_{col} = \exp\left(-k_{col}\frac{\eta_d}{\eta_c}\Omega^{3/2}\left(\frac{d_{32}}{2h^{cr}}\right)\right), \tag{7.83}$$

where k_{col} is an experimental constant ($= 0.43$) and h^{cr} is the critical rupture thickness.

The critical rupture thickness can be calculated considering the balance between Van der Waals forces and Laplace pressure,

$$h^{cr} = \left(\frac{Ad_{32}}{16\pi\sigma_s}\right)^{1/3}, \tag{7.84}$$

where A is the Hamaker constant ($= 1.059 \times 10^{-20}$ J for an oil-in-water emulsion).

RHEOLOGICAL MODEL EQUATIONS: are the same as reported by Almeida-Rivera and Bongers (2010a) in the modeling of continuous rotor-stator units for high internal phase emulsions.

MOMENTUM BALANCE: the macroscopic momentum balance in a single nozzle (Figure 7.17) suggests that shear stress and extensional forces contribute to the overall pressure (Steffe, 1996),

$$\Delta P_{in}(y) = \Delta P_{in,S}(y) + \Delta P_{in,E}(y). \tag{7.85}$$

For the case of the shear flow contribution, the pressure drop is given by the so-called Cogswell expression (Steffe, 1996),

$$\Delta P_{in,S}(y) = \frac{q^n}{r(y)^{3n+1}}\left(\frac{3n+1}{\pi n}\right)2K\,dy, \tag{7.86}$$

where q is the volumetric flow rate, K is the consistency factor and n is the flow-behavior index.

The nozzle axial and radial location are related by the angle of the conical section (θ),

$$\tan\theta = r(y)/y. \tag{7.87}$$

Thus, equation (7.86) results in

$$\Delta P_{in,S}(y) = \frac{q^n}{r(y)^{3n+1}}\left(\frac{3n+1}{\pi n}\right)\frac{2K}{\tan\theta}\,dr(y). \tag{7.88}$$

This expression can be further rewritten in terms of the apparent wall shear rate in the nozzle ($\dot{\gamma}_w$). We have

$$\Delta P_{in,S}(y) = \dot{\gamma}_w{}^n\left(\frac{3n+1}{4n}\right)^n\frac{2K}{3n\tan\theta}(1 - R/R_b)^{3n}, \tag{7.89}$$

with

$$\dot{\gamma}_w = \frac{4q}{\pi R^3}, \tag{7.90}$$

where R is the nozzle outlet diameter and R_b is the nozzle inlet diameter.

On the other hand, Cogswell expressions have been derived to estimate the extensional pressure drop,

$$\Delta P_{in,E}(y) = \dot{\gamma}_w{}^m\left(\frac{2K_E}{3m}\right)\left(\frac{\tan\theta}{2}\right)^m(1 - (R/R_b)^{3m}),$$
$$\eta_E(y) = K_E\dot{\varepsilon}(y)_E^{m-1}, \tag{7.91}$$

where m is the power law index for extensional stress.

MECHANICAL ENERGY BALANCE: a macroscopic energy balance in the nozzle leads to the following Bernoulli expression:

$$\frac{\bar{u}_2^2 - \bar{u}_1^2}{\hat{\alpha}} + g(z_2 - z_1) + \frac{P_2 - P_1}{\rho} + \sum F + W = 0 \tag{7.92}$$

with

$$\sum F = \sum \frac{2f\bar{u}^2 l}{d_h} + \sum k_f \frac{\bar{u}^2}{2}, \tag{7.93}$$

where 1 and 2 denote any position of the volume element, f is the friction factor and k_f is the friction coefficient for accessories.

The kinetic correction factor is dependent on the fluid properties and flow regime. Thus, for the case of HB fluids, $\hat{\alpha}$ is given by the following expressions (Steffe, 1996):

$$\hat{\alpha} = \begin{array}{ll} \exp(0.168c - 1.062nc + 0.954n^{0.5} - 0.115c^{0.5} + 0.831), & 0.006 < n < 0.38, \\ \exp(0.849c - 0.296nc + 0.6n^{0.5} - 0.602c^{0.5} + 0.733), & 0.38 < n < 1.6. \end{array} \tag{7.94}$$

The variable c relates the yield stress and shear stress, i. e.,

$$c = \frac{\sigma_0}{\sigma} = \frac{2\sigma_0}{f\rho\bar{u}^2}. \tag{7.95}$$

For the conical geometry of the nozzle, the friction coefficient for accessories is a function of the ratio of nozzle radii,

$$k_f = \left(1 - \left(\frac{R}{R_b}\right)^2\right)^2. \tag{7.96}$$

Thus, equation (7.92) results in

$$\frac{\delta P}{\rho} + \frac{\bar{u}_2^2 - \bar{u}_1^2}{\hat{\alpha}} + 2f\bar{u}^2\left(\frac{l}{d_h}\right) + \frac{1}{2}\left(1 - \left(\frac{R}{R_b}\right)^2\right)^2 \bar{u}^2 = 0. \tag{7.97}$$

The pressure drop of the nozzle can be alternatively expressed as a function of the local velocity, the ratio of diameters and the ratio of element length and characteristic diameter, i. e.,

$$\delta P = f(\bar{u}, R/R_b, l/d_h). \tag{7.98}$$

The friction factor required for the calculation of the overall pressure drop (equation (7.97)) is a function of the flow regime and properties of the fluid. Several correlations exist in the open domain to estimate the friction factor. In this example, the fluid is modeled as a Herschel–Bulkley material, according to the following set of correlations (Garcia and Steffe, 1986).

For *laminar flow*, $\text{Re} < \text{Re}^{\text{cr}}$, we have

$$f = \frac{16}{\Psi \text{Re}_{\text{HB}}}, \tag{7.99}$$

$$\text{Re}_{\text{HB}} = \frac{d^n u^{(2-n)} \rho}{8^{(n-1)} K}\left(\frac{4n}{1+3n}\right)^n, \tag{7.100}$$

$$\Psi = (1 + 3n)^n (1 - c)^{1+n} \left[\frac{(1 - c)^2}{(1 + 3n)} + \frac{2\xi_0(1 - c)}{(1 + 2n)} + \frac{c^2}{(1 + n)} \right]^n. \tag{7.101}$$

For *turbulent flow*, $\mathrm{Re} > \mathrm{Re}^{cr}$, we have

$$\frac{1}{\sqrt{f}} = 0.45 - \frac{2.75}{n} + \frac{1.97}{n} \ln(1 - c) + \frac{1.97}{n} \ln\left(\mathrm{Re_{HB}} \left(\frac{1 + 3n}{4n} \right)^n f^{1-n/2} \right), \tag{7.102}$$

where the variable c is given by equation (7.95).

The influence of turbulence on the population balance is accounted for in the calculation of the We number, while that on the momentum balance is accounted for in the estimation of the friction coefficient f.

MODEL IMPLEMENTATION AND VALIDATION: in building the integrated simulation model a number of assumptions and estimations have been made. First of all, it should be noted that the models are approximations of reality. In this respect, a number of assumptions have been introduced for the sake of model development: (i) droplets are spherical; (ii) daughter droplets are all identical in shape and size; (iii) an initial droplet size distribution is supposed to be occupying the whole nozzle at $t = 0$; (iv) the initial preemulsion is represented by a log-normal distribution of droplets; (v) droplets in the same size class are broken at the same frequency and only once; (vi) no backmixing is considered, so all droplets of the same class size have the same residence time in the unit; (vii) droplets are reduced in size exclusively by mechanical breakage; (viii) the number of daughter droplets produced by the inertial mechanism is supposed to be a function of the Weber number; (ix) the number of daughter droplets produced by the viscous mechanism is supposed to be a function of the capillary number and critical capillary number; (x) the density of the emulsion is a linear function of the oil content; (xi) the desired microstructure of the product is exclusively given by the droplet size distribution of the discrete phase in the product matrix; (xii) the emulsion-like fluid is modeled as a shear thinning substance according to the Herschel–Bulkley expression; (xiii) model parameters are constant, otherwise explicitly stated; (xiv) each oil and water mixture forms an O/W emulsion, so that there is not any limitation in the amount of emulsifier agent present in the system; (xv) the unit under consideration consists of a conical section, through which the material flows and simultaneously pressure drop and extensional stress are built up; the geometry of the unit resembles a nozzle with a converging angle close to 90° (Figure 7.17); (xvi) the unit is symmetrical with respect to the axial direction; and (xvii) for a nozzle-like unit the overall pressure drop, P_{in}, is composed of contributions originating from shear stress and a dominating extensional force flow.

The population balance expression has been solved numerically by discretization of the size classes in the birth term using a multi-step Gear method. The model has been developed to be flexible enough to accommodate fluid of different physical properties. The number of size classes is automatically estimated based on the upper and lower size classes and a given volume ratio between adjacent classes ($v_i = av_{i-1}$). The axial co-ordinate is divided into a large number of volume elements which are solved independently. The friction factor is calculated at each integration step and at each spatial location. The breakup mechanisms are selected based on the droplet size and residence time at a given size class. Population balances are performed for each size class at each spatial location and integrated in time.

Based on the behavior of the unit, we are able to determine the optimal point of operation in terms of volumetric flow and initial droplet size distribution. The criterion for choosing the optimal operating conditions is exclusively the finished product droplet size distribution. As no strong dependency is reported between d_{32} and $d_{32-\text{init}}$ we based the selection of the optimal operational conditions on the volumetric flow.

For a target d_{32} of approximately 3–3.5 μm we estimated the optimal operational conditions. These operating conditions were tested at pilot-plant scale, resulting in a remarkable agreement (deviation < 3 %) between these and the simulated data (Figure 7.19-left). The DSD found experimentally for the emulsion was slightly broader than the simulated predictions. This can be explained by the slight deviation between the premix DSDs for the experimental and simulated data sets. In fact, the simulation results were obtained under the approximation that the initial preemulsion is represented by a log-normal distribution of droplets. The unit used during the validation trials was a nozzle featuring high-shear mixer of known dimensions and operated in a continuous mode. A simplified recipe of an emulsion with high internal phase emulsion ($x_{\text{oil}} > 0.74$) and whose premix was produced in a stirred vessel equipped with a marine propeller was considered. The premix DSD was targeted in the range 70–75 μm. Both premix and product DSDs were determined using laser diffraction. An extended set of operating conditions were tested to further validate the predictability of the model. The results of such extended trials are shown in Figure 7.19-right, stressing the remarkable proximity of the predicted and experimental values.

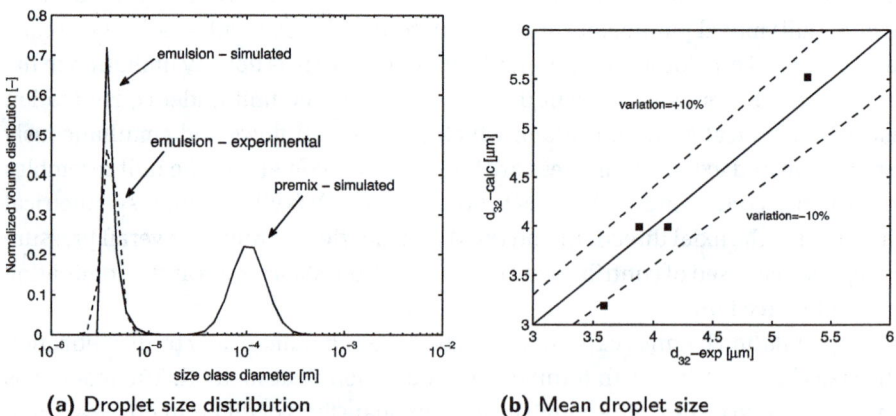

(a) Droplet size distribution

(b) Mean droplet size

Figure 7.19: Comparison between experimental and simulated data of the extensional-flow model. System features: $d_{32-\text{init}} = 70$–75 μm; $q = 1.8$–$2.0\ \text{m}^3/\text{h}$; $x_{\text{oil}} > 0.74$.

7.1.2.3 Designing process flowsheets

Flowsheeting is a key method to represent and explain processes, equipment units or steps of processes. The most simple flowsheeting representation starts with a block that shows the entering and leaving streams and interaction flows with the surroundings, with amounts and properties (Walas, 1990). Block flow diagrams (BFD) are extremely useful during the early phases of process design and particularly for orientation purposes, where preliminary decisions need to be made with limited availability of information. Material and energy flows are included together with inputs and outputs of the system and represented as a collection of rectangles and arrows. Increasing the level of complexity, flowsheeting can be carried out by means of a Process flow diagram (PFD). In addition to the material and energy balances, this representation includes information about the equipment (*e. g.*, estimates of dimensions, operation parameters) and subsystems and the flow of product(s) between them. Next to the numerical framework of the units, PFDs embody major instrumentation essential for process control and complete understanding of the flowsheet (Walas, 1990). The highest level of detail is provided by the Process and Instrumentation diagram (P&ID). Contrary to BFD and PFD, P&IDs do not show operating conditions of compositions or flow quantities, but embrace *all* equipment within the process or process step. Every mechanical aspect of the units and their interconnections are in the scope of this representation. P&IDs can be built for individual utilities to specify how the utilities are piped up to the process equipment, including the flow quantities and conditions (Walas, 1990). Generic information to be included on a process flowsheet is given in Table 7.8.

Table 7.8: Data normally included on a process flowsheet (Walas, 1990).

Nr.	Information
1.	Process lines, but including only those bypasses essential to an understanding of the process
2.	All process equipment; spares are indicated by letter symbols or notes
3.	Major instrumentation essential to process control and to understanding of the flowsheet
4.	Valves essential to an understanding of the flowsheet
5.	Design basis, including stream factor
6.	Temperatures, pressures, flow quantities
7.	Mass and/or mol balance, showing compositions, amounts and other properties of the principal streams
8.	Utility requirements summary
9.	Data included for particular equipment: (a) compressors: SCFM (60 °F, 14.7 psia); ΔP psi; HHP; number of stages; details of stages if important; (b) drives: type; connected HP; utilities such as kW, lb steam/hr, or Btu/hr; (c) drums and tanks: ID or OD, seam to seam length, important internals; (d) exchangers: Sqft, kBtu/hr, temperatures, and flow quantities in and out; shell side and tube side indicated; (e) furnaces: kBtu/hr, temperatures in and out, fuel; (f) pumps: GPM (60 °F), ΔP psi, HHP, type, drive; (g) towers: number and type of plates or height and type of packing identification of all plates at which streams enter or leave; ID or OD; seam to seam length; skirt height; (h) other equipment: sufficient data for identification of duty and size

7.2 Take-away message

This level of the PDPS approach focuses on the equipment design aspects of the overall design problem. The mechanisms selected at Level 5 are materialized (or embodied) in units. The ultimate objective of this level is the sizing of the units (key dimensions), together with the exploration of potential integration opportunities and the definition of connectivity of each unit within the final flowsheet.

Bibliography

Almeida-Rivera, C. and Bongers, P. M. M. (2009). Modelling and experimental validation of emulsification process in continuous rotor-stator units. *Computer-Aided Chemical Engineering*, 26, pp. 111–116.

Almeida-Rivera, C. and Bongers, P. (2010a). Modelling and experimental validation of emulsification processes in continuous rotor stator units. *Computers and Chemical Engineering*, 34(5), pp. 592–597.

Almeida-Rivera, C. and Bongers, P. (2010b). Modelling and simulation of extensional-flow units in emulsion formation. *Computer Aided Chemical Engineering*, 28(C), pp. 793–798.

Almeida-Rivera, C. and Bongers, P. (2012). Modelling and simulation of extensional-flow units in emulsion formation. *Computers and Chemical Engineering*, 37, pp. 33–39.

Almeida-Rivera, C., Khalloufi, S., Jansen, J. and Bongers, P. (2011). Mathematical description of mass transfer in supercritical-carbon-dioxide-drying processes. *Computer-Aided Chemical Engineering*, 29, pp. 36–40.

Almeida-Rivera, C. P., Bongers, P. and Zondervan, E. (2017). A Structured Approach for Product-Driven Process Synthesis in Foods Manufacture. In: M. Martin, M. Eden and N. Chemmangattuvalappil, eds., *Tools For Chemical Product Design: From Consumer Products to Biomedicine*. Elsevier, pp. 417–441. Chapter 15.

Almeida-Rivera, C. P., Khalloufi, S. and Bongers, P. (2010). Prediction of Supercritical Carbon Dioxide Drying of Food Products in Packed Beds. *Drying Technology*, 28(10), pp. 1157–1163.

Bindhumadhavan, G., Seville, J. P. K., Adams, M. J., Greenwood, R. W. and Fitzpatrick, S. (2005). Roll compaction of a pharmaceutical excipient: experimental validation of rolling theory for granular solids. *Chemical Engineering Science*, 60, pp. 3891–3897.

Bird, R. B., Curtiss, C. F. and Stewart, W. E. (1959). Tangential newtonian flow in annuli – II. Steady state pressure profiles. *Chemical Engineering Science*, 11, pp. 114–117.

Bongers, P. (2006). A heat transfer of a scrapped surface heat exchanger for ice cream. *Computer-Aided Chemical Engineering*, 21, pp. 539–544.

Bongers, P. (2008). Model of the Product Properties for Process Synthesis. *Computer-Aided Chemical Engineering*, 25, pp. 1–6.

Bongers, P. and Almeida-Rivera, C. (2011). Dynamic modelling of the margarine production process. *Computer-Aided Chemical Engineering*, 29, pp. 1301–1305.

Brown, Z. K., Fryer, P. J., Norton, I. T., Bakalis, S. and Bridson, R. H. (2008). Drying of foods using supercritical carbon dioxide—Investigations with carrot. *Innovative Food Science and Emergency Technologies*, 3(9), pp. 2980–2989.

Cheesters, A. K. (1991). The modelling of coalescence processes in fluid liquid dispersion – a review of current understanding. *Chemical Engineering Research and Design: Part A*, 4, pp. 259–270.

Davies, J. T. (1985). Drop Sizes in Emulsion Related to Turbulent Energy Dissipation Rates. *Chemical Engineering Science*, 40(5), pp. 839–842.

de Bruijn, R. (1989). *Deformation and breakup of drops in simple shear flow*. Ph. D. thesis, Delft University of Technology.

Fredrickson, A. and Bird, R. B. (1958). Non-Newtonian flow in annuli. *Industrial and Engineering Chemistry Research*, 50(3), pp. 347–352.

Garcia, E. J. and Steffe, J. F. (1986). Comparison of friction factor equations for non-newtonian fluids in pipe flow. *Journal of Food Process Engineering*, 9, pp. 93–120.

Grace, H. (1982). Dispersion Phenomena in High Viscosity Immiscible Fluid Systems and Application of Static Mixers as Dispersion Devices in Such Systems. *Chemical Engineering Communications*, 14, pp. 225–277.

Grossmann, I. and Westerberg, A. W. (2000). Research Challenges in Process Systems Engineering. *AIChE Journal*, 46(9), pp. 1700–1703.

Guigon, P. and Simon, O. (2003). Roll press design – influence of force feed system on compaction. *Powder Technology*, 130, pp. 41–48.

Harmsen, J., de Haan, A. B. and Swinkels, P. L. J. (2018). *Product and Process Design – Driving Innovation*. De Gruyter.

Holdsworth, S. D. (1993). Rheological models used for the prediction of the flow properties of food products: a literature review. *Transactions of the Institute of Chemical Engineers – Part C*, 71, pp. 139–179.

Hsia, M. A. and Tavlarides, L. L. (1983). Simulation and analysis of drop breakage, coalescence and micromixing in liquid–liquid stirred tanks. *Chemical Engineering Journal*, 26, pp. 189.

Johanson, J. R. (1965). A rolling theory for granular solids. *Transactions of the American Institute of Mechanical Engineers*, pp. 842–848.

Johanson, J. R. (1984). Feeding roll pressing for stable roll operations. *Bulk Sol. Handl. 4*, pp. 417–422.

Karbstein, H. (1994). *Investigation on the preparation and stabilization of oil-in-water emulsions*. Thesis/dissertation, University of Karlsruhe.

Karbstein, H. and Schubert, H. (1995). Developments in the continuous mechanical production of oil-in-water macro-emulsions. *Chemical Engineering and Processing*, 34, pp. 205–211.

Khalloufi, S., Almeida-Rivera, C. P. and Bongers, P. (2009). Prediction of supercritical-carbon-dioxide drying kinetics of foodstuffs in packed beds. In: *Proceedings of the World Congress of Chemical Engineering*, pp. 001–006.

Khalloufi, S., Almeida-Rivera, C. P. and Bongers, P. (2010). Supercritical-CO_2 drying of foodstuffs in packed beds: Experimental validation of a mathematical model and sensitive analysis. *Journal of Food Engineering*, 96(1), pp. 141–150.

Kleinebudde, P. (2004). Roll compaction/dry granulation: pharmaceutical applications. *European Journal of Pharmaceutics and Biopharmaceutics*, 58, pp. 317–326.

Krokida, M. and Maroulis, Z. B. (1997). Effect of drying method on shrinkage and porosity. *Drying Technology*, 10(15), pp. 2441–2458.

Krokida, M. and Maroulis, Z. B. (1999). Effect of microwave drying on some quality properties of dehydrated products. *Drying Technology*, 3(17), pp. 449–466.

Krynke, K. and Sek, J. P. (2004). Predicting viscosity of emulsions in the broad range of inner phase concentrations. *Colloids and Surfaces A: Physicochemical and Engineering Aspects*, 245, pp. 81–92.

Nere, N. and Ramkrishna, D. (2005). Evolution of drop size distributions in fully developed turbulent pipe flow of a liquid–liquid dispersion by breakage. *Industrial and Engineering Chemistry Research*, 44, pp. 1187–1193.

Nienow, A. (2004). Break-up, coalescence and catastrophic phase inversion in turbulent contactors. *Advances in Colloid and Interface Science*, 108, pp. 95–103.

Pal, R. (2001a). Evaluation of theoretical viscosity models for concentrated emulsions at low capillary numbers. *Chemical Engineering Journal*, 81, pp. 15–21.

Pal, R. (2001b). Single-parameter and two-parameter rheological equations of state for nondilute emulsions. *Industrial and Engineering Chemistry Research*, 40, pp. 5666–5674.

Park, J. Y. and Blair, L. M. (1975). The effect of coalescence on drop size distribution in an agitated liquid–liquid dispersion. *Chemical Engineering Journal*, 20, pp. 225.

Peter, S., Lammens, R. and Steffens, K. J. (2010). Roller compaction/dry granulation: use of the thin layer model for predicting densities and forces during roller compaction. *Powder Technology*, pp. 165–175.

Princen, H. M. and Kiss, A. D. (1989). Rheology of foams and highly concentrated emulsions. *Journal of Colloid Interface Science*, 128(1), pp. 176–187.

Rahman, M. S. (2001). Toward prediction of porosity in foods during drying: A brief review. *Drying Technology*, 1(19), pp. 1–13.

Raikar, N., Bhatia, S., Malone, M. F. and Henson, M. (2006). Self-similar inverse population balance modeling for turbulently prepared batch emulsions: Sensitivity to measurement errors. *Chemical Engineering Science*, 61, pp. 7421–7435.

Ratti, C. (1994). Shrinkage during drying of foodstuffs. *Journal of Food Engineering*, 32, pp. 91–105.

Reynolds, G., Ingale, R., Roberts, R., Kothari, S. and Gururajan, B. (2010). Practical application of roller compaction process modelling. *Computers and Chemical Engineering*, 34(7), pp. 1049–1057.

Saboni, A., Alexandrova, S., Gourdon, C. and Cheesters, A. K. (2002). Interdrop coalescence with mass transfer: comparison of the approximate drainage models with numerical results. *Chemical Engineering Journal*, 88, pp. 127–139.

Sherwood, J. D. and Durban, D. (1998). Squeeze-flow of a Herschel-Bulkey fluid. *Journal of Non-Newtonian Fluid Mechanics*, 77, pp. 115–121.

Silla, H. (2003). *Chemical Process Engineering – Design and Economics*. Marcel Dekker Inc.

Stankiewicz, A. and Moulijn, J. (2002). Process intensification. *Industrial and Engineering Chemistry Research*, 41(8), pp. 1920–1924.

Steffe, J. F. (1996). *Rheological Methods in Food Process Engineering*, 2nd ed. Michigan: Freeman Press.

Steiner, H., Teppner, R., Brenn, G., Vankova, N., Tcholakova, S. and Denkov, N. (2006). Numerical Simulation and Experimental Study of Emulsification in a Narrow-Gap Homogenizer. *Chemical Engineering Science*, 61, pp. 5841–5855.

Teng, Y., Qiu, Z. and Wen, H. (2009). Systematic approach of formulation and process development using roller compaction. *European Journal of Pharmaceutics and Biopharmaceutics*, 73, pp. 219–229.

Tjaberinga, W. J., Boon, A. and Cheesters, A. K. (1993). Model Experiments and Numerical Simualtions on Emulsification Under Turbulent Conditions. *Chemical Engineering Science*, 48(2), pp. 285–293.

Trommelen, A. (1967). Heat transfer in a scraped surface heat exchanger. *Trans. Inst. Chem. Engrs.* 45, pp. T176–T178.

Walas, S. M. (1990). *Chemical Process Equipment – Selection and Design*. Butterworth-Heinemann.

Walstra, P. (1993). Principles of Emulsion Formation. *Chemical Engineering Science*, 48(2), pp. 333–349.

Yusof, Y., Smith, A. and Briscoe, B. (2005). Roll compaction of maize powder. *Chemical Engineering Science*, 60, pp. 3919–3931.

8 Multi-product and multi-product-equipment integration

8.1 Level 6: multi-product integration

This level of the methodology is characterized by the exploration of the possibilities of combining manufacturing processes of relatively similar products (Figure 8.1). Such products, or Stock Keeping Units (SKUs), are the lowest-detailed product description and differ from each other in attribute or product properties, including size, flavor, packaging format and concentration of active ingredient. This level is of relevance only if the design problem involves the manufacture of a set of products with different attributes or properties, which implies the requirement of additional fundamental tasks. At this level the outcomes of Levels 3 (input–output level) to 5 (mechanism and operational window) for each different product or SKU are compared and possibilities to combine production are explored.

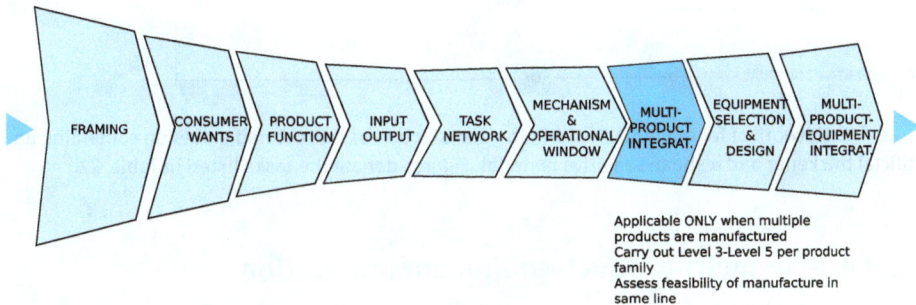

Figure 8.1: Multi-product integration as part of the PDPS strategy.

During the synthesis of multi-product processes specific requirements are identified for each type (or family) of products. These requirements might involve the introduction or removal of tasks from a generic synthesis process. The underlying principle behind this level aims at assuring that the process is designed to be robust enough to accommodate for and deliver each type (or family) of products. A set of task-based designs is created separately for each type (or family) of products following the sequence of levels of the approach. Upon addressing all products of relevance to the design problem, the corresponding processes are integrated within the generic processes. By developing a compatibility matrix of products, the practitioner explores the opportunity of combining the manufacture of multiple products in the same process line. As an example of the scope of this level, let us consider the manufacture of a structured emulsion comprising a shear-sensitive material. The baseline task network (*i. e.,* the one without the shear-sensitive material) is depicted in Figure 6.4. Expanding the

https://doi.org/10.1515/9783110570137-008

design scope implies the addition of specific tasks, which include the conditioning of the shear-sensitive material and its incorporation into the main stream. The resulting task network is included in Figure 8.2.

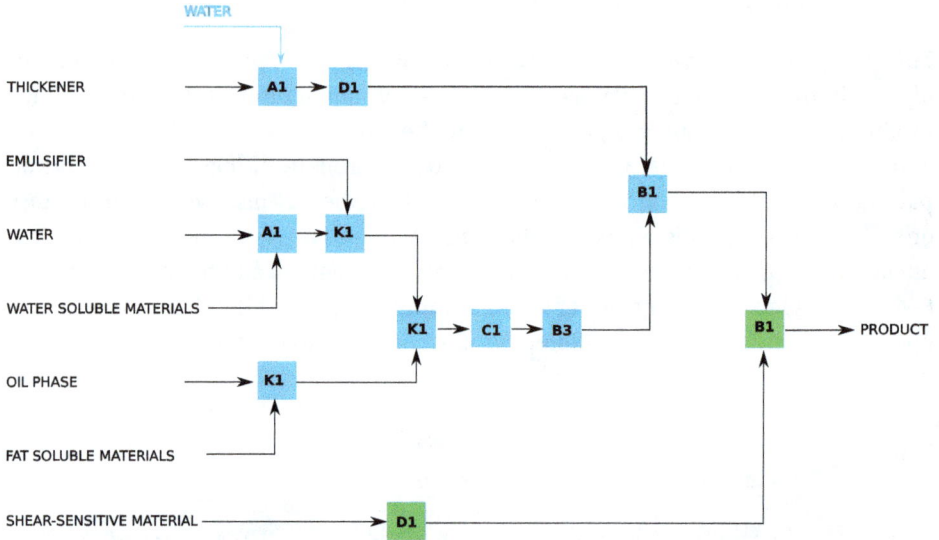

Figure 8.2: Network of fundamental tasks in the manufacture of a structured emulsion containing an artificial thickener and a shear-sensitive material. Figures denote the tasks listed in Table 6.6.

8.2 Level 8: multi-product-equipment integration

This last level of the PDPS approach (Figure 8.3) expands the scope of the approach towards plant-wide operations, as depicted in Figure 2.36. It focuses on the optimization of the interaction of the various unit operations in the flowsheet (plant-wide control) and the production planning. To support this activity, multi-stage scheduling of the multiple products is applied, fed by the run-strategy based on the product demand and portfolio defined at Level 0.

This level comprises the definition of realistic disturbance scenario and the design of the control structure and control strategy to mitigate the considered disturbances. A multi-stage scheduling is generated and for each control structure and strategy a controllability performance index is determined. The proposed multi-stage scheduling is preferably validated in a live environment. While Supply Chain Modeling and process control are considered in detail in Chapters 20 and 15, the challenge at this level of the approach is given by the realization of the trade-off between marketing and manufacturing functions. In the context of Supply Chain Management (SCM), the nature of the relationship between both functions is a key driver for the growth opportunities of a corporation (Section 8.2.1).

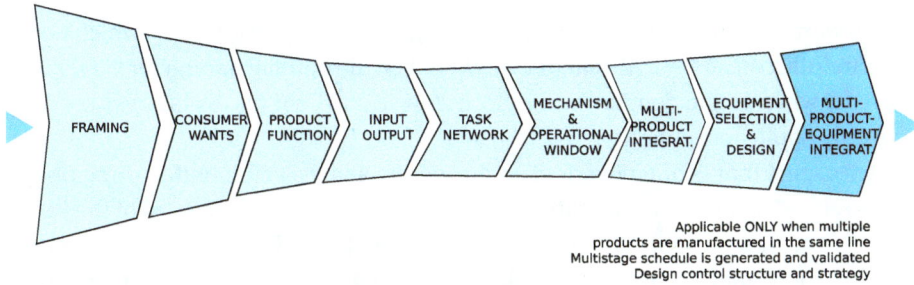

Figure 8.3: Multi-product-equipment integration as part of the PDPS strategy.

As mentioned in Chapter 20, while SCM (Supply Chain Management) and EWO (Enterprise-wide Optimization) are often considered interchangeable terms, they do not address the same problems. SCM focuses on a broader set of real-world applications with an emphasis on logistics and distribution, involving usually linear models. EWO, on the other hand, is concerned with manufacturing facilities with a strong focus on their planning, scheduling and control, involving the use of non-linear process models.

Within supply chain, procurement, manufacture, distribution and sales (or marketing) functions (or entities) in an organization have operated traditionally in an independent fashion. The main rationale behind this mode of operation might be explained by the different and often conflicting objectives of each function. For example, a manufacturing function is designed to maximize throughput with little focus on the impact on the inventory level or distribution capabilities.

The dynamics of the FMCG sector require the manufacturer to have their products or services available when the consumer wants to buy them (on-shelf availability) and to have product propositions affordable for the consumers (value for money), while maintaining an innovative product portfolio (Bongers and Almeida-Rivera, 2011). On-shelf availability requires a large inventory of products, which is penalized by substantial costs. At large product inventories, novel product innovations are costly due to a write-off of obsolete products, or due to the long lead time before the old inventory is sold. On the other hand, manufacturing sites need to be efficient, which is achieved by manufacturing the products continuously for longer time (*i. e.*, long run length) and keeping the number of changeovers from one product to the next to the minimum. This apparent conflict of objectives (manufacturing: long run length and high inventories; marketing: low inventories and short run lengths) is captured in the Economic Batch Quantity (EBQ) parameter, which is commonly used to determine the minimal production quantity of a product or SKU (Grubbstrom, 1995). It is calculated as

$$\text{EBQ} = \sqrt{\frac{2 \times \text{average weekly demand} \times \text{changeover cost}}{\text{storage cost}}}. \tag{8.1}$$

It is worth mentioning that the EBQ is a compromise-driven quantity, which needs to be carefully considered. Although it might not be the optimal parameter to rely on – especially when actual demand varies from 25 % to 300 % of the average value – EBQ is the only applicable quantity.

One aspect that is often overlooked during the analysis of manufacturing operations is related to the local constraints at the manufacturing sites. In most of the FMCG factories, the final product is made from intermediate product(s). The final product manufacture is often continuous, whilst the intermediate products are made batch-wise. Thus, a minimum batch size is needed when manufacturing the product. Batch vessels have a limited capacity, as well as a minimum quantity that should be inside the vessels to ensure proper processing (*e. g.*, mixing, heating or cooling). Excess intermediate products go to waste, resulting in increased manufacturing costs. Bongers and Almeida-Rivera (2011) formulate a MILP model to analyze the trade-off between short run length and frequent changeovers and fixed batch sizes in the factory, inventory level, on-shelf availability and overall costs.

Supply Chain Management (SCM) can be regarded as a process to reach a level of integration within the organizations of the value chain.

8.2.1 Supply Chain Management

SCM is a relatively new concept in the business and industrial worlds and has grown in popularity over time, especially among industrial companies within the FMCG sector. In simple terms, the ultimate goal of SCM is customer satisfaction by having the required product in the right place at the right time and with the lowest possible price. As mentioned by Shah (2005), SCM comprises a series of logistic internal and external activities between companies to increase the efficiency by sharing information between entities to gain customers' satisfaction at the lowest possible cost. It includes all movement and storage of raw materials, work-in-process inventory and finished goods from point-of-origin to point-of-consumption. If the activities within an organization are mapped onto the generic value creation model SOURCE ▷ MAKE ▷ DISTRIBUTE ▷ SELL and projected onto a temporal layering framework (STRATEGIC ▷ TACTICAL ▷ OPERATIONAL), a comprehensive overview of the internal and external SCM activities can be depicted (Figure 8.4). The temporal layers represent the different time horizons for decision making.

In this model, source can be defined as the processes to procure goods and services to meet planned or actual demand; make is concerned with the processes that transform goods into a finished state to meet planned or actual demand; deliver focuses on the processes that provide finished goods and services to meet planned or actual demand; and sell looks after the processes that transfer the goods or services to the customers that have the demand. The description of the key elements of the supply chain matrix are depicted in Figure 8.4 and the relevant terms of supply chain value creation are listed in Table 8.1.

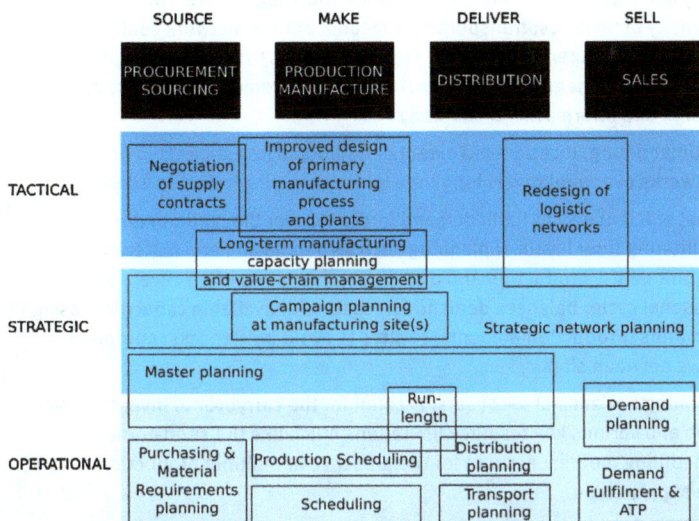

Figure 8.4: Scope of supply chain in manufacturing industry: SOURCE ▷ MAKE ▷ DISTRIBUTE ▷ SELL at several temporal layers.

To support the optimization of all the steps from supplier to consumers, SCM comprises the development of modeling tools and methods in two specific areas: (i) planning and scheduling and (ii) managerial issues.

Planning and scheduling has been established in order to find the right quantity and quality of the product. Planning and scheduling is the decision making process and is used on a regular basis by the process industries. Planning and scheduling designs the plants, allocates the resources over time and decides on the quality and quantity of the product as the customer is expecting it. In order to make these decisions, a time-based model has to be built. At the beginning all the decisions are made based on forecasting, but over time new information arrives and the model becomes more accurate. Novel methods and approaches have been recently developed to address the complexity and uniqueness of FMCGs (van Elzakker *et al.*, 2012, 2014b, 2017). Specific features included in the formulation of the planning problem are shelf-life constraints (van Elzakker *et al.*, 2013b, 2014a), demand uncertainty (van Elzakker *et al.*, 2013a), intermediate storage (van Elzakker *et al.*, 2011; Bongers and Almeida-Rivera, 2011) and seasonality of raw materials (Mehdizadeh *et al.*, 2011; Bongers and Almeida-Rivera, 2011).

Managerial issues can be solved by applying mathematical programming. However, there is a number of problems, such as location/allocation, inventory control,

Table 8.1: Relevant terms and definitions in supply chain value creation activities.

Activity	Scope
Strategic network planning	location of production sites, warehouses and geographical customer areas; capacity of facilities; transportation means; optimal network configuration, inventory management policies, supply contracts, distribution strategies, supply chain integration, outsourcing and procurement strategies, product design and information technology
Strategic or long-term planning	optimal timing, location and extent of additional investments in processing networks over a relatively long time horizon ranging from 5 to 10 years
Master planning	most efficient way to fulfill demand forecasts over the medium term, facilitating finer levels of planning such as purchasing and material requirements, production and distribution planning; often covers a full seasonal cycle; balances demand forecasts with available capacities; assigns demands (production amounts) to sites to avoid bottlenecks; coordinates flows between sites
Midterm tactical models	planning horizons of 1–2 years; account for the carryover of inventory over time and various key resource limitations much like the short-term scheduling models; account for the presence of multiple production sites in the supply chain
Short-term (operational) models	characterized by short time frames, such as 1–2 weeks, over which they address the exact sequencing of the manufacturing tasks while accounting for the various resource and time constraints
Operation models	include planning, scheduling, real-time optimization and inventory control

production planning, transportation mode selection and supplier selection, that are not solved by traditional analytical tools. Thus, Supply Chain Modeling is a very relevant concept to address.

8.2.2 Supply Chain Modeling

In the context of SCM, the information flow and accuracy among all entities of the supply chain are crucial for a successful integration. Supply Chain Modeling is concerned with capturing this information, which is shared in the whole supply chain, by going through all the nodes (from source to sell) and finding the optimal solution for each step. The first step of building a model is to define the scope of the supply chain model. According to Chopra and Meindl (2002), the scope can be defined by addressing three decision making levels: (i) competitive strategy, (ii) tactical plans and (iii) operational routines.

The competitive strategy is normally encountered in problems such as location-allocation decisions, demand planning, new product development, information technology selection and pricing.

Tactical plans are encountered in problems such as production/distribution coordination, material handling, inventory control and equipment selection.

Operational routines are encountered in problems such as vehicle routing/scheduling, workforce scheduling, record keeping and packaging.

Next to the scope definition, supply chain models require specific goals, which can be expressed in terms of four classes: customer service, monetary value, information/knowledge transactions and risk elements. These factors define the model driving force. Next to the scope and goal, constraints need to be formulated, which normally originate from three classes: capacity, service compliance and the extent of demand (Min and Zhou, 2002). Decision variables are crucial elements of the model and include, among others, production/distribution scheduling, inventory level, number of stages (echelons), the distribution center, plant-product assignments, buyer-supplier relationships, product differentiation step specification and number of product types held in inventory (Beamon, 1998). The model requires measures to assess its performance. Performance measures determine not only the effectiveness and efficiency of the model, but also the values of the decision variables of the system to yield the most desired level of performance. Performance measures are categorized as quantitative or qualitative. Quantitative performance measures are either measures based on costs (cost minimization, sales maximization, profit maximization, inventory investment minimization, return on investment maximization) or measures based on customer responsiveness (fill-rate maximization, product lateness minimization, customer response time minimization, lead time minimization, function duplication minimization). A relevant performance indicator of how well a manufacturing operation is utilized compared to its full potential at operational level is the Overall Equipment Effectiveness (OEE). The associated terms for its estimation are listed in Table 8.2. On the other hand, qualitative performance measures are customer satisfaction, flexibility, information and material flow integration, effective risk management, supplier performance, etc. (Beamon, 1998).

To create a corporation-wide model, integration of the information and the decision making among the various entities is essential. To fully realize the potential of transactional IT tools, the development of sophisticated deterministic and stochastic linear/non-linear optimization models and algorithms is needed. These models allow for the exploration and analysis of alternatives of the supply chain to yield overall optimum economic performance and high levels of customer satisfaction. Several challenges have been identified in the development of an integrated corporation-wide supply chain model. As mentioned by Grossmann (2005), these challenges include (i) functional integration (Shapiro, 2004), (ii) geographical integration and (iii) intertemporal integration.

Without good cross-functional integration, supply chain planning will not only suffer from poor coordination, but may also create large demand/supply mismatches and their attendant inventory holding, production and opportunity costs (Silver *et al.*,

Table 8.2: Relevant terms and definitions in supply chain for the estimation of OEE.

Term	Definition
OEE	Overall Equipment Effectiveness
	availability × performance × quality
	$\dfrac{\text{loading time} - \text{downtime}}{\text{loading time}} \times \dfrac{\text{standard cycle time} \times \text{units processed}}{\text{operating time}} \times \dfrac{\text{units processed} - \text{defect units}}{\text{units processed}}$
Maximum capacity	line nominal speed × OEE × available time
MCU	Maximum Capacity Utilization; measure of how much of capacity was utilized over the available time
	$\dfrac{\text{loading time}}{\text{available time}} = \dfrac{\text{production volume}}{\text{maximum capacity}}$
Maximum time	maximum time within a reporting period
Unavailable time	time when the factory cannot operate due to imposed national or local legal restrictions and time for planned annual maintenance and shut-downs
Available time	maximum time when the line is available for production and related downtime activities
Loading time	time during which the line could be occupied with some type of activity (production or downtime), whether or not the machine is actually producing
Operating time	time during which the line is planned to run for production purposes.
Net operating time	maximum time during which the line could be expected to be operated productively.
Value operating time	theoretical minimum hours required to produce a given production.
Shut-down losses	time lost due to legal restrictions, public holidays, annual maintenance and modifications planned in advance; time during which the line is physically capable to produce but no production is occurring
Downtime losses	time lost due to equipment breakdowns, changeovers, start-up, ramp-down, shut-down; losses due to inefficient layout, periodic maintenance, waiting time losses generated by management problems
Performance losses	time lost due to minor stoppages (less than 10 minutes), speed loss, loss due to shortage of operators on the line, inefficient delivery of materials to the lines and removal of finished products from the lines
Defect losses	time lost due to product quality defects, cleaning routine, machine adjustment to prevent the recurrence of quality problems
Standard cycle time	time based on the design speed (or maximum speed), considering the bottleneck in the line.

1998). Inefficient integration might originate from the limited cross-functional inter-action in a corporation. In fact, functional excellence without cross-functional integration is often a waste of resources.

According to Oliva and Watson (2006), achieving alignment in the execution of plans between different functions, geographical locations or different time scales within a company can be more important than the quality of the information and the quality of the inferences made from available information.

8.3 Take-away message

Levels 6 and 8 of the PDPS approach focus on the multi-product and multi-product-equipment integration, respectively. Level 6 explores the possibility of combining manufacturing processes of several types (or families) of products. The requirement of each type (or family) of products is considered as a potential extension of the current process structure. Level 8 is concerned with the plant-wide operation of the process, where the demand of a portfolio of products needs to be satisfied. Both Level 6 and Level 8 of the approach are exclusively of relevance in the cases where multi-products and multi-stage scheduling are included in the scope of the design problem. Plant-wide control and multi-stage scheduling are described in detail in Chapters 15 and 20.

Exercises
- Enumerate the conflicting interests that could arise from manufacturing and marketing functions when it comes to run length and stock levels.
- The Economic Batch Quantity (EBQ) is a key variable used to compromise the conflicting requirements of manufacture and marketing functions. Derive this expression analytically, listing the assumptions from which it is derived.

Bibliography

Beamon, B. M. (1998). Supply chain design and analysis: models and methods. *International Journal of Production Economics*, 55, pp. 281–294.

Bongers, P. and Almeida-Rivera, C. (2011). Optimal run length in factory operations to reduce overall costs. *Computer-Aided Chemical Engineering*, 29, pp. 900–904.

Chopra, S. and Meindl, P. (2002). Supply chain management: strategy, planning and operation. *IIE Transactions*, 2, pp. 221–222.

Grossmann, I. (2005). Enterprise wide Optimization: A New Frontier in Process Systems Engineering. *AIChE Journal*, 51(7), pp. 1846–1857.

Grubbstrom, R. (1995). Modelling production opportunities – a historical overview. *International Journal of Production Economics*, 41, pp. 1–14.

Mehdizadeh, A., Shah, N., Bongers, P. M. M. and Almeida-Rivera, C. P. (2011). Complex Network Optimization in FMCG. *Computer-Aided Chemical Engineering*, 29, pp. 890–894.

Min, H. and Zhou, G. (2002). Supply chain modelling: past, present and future. *Computers and Industrial Engineering*, 43, pp. 231–249.

Oliva, R. and Watson, N. H. (2006). Cross Functional Alignment in Supply Chain Planning: A Case Study of Sales & Operations Planning. *Harvard Business Review* No. 07-001.

Shah, N. (2005). Process industry supply chains: Advances and challenges. *Computers and Chemical Engineering*, 29, pp. 1225–1235.

Shapiro, J. F. (2004). Challenges of strategic supply chain planning and modeling. *Computers and Chemical Engineering*, 28, pp. 855–861.

Silver, E. A., Pyke, D. F. and Peterson, R. (1998). *Inventory Management and Production Planning and Scheduling*. John Wiley and Sons.

van Elzakker, M., La-Marie, B., Zondervan, E. and Raikar, N. (2013a). Tactical planning for the fast moving consumer goods industry under demand uncertainty. In: *Proceedings of ECCE9*.

van Elzakker, M., Maia, L., Grossmann, I. and Zondervan, E. (2017). Optimizing environmental and economic impacts in supply chains in the FMCG industry. *Sustainable Production and Consumption*, 11, pp. 68–79.

van Elzakker, M., Zondervan, E., Almeida-Rivera, C., Grossmann, I. and Bongers, P. (2011). Ice Cream Scheduling: Modeling the Intermediate Storage. *Computer-Aided Chemical Engineering*, 29, pp. 915–919.

van Elzakker, M., Zondervan, E., Raikar, N., Grossmann, I. and Bongers, P. (2012). Tactical planning in the fast moving consumer goods industries: an SKU decomposition algorithm. In: *Proceedings of FOCAPO 2012*.

van Elzakker, M., Zondervan, E., Raikar, N., Hoogland, H. and Grossmann, I. (2013b). Tactical Planning with Shelf-Life Constraints in the FMCG Industry. *Computer-Aided Chemical Engineering*, 32, pp. 517–522.

van Elzakker, M., Zondervan, E., Raikar, N., Hoogland, H. and Grossmann, I. (2014a). Optimizing the tactical planning in the Fast Moving Consumer Goods industry considering shelf-life restrictions. *Computers and Chemical Engineering*, 66(4), pp. 98–109.

van Elzakker, M., Zondervan, E., Raikar, N., Hoogland, H. and Grossmann, I. (2014b). An SKU decomposition algorithm for the tactical planning in the FMCG industry. *Computers and Chemical Engineering*, 62(5), pp. 80–95.

9 Molecular product design

9.1 Introduction and motivation

One of the key tasks in process design is the selection of chemicals used within a given production process: solvents, heat transfer media, diluents, reaction media and fuels for boilers, furnaces and engines. Chemical engineers also are involved in the design of final products, especially those products which are mixtures. Thus, a large fraction of the research currently performed today in the chemical industries is devoted to product design, that is, the search for new materials with specifically tailored properties. High-performance polymers for aircraft applications, new environmentally benign fuel additives and more efficient refrigerants are examples of novel products which are in demand and currently being investigated. Due to the tedious trial-and-error approach employed in most cases and the high cost of experimentation, most new designs for such molecules tend to have similar structures to those previously used. In 2001, the American Chemical Society convened a panel of industry and government scientists, who set as a goal a 15 % increase in the number of new chemical products and applications which are developed annually (ACS, 2001). It is clear that the chemical industry requires new materials which will provide the capability to produce chemicals more efficiently, and with less impact on the environment. In order to reach this goal, the panelists specifically mention the use of combinatorial techniques as "enabling research tools" to provide a more efficient search technique than the traditional trial-and-error approach currently employed in most cases. This use of molecular systems engineering to improve the product design process holds great promise in terms of finding truly novel chemical products in an efficient manner.

In order to accelerate the molecular discovery and development process, computational approaches are needed which can help guide synthesis and testing procedures. This has been noted by a number of major chemical and pharmaceutical concerns, many of which are now employing Computer-Aided Molecular Design (CAMD) techniques to develop new solvents, medicines and refrigerants (Gane and Dean, 2000; Sahinidis *et al.*, 2003). The current methodologies in CAMD are starting to bear fruit, but the rate of discovery is still lower than is desirable. Two major challenges arise in the development of such a computational molecular design procedure: the ability to predict the physical and chemical properties of a given molecule and the ability to solve the large optimization problem which is derived from the search for the best molecule for a given application. Advances in both deterministic and stochastic optimization algorithms, along with the ever-increasing computational power of modern parallel computers, provide an opportunity to attack complex molecular design problems effectively (Achenie *et al.*, 2002). Furthermore, research in molecular characterization has reached a point at which many physical and chemical properties can be predicted to a reasonable accuracy for common molecules.

https://doi.org/10.1515/9783110570137-009

9.2 CAMD methodology

Figure 9.1 provides an overview of the methodology for CAMD.

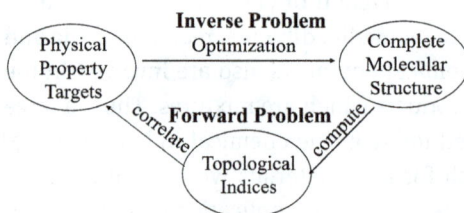

Figure 9.1: Overall methodology for computational molecular design.

In this figure, the solution to the molecular design problem is shown at the upper left, namely, the chemical structure and formulation conditions for a new candidate molecular product. The top section of this figure is known as the *forward problem*, which involves the prediction of the macroscopic properties of a molecule based on molecular properties, namely, the atoms and bonds in their specific topological arrangement. These macroscopic properties can include anything from simple properties such as viscosity or density to highly complex properties such as catalytic activity, toxicity and long-term resistance to wearing. Except for the simplest of properties, accurate correlations cannot be directly made between properties and the atoms and bonds in a molecule. Numerical descriptors which quantify important aspects of a molecular structure, such as the hydrophobic surface area of a protein or the three-dimensional shape of a ligand, are fundamental properties of a molecule, and as a whole do correlate with macroscopic properties, especially when combined with formulation conditions for complex molecular systems. In any CAMD problem formulation, a methodology must be developed to predict the critical properties of the candidate molecules. These methods must be fast enough to allow for a reasonable number of candidates to be sampled, yet accurate enough such that this type of screening algorithm does not generate too many false positives.

In molecular design, the main goal is to solve the *inverse problem*, which means that critical macroscopic properties are provided as targets, and the molecular structure and formulation conditions are the unknowns to be solved for (Venkatasubramanian *et al.*, 1994). Optimization techniques are used to solve this problem, as is shown in the bottom section of Figure 9.1. The techniques used range from enumeration approaches where molecules are sequentially culled from all those which are initially part of the search space (Harper and Gani, 2000), to stochastic algorithms which perform directed, randomized searches though the molecular space in search of locally optimal candidates (Venkatasubramanian *et al.*, 1994), to deterministic algorithms which seek to find the globally optimal solution to the optimization problem by successive bound improvements (Sahinidis *et al.*, 2003).

The optimization problem to be solved consists of an objective function, which describes the variance of a candidate molecule's property values from prespecified targets, along with a number of constraints, which ensure that a stable, connected molecule is formed. In many cases, the optimization method is designed to start with chemically feasible molecules, and only makes moves through the search space which cause the candidate solutions to remain feasible, thus avoiding the need for explicit enforcement of structural constraints (Lin *et al.*, 2005). Since the number of atoms of a given type in a molecule and a molecule's topology are not continuous variables, the final optimization problem must include both discrete and continuous variables. Furthermore, it is often found that non-linear correlations provide better predictive power for the properties of interest when compared to linear ones. The resulting CAMD optimization problems are therefore often mixed-integer non-linear programs (MINLPs). While these types of problems are known to be NP-complete and thus computationally very challenging, research since the 1970s has shown that many instances of MINLP problems are tractable in a reasonable time frame, particularly when the globally optimal solution is not required. In CAMD problems, inaccuracies in property prediction techniques combined with a fairly large array of possible molecules mean that for many cases, locally optimal solutions provide candidate molecules which are equally likely to be practically effective. Thus methods which cannot find the global optimum with mathematical certainty, but consider more feasible solutions in the same computational time and have other features such as parallelizablility, are often employed to solve CAMD optimization problems.

9.3 Quantitative structure-property relationships

All CAMD methodologies require the generation of a predictive model to provide estimated property values for candidate molecules. A common way to quantify chemical structures is to use group contribution methods, in which the number of each type of functional group or larger chemical unit in a given molecule describes the chemical structure. UNIQUAC and UNIFAC are group contribution methods that are used to predict thermodynamic properties in liquid mixtures (Fredenslund *et al.*, 1975), and are used in nearly every simulation program for determination of physical properties of compounds not in the program's database. Within CAMD problems, the group contribution approach has been widely used (Gani *et al.*, 1989; Venkatasubramanian *et al.*, 1994; Harper and Gani, 2000).

Because most group contribution methods cannot adequately account for steric effects (Wang and Milne, 1994) and cannot distinguish structural isomers, several researchers have begun using topological descriptors involving more structural information to take larger-scale structures into account when predicting physical and chemical properties. Connectivity indices, first developed by Randić (1975) and implemented by Trinajstic (1975), are numerical topological descriptors which are easy to compute and contain a great deal of information about bonding structure, local steric environments and electron density. These indices have been used extensively by Bicerano (2002) for prediction of the properties of straight-chained, non-cross-linked, limited-molecular weight polymers. Raman and Maranas (1998) first employed these indices within an optimization framework, and Camarda and Maranas (1999) used connectivity indices to design polymers which prespecified values of specific properties. Similar structural descriptors have been considered by Kier and Hall (1986) to design small drug molecules. The property predictions generated from these indices are more accurate than those from group contributions, and furthermore, when a molecular design problem is solved using these indices, a complete molecular structure results, and no secondary problem has to be solved to recover the final molecular structure (Siddhaye *et al.*, 2000; Chavali *et al.*, 2004).

More recently, approaches have focused on the use of quantum-based calculations, especially for solvation properties of mixtures. Klamt (2011) reviews the use of COSMO-RS for property predictions, and gives some ideas as to the applicability of the method within a molecular design framework. A simpler and more practical approach to the prediction of solvation properties is PC_SAFT (Burger *et al.*, 2015; Lampe *et al.*, 2014), which treats the molecules as a flexible chain, and is now being implemented within the process simulation software gPROMS (Process Systems Enterprise, 1997–2015) for the prediction of both pure-component and mixture properties.

To create a predictive model, numerical descriptors of chemical structure are related to physical or chemical properties of the molecular system of interest. While phenomenological models are more accurate in terms of property prediction, QSPRs based on structural descriptors allow rapid estimation of properties directly from the chemical structure, and are often straightforward to include within an optimization framework for CAMD. Most QSPR are created using linear regression, but many optimization algorithms are now available which can handle non-linear QSPR expressions. Partial Least Squares regression (PLS) is another option for the generation of QSPR models, and is an effective method when a large number of descriptors are to be used (Oprea, 2005).

There are several methods which may be used to determine the best number of descriptors to use in a model. Measures which weigh the quality of fit against the degrees of freedom, such as the adjusted correlation coefficient, are often employed; however, these values tend to overestimate the number of descriptors that are useful. It is also important to consider the statistical significance of the model and of the parameters. The most effective means of determining model size is to use cross-validation, in which part of the experimental data is left out when making the model (Wasserman, 2004). The error is then measured for the predicted values of data not used to create the model. If too many descriptors are used, large errors occur in the predictions.

9.4 Optimization formulations for CAMD

An explicit, general formulation of a CAMD optimization problem may be written as

$$\min s = \sum_m \frac{1}{P_m^{\text{scale}}} |P_m - P_m^{\text{target}}|,$$

$$P_m = f_m(y),$$

$$y = g(a_{ijk}, w_i),$$

$$h_c(a_{ijk}, w_i) \geq 0, \tag{9.1}$$

where P_m is the value of the mth physical or chemical property, P_m^{target} is a preselected target value for the mth property, P_m^{scale} is a scaling factor for the mth property, $f_m(y)$ are the m QSPR relations which are functions of the structural descriptors y and g_n are the defining equations for the structural descriptors. These descriptors are computed via a data structure a_{ijk} which is used to store bonding and w_i which stores the identity of each group i in the molecule; $h_c(a_{ijk}, w_i)$ are structural constraints used to ensure that a stable, connected molecule is designed. Specific solution algorithms may ensure structural feasibility in all cases (by enforcing rules on generation of a new candidate, for example), and thus these constraints may not be written explicitly. In the general case, the CAMD problem is an MINLP, since the property prediction equations or the structural constraints may be non-linear. The solution algorithm chosen is thus dependent on the QSPR model. As is always the case in the solution of NP-hard problems, the structure of specific instances of the problem must be exploited, whether an enumerative, deterministic or stochastic method is used. In fact, the formulation of the problem is normally quite specific to the solver which is to be employed. The further subsections of this section focus on reformulations of the general CAMD problem for specific problem instances and solution techniques.

One of the earliest attempts to solve a molecular design problem was the work of Joback (1987), who employed an enumerative approach with a group contribution-based QSPR model to find novel refrigerants. This work was further developed by Gani and co-workers, and is perhaps best described in Harper and Gani (2000), where a generate-and-test approach for molecular design is included with a predesign phase for property target identification and a verification phase for result analysis. The QSPR model varies with the target properties desired, and group contribution algorithms such as UNIFAC are employed. The building blocks (molecular fragments) to be combined are selected based on the range of applicability of the QSPR model. A number of other CAMD formulations for enumeration-based design methods are provided in Achenie *et al.* (2002).

Deterministic algorithms, which bound the optimal solution from above and below and then refine those bounds to guarantee a globally optimal solution within a finite number of steps, have been used for CAMD optimization problems when the mathematical structure is known to be tractable. When all of the property prediction expressions can be formulated as linear equations; for example, the general CAMD optimization problem shown in equation (9.1) reduces to a mixed-integer linear program (MILP) (Siddhaye *et al.*, 2000, 2004), which can then be solved to optimality in order to find the globally optimal candidate molecule. In Camarda and Maranas (1999), a connectivity index-based formulation was transformed to an MILP and applied to polymer design, while Adjiman *et al.* (2000) applied specific improvements to the MINLP solver αBB to solve a specific instance of a CAMD problem. Sahinidis *et al.* (2003) solved an MINLP formulation directly via a branch-and-reduce algorithm implemented in the solver BARON. In general, global optimization algorithms are specifically designed for a particular MINLP structure, and thus the CAMD problem must be reformulated for the given algorithm, unless a novel algorithm is to be designed.

When a stochastic algorithm is to be applied to solve a CAMD optimization problem formulation, requirements on linearity, continuity and even direct access to the QSPR model need not be enforced. Such algorithms start with a set of initial feasible solutions (molecular candidates), and improve these solutions by making directed random moves through the molecular search space. A good example of this is the polymer design work of Venkatasubramanian *et al.* (1994), in which a genetic algorithm was applied to the generation of molecular candidates. Since no rigorous bound calculations are made, no guarantees exist as to the optimality of the solutions; however, the examples shown in that work validate that the genetic algorithm is able to recover molecules for which known properties were input. Another example may be found in Chavali *et al.* (2004), in which a CAMD problem for the design of homogeneous transition-metal catalysts is formulated for solution via Tabu Search, a stochastic optimization method applying adaptive memory to escape local optima. Any type of QSPR model, including highly non-linear or neural network-type models, may be used with these heuristic methods. Thus the challenge in terms of problem formulation is certainly lessened in comparison to deterministic algorithms. Unfortunately, no well-defined method exists for the tuning of such stochastic algorithms for effective performance, as will be described in the following section.

9.5 Mathematical techniques for the solution of CAMD optimization problems

As described in the previous section, a variety of combinatorial optimization algorithms are currently employed for CAMD. These include enumerative techniques, which use multi-level screening approaches to attempt to narrow the search space down to a few final useful candidates; deterministic methods, which bound the optimal solution from above and below and then refine those bounds sequentially, thus giving a guarantee of convergence within a finite number of steps; and stochastic algorithms, which perform iterative improvements to a set of initial solutions. The formulation and QSPR model chosen should match with the solution algorithm, such that the entire problem can be solved in a reasonable period of time. This allows for a cycle in which experimental testing of novel alternative molecules can be included, and the QSPR models can be refined if the newly designed molecules do not possess the desired physical and chemical property values.

The excellent compilation by Achenie *et al.* (2002) includes a number of case studies using generate-and-test methods such as those of Harper and Gani (2000), who apply the multi-level approach of Hostrup *et al.* (1999), in which successive steps of generation and screening are implemented and the level of molecular detail is increased in subsequent steps as the candidate molecules become fewer in number and more likely to satisfy the entire set of property targets. Enumerative algorithms are advantageous in that they do not limit the functional form of the QSPR expressions and do not require derivatives or convex underestimators to be derived. As the molecular search space increases, the QSPR relations must be evaluated more frequently, so they must be kept fairly simple.

Deterministic (global) optimization algorithms can be effective for the solution of CAMD optimization problems despite the obvious issue concerning the combinatorial nature of the search space. In general, the problem is formulated in a specific manner such that a known deterministic algorithm may be applied. Maranas (1996) transform the non-convex MINLP formulation into a tractable MILP model by expressing integer variables as a linear combination of binary variables and replacing the products of continuous and binary variables with linear inequality constraints. Using connectivity indices, Camarda and Maranas (1999) describe a convex MINLP representation for solving several polymer design problems. Sahinidis and Tawarmalani (2000), as well as Sahinidis *et al.* (2003), report a branch-and-reduce algorithm for identifying a replacement for Freon.

Stochastic optimization approaches have also been developed as alternative strategies to rigorous deterministic methods. These algorithms show great promise in terms of parallelization, since each member of the initial solution set can be updated independently. Venkatasubramanian *et al.* (1994) employed a genetic algorithm for polymer design in which properties are estimated via group contribution methods. Marcoulaki and Kokossis (1998) describe a simulated annealing (SA) approach for the design of refrigerants and liquid–liquid extraction solvents. Wang and Achenie (2002) present a hybrid global optimization approach that combines the OA and SA algorithms for several solvent design problems.

9.6 Example

In the example problem provided in this chapter, an optimization formulation of the CAMD problem is solved via Tabu Search, a heuristic algorithm developed by Glover and Laguna (1997). The method operates by generating a list of neighbor solutions similar to a current solution and keeping a record of recent solutions in a Tabu list.

This memory helps prevent cycling near local optima and encourages exploration of the entire search space. Tabu Search has been used for many applications of combinatorial optimization, including chemical process optimization (Lin and Miller, 2004), planning and scheduling (Kimms, 1996) and molecular design (Eslick and Camarda, 2006). In this example, the method was selected for a polymer design formulation because the constraints are highly non-linear and because Tabu Search rapidly finds and evaluates many near-optimal solutions. This is of great advantage within a molecular design context since the property correlations and target property values are normally approximate. Structure-property correlations created using connectivity indices typically involve an error of 3 %–10 %. Given this type of uncertainty, it is unlikely that the one "best" molecule from a mathematical standpoint is actually the best in practice. What the designer needs is a short list of candidate molecules, which he or she can parse using a well-developed chemical intuition, to determine which of the structures is worth the effort and expense of a chemical synthesis and testing procedure.

9.6.1 Initial steps

The initial solution is created using the following steps. First, one of the basic groups is selected as a root group. Then, another group is added to the root group, which must have some number of free bonds. Additional groups are then added to connect to the empty bonds. For polymers, both main chain and side chain groups may be added, and a limit is placed on the length of the sidechains and on the overall size of the repeat unit.

After generating any candidate molecule, the complete two-dimensional molecular structure is known, so the QSPR relations (in this case applying molecular connectivity indices) can be applied to estimate physical properties. At each iteration, certain solutions are classified as Tabu (forbidden) and added to Tabu lists. At the same time, the Tabu property of other solutions expires, and those solutions will be removed from the Tabu lists. In this way, Tabu lists are updated continuously and adapt to the current state of the search. The use of adaptive memory enables TS to exhibit learning and creates a more flexible and effective search.

In this example, a molecular design problem requiring the generation of candidate structures for dental adhesives is considered (Eslick *et al.*, 2009). Polymeric materials, such as dental composites, are increasingly being used to replace mercury containing dental amalgam, and new adhesives are required which chemically match with these new materials. The overall approach to the development of durable dentin adhesives pursued by Eslick *et al.* (2009) combines CAMD techniques and mechanical modeling, followed by synthesis and testing of modified methacrylate resins, which includes multi-scale *in situ* adhesive/dentin interfacial characterization. The critical design parameter is the clinical longevity of the adhesive/dentin bond, which involves both physical and chemical factors and is strongly dependent on the rate of cleavage of ester bonds by salivary esterases. However, many other property values, such as viscosity, cure time and monomer toxicity, affect the practical usage of such materials. A design method which estimates many properties at the same time and seeks to optimize these property values simultaneously, such as CAMD, can greatly improve the efficiency of such research. However, the cross-linked random co-polymer structure of dental polymers presents a challenge to any CAMD approach, since the exact molecular structure of these polymers is difficult to determine experimentally or describe in a manner useful for calculation. Both physical and chemical degradation pathways are considered within a multi-scale characterization and modeling approach. The application of CAMD generates reasonable, novel candidate polymer structures, and therefore allows the rational design of durable, water-compatible, esterase-resistant dentin adhesives.

9.6.2 Generation of QSPR models

The next step in any CAMD study is to create QSPRs for the physical and chemical properties of interest. Experimental property data are collected to develop QSPRs specific to the current type of polymeric materials in use as dental adhesives. In the dental polymer example used here, QSPRs are needed for cross-linked polymethacrylates. Polymer properties are often dependent on processing conditions, so it is best to design and conduct a set of consistent experiments to measure important properties, which here include tensile strength, elastic modulus, esterase resistance and glass transition temperature. Once the experimental data are collected, the property information and polymer structures are entered into a database. A specific data structure is employed to ease the computation of structural descriptors. The particular descriptors chosen for use in the QSPRs developed here are connectivity indices, which are described in Section 2.2. These numerical descriptors and property data are then exported to statistical software to develop QSPRs. The QSPRs are then embedded within an optimization framework to design new materials which are likely to have desirable properties. The candidate materials can later be synthesized and tested experimentally, and the results can be used to refine the QSPRs as necessary. After some iterations of the design process, synthesis and characterization of new materials is achieved.

Cross-linked polymers are an example of a molecular system which provides a significant challenge for structure-based property prediction, since group contributions and connectivity indices are unable to describe polymer cross-linking. While these descriptors are still useful for cross-linked polymers, additional information must also be included, since cross-linking has a large effect on many polymer properties, such as glass transition temperature (Van Krevelen, 1997). For the dental polymer design example provided here, descriptors which consider a larger, more representative section of the polymer network must be used, since both random cross-links and small rings not contributing to cross-linking may form, and there are areas in the polymer network of varying cross-link density (Ye *et al.*, 2007a,b).

To generate desirable polymer structures, a number of assumptions are made. As a starting point, it is assumed that all double bonds in the methacrylate monomers have equal reactivity and that no intramolecular reactions occur during polymerization.

The simplicity of using a fixed representative section of polymer greatly accelerates computation of structural descriptors needed for property prediction. A small preliminary set of descriptors is selected, including connectivity indices, degree of conversion and cross-link density. Addition of more descriptors would only be useful if additional experimental data were available; otherwise overfitting can occur. After collecting experimental property data for a set of dental polymers, the structural descriptors are calculated.

Multiple linear regression can be used to develop models relating these descriptors to the measured property values, and no transformations of the data were considered. It is possible to use linear regression to find polynomial, logarithmic or any other models where the parameters have a linear relationship to each other. Combinations of descriptors were obtained which provide the best correlation coefficient for QSPRs containing different numbers of descriptors, as shown in Table 9.1. Once these models were found, it was necessary to select those with the most appropriate number of descriptors.

Table 9.1: QSPRs.

Property model	Correlation Coeff.
Tensile strength, $\sigma = 1406.6 - 7484.5\xi^1 + 6611.6\xi_v^1 + 78231.7CD_{max} - 149268.6CD$	0.94
Modulus, $E = 257.362 - 135.89\xi^0 - 276.37\xi^1 - 78.24\xi_v^1 + 0.02336DC - 0.03146WC$	0.97
Glass transition temperature, $T^g = 11664.9 - 14036.5\xi^1 - 15286.9\xi_v^1 + 94671.6CD$	0.82
Initial polymerization rate (IPR) $= -4028.55 + 6510.10\xi^0 - 2394.13\xi_v^1$	0.82

A short list of structural descriptors is compiled for use in the development of the QSPRs. The degree of conversion (DC) is used as a descriptor even though it is experimentally measured, since it is needed to characterize the cross-link density. The maximum cross-link density (CD_{max}) and estimated cross-link density (CD) as described by Cook *et al.* (2003) are also used. The zeroth- and first-order valence and simple connectivity indices (ξ^0, ξ_v^0, ξ^1 and ξ_v^1) are considered in their-size independent form (Bicerano, 2002), since there are no obvious repeat units for the polymers in this example, and all properties of interest are intensive. The appropriate number of descriptors to include in each model is selected by evaluating statistical signifi-cance. Cross-validation is then performed for all models, and this evaluation confirms the model size choice. The property models for this example are shown in Table 9.1. Note that these expressions are based on a very small experimental data set, and are not intended for use in prediction of properties of cross-linked polymers involv-ing groups not included within that small set. However, they serve as an example of the types of models which can be included within a molecular design optimization problem.

9.6.3 Problem formulation

The goal of this CAMD example is to determine new polymer structures which are likely to possess desirable properties for use as dental adhesives. To find these new polymers, the QSPRs were included in an optimization framework. Structural require-ments, such as valency restrictions, were combined with the QSPR model to form the constraint set for this problem. The objective function seeks to minimize variation in the predicted properties from the user-specified target values, as described earlier in the formulation section. This formulation of the CAMD optimization problem is then solved using Tabu Search, but with a novel implementation. Neighboring structures are created by modifying monomer structures or changing composition. For optimiza-tion, the monomers are built from several small groups of atoms. Structures of the monomers are modified by adding, deleting or replacing groups. Sufficient groups are provided such that nearly any reasonable monomer can be made. Monomers are modified in such a way that only feasible structures will result. The method of struc-ture modification eliminates the need for most formal mathematical constraints. Some constraints may still be required to maintain a valid resin composition.

For random block co-polymers, specialized data structures are needed to store monomer information, as well as information about the overall network of monomer units (Eslick *et al.*, 2009). In monomer graphs, each vertex represents an atom or con-nection point between monomers, and each edge represents a chemical bond. These graphs are useful in forming the polymer structure and may be used to predict prop-erties of unreacted monomers, such as viscosity. Each functional group can have one or more states, which represent ways in which that functional group can react. These monomer graphs are well suited to systematic generation of polymer structure, due to their ability to represent several bonding states. When the state of a monomer graph changes, the vertex and edge indices are resorted so that all of the vertices and edges in the current state have a lower index than the ones that are out-of-state. Graph al-gorithms may then ignore the out-of-state vertices and edges. Figure 9.2 shows an example of two monomer graph states for the 2-hydroxyethyl methacrylate (HEMA) monomer, often used in dental polymer formulations. The dummy Xx atoms represent connection points to other monomers.

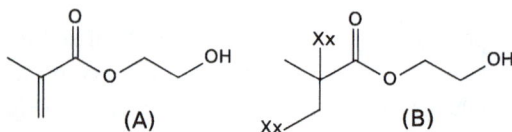

Figure 9.2: Graph states for HEMA monomer.

To determine whether a monomer matches with one on the Tabu list, the structural descriptors of a solution are compared to solutions in the Tabu list. The structures cannot be compared by directly matching the groups they contain, since this would not disallow structural isomers with nearly identical properties. In fact, a certain amount of structural difference is allowed, so structures do not have to exactly match those on the list to be considered the same as a previous structure. This allowed structural difference is an adjustable parameter. Solutions may be refined by allowing them to be close to those in the Tabu list, while diversification of the candidate solutions considered may be encouraged by requiring a large difference.

These property models are then used to evaluate the quality of each molecular candidate, compared with target property values selected to yield a more durable dental polymer. In cases where a property is to be maximized, a target near the upper limit of applicability of the property model may be set as the target. The first row of Table 9.2 shows the target property values used in the example.

Table 9.2: Objective function values.

Monomer	σ (MPa)	E (GPa)	IPR (mol/l/s)	T_g (°C)	Objective
Target	80	3.0	130	120	0.000
bisGMA	68	1.0	107	115	0.497
Candidate 1	79	1.4	98	132	0.372
Candidate 2	77	1.3	105	134	0.365
Candidate 3	88	1.7	119	148	0.262
Candidate 4	85	1.3	116	143	0.367
Candidate 5	86	1.4	121	134	0.299

The objective function used simply minimizes the total scaled deviation of the properties from the prespecified property targets.

9.6.4 Example results

The goal of the optimization problem was to design an improved monomer, while keeping the composition ratio of new monomers consistent with that used in the experiments. In this example, solutions to the optimization problem are monomer structures. The composition and structure of the first two monomers are fixed. The monomer structures are constructed from small groups of atoms. The groups include a methacrylate functional group, ring structures such as aromatic rings and small basic groups such as $-CH_3$. Sufficient groups are supplied such that almost any methacrylate monomer of interest can be generated. Groups can be added, deleted or replaced to generate neighbor solutions to a current solution, but only in a manner such that all neighbor solutions are feasible. The number of functional groups is fixed by the initial structure, but variation is included by using starting points with different numbers of functional groups.

Several starting points are selected for Tabu Search, including currently used monomers such as bisGMA, as well as randomly generated monomers with different numbers of functional groups. The Tabu Search algorithm is applied, and it is terminated when a fixed number of non-improving iterations is completed, in this case 300. A list of the best solutions found is compiled and is provided to polymer chemists for further consideration and refinement. Figure 9.3 shows two of the resulting structures, and Table 9.2 shows objective function values for those two structures (candidates 1 and 2), as well as three other candidate monomers, as compared to bisGMA, which is typically used in commercial dentin adhesives. The solutions shown all give a better objective function value than bisGMA.

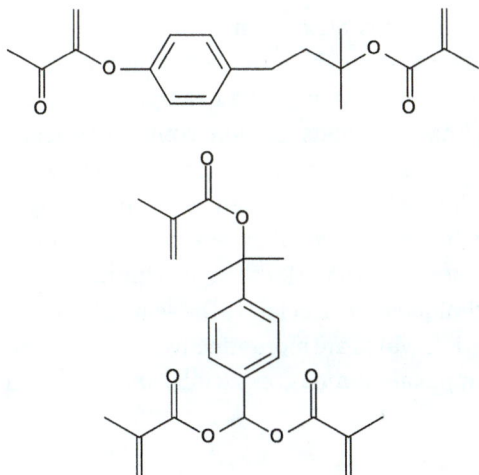

Figure 9.3: Candidate monomers 1 and 2.

The results provide numerous structures which can be further evaluated for use within polymethacrylate dentin adhesives.

The Tabu Search algorithm chooses neighbors to generate randomly, and therefore does not always yield the same solution from the same initial structure. While there is no guarantee that the globally optimal solution was found, each of these strong candidate monomers was generated in about 30 seconds on average. The computation time depends on a number of Tabu Search parameters, including the length of the Tabu list and the number of non-improving iterations allowed prior to termination. Further refinement of the algorithm parameters is expected to improve solution time and solution quality.

9.7 Take-away message

The example describes an implementation of a CAMD methodology for complex polymer systems. With the framework in place, preliminary work for the design of dentin adhesives can progress rapidly as more experimental data are collected and more accurate property models are created. With additional data, further evaluation of new structural descriptors and QSPRs and the addition of more properties, the newly designed polymers are more likely to have practical value. The optimal dentin adhesive will be produced by balancing the desired physical, chemical and mechanical properties with the need for esterase resistance and water compatibility. The combinatorial optimization allows the relative importance of each property to be varied, and predicts novel methacrylate structures for further evaluation.

This chapter provides an overview of CAMD formulations and the algorithms which have been used to solve them. Within the context of molecular systems engineering, CAMD provides design methodologies, much as process design methodologies are widely applied to process systems engineering. The example given here attempts to provide insight into the use of CAMD methods for more complex systems, such as the design of biomaterials, for which a rational design methodology is greatly in demand. Many researchers are attempting to build integrated software packages which will allow industrial practitioners without significant experience in optimization problem formulation to apply the techniques of CAMD for their specific molecular selection issues. The hope is that these techniques will accelerate the search for novel molecules, and thus permit the use of materials which are more effective and less damaging to our environment, within many application areas, especially those beyond traditional chemical processing.

Bibliography

Achenie, L. E. K., Gani, R. and Venkatasubramanian, V. (Eds.) (2002). Computer Aided Molecular Design. In: *Theory and Practice*. Amsterdam: Elsevier.

Adjiman, C. S., Androulakis, I. P. and Floudas, C. A. (2000). Global Optimization of mixed-integer nonlinear problems, *AIChE J.*, 46, pp. 1769.

American Chemical Society (2001). Technology Roadmap for New Process Chemistry. Available at http://www.oit.doe.gov/chemicals/visions_new_chemistry.shtml.

Bicerano, J. (2002). *Prediction of Polymer Properties*, 3rd ed. New York: Marcel Dekker.

Burger, J., Papaioannou, V., Gopinath, S., Jackson, G., Galindo, A. and Adjiman, C. S. (2015). A hierarchical method to integrated solvent and process design of physical CO_2 absorption using the SAFT-γ Mie approach. *AIChE J.*, 61, pp. 3249–3269.

Camarda, K. V. and Maranas, C. D. (1999). Optimization in polymer design using connectivity indices. *Ind. Eng. Chem. Res.*, 38, pp. 1884.

Chavali, S., Lin, B., Miller, D. C. and Camarda, K. V. (2004). Environmentally-benign transition-metal catalyst design using optimization techniques. *Comp. Chem. Eng.*, 28(5), pp. 605.

Cook, W. D., Forsythe, J. S., Irawati, N., Scott, T. F., Xia, W. Z. (2003). Cure kinetics and thermomechanical properties of thermally stable photopolymerized dimethacrylates. *J. Appl. Polym. Sci.*, 90, pp. 3753.

Eslick, J. C. and Camarda, K. V. (2006). Polyurethane design using stochastic optimization. In: W. Marquart and C. Pantelides, eds., *Proc. 9th International Symposium on Process Systems Engineering*. Amsterdam: Elsevier, pp. 769–774.

Eslick, J. C., Ye, Q., Park, J., Topp, E. M., Spencer, P. and Camarda, K. V. (2009). A computational molecular design framework for crosslinked polymer networks. *Comput Chem Eng.*, 33(5), pp. 954–963.

Fredenslund, A., Jones, R. L., Praunsnitz, J. M. (1975). Group-contribution estimation of activity coefficients in nonideal liquid mixtures. *AIChE J.*, 21, pp. 1086.

Gane, P. J. and Dean, P. M. (2000). Recent advances in structure-based rational drug design. *Curr. Op. Struct. Bio.*, 10, pp. 401.

Gani, R., Tzouvaras, N., Rasmussen, P. and Fredenslund, A. (1989). Prediction of gas solubility and vapor-liquid equilibria by group contribution. *Fluid Phase Equilib.*, 47, pp. 133.

Glover, F. and Laguna, M. (1997) *Tabu Search*. Boston: Kluwer Academic Publishers.

Harper, P. M. and Gani, R. (2000). A multi-step and multi-level approach for computer aided molecular design. *Comp. Chem Eng.*, 24, pp. 677.

Hostrup, M., Harper, P. M., Gani, R. (1999). Design of environmentally benign processes: Integration of solvent design and separation process synthesis. *Computers and Chemical Engineering*, 23(10), pp. 1395–1414.

Joback, K. G. (1987). *Designing molecules possessing desired physical property values*, Ph. D. Thesis, MIT, Cambridge.

Kier, L. B. and Hall, L. H. (1986). *Molecular Connectivity in Structure-Activity Analysis*. Letchworth, England: Research Studies Press.

Kimms, A. (1996). Competitive methods for multi-level lot sizing and scheduling: Tabu Search and randomized regrets. *Int. J. Prod. Res.*, 34, pp. 2279.

Klamt, A. (2011). The COSMO and COSMO-RS solvation models. *WIREs Comput Mol Sci*, 1, pp. 699–709.

Lampe, M., Kirmse, C., Sauer, E., Stavrou, M., Gross, J. and Bardow, A. (2014). Computer-aided Molecular Design of ORC Working Fluids using PC-SAFT, *Computer Aided Chemical Engineering*, 34, pp. 357–362.

Lin, B. and Miller, D. C. (2004). Tabu search algorithm for chemical process optimization. *Comp. Chem. Eng.*, 28, pp. 2287.

Lin, B., Chavali, S., Camarda, K. and Miller, D. C. (2005). Computer-aided molecular design using Tabu Search. *Comp. Chem. Eng.*, 29, pp. 337.

Maranas, C. D. (1996). Optimal computer-aided molecular design: a polymer design case study. *Ind. Eng. Chem. Res.*, 35, pp. 3403.

Marcoulaki, E. C. and Kokossis, A. C. (1998). Molecular design synthesis using stochastic optimization as a tool for scoping and screening. *Comp. Chem. Eng.*, 22, pp. S11–S18.

Oprea, T. I. (Ed.) (2005). *Chemoinformatics in Drug Discovery*. Weinheim: Wiley-VCH.

Process Systems Enterprise, gPROMS, www.psenterprise.com/gproms, 1997–2015.

Raman, V. S. and Maranas, C. D. (1998). Optimization in product design with properties correlated with topological indices. *Comp. Chem. Eng.*, 22, pp. 747.

Randić, M. (1975). On the characterization of molecular branching. *J. ACS*, 97, pp. 6609.

Sahinidis, N. V. and Tawarmalani, M. (2000). Applications of global optimization to process and molecular design. *Comp. Chem. Eng.*, 24, pp. 2157.

Sahinidis, N. V., Tawarmalani, M. and Yu, M. (2003). Design of alternative refrigerants via global optimization. *AIChE J.*, 49, pp. 1761.

Siddhaye, S., Camarda, K. V., Topp, E. and Southard, M. Z. (2000). Design of novel pharmaceutical products via combinatorial optimization. *Comp. Chem. Eng.*, 24, pp. 701.

Siddhaye, S., Camarda, K., Southard, M., Topp, E. (2004). Pharmaceutical product design using combinatorial optimization. *Computers and Chemical Engineering*, 28(3), pp. 425–434.

Trinajstic, M. (1975). Chemical Graph Theory. Boca Raton, FL: CRC Press.

Van Krevelen, D. W. (1997). *Properties of Polymers*. Amsterdam: Elsevier.

Venkatasubramanian, V., Chan, K. and Caruthers, J. M. (1994). Computer aided molecular design using genetic algorithms. *Comput. Chem. Eng.*, 18, pp. 833.

Wang, S. and Milne, G. W. A. (1994). Graph theory and group contributions in the estimation of boiling points. *J. Chem. Info. Comp. Sci.*, 34, pp. 1232.

Wang, Y. and Achenie, L. E. K. (2002). A hybrid global optimization approach for solvent design. *Comp. Chem. Eng.*, 26, pp. 1415.

Wasserman, L. (2004). A Concise Course in Statistical Inference. New York: Springer.

Ye, Q., Spencer, P., Wang, Y. and Misra, A. (2007a). Relationship of solvent to the photopolymerization process, properties, and structure in model dentin adhesives. *J. Biomed. Mater. Res. Part A*, 80, pp. 342.

Ye, Q., Wang, Y., Williams, K. and Spencer, P. (2007b). Characterization of photopolymerization of dentin adhesives as a function of light source and irradiance. *J. Biomed Mater. Res Part B*, 80, pp. 440.

Part II: **Process Design Principles**

10 Process synthesis

Process synthesis is the art and science of the development of new flowsheets, or more generally, the selection of equipment and material flows to produce a chemical product. For most of the traditional chemical industries, these products are fluids, so moving the materials from one unit to another is accomplished with pumps and compressors. The challenges associated with process synthesis stem from the fact that we must develop something new, and that there are conflicting objectives which drive the effort. Synthesis should be contrasted with design, which includes defining the specific material and energy flows, the detailed descriptions of each piece of equipment and the physical layout of the plant. Our focus here is at a more conceptual level, whereby we must consider different raw materials, reaction chemistries and separation technologies. While we can return to the synthesis stage after some of the detailed design work is complete, in practice this is rare: the specifics of the design are usually altered to make the synthesis idea work. Next we will consider where we look to find the novel ideas needed for process synthesis, as well as how we can evaluate those ideas on a conceptual level.

10.1 Introductory concepts

In industrial practice, a process synthesis task usually begins with a prompt from management. An idea arises for the production of a new product, perhaps based on market analysis or a development from the R&D department. In a consulting design firm, a bid has been accepted to perform a synthesis and design task for another company. In either case, the prompt is usually vague, rarely containing necessary data about the system of interest. One key here is to keep the communications open between those requesting the new process and those synthesizing and designing it. Expectations may be set far beyond what is physically or economically reasonable, and so open discussions between management or clients and process engineers is necessary to arrive at a satisfactory conclusion.

In any synthesis task, it is critical to make a full consideration of the *space of design alternatives*. This is not to say that every possibility must be considered or even generated, but a systematic method is needed to ensure that high-quality designs are not ignored simply through omission. Section 10.4 will discuss mathematical methods to perform this consideration of alternatives. The use of systematic methods also avoids the tendency to simply repeat previous designs found in the patent literature. While it is important to be aware of production pathways used by others, an overreliance on previous practice tends to exclude truly novel designs which could in fact greatly improve profitability, environmental performance or safety. The fact that new chemical plants are extremely expensive tends to exacerbate the conservative approach. But modern chemical process simulation, pilot plant analysis and testing and

https://doi.org/10.1515/9783110570137-010

more robust control strategies mean that better information can be obtained prior to constructing a full plant, and designs which are more highly integrated can be controlled effectively. A classic example of this is reactive distillation, in which a reaction occurs on the trays of a distillation column, and the separation of the product from the reactants drives the reaction extent further than would occur in a traditional reactor. When this novel system was first considered, conservative thinkers suggested that controllability would be a huge issue and that the loss of degrees of freedom brought on by the integration would greatly limit the operability of the system. However, reactive distillation systems have been successfully implemented in chemical plants around the world, and modern controllers keep the processes running at a stable steady state.

Once we have a reasonable understanding of the design alternatives, the next step in synthesis is to evaluate those alternatives. Economic analysis is nearly always the first consideration, except in special cases where time-to-market is critical (such as certain pharmaceutical products), and no income at all can be generated if the product is not produced by a given date. But for industrial bulk chemicals, being able to produce the product at a low enough cost such that a profit can be made is foremost. Environmental considerations must also be considered here, and should be an important objective from early on in the synthesis. In some cases, including highly energy-intensive processes, the objective of decreasing energy use is both an environmental and an economic goal. Sustainability is a third objective which should be considered at the conceptual stage of process synthesis, since it depends on the fundamental decisions being made at this stage: the choices of raw materials, catalysts and intermediates, and even the structure of the product in some cases. Inherent safety, which is defined as the risk involved with the materials and processes chosen in the synthesis, should be considered at this level as well. Once the decision has been made to use a highly toxic intermediate, for example, later design and operating variable choices can only mitigate the risks involved with that intermediate, not eliminate them. The specific choices for layers of protection and redundant safety systems occur far later in the design process; but the big decisions made in the synthesis and design phases lay the foundation under which those systems are constrained. Finally, plant flexibility and controllability should be considered as an objective in many plant synthesis tasks. Many batch plants, for example, need to be able to produce multiple similar products. Some products, like an influenza vaccine, must change yearly, and a flexible plant is needed to accommodate these changes. Commodity chemical plants normally make a single product or set of products, but even there, the concentration of a product in solution may need to be changed for different customers, or the production capacity may need to vary based on demand. As we switch to renewable energy sources, a chemical plant designed to operate with widely varying capacity could take advantage of energy sources which are intermittent, and avoid the need to use more constant sources of energy which emit carbon dioxide. A process design team must consider all of these objectives to some degree when evaluating a given process alternative, and make value judgements to balance these conflicting factors.

10.2 Collection of relevant information

As mentioned in the previous section, the first pieces of information to be gathered for the basis of the design are obtained from the entity requesting the synthesis work, so good communication with that entity is critical. The next set of information to be considered is the prior art. How have other companies made this product and/or similar ones in the past? This information is found primarily in the patent literature. A thorough search of all relevant patents (which may require some translation) is very helpful to understand the history of a given chemical's production. It should be remembered that patents are written to protect the inventions of companies, not to provide sufficient details for other entities to build the process. Often, patents are created to prevent other companies from using a process, rather than to protect a process to be operated by a company. So while processes in patents give us a good roadmap as to what is currently possible, they often overstate the true capability of a process. The trade literature also can guide us to an understanding of what plants are currently operating making our product, and with which technologies. Many plants make multiple products, and the patent literature gives us an understanding of what co-products are often associated with the product we have been tasked to produce. Based on this industrial information, a decision is often made as to whether technology to produce the product should simply be licensed from another company, or a new plant should be built which does not infringe on the patents of other companies.

Example (Production of ethylene glycol). Consider the task of producing 150,000 metric tons of ethylene glycol (EG) per year. Management considers that unmet demand for EG exists, and that this production amount could be sold. Furthermore, they believe that modern production technologies could be employed to undercut other firms and make significant profits. However, they do not know which raw material should be chosen, nor which process should be used.

As the process designer, you first conduct a thorough literature search. Patents and reference books on chemical processing tell you that there are many ways to produce EG. By far the most prevalent raw material in use is ethylene, which is produced in large quantities in refineries. But newer patents and papers suggest that EG can be produced from glycerol or xylitol, which might exist in large biorefineries; from syngas (carbon monoxide and hydrogen) produced from many sources; or from ethylene and hydrogen peroxide in a new "green" process. The process chemistries vary widely, and in fact when starting from ethylene, there are many catalyst types, and many processes which can be considered. Shell Chemicals licenses the OMEGA process, which involves ethylene carbonate as an intermediate, so that technology could be purchased and used directly. The set of alternatives is large, and the objectives unclear are. Should the lowest-cost production route be selected, or is there an advantage to a process which could be considered "green," yet might have a higher production cost?

Once a complete understanding of the technologies currently in use and in development for the production of a given product is developed from the patent literature, the next step is to gather specific information about all chemicals which will be present in the plant. It is important to gather all of this information at this stage, to avoid repeating steps later on in the synthesis process. For example, we might determine that a given reaction process is highly efficient, but produces a by-product which cannot be separated from the main product. This thermodynamic information needs to be found as early as possible, to avoid effort being wasted to design a reaction scheme which in the end cannot be chosen. Physical property data are needed for every chemical in the system: products and raw materials, intermediates, solvents, catalysts and even fuels for the pumps and cleaning solutions. This information should be as extensive as possible, and must include not only properties directly needed for the design, like heat capacities, densities and vapor pressures, but also information which can be used for later environmental, sustainability and safety assessments. The source of each property value should be carefully logged, and when possible, the error in the value should be listed as well. This allows for an estimation of the precision of future calculations. Most large companies have physical property databanks, and some information can be found in textbooks, chemical encyclopedias and handbooks. The use of internet sources for physical property data is very risky, unless those sources have a print version which is peer-reviewed.

Thermodynamic data, especially those involving mixture properties, are clearly important for designing separation systems, and can be difficult to obtain. Here the academic literature is a good start, since many hundreds of journal papers containing thermodynamic data are published yearly. Many large companies have labs in which values such as binary interaction parameters, solubilities and vapor pressures can be measured over a wide range of temperatures, pressures and concentrations. This is always preferred unless all necessary data can be found in the peer-reviewed literature. Modern chemical process simulators also have physical and thermodynamic property databanks, which should be used with caution. Again, all data found should be referenced to the academic literature and checked when possible with other sources. It is obviously better to put in extra work to check data consistency at this phase, rather than discovering during later design of even plant construction that the separation which has been developed will not work.

A key facet of thermodynamic data which should be sought at this stage is the existence of azeotropic mixtures within our system. The separation of azeotropes is of course quite difficult, and usually requires additional solvents, which must then be separated out and recycled. Information about potential azeotropes involving reaction by-products helps guide us to choose reaction schemes which avoid those by-products, even if we end up making more of other by-products which are not so difficult to remove. If azeotropic mixtures cannot be avoided, then complete physical and environmental property data regarding appropriate solvents for extractive distillation processes, or extraction processes, should be collected in this step.

When everything else fails, property values can be estimated from structure. This should only be done for novel molecules, and only when experimental values cannot be obtained. While it is tempting to use estimation methods for a "back-of-the-envelope" conceptual process synthesis, the tendency is to assume that the values we then obtain are valid, and not spend time and resources to check those values later. Most process simulators include group contribution-based methods like UNIFAC to estimate missing physical and thermodynamic property values. These methods predict properties by correlation with the number of each type of functional group found in the molecule. They therefore do not take into account the internal structure of that molecule, and cannot distinguish between structural isomers. Methods which use more detailed structural information, such as the family of approaches known as PC-SAFT, are currently being implemented within simulators like gPROMS. However, despite recent improvements in structure-property estimation, high-quality experimental data are always preferred.

Kinetic data are perhaps the hardest category of information about a chemical process to find. Reaction rates, activation energies and preexponential factors define the size of our reactors, the energy required to form the products and the need for a catalyst. These values are of course different for each catalyst, and can be complicated by mass transfer effects in different reactor configurations. A good place to start the search for kinetic data is the patent and academic literature. This must of course be done for each set of raw materials to be considered and each reaction pathway which may be used. Companies which produce catalysts for bulk chemical production are happy to provide data, since of course they need to sell their product. Large companies may be able to synthesize their own catalysts in-house, but most chemical producers buy them from specialized firms. In many cases, experimental data are collected to obtain the required kinetic information. Either the catalyst producer runs the experiments to convince their customer of the efficacy of the material, or the design firm must conduct its own experimental investigations. Without accurate kinetic data about not only the main reaction but also the side reactions, conceptual design tends to lead to poor results. Note that kinetic parameters are highly temperature-dependent, so accurate data at a range of temperatures are needed. By-product formation is critical to understand, since it strongly affects downstream processing. The overall conversion to the product is often less important than the selectivity of the reaction, since unconverted raw materials are usually recycled into the feed, but once by-products form, it is usually difficult to recover the raw materials, and most separation processes are energy-intensive. So information about the distribution of by-products is crucial within conceptual process synthesis.

Environmental and safety data should also be collected at this early stage of process synthesis. Such data are not needed at a high level of accuracy, but considerations of environmental risks can help with material selection within the design. For example, an understanding of the environmental impact of various by-products may help us choose a synthesis path. The path which uses somewhat less energy in the reaction section may seem like a better choice, but if the by-products thus formed are highly regulated in terms of environmental impact, then the separation and disposal of those by-products may increase the cost and carbon footprint beyond that of the process alternative which uses more energy at the initial reaction step. At this phase, we do not consider the probabilities of escape of the various chemicals into the environment, since those are dependent on specific details of process equipment, as well as flow rates and temperatures. But we need a basic idea of the impacts of each chemical in our process, in terms of toxicity, smog formation, ozone depletion potential, global warming potential and other short- and long-term effects on humans and the environment.

Evaluation of the sustainability of a chemical process is described in detail in Chapter 18. In order to make such an evaluation, properties related to the supply and the long-term fate of chemicals in our process are needed. For a process to be completely sustainable, all of the limited resources we use need to be recycled back to their sources, so we need information about the supply of various raw materials (including those needed to produce our catalysts, solvents and heat transfer media) and about whether our product is recyclable.

Finally, significant economic information is required for even preliminary process synthesis. Prices of raw materials, energy, and subsidiary materials like catalysts and solvents must be found, or even negotiated. Since the initial evaluation of a process alternative is always economic, a simple cost calculation based on the input–output structure of a potential flowsheet is one of the very first steps in process synthesis. For commodity chemicals, a market analysis helps judge the stability of the product price, and a consideration must be given to market share and to whether the new amount of product will significantly change the current price structure. While the cost of energy may be the most critical factor for profitability, fixed costs such as equipment purchases almost never play that role, so at the information gathering stage, we simply need to find prices relating to operating costs, so that we can compare these costs to the annual income from sales and judge economic feasibility.

10.3 The hierarchy of decisions

Process design is essentially a sequence of decisions which lead to a final plan of how a chemical product is to be produced. Each decision narrows the set of alternatives, leading us to a final detailed design which can be built. While design is an iterative process, and many of these decisions can be reconsidered after further information is obtained, the process synthesis phase is usually not reconsidered. Most of the decisions made here are fundamental, meaning changing them later would involve a complete redesign of the process.

Before we start with the overall hierarchy, some thought should be given to the production capacity we are striving for. This decision should be made based first of all on economic considerations. How much product can we sell, and how will selling that product affect the overall market? There is also the question of raw material availability. If sufficient raw materials are not available to produce at a given product capacity, then sometimes negotiations with suppliers must take place prior to the process synthesis to determine whether the needed raw materials can be procured, and if so at what price. Another consideration for smaller concerns is the overall process construction cost. If the capacity is large, it may be that financing for a project of that size cannot be obtained. A smaller capacity plant might be a good first step, with a future expansion in view after profitability has been established at the smaller scale. Of course, economies of scale exist such that smaller plants may be less profitable per kilogram of product, in some industries. Our initial search of market information should tell us the average capacity of current plants (or plants producing similar products) and give us guidance in how to set the capacity. Judging market demand is much more difficult, and should be left to management, especially for consumer products, the demand for which is highly variable and dependent on advertising, marketing and psychological factors.

Our first big decision to be made involves whether to operate a plant in batch mode or as a continuous operation. In general, those not familiar with large-scale chemical processing tend to favor batch processes. Often, we think of industrial manufacture as requiring a sequence of tasks, whether we aim to produce a device or a chemical product. The assembly line mechanized automobile production, for example, by putting each task in a physical as well as temporal sequence. This also allowed for the eventual automation of each task, with specialized robots being designed to perform each repetitive operation. For chemical production, we can think of raw material preparation, reaction, separation and product storage as independent tasks to be performed. But the batch mode of production is usually not the most efficient route for bulk chemical production. When we make chemicals in a batch, we essentially must start and stop each process, reacting or separating for a given amount of time. Inevitably, tasks require different amounts of time to complete, and unless we provide extra equipment, bottlenecks occur which slow our processing and limit our total throughput. Furthermore, chemical process equipment such as furnaces, boilers and distillation columns does not respond well to repeated start-up and shut-down cycles, due to thermal stresses.

So when is it appropriate to operate in batch mode? The following considerations should be applied to make this decision.

1. What is the residence time required in reactors? If these times are very long, then the size of reactors operated in continuous mode will be unmanageable. A simple example of this is winemaking, in which product must react (ferment) for months to years. Systems with long residence time requirements are always operated in batch mode.

2. What is the capacity required, and how much flexibility is needed? If many products are required to be produced on the same equipment, then a batch or semi-

batch mode makes sense. Small capacities also suggest batch operation. An example of this is yogurt production, which requires fairly small amounts of different flavors to be produced frequently. The production profile for different flavors varies seasonally, matching consumer demand. Normally, yogurt is produced in a semi-batch mode, with a single flavor being produced (continuously) for some number of hours, followed by an equipment cleaning phase, and then switchover to the next flavor batch.

3. What are the challenges of start-up and shut-down? Large-scale chemical facilities such as oil refineries operate 24 hours a day, seven days a week, and the start-up and shut-down of such plants can take weeks. Clearly, a batch process would not make sense here. Regulatory issues such as sterility testing as part of start-up may require a batch mode for smaller quantities of pharmaceutical products. In general, start-up and shut-down are dynamic, challenging processes, and thus large-scale chemical production seeks to avoid them.

4. How sensitive is the product to changes over time? If we are producing a live vaccine, a short-lived radioactive isotope, or a highly reactive specialty chemical, then clearly batch processing is the mode of operation which works best. Note that batch processing also requires more operator interaction, which in general entails more risk, and higher labor costs, than a computer-controlled continuous process. But for specialty products, this level of oversight and direct operator interaction may be necessary.

A further consideration in terms of batch *versus* continuous processing is the opportunity for heat integration. For energy-intensive processes, operating in a continuous manner allows for energy recycling, which can be extremely important for both profitability and environmental impact. Continuously flowing waste streams which are at high temperature are fine sources of energy for inlet streams needing to be heated to reaction temperature. The product streams from exothermic reactors are often cooled to permit easier separation, and again, these streams can be contacted with a stream which needs to be heated, or perhaps a utility stream of steam or a molten salt, in order to effectively reuse some of that energy. Such integration is rarely possible in a batch process, and therefore nearly all commodity chemical production processes requiring distillation operate in a continuous manner.

Example (EG production). Consider again the yearly production of 150,000 tons of ethylene glycol. Is there any logic to perform this operation in batch mode? Not really. When the reaction rate data are collected for all reaction schemes listed in Section 10.2, one finds that none of the reactions have extremely large residence times. Furthermore, the only EG production described in the literature discusses laboratory experiments, while all of the processes operating in industry are continuous. The large capacity also strongly suggests that continuous operation is appropriate. Finally, one can determine from the reaction mechanisms the none of the reactions give complete conversion to EG, and so distillation or extraction processes will be required. These will require energy input, and thus heat integration can be a valuable facet of the design. Only continuous processing will allow for this integration and for the easy recycling of raw materials.

Once the basic decision about continuous or batch processing has been made, the next step is to define the input–output structure of the flowsheet. Focusing on continuous processing for a moment, let us consider, *e. g.*, the production of ethylene glycol. The input–output structure contains the raw materials we choose (with their associated impurities) and all products and by-products. If there are alternative chemistries to lead to our products (based on different raw materials, different catalysts or both), then we should compare the different input–output structures. These structures are extremely simple, and yet with a basic idea of the process chemistry, we can work out the initial feasibility of the process. The stoichiometry of the reactions, along with the conversion expected and the feasibility of recycling unconsumed raw material, tells us how much raw material must be purchased to produce our capacity of product. One can then consider the purchase prices of raw materials, the sales prices of our product and any valuable by-products and the disposal costs of non-valuable by-products, and get a very rough estimate of the profitability of our process. If the process loses money at this step, then clearly it will not be profitable once energy and equipment costs are considered. So making this quick calculation for each process alternative gives a quick guideline as to which possible designs should be further considered.

Example (EG production). In this case, we can sketch the input–output structure for each of the alternatives to be considered. Each alternative will have different material flows, and we need to decide how each exit stream is to be handled: should we recycle, sell or simply waste each stream? Based on the prices we determined earlier, we can now perform this first-pass economic evaluation. Getting an estimated sales price from management allows us to determine whether there is any profit to be made for each alternative. For those that look at least possibly profitable, we continue the evaluations. Anything which does not make money at this stage is discarded, since costs will only rise when energy and internal flows are considered later.

Once we select an input–output structure for further consideration, our next decision involves recycling. Many continuous chemical processes operate with a low conversion (but a high selectivity), since the separation and recycling of the raw materials from the products and by-products is not terribly expensive. A review of the patent literature and of the thermodynamic data can tell us whether recycling is useful in our process. This is a fundamental step where chemical engineers think differently than other product producers. When we produce automobiles in a factory, lasagna in the kitchen, or paintings for our walls, there is no concept of recycling of raw materials. We process the product until all the raw materials are consumed, and if we have excess raw materials, they are stored until we make the next batch. But in continuous chemical processing, it is very rare that all of the raw materials are consumed. And often when that does occur, rather than producing the product we desire, we get by-products which are not useful. For example, chemical products produced by oxidation reactions are never allowed to completely react, since 100 % conversion of reactants would lead to the production of just carbon dioxide and water. Instead, we run the reaction at a controlled temperature and at a shorter residence time, and we separate out the unoxidized raw materials from the products and recycle them. There are of course cases where energy costs of separation are so high as to preclude recycling, and in those cases, reaction conditions must be tuned to avoid by-product formation. But most chemical processing involves recycling, and large, integrated plants which produce many products often share recycling between processes for maximum mass efficiency. Defining the recycle structure of a process alternative is thus needed for a reasonable evaluation.

Each alternative process structure should then be defined via the reaction and separation steps which take place. While these often occur in separate unit operations, we should always be cognizant of the possibility of integration of these, such as in reactive distillation. If we do not allow for such an alternative here, the possibility will be ruled out in future decision making steps, thus discarding one of our best alternatives. With regard to the reaction section, our decision at this point concerns the choice of catalysts and the specific chemistry to be operated. This may already be defined by the input–output structure, but there are also cases in which reaction intermediates, inert reaction media or other internal variations may need to be defined, leading to different process alternatives for the same input–output structure.

Given an understanding of the reaction section, the next task in formulating a process alternative is to define the separation scheme. Nearly all large-scale chemical production involves separation, since most reactions produce by-products, and most raw materials are not pure to begin with. The first question involves the order of the separations and their placement within the flowsheet. Impurities within raw materials may lead to separations prior to the reaction section of the plant. In some cases, it may be a cheaper alternative to buy a less pure raw material and perform the separations at our plant. Another option is to simply leave inert impurities in the reactant flows, and in fact sometimes these inert chemicals can serve as heat buffers to mitigate overheating, especially when the reactions are highly exothermic. This will of course depend on the specifics of the reaction and the purity of the raw materials. If the impurities are not inert, they will likely form by-products, which then must be separated out after the reactions have taken place. In this case, separation prior to the reaction is preferred. If multiple impurities must be removed, then the order of separation and the technologies to be employed must be considered. This can lead to a large number of feasible alternatives.

The post-reaction separations are usually more extensive than those prior to the reaction section. Again, the best way to separate unwanted chemicals from our product is to never produce them at all, so a careful consideration of reaction selectivity is important. But even when conditions are chosen optimally to minimize by-product formation, many reactions simply produce a variety of products, and these must be separated. Also, often our raw material is not completely consumed, and so removing that early in the separation steps and recycling it leads to high mass efficiency. Again, the alternatives are formulated by considering the order of separations, and then the technologies which can be employed. The patent literature can help to define the scope of possible separation technologies, but of course distillation is preferred for most large-scale fluids processing. That being said, liquid–liquid phase separation should be performed whenever possible, unless the desired product distributes between the two phases. This is of course due to the fact that decantation requires very little energy, and also relates to the idea that the formation of two liquid phases within a distillation column is rarely preferred, since the separation efficiency can be severely compromised.

10.4 Data structures for the space of alternative designs

In order to perform a systematic optimization over the many alternative designs which were generated in the previous step, the information within those flowsheets needs to be stored in a data structure. The most common form of these is a graph, in which unit operations are stored as labeled vertices, and directed edges are used to define flows within pipes connecting those unit operations (Figure 10.1).

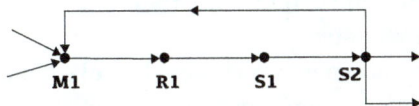

Figure 10.1: A graph representation of a simple flow sheet. Vertices correspond to unit operations, edges to material flows.

By storing a list of the vertices with their labels, and the end-points connected to each edge, a computer-readable representation of a flowsheet is created. Each type of vertex, corresponding to a unit operation, can be assigned a set of mass balance equations, which are based on the reaction and separation processes going on in that unit. These simplified, linear mass balances are not temperature-dependent, and are only used for preliminary estimation of flow rates. But they are very helpful in creating a simplified picture of the process, more complex than the input–output structure, but still basic enough to allow for the consideration of many possible alternatives.

Often, we build a flowsheet which contains all of the possible alternative technologies, and all of the possible connections, which could lead to the desired product. This representation is known as a *superstructure* (Figure 10.2), and permits the

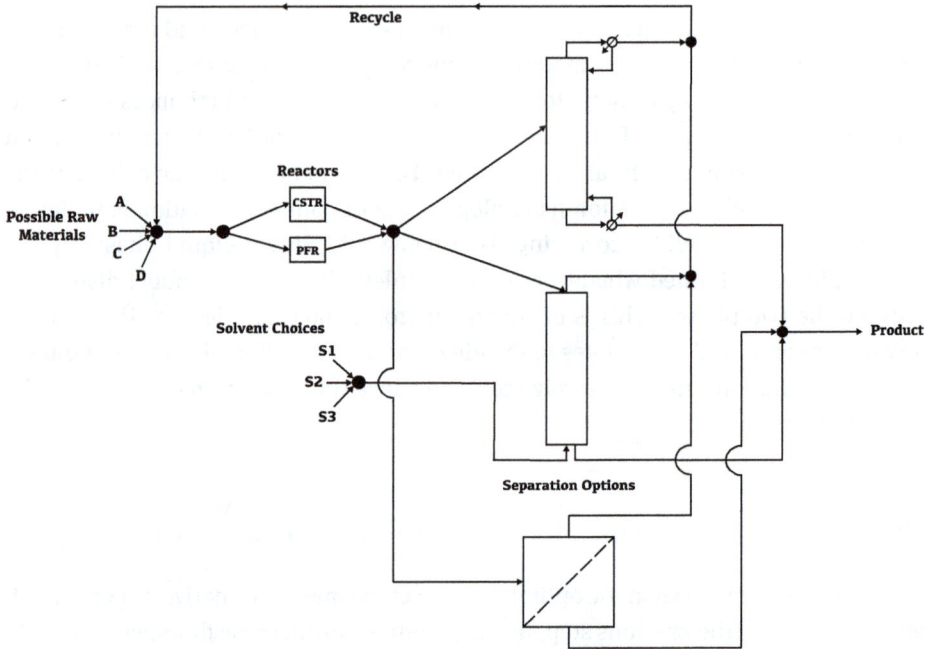

Figure 10.2: A superstructure representation of flowsheet alternatives.

formulation of an optimization problem, when the mass and energy balances defining the units are combined with equations for selection of units, constraints based on physical limitations and one or more objective functions.

A superstructure does not represent a list of equipment which would ever be built or operated, but simply combines all structural information about all alternatives into a single diagram.

When the set of alternatives is fully generated, we can then consider how they are to be evaluated. First of all, we look at the total size of the space of alternatives. If it is just a few designs, we can evaluate all of them at a coarse level, based on the criteria listed in the next section. When there are many hundreds of alternatives, however, a systematic approach based on reliable optimization strategies should be used. A systematic approach avoids the tendency to avoid innovation, and select alternatives similar to currently operating systems. While this certainly minimizes risk, it also tends to discount novel designs which may lead to significant improvements in terms of energy usage, material efficiency or safety.

A word should be given here about the use of process simulation in the evaluation of alternatives. Some would argue that the accuracy and ease of use of modern process simulators would suggest that they should be employed early in the alternative evaluation process for process designs. However, a simulator requires many details to be set in advance, such as choices for solvents and specific reaction data, and this tends to impede the search for novel process formulations. A designer can get lost in the specific details of one alternative, and never return to a consideration of the entire space of alternative designs. This author prefers to avoid the use of simulators (except when their databanks hold useful, easy-to-obtain and reliable data) until the space of alternatives has been narrowed to a small number. Only then should those designs be refined and simulated.

10.5 Evaluation of alternative designs

When the full set of alternative designs has been formulated, and simplified mass balances have been used to determine approximate flow rates between each unit operation, then those designs can be evaluated for suitability. Clearly, the economic viability of the process is foremost in this evaluation process. The simple mass flow-based consideration of the input–output structure has already been used to make a first-pass economic viability check. Next we need a rough estimation of total energy use in the potential plant, since some processes are dominated by energy costs. To do this, a preliminary energy balance is required, and so here the use of a process simulator is recommended. If there are still a large number of possible flowsheets, a heuristic approach which eliminates those involving obviously high energy usage can be employed. For example, processes which form azeotropic mixtures and then break them via extractive distillation are almost always inferior to those in which such an azeotropic mixture is never formed. In order to perform the preliminary simulation, temperatures and pressures must be estimated for all units. Consideration should be taken here for extremely high pressure or vacuum operation, as these lead to higher energy and equipment costs and entail safety hazards. The preliminary energy balances should only be used to compare alternatives, as the errors in the various computations tend to be consistent across the computations for all alternatives. The energy and material costs can then be compared across designs, since the sales income from the product should be similar for all alternatives produced at the same capacity.

If there is a need for the plant to be designed flexibly, that is, to allow for future changes in production amounts, product concentration or even product identity, then this must be considered alongside the economics. Obviously, there will be costs associated with the added flexibility, perhaps in terms of extra equipment, switching valves or alternative solvents which are needed. Equipment capacities must be larger, such that an increase in production capacity can be handled by the system. The flexibility can be enforced mathematically by placing constraints within the optimization problem to be solved over the superstructure, or within a screening-based approach simply by assigning a cost penalty to those designs with less flexibility.

The consideration of cost of production serves as a screening tool which eliminates many feasible but less efficient designs. However, a few designs should be saved to be considered in further phases of evaluation, so that options exist in terms of environmental and safety considerations. Numerous metrics exist for environmental and sustainability evaluation, and these can be used in a comparative fashion to weigh the effects of our designs. While environmental impact can be enforced simply by requiring that regulations be met, treating environmental and sustainability concerns as separate objectives can often identify logical trade-offs, whereby little extra cost is incurred but a significant environmental benefit is obtained. Chapter 18 will discuss some of the ideas surrounding sustainability-based process design.

Inherent safety is a separate objective which cannot be simply assigned a cost value and combined with an economic objective. The designer must consider not only the difference in cost between two processes exhibiting different levels of inherent safety, but also the potential future risk of accident. Remember that risk is the product of the probability of an event occurring times the impact of that event. If a highly dangerous reactant is not used, the probability of that chemical causing an accident is zero. So the consideration of inherent safety needs to be made at this phase of design, while choices for raw materials, solvents, reaction mechanisms and separation processes are still in play. A naïve approach which only considers safety as a cost is short-sighted and dangerous.

Given that some of these objectives cannot be neatly combined into a single objective function, the question remains as to how the different objectives can be considered simultaneously. In a purely mathematical approach, the selection of a good alternative can be formulated as a multi-objective optimization problem. In this case, we seek solutions which are non-dominated, meaning those for which no objective can be improved in a different design without some other objective being worsened. The set of all of these alternative designs is called the *pareto-optimal* set and represents a subset of the alternative designs which reach this stage. The narrow remaining set of designs can then be considered, looking for options which meet the specific goals of the company, while still giving an acceptable performance in terms of all other objectives. When conflicting objectives must be considered, only a designer can choose the compromise solution which seems to best balance the economic, environmental and safety performance of the design. The pareto-optimal alternatives form a nice set of designs, which can be reconsidered sequentially if one design turns out to not be quite as effective as was originally predicted. But the best of these alternatives, based on the weighting of the objectives by the designer, can then move on to become the base-case design.

Example (EG production). For the designs that remain after basic screening, we compute the mass flows based on simplified material balances and estimate energy usage. At this point, many processes can be shown to be inferior under multiple objectives. The non-catalytic hydration of ethylene oxide, produced from ethylene, is shown to be less profitable, more energy intensive and more environmentally damaging than the other processes, and is discarded. The method involving syngas is similarly rejected. Routes coming from xylitol and glycerol are economically infeasible at current prices, but could be retained if we consider that future biorefineries may be producing these chemicals in excess, thus decreasing costs. Plants employing these raw materials are much more sustainable than those using hydrocarbon-resources ethylene. The direct reaction of ethylene and hydrogen peroxide is an interesting choice. It is shown to be somewhat less economically attractive than processes involving ethylene oxide and carbonate, but it is more energy-efficient due to simpler downstream separation processes. Safety is of course a consideration for this process, given the highly concentrated hydrogen peroxide stream required, but the more complex designs involved with the ethylene oxide/ethylene carbonate route require many more pieces of equipment and solvents which could leak or fail. Now the designer has a smaller set (perhaps two or three) of designs to choose from, to form the base-case which balances the various objectives.

10.6 Take-away message

This chapter has discussed the basic ideas surrounding process synthesis, the design task in which novel flowsheet structures are developed to produce a chemical product. The importance of the collection of as much information as possible about the design, including the history of production processes for this product and the kinetic data for every reaction scheme which could be used to produce the product, cannot be stressed enough. A logical sequence of decisions can be made regarding potential flowsheets, which narrows the initially wide set of alternatives into a more manageable set. This set can be stored as a graph or a superstructure, and these data structures can then be employed to allow for optimization-based evaluation of the remaining alternatives. The remaining set should be evaluated using metrics for economic, environmental and safety metrics. At this stage of design, these are only preliminary evaluations, but they can be used comparatively to further narrow the alternative set. The designer must eventually make a weighting of the various objectives and select one alternative to move on to further simulation, optimization and detailed design, which is known as the base-case.

Exercises

1. Consider the production of 100,000 tons per year of vinyl chloride monomer. Perform a literature search to investigate possible raw materials and process chemistries. What do you think the major safety and environmental hazards would be for each process alternative?
2. The production of propylene oxide normally occurs within plants making other co-products. Looking at the patent literature, do you believe that a plant designed solely to produce propylene oxide would be economically feasible? Why or why not?
3. The management of a firm producing ethanol from biomass would like to consider the production of 1,3-butadiene at a plant adjacent to their ethanol production facility. Use the literature to evaluate the soundness of this proposal, based on current prices for ethanol and 1,3-butadiene and various reaction chemistries which could be used.
4. Paracetamol (acetaminophen) is one of the most widely used pharmaceutical products in the world. Consider the advantages and disadvantages of joining this market as a producer. The

molecule can be synthesized from various raw materials, and while the price is low, newer technologies might make production via a new plant cheaper than older designs. Investigate a set of alternative designs, and make an estimation of which one you believe is likely to meet your objectives the best. State your objectives carefully, and then justify your choice.

10.7 Further reading

Biegler, L. T., Grossmann, I. E. and Westerberg, A. W. (1997). *Systematic methods for chemical process design*. Prentice-Hall.

Douglas, J. M. (1988). *Conceptual design of chemical processes*. McGraw-Hill.

Duncan, T. M. and Reimer, J. A. (2019). *Chemical Engineering Design and Analysis*. Cambridge University Press.

Seider, W. D., Seader, M. and Lewin, D. (2009). *Product and Process Design Principles*. Wiley.

Kleiber, M. (2016). *Process Engineering*. Berlin: De Gruyter.

11 Process simulation

11.1 Process simulators

When finalizing the process creation and process synthesis stages, one would like to analyze several things in the process design. For example, one would like to obtain accurate information for the mass and energy balances, phase equilibria, mass transfer and kinetics, all of this with the aim of finding suitable operating conditions for the process, for example at which pressures and temperatures the equipment should be operated. The main aim of process simulation via (commercially) available software is to obtain all of the lacking information that one needs for further decision making. It is noted that process simulators are also developed at an increasing rate and used extensively to train operators of chemical plants.

Process simulators are mostly used for steady-state and scheduling calculations. However, process simulators can also be used for process dynamics and control, for economic evaluation and profitability analysis or for process optimization.

We need process simulators, because they allow us to find solutions of mass and energy balances in a simultaneous manner. Often the process simulators have large libraries of non-ideal thermodynamic models that we can use and for many unit operations detailed (rigorous) models are included in the simulators. Process simulators have solvers that allow us to solve large sets of equations resulting from setting up all balance equations (typically around 10,000 to 100,000 equations).

Over the years many process simulator packages have been developed. Those packages can be divided into two main types, *i. e.*, the modular type, with packages from AspenPlus, HYSIS, Chemcad, Pro II and Honeywell's UniSim, and the equation-oriented type, with packages from MATLAB®, GProms, GAMS and many more.

11.2 Modular- and equation-oriented modes

Process simulators are available in two "modes," *i. e.*, modular packages and equation-oriented packages. In modular packages, the unit operation and thermodynamic models are self-contained subprograms (modules). The generated flowsheets are called at a higher level to converge the stream connectivity of the flowsheet (recycles). Modular packages have a long history and are in general more popular for design work; they are easy to construct and debug, but in some cases inflexible for various user specifications. Figure 11.1 shows a screenshot of a modular-based simulation package.

Alternatively there are equation-oriented simulation packages, in which the user has to program the equations for unit operations, thermodynamics and connectivity manually. Often also sophisticated numerical methods and software engineering skills are needed to create suitable process simulation models. The equation-oriented

https://doi.org/10.1515/9783110570137-011

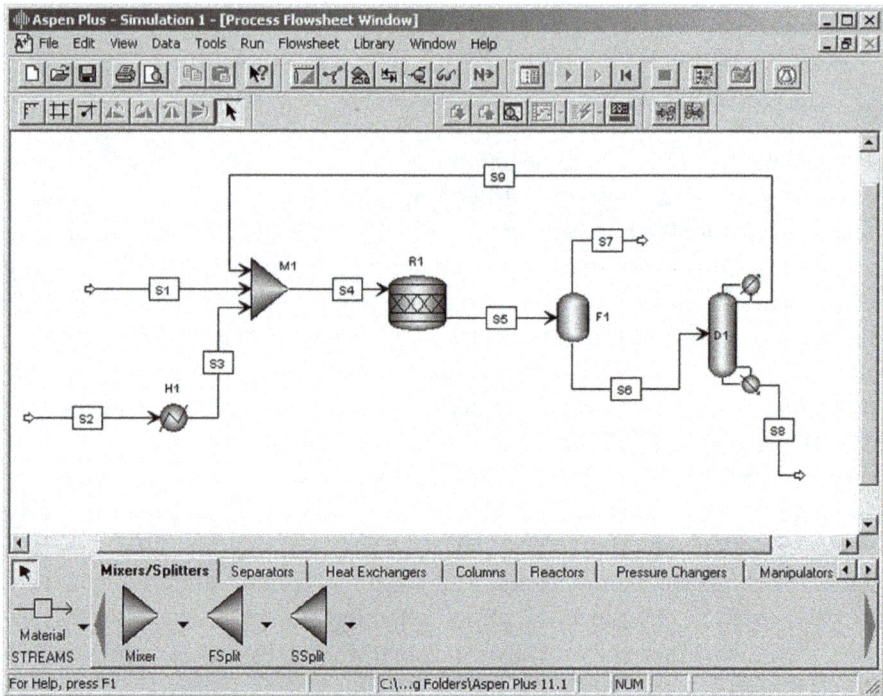

Figure 11.1: Screenshot of the AspenPlus interface.

simulation packages are primarily applied to online modeling and optimization. Figure 11.2 shows a screenshot of an equation-oriented software package.

The first simulation packages were developed in the 1950s with unit operation stand-alone models. The equation and solution of the models was done sequentially to form a flowsheet. This was the origin of the modular software packages.

In the 1960s, especially the petrochemicals companies developed the sequential modular simulation packages in-house, while academia focused on the fundamentals of equation-oriented simulators.

In the 1970s advanced methods for modular flowsheet packages were developed, leading to the simultaneous solution of flowsheets. Also general unit operation models and advanced numerical methods were embedded in the software. This led to a very fruitful collaboration between MIT and Aspen.

When in the 1980s and 1990s computers became accessible to a broader audience, there was also increased interest from industry in equation-oriented simulation packages. Vendor-supported software companies worked on user-friendly interfaces and further optimized solution algorithms.

In the resent past there has been a considerable interest to include additional functionality in the simulation packages, for example to solve supply chain problems and to execute advanced economic analysis of flowsheets.

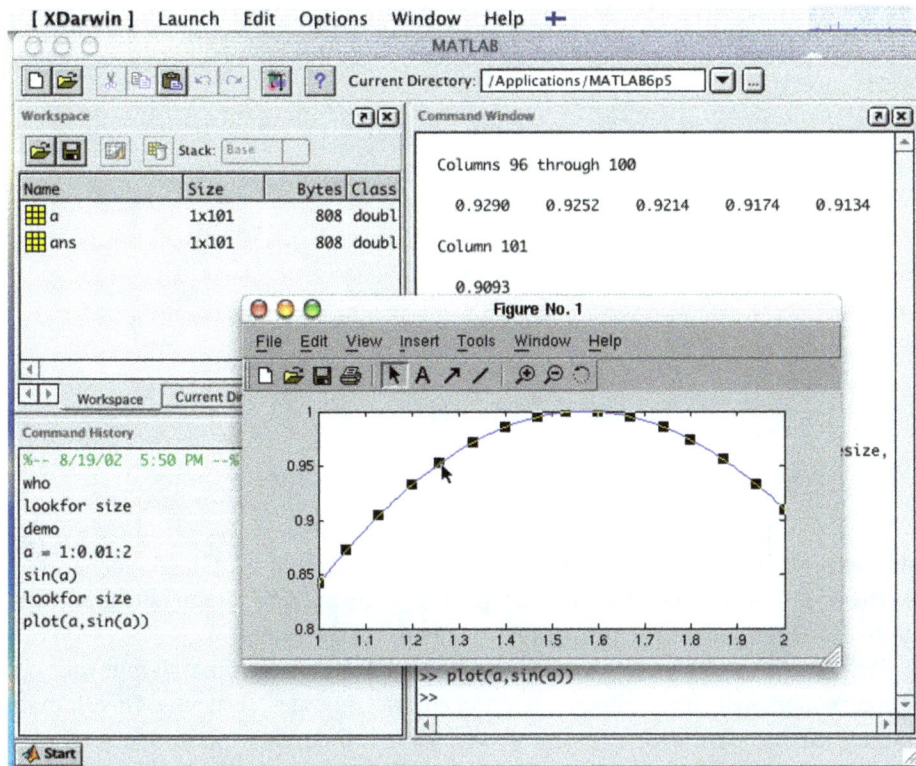

Figure 11.2: Screenshot of the MATLAB user interface.

11.3 Analysis, process and simulation flowsheets

From the first stages of a conceptual process design follows a process flow diagram or initial flowsheet. As an example, a flowsheet for the production of vinyl chloride is shown in Figure 11.3.

Although the flowsheet shows which equipment is needed, it does not provide any insight in which temperatures, pressures and flow rates are needed in the various units. Process simulators can be used to obtain such data and to test potential changes in operating conditions and do retrofitting.

During process analysis we have to deal with increasing mathematical complexity at each level of the design. That means that we will have more equations to solve and that we require more sophisticated solvers for solving the material and energy balances, equipment sizing and profitability analysis.

Modular simulation packages can be very useful in finding the answers to our design questions. For classical conceptual design problems we mostly want to use for our analysis conventional equipment (such as heat exchangers, pumps, distillation

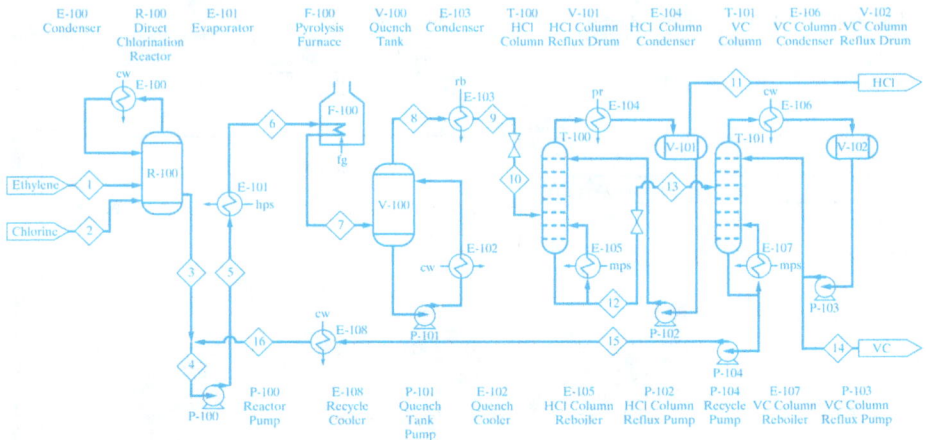

Figure 11.3: Process flowsheet of the vinyl chloride process (from Seider *et al.*).

columns and absorbers). These units are programmed in self-contained blocks. Mind that the equations for the different units themselves do not differ among the processes, but that the thermodynamic properties and chemical constants do.

With this it is of course possible to write subroutines for the different unit operation models, which are often called procedures, modules or blocks. The different modules can be connected to each other and solved simultaneously with an equation solver, for example by Newton's method.

Modular software packages (such as AspenPlus) can often also show two different simulation environments: the simulation flowsheet and the process flowsheet. In the process flowsheet the emphasis is on the flows of material and/or energy, while in the simulation flowsheet the emphasis is on the flow of information. The arcs in a simulation flowsheet represent information (flow rates, temperature, enthalpy, vapor/liquid fractions, etc), the solid blocks in the simulation flowsheet denote the simulation units (unit operation models) while dashed line rectangles are mathematical convergence units (solvers). A typical example of a simulation flowsheet is shown in Figure 11.4.

Modular-based simulators have many preprogrammed subroutines available. The process models for the units can often be addressed at different levels of detail, from approximately detailed (at the beginning of the design process) to rigorous (fewer options left).

11.4 Degrees of freedom analysis

A very important concepts in flowsheet simulation is the concept of the degrees of freedom analysis. A block solves N_{eq} equations with N_{var} variables when

$$N_{eq} < N_{var}. \tag{11.1}$$

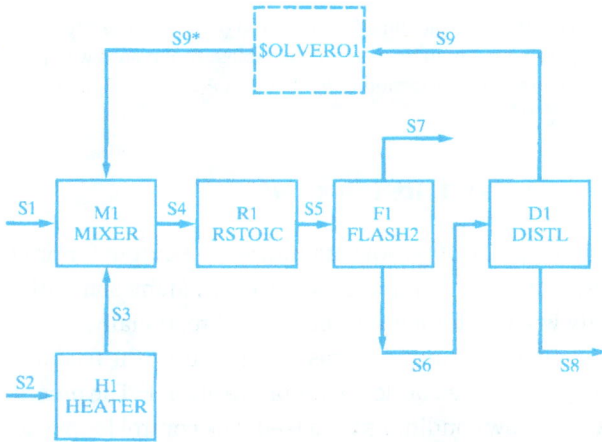

Figure 11.4: Simulation flowsheet with information flows, simulation units and convergence units.

There are in this case N_D degrees of freedom (or input decision variables), where N_D is calculated as

$$N_D = N_{var} - N_{eq}. \qquad (11.2)$$

If the number of equations does not match the number of variables the resulting system cannot be solved. That is why simulation packages often give the warning message "required inputs are not complete" when the degrees of freedom have not been provided. Often the input data (flows, concentration, temperatures, etc) are specified, and the output data can then be calculated.

Example 1 (Degrees of freedom analysis for a Cooler). A binary system where S1 contains benzene and toluene at a vapor fraction of $F = 0.5$ is considered in the figure below.

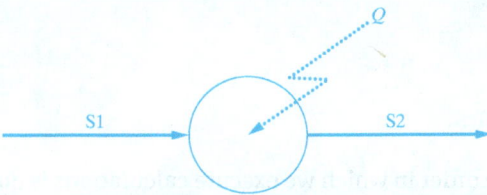

Carry out a degree of freedom analysis. At steady state the material and energy balances are

$$F_1 x_{B1} = F_2 x_{B2},$$

$$F_1 x_{T1} = F_2 x_{T2},$$

$$F_1 h_1 + Q = F_2 h_2,$$

where F are the molar flow rates and x are the mole factions, *i. e.*, $x_{Ti} = 1 - x_{Bi}$, $i = 1, 2$. The enthalpy can be expressed as $h_i = h_i(P_i, F_i, x_i)$. We have 7 equations and 13 variables. That means we have $13 - 7 = 6$ degrees of freedom. A common set of specifications would be $F_1, P_1, F_2, x_{B1}, P_2$ and Q.

11.5 Bidirectional information and design specs

Often a process designer wants to work the other way around, set the output (or product) data and calculate what are the required input specifications. Many simulation packages have such functionality where information can flow bidirectionally.

Suppose that for a mixer that mixes two inlet streams into a product the feed flow conditions have to be found that give a desired outlet temperature of $85\,°F$. The process simulator can now estimate the feed flow conditions via a feedback control loop (with a solver block), as shown Figure 11.5.

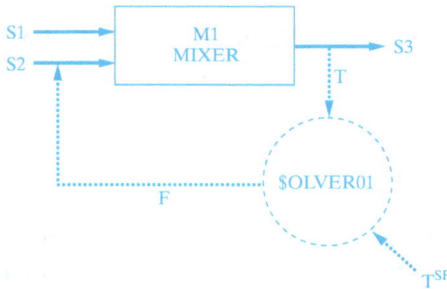

Figure 11.5: Finding the right feed flow of stream S2 that ensures that the outlet temperature of the product stream reaches a desired set point.

The solution for bidirectional information flow problems, where a desired output is specified and the corresponding inlet conditions are varied, is called a design specification.

11.6 Recycles and tear streams

As we saw in the previous section, the order in which we execute calculations is quite important. Actually the calculation order has to be consistent with the flow of information (often parallel to the material/energy flows). However, when we deal with recycles, reversible or competing reactions or purge streams the information flow can become problematic, calculations can no longer be done in a direct way and iterations are required.

Simulation packages have iterative schemes to solve such tear variables. This is best illustrated with the example shown in Figure 11.6.

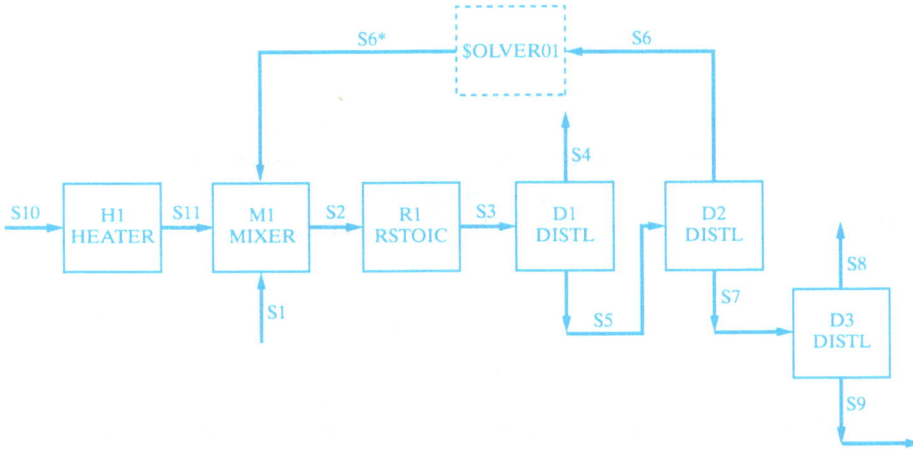

Figure 11.6: Simulation flowsheet with recycle.

Stream S11 can be calculated directly from stream S10. But stream S2 depends on S11, S1 and S6, and S6 is calculated only further in the flowsheet. That is why we tear a stream in the recycle loop; in other words, we will provide a guess value S6*. With this guess value we can calculate S2. We then iterate until the difference between S6* and S6 is smaller than a preset tolerance.

11.7 Convergence algorithms

The convergence units or solver blocks contain numerical methods for solving systems of (non-)linear equations. The simplest algorithm is the Wegstein method. There are also more advanced solver algorithms, for example Newton–Raphson's method, the quasi-Newton method or Broyden's method.

Suppose we want to solve a system of non-linear equations, *i. e.*,

$$f_1(x) = 0,$$
$$f_2(x) = 0,$$
$$\vdots$$
$$f_n(x) = 0, \tag{11.3}$$

with Newton's method. Newton's method makes use of a Taylor series expansion (only the first term):

$$
\begin{bmatrix} f_1(x + dx) \\ f_2(x + dx) \\ \vdots \end{bmatrix}
=
\begin{bmatrix} f_1(x) \\ f_2(x) \\ \vdots \end{bmatrix}
+
\begin{bmatrix} \frac{\partial f_1}{\partial x_1} & \frac{\partial f_2}{\partial x_1} & \cdots \\ \frac{\partial f_1}{\partial x_2} & \frac{\partial f_2}{\partial x_2} & \cdots \\ \vdots & \vdots & \ddots \end{bmatrix}
\begin{bmatrix} dx_1 \\ dx_2 \\ \vdots \end{bmatrix},
\tag{11.4}
$$

or in a more compact format,

$$F(x_i + dx_i) = F(x_i) + Jdx_i, \tag{11.5}$$

with $x_{i+1} = x_i + dx_i$ being a better approximation to the solution, which means that $F(x_{i+1}) = 0$, implying $-F(x_i) = Jdx_i$ or alternatively $-F(x_i)J^{-1} = dx_i$. The dx is the Newton step that is taken until a stopping criterion applies. Commercial simulation packages have of course implemented these numerical methods and a user can leave it up to the software or decide which of the numerical methods to use.

11.8 Open-loop simulation of an ammonia synthesis plant in UniSim

In the following sections we are going to demonstrate the use of the professional process simulator Honeywell's UniSim. The UniSim Design Suite (https://www.honeywellprocess.com/library/marketing/brochures/Academic%20Program%20-%20UniSim%20Design.pdf) is free for use within universities.

In the first part of this tutorial we will set up an open-loop simulation of an ammonia synthesis plant, and in the second part we will also include a recycle.

First one should become comfortable and familiar with the UniSim Design graphical user interface, explore UniSim Design flowsheet handling techniques and understand the basic input required to run a UniSim Design simulation. Secondly one should learn how to approximate the "real" kinetic-based, heterogeneous reaction into an equilibrium reaction and determine the appropriate reactor model type. Additionally we will determine a physical properties method for ammonia synthesis and apply the acquired skills to build an open-loop ammonia synthesis process simulation, to enter the minimum input required for a simplified ammonia synthesis model and to examine the open-loop simulation results.

11.8.1 Background

Ammonia is one of the most produced chemicals in the world and is often used in fertilizers. In 1913 Fritz Haber and Carl Bosch developed a process for the manufacture of ammonia on an industrial scale (the Haber–Bosch process). This process is known for the extremely high pressures which are required to maintain a reasonable equilibrium constant. Today, 500 million tons of nitrogen fertilizer are produced by this process per year and the process is responsible for sustaining one-third of the Earth's population.

Ammonia is produced by reacting nitrogen from air with hydrogen. Hydrogen is usually obtained from steam reformation of methane, and nitrogen is obtained from deoxygenated air. The chemical reaction is the following:

$$N_2 + 3H_2 \leftrightarrow 2NH_3.$$

Our goal is to simulate the production of ammonia using the UniSim Design Suite. We will create a simplified version of this process in order to learn the basics of how to create a flowsheet in the UniSim Design R443 user interface. A process flow diagram for this process is shown in Figure 11.7.

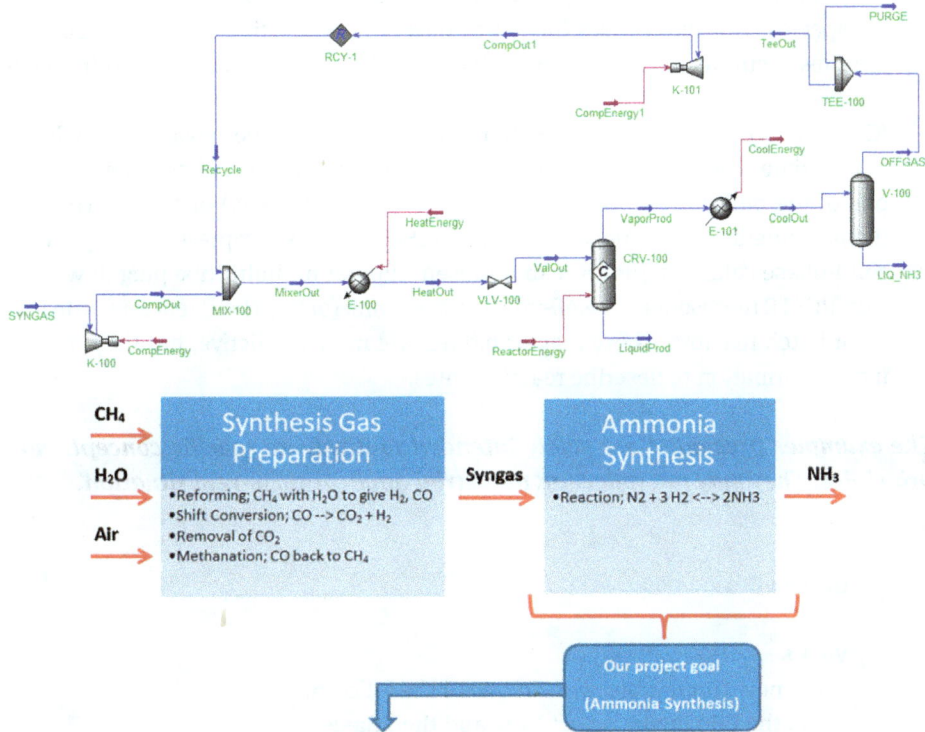

Figure 11.7: Process flowsheet of the ammonia synthesis process.

Knowledge base: physical properties for ammonia synthesis process

Equation-of-state models provide an accurate description of the thermodynamic properties of the high-temperature, high-pressure conditions encountered in ammonia plants. The Peng–Robinson model was chosen for this application.

Knowledge base: reactor model types

The UniSim Design Suite includes three types of reactor models. The type one chooses depends on the level of rigor one wants to use and the amount of information one has available.

- Balance-based (RYIELD, RSTOIC) – These reactors are for mass and energy balance purposes. One specifies the conversion or yield and the reaction stoichiometry. In essence, one provides UniSim with the expected result and it handles the details of the mass, energy and species balances.

– Equilibrium-based (REQUIL, RGIBBS) – These reactor models are appropriate for fast reactions that reach equilibrium quickly (although there are ways to specify the approach to equilibrium for non-ideal cases). RGIBBS is the most flexible model. It allows for multiple phases (including multiple solid phases) and multiple species. This model uses Gibbs free energy minimization to predict results. It requires accurate thermodynamics since the Gibbs energy is calculated from enthalpy and entropy.

– Kinetic-based or rate-based (RCSTR, RPLUG, RBATCH) – These reactor models are appropriate when one knows the reaction kinetics. One describes kinetics using one of the built-in reaction models (power law, LHHW, etc.) or one's own user-defined kinetic subroutine. RBATCH and RCSTR are able to represent reactors with solid-phase catalysts. RPLUG can represent tubular or multi-tube plug flow reactors. RCSTR represents any well-mixed stirred tank (or fluid bed) reactors. RBATCH is for batch reactors. These reactor models are more predictive, but they require more information to describe reaction rates.

The examples presented are solely intended to illustrate specific concepts and principles. They may not reflect an industrial application or real situation.

11.8.2 UniSim Design solution

Setting your session preferences

1. To start a new simulation case, do one of the following:
   – From the **File** menu, select **New** and then **Case**.
   – Click on the **New Case** icon.
   The **Simulation Basis Manager** appears:

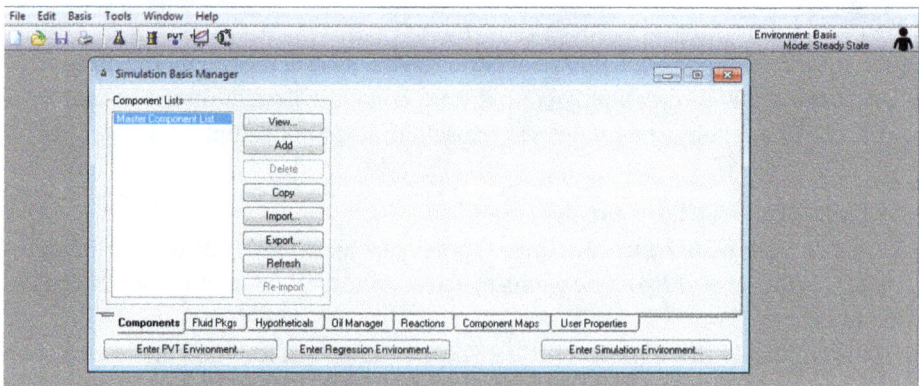

The Simulation Basis Manager view allows you to create, modify and manipulate Fluid Packages in your simulation case. Most of the time, as with this example, you will require only one Fluid Package for your entire simulation. Next, you will set your session preferences before building a case.

2. From the **Tools** menu, select **Preferences.** The Session Preferences view appears. You should be on the **Options** page of the **Simulation** tab.

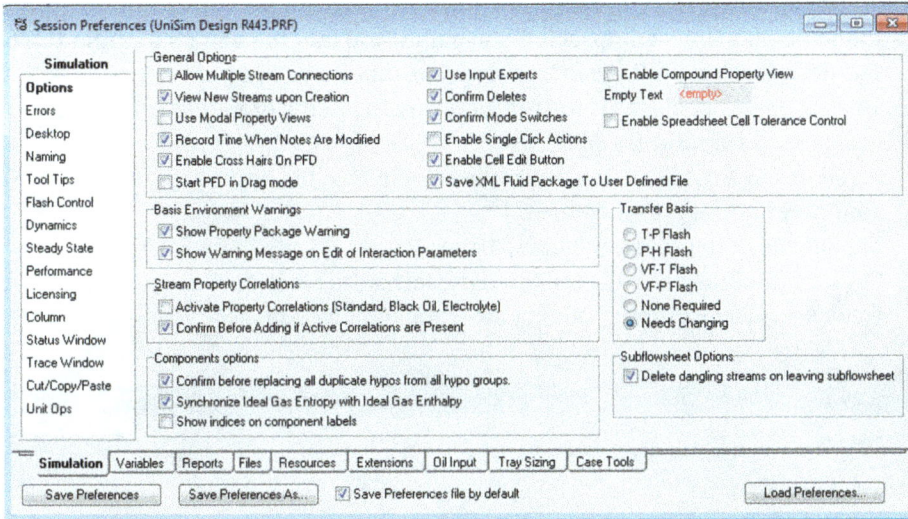

The UniSim Design default session settings are stored in a Preference file called **Unisimdesign Rxxx.prf.** When you modify any of the preferences, you can save the changes in a new Preference file by clicking either **Save Preferences**, which will overwrite the default file, or **Save Preferences As...** and UniSim Design will prompt you to provide a name for the new Preference file. Click on the **Load Preferences** button to load any simulation case. Check the **Save Preferences File by default** to automatically save the preference file when exiting UniSim Design. In the General Options group, ensure the **Use Modal Property Views** checkbox is unchecked.

11.8.3 Creating a new unit set

The first step in building the simulation case is choosing a unit set. Since UniSim Design does not allow you to change any of the three default unit sets listed, you will create a new unit set by cloning an existing one. For this example, a new unit set will be made based on the UniSim Design **Field** set, which you will then customize.

To create a new unit set, do the following:

1. In the Session Preferences view, click on the **Variables** tab.
2. Select the **Units** page if it is not already selected.
3. In the **Available Unit Sets** group, select **Field** to make it the active set, choose SI.
4. Click on the **Clone** button. A new unit set named **NewUser** appears. This unit set becomes the currently Available Unit Set.
5. In the **Unit Set Name** field, enter a name for the new unit set, for example "Ammonia." You can now change the units for any variable associated with this new unit set.
6. In the Display Units group, scroll down until you find the unit for **Temperature**. The default setting is C. A more appropriate unit for this process is **K**.
7. To view the available units for **Temperature**, open the drop-down list in the cell beside the **Temperature** cell.
8. Scroll through the list using either the scroll bar or the arrow keys, and select **K**. Your new unit set is now defined.
9. Change the unit for Pressure from kPa to **atm**.
10. Click on the **Close** icon (in the top right corner) to close the Session Preferences view. Next, you will start building the simulation case.

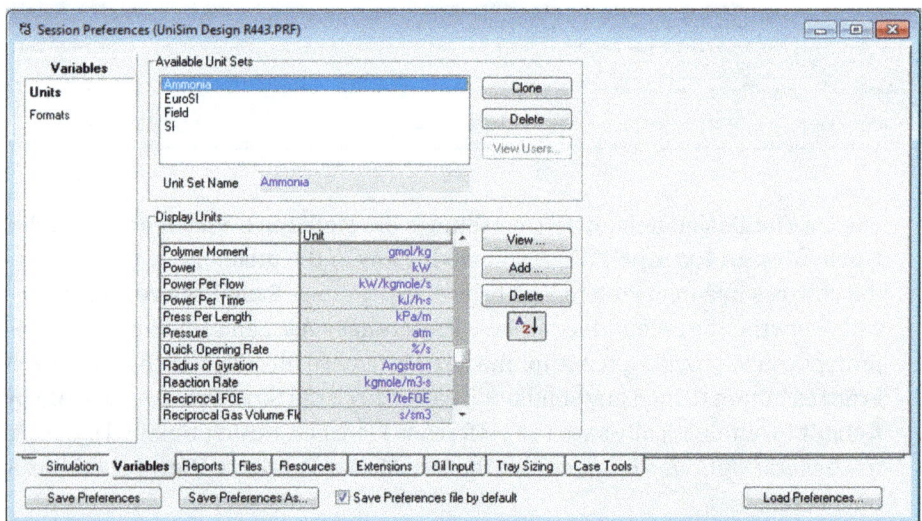

11.8.4 Building the simulation

In building a simulation case UniSim splits the configuration options into two different environments:

- The basis environment enables you to specify the basic information, like components, property package and reactions, associated with the simulation.

– The simulation environment enables you to specify the streams and operation equipment associated with the simulation and view the calculated results from the simulation.

Creating a Fluid Package

The next step is to add a Fluid Package. As a minimum, a Fluid Package contains the components and a property method (for example an equation of state) that UniSim Design uses in its calculations for a particular flowsheet. Depending on what is required in a specific flowsheet, a Fluid Package may also contain other information, such as reactions and interaction parameters.

1. In the Simulation Basis Manager view, click on the **Fluid Pkgs** tab.
2. Click on the **Add** button, and the property view for your new Fluid Package appears.

   The property view is divided into a number of tabs. Each tab contains the options that enable you to completely define the Fluid Package.

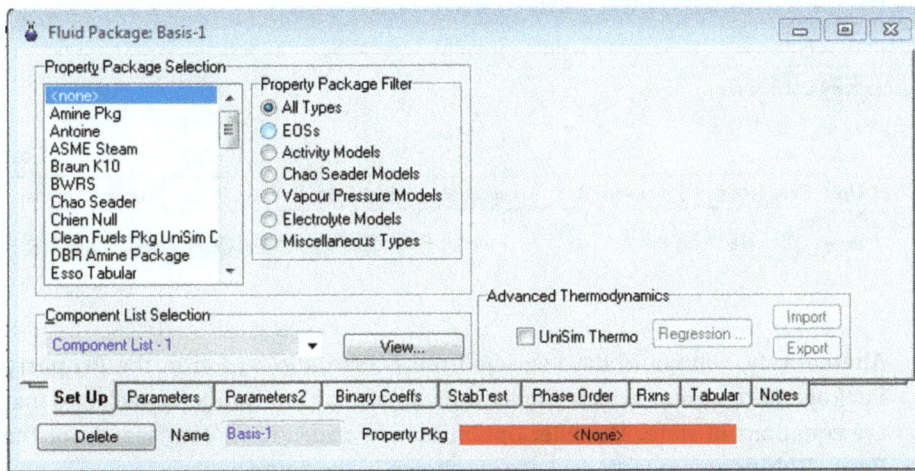

The first step in configuring a Fluid Package is to choose a Property Package on the **Set Up** tab. The current selection is **<none>**. For this tutorial, you will select the Peng–Robinson property package.

3. Do one of the following:
   – Select **<none>** in the Property Package Selection list and type **Peng–Robinson**. UniSim Design automatically finds the match to your input.
   – Select **<none>** in the Property Package Selection list and the up and down keys to scroll through the Property Package Selection list until **Peng–Robinson** is selected.

– Use the vertical scroll bar to move up and down the list until **Peng–Robinson** becomes visible, then select it.

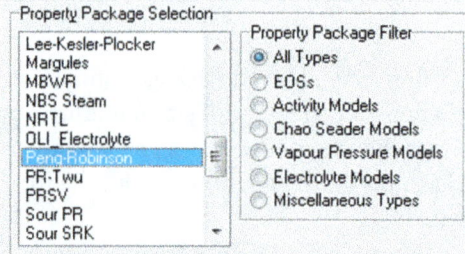

The **Property Pkg** indicator at the bottom of the Fluid Package view now indicates that **Peng–Robinson** is the current property package for this Fluid Package. UniSim Design has also automatically created an empty component list to be associated with the Fluid Package.

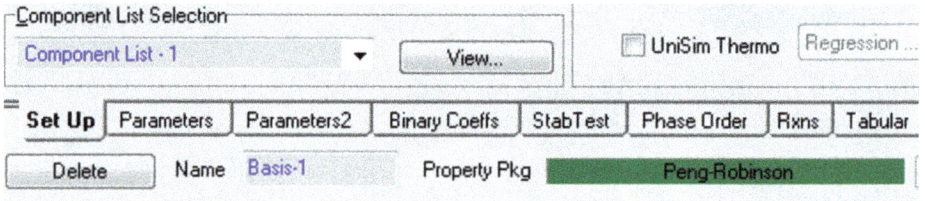

Alternatively, you could have selected the **EOSs** radio button in the **Property Package Filter** group, which filters the list to display only property packages that are equations of state. The filter option helps narrow down your search for the **Peng–Robinson** property package, as shown in the figure below.

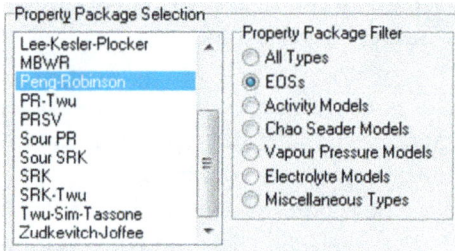

Creating a component list

Now that you have selected a property package to be used in the simulation, the next step is to select the components. You can create a list of components using the options on the Components tab of the Simulation Basis Manager view or from the Set Up tab of the Fluid Package view.

In this tutorial, we will create the component list using the option in the Fluid Package view.

1. In the **Set Up** tab, select Component List - 1 from the Component List Selection drop-down list.
2. Click on the **View** button.
   The Component List View appears.

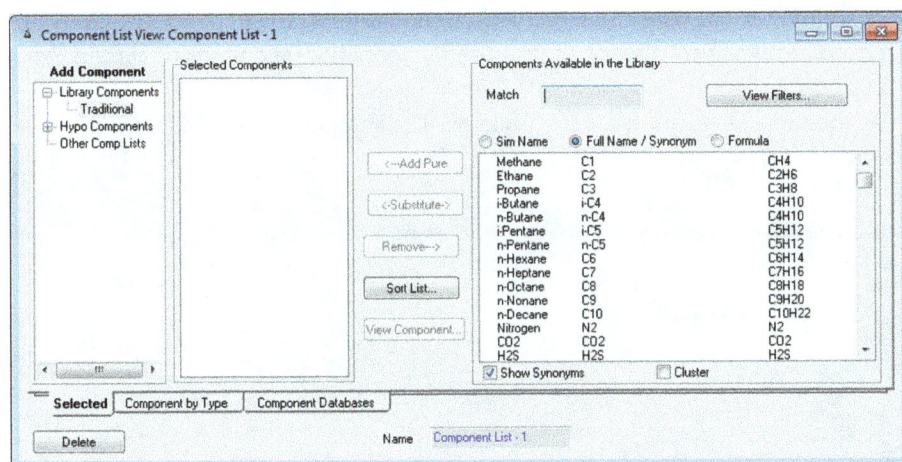

There are a number of ways to select components for your simulation. One method is to use the matching feature. Each component is listed in three ways on the Selected tab:

Matching Method	Description
Sim Name	The name appearing within the simulation.
Full Name/ Synonym	IUPAC name (or similar), and synonyms for many components.
Formula	The chemical formula of the component. This is useful when you are unsure of the library name of a component, but know its formula.

At the top of each of these three columns is a corresponding radio button. Based on the selected radio button, UniSim Design locates the component(s) that best matches the input you type in the **Match** cell.

For this tutorial, you will add the following components: **H2, N2, CH4, AR, CO** and **NH3**. First, you will add Hydrogen using the match feature.

3. Ensure the **Full Name/Synonym** radio button is selected, and the **Show Synonyms** checkbox is checked.
4. Move to the **Match** field by clicking on the field, or by pressing **Alt + M**.
5. Type **Hydrogen**. UniSim Design filters as you type and displays only components that match your input.
6. With **Hydrogen** selected, add it to the current composition list by doing one of the following:
   - Press the **ENTER** key.
   - Click on the **Add Pure** button.
   - Double-click on **Hydrogen**.

   In addition to the three match criteria radio buttons, you can also use the **View Filters...** button to display only those components belonging to certain families.
7. Do the same steps for other components. The completed component list appears below.

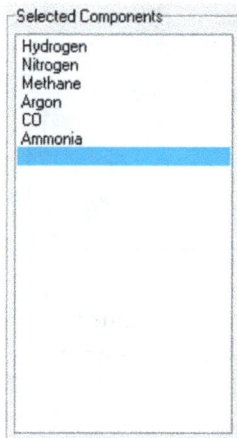

11.8.5 Defining Reaction Sets

1. Return to the Simulation Basis Manager view by clicking on its title bar, or by clicking the **Home View** icon.
2. Click on the **Reactions** tab. This tab allows you to define all the reactions for the flowsheet.

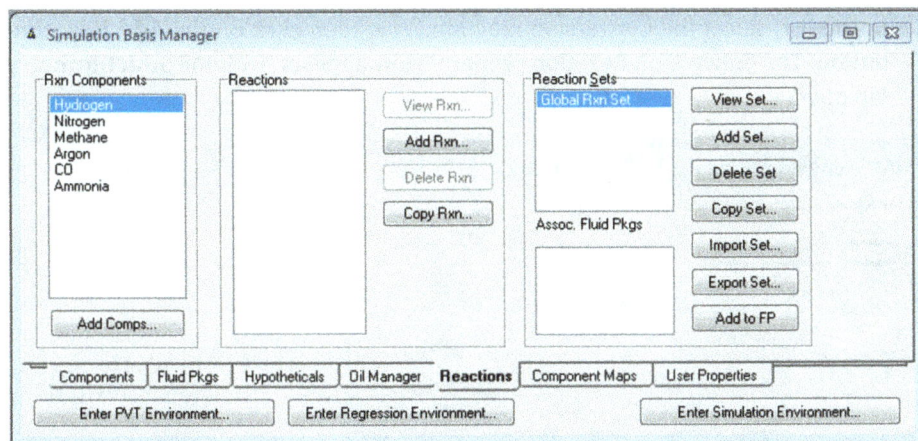

The reaction between nitrogen and hydrogen to produce ammonia is as follows:

$$N_2 + 3H_2 \leftrightarrow 2NH_3.$$

3. The first task in defining the reaction is choosing the components that will be participating in the reaction. In this tutorial, all the components that were selected in the Fluid Package are participating in the reaction, so you do not have to modify this list. For a more complicated system, however, you would add or remove components from the list.

4. Once the reaction components have been chosen, the next task is to create the reaction.

5. In the Reactions group, click on the **Add Rxn** button. The Reactions view appears.

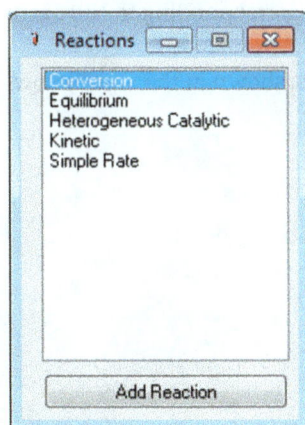

6. In the list, select the Conversion reaction type, then click on the **Add Reaction** button. The Conversion Reaction property view appears, with the **Stoichiometry** tab open.

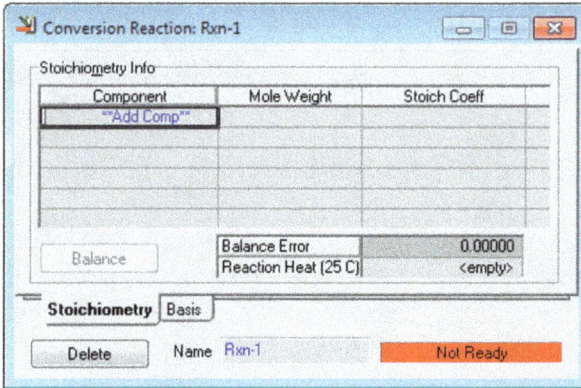

7. In the **Component** column, click in the cell labeled ****Add Comp****.
8. Select **Nitrogen** as a reaction component by doing one of the following:
   - Open the drop-down list and select Nitrogen from the list of available reaction components.
   - Type "Nitrogen." UniSim Design filters as you type, searching for the component which matches your input. When Nitrogen is selected, press the **ENTER** key to add it to the Component list.
9. Repeat this procedure to add **Hydrogen** and **Ammonia** to the reaction table.
   The next task is to enter the stoichiometric information. A negative stoichiometric coefficient indicates that the component is consumed in the reaction, while a positive coefficient indicates the component is produced.
10. In the **Stoich Coeff** column, click in the **<empty>** cell corresponding to Nitrogen.
11. Type "**-1**" and press the **ENTER** key.
12. Enter the coefficients for the remaining components as shown in the view below.

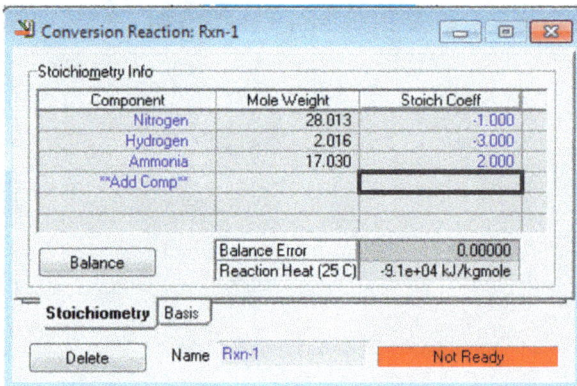

Once the stoichiometric coefficients are supplied, the **Balance Error** cell will show 0 (zero), indicating that the reaction is mass-balanced. UniSim Design will also calculate and display the heat of reaction in the **Reaction Heat** cell. In this case, the **Reaction Heat** is negative, indicating that the reaction produces heat (it is exothermic).

13. Click on the **Basic** tab. Specify **Base component** and **Conversion** as shown in the view below.

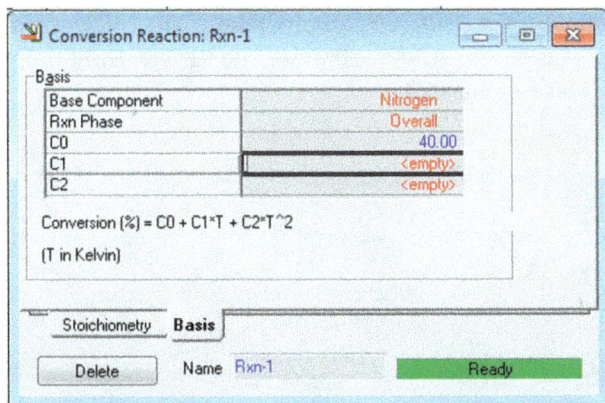

14. Close both the **Conversion Reaction** property view and the **Reactions** view.
15. Click on the **Home View** icon to ensure the **Simulation Basis Manager** view is active. In the **Reactions** tab, the new reaction, **Rxn-1**, now appears in the **Reactions** group.

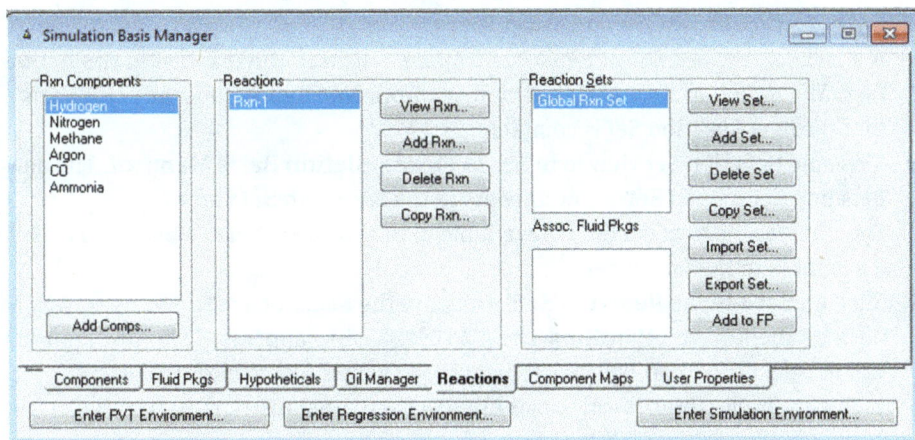

The next task is to create a Reaction Set that will contain the new reaction. In the Reaction Sets list, UniSim Design provides the Global Rxn Set (Global

Reaction Set) which contains all of the reactions you have defined. In this tutorial, since there is only one REACTOR, the default Global Rxn Set could be attached to it; however, for illustration purposes, a new Reaction Set will be created.

16. Reaction Sets provide a convenient way of grouping related reactions. For example, consider a flowsheet in which a total of five reactions are taking place. In one **REACTOR** operation, only three of the reactions are occurring (one main reaction and two side reactions). You can group the three reactions into a Reaction Set and then attach the set to the appropriate REACTOR unit operation.

17. In the Reaction Sets group, click on the **Add Set** button. The **Reaction Set** property view appears with the default name **Set-1**.

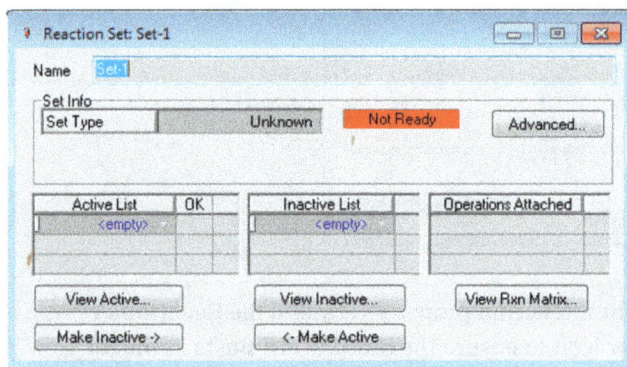

18. In the **Active List**, click in the cell labeled **<empty>**.
19. Open the drop-down list and select **Rxn-1**.
   A checkbox labeled **OK** automatically appears next to the reaction in the **Active List**. The Reaction Set status bar changes from **Not Ready** to **Ready**, indicating that the new Reaction Set is complete.
20. Close the Reaction Set view to return to the **Simulation Basis Manager**. The new Reaction Set named **Set-1** now appears in the **Reaction Sets** group.
21. The final task is to make the set available to the Fluid Package, which also makes it available in the flowsheet.
22. Click on **Set-1** in the **Reaction Sets** group on the Reactions tab.
23. Click on the **Add to FP button**. The **Add 'Set-1'** view appears. This view prompts you to select the Fluid Package to which you would like to add the Reaction Set. In this example, there is only one Fluid Package, **Basis-1**.

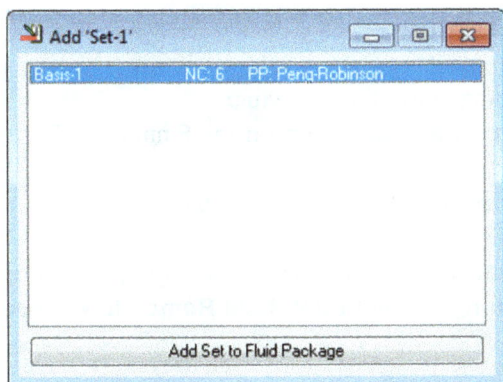

24. Select **Basis-1** and click on the **Add Set to Fluid Package** button.
25. Click on the **Fluid Pkgs** tab to view a summary of the completed Fluid Package.

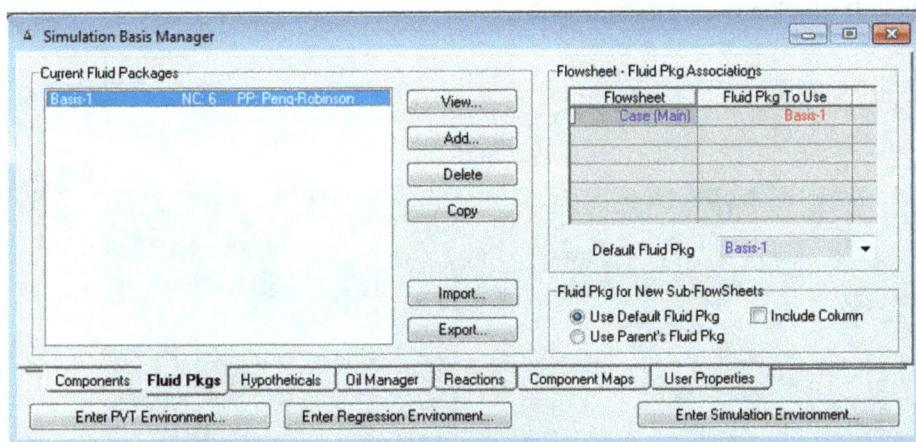

The list of **Current Fluid Packages** displays the new Fluid Package, **Basis-1**, showing the number of components (NC) and property package (PP). The new Fluid Package is assigned by default to the Main Simulation, as shown in the **Flowsheet - Fluid Pkg Associations** group. Now that the Basis is defined, you can install streams and operations in the Simulation environment (also referred to as the Parent Simulation environment or Main Simulation environment).

11.8.6 Entering the simulation environment

Go to the simulation environment by doing one of the following:
-   Click on the **Enter Simulation Environment** button on the Simulation Basis Manager.
-   Click on the **Enter Simulation Environment** icon on the toolbar.

When you enter the Simulation environment, the initial view that appears depends on your current Session Preferences setting for the **Initial Build Home View**. Three initial views are available:
-   PFD;
-   Workbook;
-   Summary.

Any or all of these can be displayed at any time; however, when you first enter the Simulation environment, only one appears. In this example, the initial Home View is the **PFD** (UniSim Design default setting).

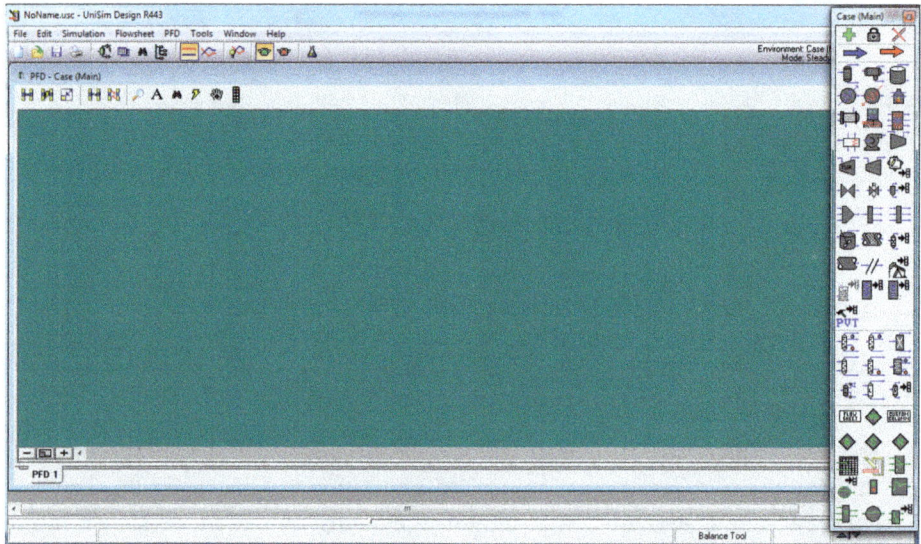

There are several things to note about the Main Simulation environment. In the upper right corner, the Environment has changed from **Basis** to **Case (Main)**. A number of new items are now available in the menu bar and tool bar, and the PFD and Object Palette are open on the Desktop. These latter two objects are described below.

Objects	Description
PFD	The PFD is a graphical representation of the flowsheet topology for a simulation case. The PFD view shows operations and streams and the connections between the objects. You can also attach information tables or annotations to the PFD. By default, the view has a single tab. If required, you can add additional PFD pages to the view to focus in on the different areas of interest.
Object Palette	A floating palette of buttons that can be used to add streams and unit operations.
	You can toggle the palette open or closed by pressing **F4**, or by selecting the **Open/Close Object Palette** command from the **Flowsheet** menu.

1. Before proceeding any further, save your case by doing one of the following:
   - From the **File** menu, select **Save**.
   - Press **CTRL + S**.
   - Click on the **Save** icon on the toolbar.
   If this is the first time you have saved your case, the Save Simulation Case As view appears.
   By default, the File Path is the **Cases** subdirectory in your **UniSim Design** directory.
2. In the **File Name** cell, type a name for the case, for example **Ammonia**. You do not have to enter the **.usc** extension; UniSim Design automatically adds it for you.
3. Once you have entered a file name, press the **ENTER** key or click on the **Save** button. UniSim Design saves the case under the name you have given it when you **save** in the future. The **Save As** view will not appear again unless you choose to give it a new name using the **Save As** command. If you enter a name that already exists in the current directory, UniSim Design will ask you for confirmation before overwriting the existing file.

11.8.7 Using the Workbook

The Workbook displays information about streams and unit operations in a tabular format, while the PFD is a graphical representation of the flowsheet.
1. Click on the **Workbook** icon on the toolbar to access the Workbook view. You can also access the Workbook from the **Tools** menu.

11.8.8 Installing the feed streams

In general, the first action you perform when you enter the Simulation environment is installing one or more feed streams. The following procedure explains how to create a new stream.

1. On the **Material Streams** tab of the Workbook, type the stream name **SYNGAS** in the cell labeled ****New****, and press **ENTER**. UniSim Design will automatically create the new stream with the name defined above. Your Workbook should appear as shown below.

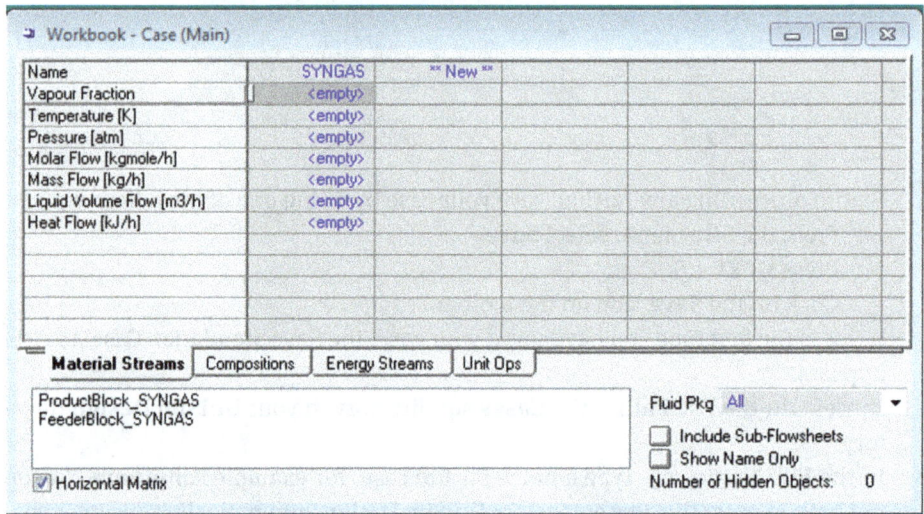

When you pressed **ENTER** after typing in the stream name, UniSim Design automatically advanced the active cell down one to Vapour Fraction. Next you will define the feed conditions.

2. Move to the **Temperature** cell for SYNGAS by clicking it, or by pressing the **DOWN** arrow key.

3. Type **553.15** in the **Temperature** cell. In the Unit drop-down list, UniSim Design displays the default units for temperature, in this case **K**. This is the correct unit for this exercise.

4. Press the **ENTER** key.
   Your active location should now be the Pressure cell for SYNGAS. If you know the stream pressure in another unit besides the default unit of atm, UniSim Design will accept your input in any one of the available different units and automatically convert the supplied value to the default unit for you. For this example, the pressure of SYNGAS is 26.17 atm.

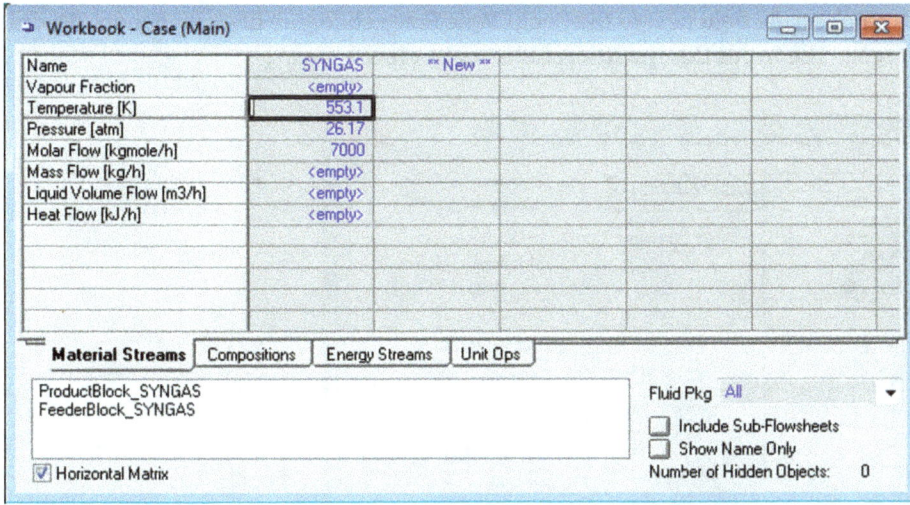

5. To finish the stream definition, we need to specify the composition. Click on the **Compositions** tab. By default, the components are listed by **Mole Fractions**.

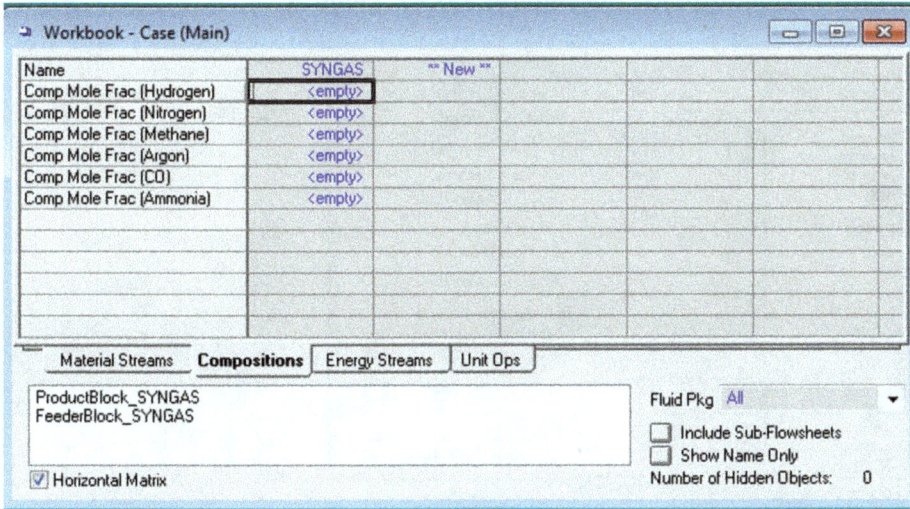

6. Click on the field next to the Comp Mole Frac cell for the first component, Hydrogen. **The Input Composition for Stream view** appears.

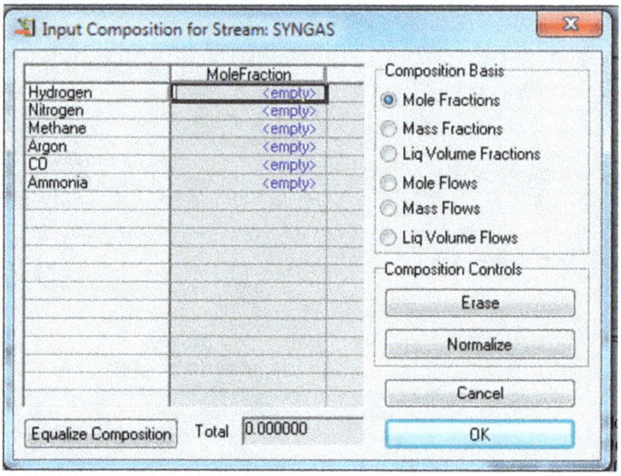

7. Choose **Mole Flows** in **Composition Basic**. Click on the **CompMoleFlow** cell for each component, and fill in as follows: **H2 = 5160 kmol/hr, N2 = 1732 kmol/hr, CH4 = 72 kmol/hr, AR = 19 kmol/hr, CO = 17 kmol/hr.**

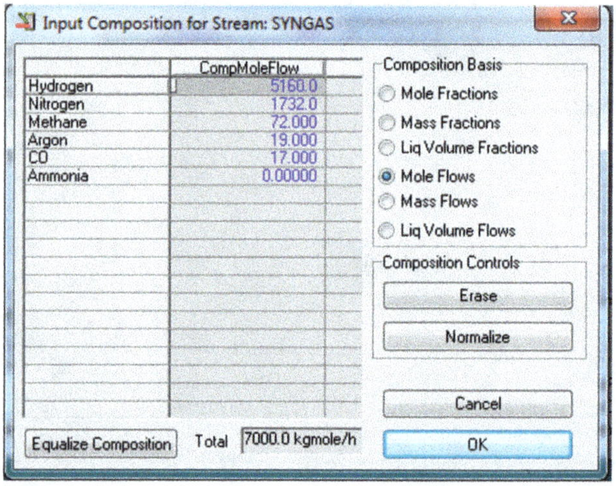

8. Click on the **OK** button, and UniSim Design accepts the composition. The stream is now completely defined, so UniSim Design flashes it at the conditions given to determine its remaining properties. Go back to the Material Streams tab; the properties of SYNGAS appear below. The values you specified are **blue** and the calculated values are **black**.

Name	SYNGAS	** New **				
Vapour Fraction	1.0000					
Temperature [K]	553.1					
Pressure [atm]	26.17					
Molar Flow [kgmole/h]	7000					
Mass Flow [kg/h]	6.131e+004					
Liquid Volume Flow [m3/h]	214.1					
Heat Flow [kJ/h]	4.482e+007					

Workbook - Case (Main)

Material Streams Compositions Energy Streams Unit Ops

ProductBlock_SYNGAS
FeederBlock_SYNGAS

Fluid Pkg All

☐ Include Sub-Flowsheets
☐ Show Name Only
Number of Hidden Objects: 0

☑ Horizontal Matrix

11.8.9 Installing unit operations

In the last section you defined the feed streams. Now you will install the necessary unit operations for the ammonia synthesis process.

11.8.9.1 Adding a Compressor

1. Double click on the **Compressor icon** in the operations palette. The **Compressor property** view appears.

K-100

Design

Connections
Parameters
Links
User Variables
Notes

Name   K-100

Inlet

Fluid Package
Basis-1

Energy

Outlet

Design Rating Worksheet Performance Dynamics Cost

Delete   Requires a feed stream   ☑ On   ☐ Ignored

As with a stream, a unit operation's property view contains all the information defining the operation, organized in tabs and pages. The status bar at the bottom of the view shows that the operation requires a feed stream.

2. Click on the **Inlet** field, choose **SYNGAS** from the list. The status indicator now displays "Requires a product stream."
3. Click in the **Outlet** field. Type **CompOut** in the cell and press **ENTER**.
   UniSim Design recognizes that there is no existing stream named **CompOut**, so it will create the new stream with this name. The status indicator now displays "Requires an energy stream."
4. Click in the **Energy** field. Type **CompEnergy** in the cell and press **ENTER**. The status indicator now displays "Unknown Duty."
5. Click on the **Worksheet** tab. Choose the **Conditions** page. Set the pressure of the **CompOut** stream to **271.4 atm**. Now, the status indicator displays a green **OK**, indicating that the operation and attached streams are completely calculated.

Name	SYNGAS	CompOut	CompEnergy
Vapour	1.0000	1.0000	<empty>
Temperature [K]	553.1	1208	<empty>
Pressure [atm]	26.17	271.4	<empty>
Molar Flow [kgmole/h]	7000	7000	<empty>
Mass Flow [kg/h]	6.131e+004	6.131e+004	<empty>
LiqVol Flow [m3/h]	214.1	214.1	<empty>
Molar Enthalpy [kJ/kgmole]	6403	2.706e+004	<empty>
Molar Entropy [kJ/kgmole-C]	126.1	130.7	<empty>
Heat Flow [kJ/h]	4.482e+007	1.894e+008	1.446e+008

6. Now that the Compressor is completely known, close the view to return to the FPD.

11.8.9.2 Adding a Mixer

As with most commands in UniSim Design, installing an operation can be accomplished in a number of ways. One method is through the **Unit Ops** tab of the **Workbook**.

1. Click on the **Workbook** icon to access the Workbook view.
2. Click on the **Unit Ops** tab of the Workbook.
3. Click on the **Add UnitOp** button. The **UnitOps** view appears, listing all available unit operations.

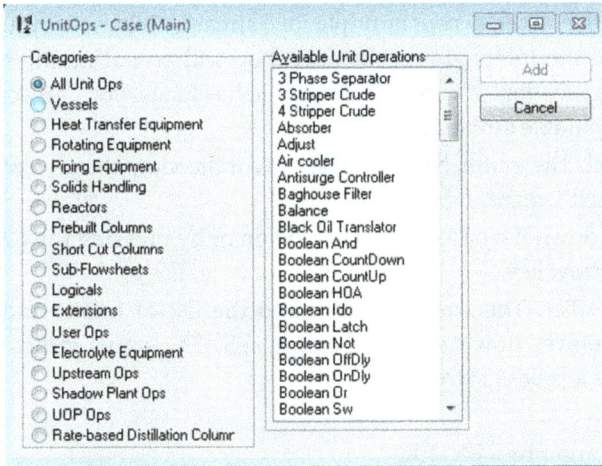

4. Select **Mixer** by doing one of the following:
   - Click on an operation in the Available Unit Operations list and type **Mixer**.
   - Click on an operation in the Available Unit Operations list and press the **DOWN** arrow key to scroll down the list of available operations to Mixer.
   - Scroll down the list using the vertical scroll bar and click on **Mixer**.
5. With Mixer selected, click on the **Add** button or press the **ENTER** key.
   You can also use the filters to find and add an operation; for example, select the **Piping Equipment** radio button under Categories. A filtered list containing just piping operations appears in the Available Unit Operations group. The Mixer property view appears.

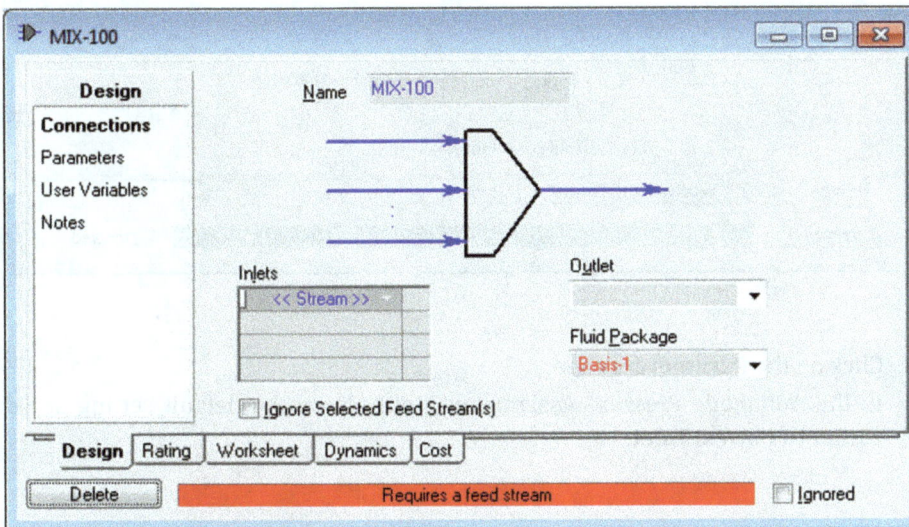

Many operations, such as the Mixer, accept multiple feed streams. Whenever you see a table like the one in the Inlets group, the operation will accept multiple stream connections at that location. When the Inlets table has focus, you can access a drop-down list of available streams.

6. Click on the **«Stream»** cell. The status bar at the bottom of the view shows that the operation requires a feed stream.

7. Open the **«Stream»** drop-down list of feeds by clicking on or by pressing the **F2** key and then the **DOWN** arrow key.

8. Select **CompOut** from the list. The stream is added to the list of Inlets, and **«Stream»** automatically moves down to a new empty cell. The status indicator now displays "Requires a product stream."

9. Click in the **Outlet** field.

10. Type **MixerOut** in the cell and press **ENTER**.
    UniSim Design recognizes that there is no existing stream named MixerOut, so it will create a new stream with this name. The status indicator now displays a green **OK**, indicating that the operation and attached streams are completely calculated.

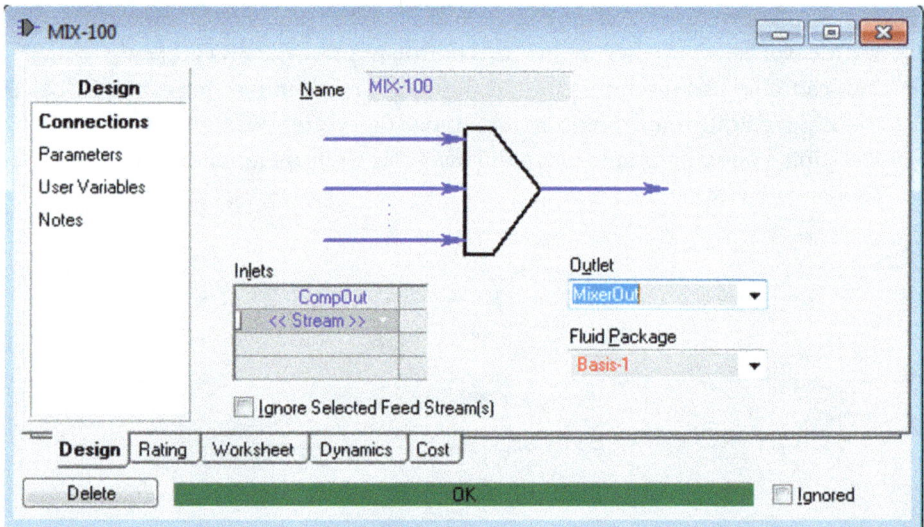

11. Click on the **Parameters** page.

12. In the Automatic Pressure Assignment group, leave the default setting at **Set Outlet to Lowest Inlet**.

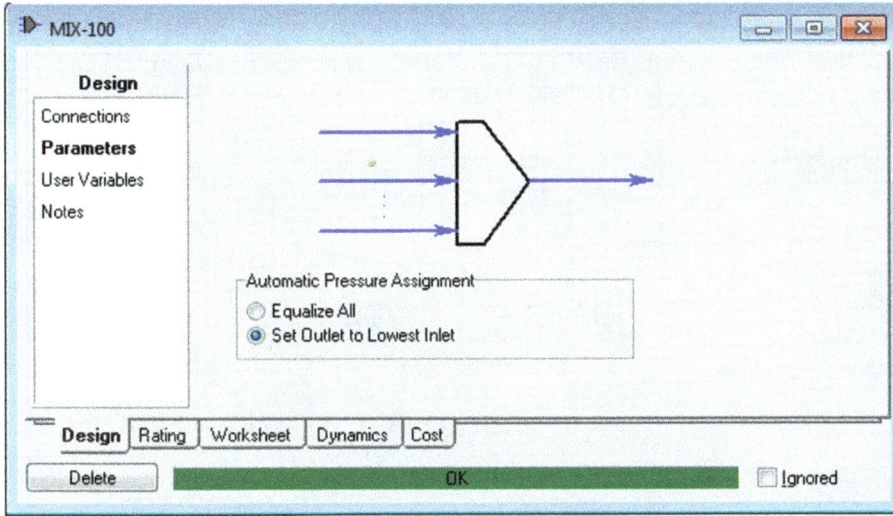

UniSim Design has calculated the outlet stream by combining the two inlets and flashing the mixture at the lowest pressure of the inlet streams.

13. To view the calculated outlet stream, click on the **Worksheet** tab and select the **Conditions** page.
14. Now that the Mixer is completely known, close the view to return to the Workbook. The new operation appears in the table on the **Unit Ops** tab of the Workbook.

11.8.9.3 Adding a Heater

1. Double click on the **Heater icon** ⚙ on the operations palette. This will bring up a screen where you can specify the Heater design.

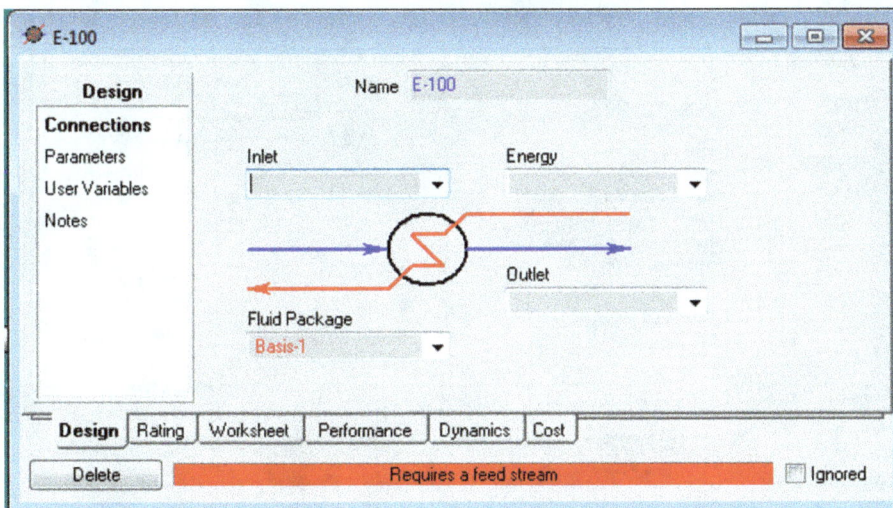

2. Attach the **MixerOut** stream by clicking on the box below **Inlet** and selecting it. In the **Outlet** box, type **HeatOut**. In the **Energy** box, type **HeatEnergy**. The status indicator now displays "Unknown Delta P."

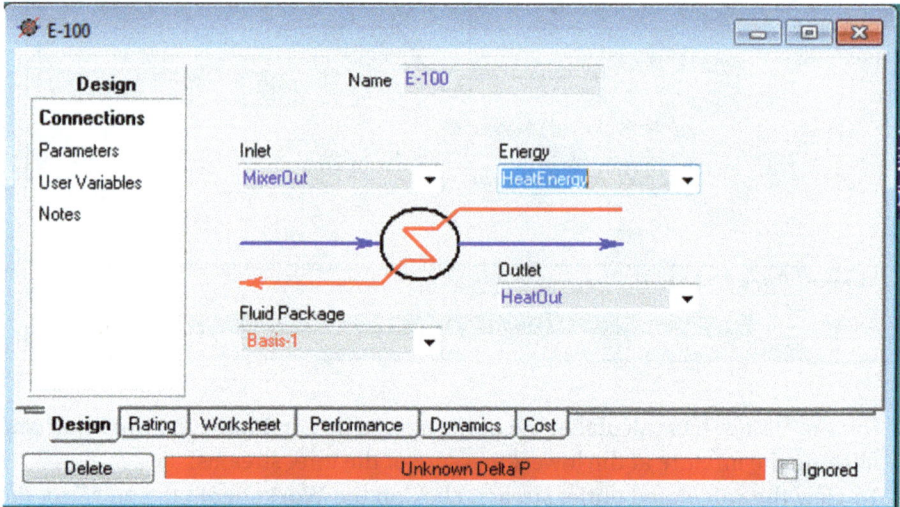

3. Select the **Parameters** page. In the **Delta P** field, specify a pressure drop of **0 atm**. The status indicator now displays "Unknown Duty."
4. Select the **Worksheet** tab, fill in a temperature of **755 K** for the HeatOut stream. The status indicator now displays a green OK, indicating that the operation and attached streams are completely calculated.

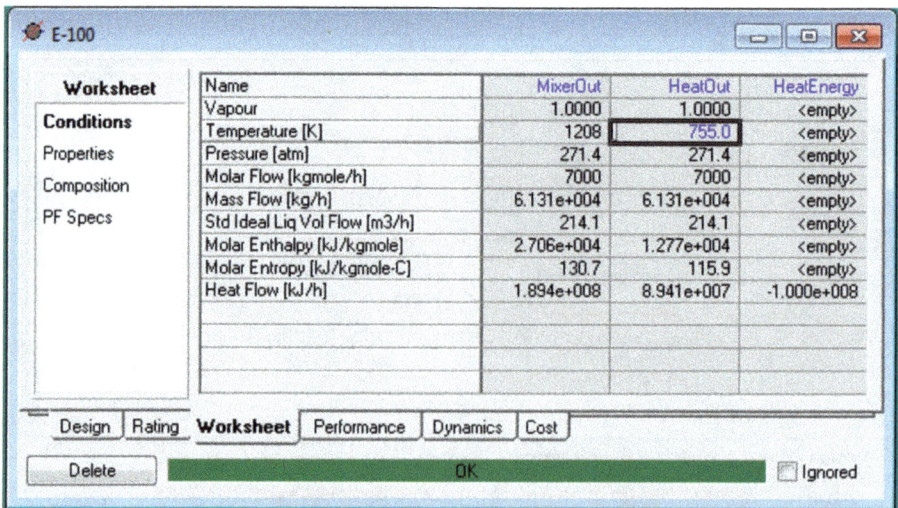

Name	MixerOut	HeatOut	HeatEnergy
Vapour	1.0000	1.0000	<empty>
Temperature [K]	1208	755.0	<empty>
Pressure [atm]	271.4	271.4	<empty>
Molar Flow [kgmole/h]	7000	7000	<empty>
Mass Flow [kg/h]	6.131e+004	6.131e+004	<empty>
Std Ideal Liq Vol Flow [m3/h]	214.1	214.1	<empty>
Molar Enthalpy [kJ/kgmole]	2.706e+004	1.277e+004	<empty>
Molar Entropy [kJ/kgmole-C]	130.7	115.9	<empty>
Heat Flow [kJ/h]	1.894e+008	8.941e+007	-1.000e+008

5. Now that the Heater is completely known, close the view to return to the FPD.

11.8.9.4 Adding a Valve

1. On the Object Palette, right-click and hold the **Valve** icon.
2. Drag the cursor over the PFD. The mouse pointer becomes a bullseye.
3. Position the bullseye pointer beside the HeatOut stream and release the mouse button.
4. A Valve icon named VLV-100 appears.
5. Double-click the VLV-100 icon on the PFD to open its property view.
6. In the Valve property view, specify the following connections:

Tab [Page]	In this cell...	In this cell...
Design [Connections]	Inlet	HeatOut
	Outlet	ValOut
Design [Parameters]	Delta P	1.4 atm

7. Click on the **Close** icon to close the Valve property view.

11.8.9.5 Adding a Conversion Reactor

1. Ensure that the Object Palette is displayed; if it is not, press **F4**.
2. You will add the Conversion Reactor (CR) to the right of the Valve, so if you need to make some empty space available in the PFD, scroll to the right using the horizontal scroll bar.
3. In the Object Palette, click on the **General reactors** icon, choose the **Conversion reactor** .
4. Position the cursor in the **PFD** to the right of the **ValOut** stream. The cursor changes to a special cursor with a plus (+) symbol attached to it. The symbol indicates the location of the operation icon.
5. Click to "drop" the **Reactor** onto the **PFD**. UniSim Design creates a new **Reactor** with a default name, **CRV-100**. The **Reactor** has a red status (color), indicating that it requires feed and product streams.
6. Double-click the **CRV-100** icon to open its property view.
7. Click on the **Design** tab, then select the **Connections** page. The names of the **Inlet, Vapor Outlet**, Liquid Outlet and Energy streams are specified as **ValOut, VaporProd, LiquidProd** and **ReactorEnergy**, respectively.

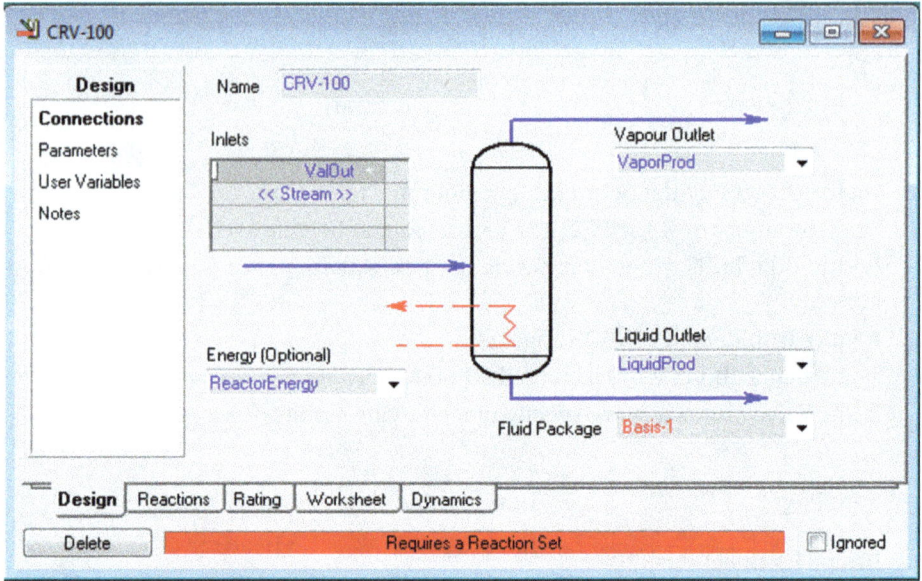

8. Click on the **Reactions** tab. Next you will attach the Reaction Set that you created in the Basis Environment.
9. From the **Reaction Set** drop-down list, select **Set-1**. The completed **Reactions** tab appears below.

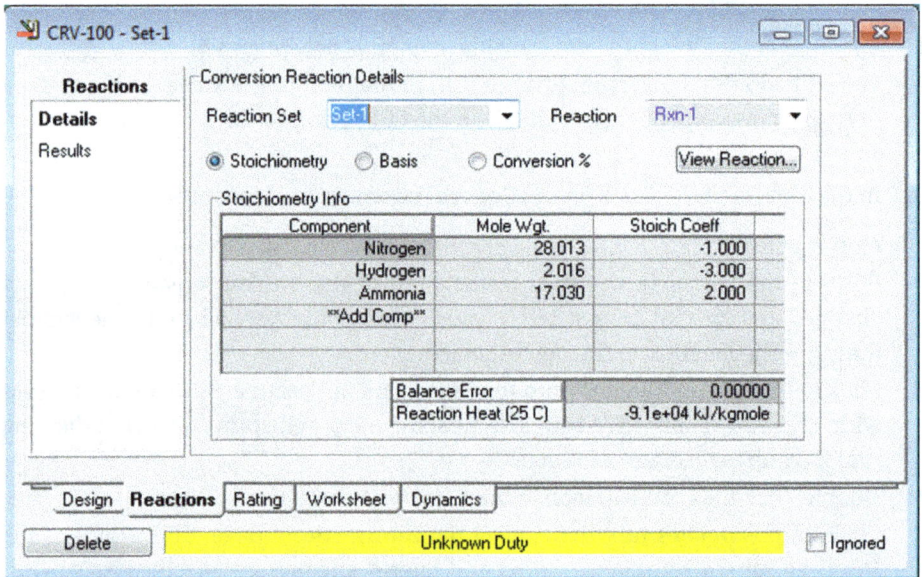

10. Click on the **Worksheet** tab.

CRV-100 - Set-1

Name	ValOut	LiquidProd	VaporProd	F
Vapour	1.0000	0.0000	1.0000	
Temperature [K]	755.1	<empty>	<empty>	
Pressure [atm]	270.0	270.0	270.0	
Molar Flow [kgmole/h]	7000	<empty>	<empty>	
Mass Flow [kg/h]	6.131e+004	<empty>	<empty>	
Std Ideal Liq Vol Flow [m3/h]	214.1	<empty>	<empty>	
Molar Enthalpy [kJ/kgmole]	1.277e+004	<empty>	<empty>	
Molar Entropy [kJ/kgmole-C]	115.9	<empty>	<empty>	
Heat Flow [kJ/h]	8.941e+007	<empty>	<empty>	

Worksheet | Conditions | Properties | Composition | PF Specs

Design | Reactions | Rating | **Worksheet** | Dynamics

Delete | Unknown Duty | Ignored

At this point, the **Reactor** product streams and the energy stream **ReactorEnergy** are unknown because the **Reactor** has one degree of freedom. At this point, either the outlet stream temperature or the Reactor duty can be specified. For this example, you will specify the outlet temperature. Initially the **Reactor** is assumed to be operating at 755 K.

11. In the LiquidProd column, click in the **Temperature** cell. Type 755, then press **ENTER**. UniSim Design solves the Reactor.

There is no phase change in the Reactor under isothermal conditions since the flow of the liquid product stream is zero.

CRV-100 - Set-1

Name	ValOut	LiquidProd	VaporProd	ReactorEnergy
Vapour	1.0000	0.0000	1.0000	<empty>
Temperature [K]	755.1	755.0	755.0	<empty>
Pressure [atm]	270.0	270.0	270.0	<empty>
Molar Flow [kgmole/h]	7000	0.0000	5614	<empty>
Mass Flow [kg/h]	6.131e+004	0.0000	6.131e+004	<empty>
Std Ideal Liq Vol Flow [m3/h]	214.1	0.0000	168.3	<empty>
Molar Enthalpy [kJ/kgmole]	1.277e+004	2338	2419	<empty>
Molar Entropy [kJ/kgmole-C]	115.9	132.5	132.3	<empty>
Heat Flow [kJ/h]	8.941e+007	0.0000	1.358e+007	-7.582e+007

Worksheet | Conditions | Properties | Composition | PF Specs

Design | Reactions | Rating | **Worksheet** | Dynamics

Delete | OK | Ignored

12. Close the **Reactor property** view.

11.8.9.6 Adding a Cooler

1. Double click on the **Cooler icon** ⚙ on the operations palette. This will bring up a screen where you can specify the Cooler design.

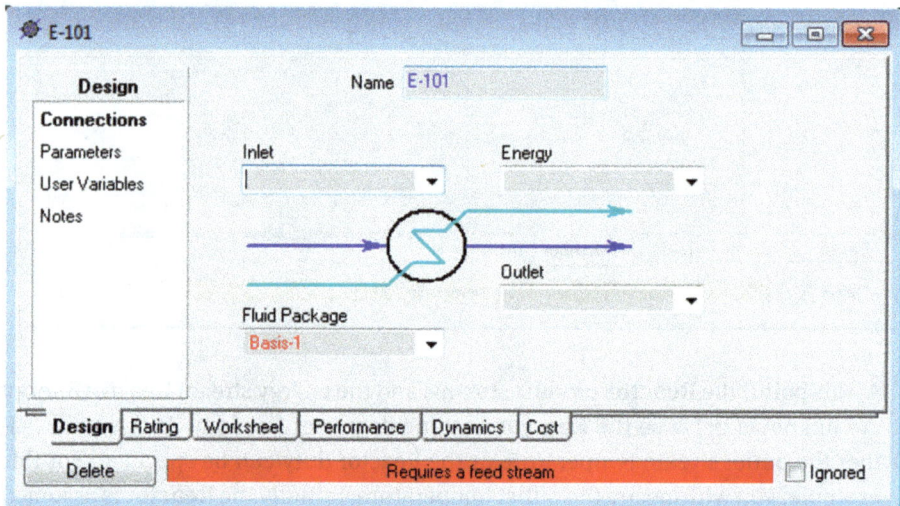

2. Attach the **VaporProd** stream by clicking on the box below **Inlet** and selecting it. In the **Outlet** box, type **CoolOut**. In the Energy box, type **CoolEnergy**. The status indicator now displays "Unknown Delta P."

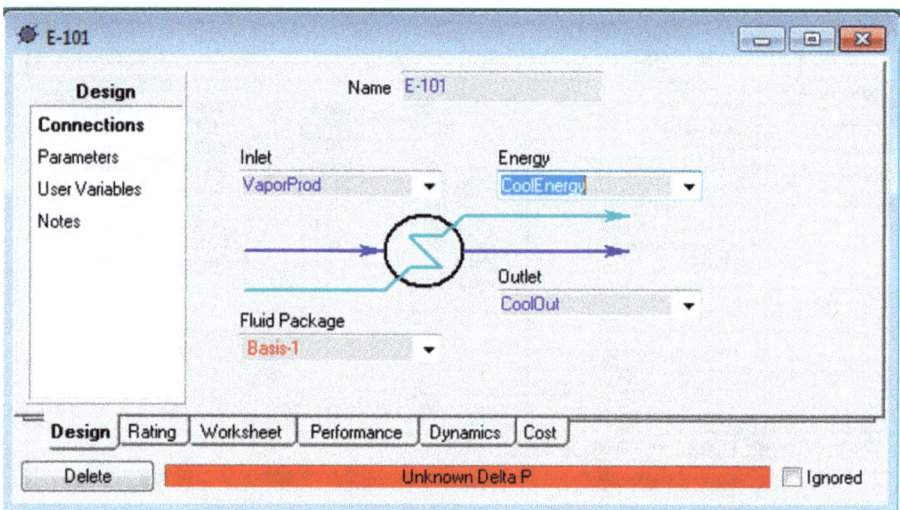

3. Select the **Parameters** page. In the **Delta P** field, specify a pressure drop of **0 atm**. The status indicator now displays "Unknown Duty."
4. Select the **Worksheet** tab, fill in a temperature of **280 K** for the CoolOut stream. The status indicator now displays a green OK, indicating that the operation and attached streams are completely calculated.

E-101				
Worksheet	Name	VaporProd	CoolOut	CoolEnergy
	Vapour	1.0000	0.7881	<empty>
Conditions	Temperature [K]	755.0	280.0	<empty>
Properties	Pressure [atm]	270.0	270.0	<empty>
Composition	Molar Flow [kgmole/h]	5614	5614	<empty>
	Mass Flow [kg/h]	6.131e+004	6.131e+004	<empty>
PF Specs	Std Ideal Liq Vol Flow [m3/h]	168.3	168.3	<empty>
	Molar Enthalpy [kJ/kgmole]	2419	-1.788e+004	<empty>
	Molar Entropy [kJ/kgmole-C]	132.3	86.43	<empty>
	Heat Flow [kJ/h]	1.358e+007	-1.004e+008	1.140e+008

Design | Rating | **Worksheet** | Performance | Dynamics | Cost

Delete | OK | Ignored

5. Now that the Cooler is completely known, close the view to return to the PFD.

11.8.9.7 Adding a Separator
Next you will install and define the inlet Separator, which splits the two-phase CoolOut stream into its vapor and liquid phases.
1. In the Workbook property view, click on the **Unit Ops** tab.
2. Click on the **Add UnitOp** button. The UnitOps view appears. You can also access the Unit Ops view by pressing **F12**.
3. In the Categories group, select the **Vessels** radio button.
4. In the list of Available Unit Operations, select **Separator**.
5. Click on the **Add** button. The Separator property view appears, displaying the **Connections** page on the **Design** tab.
6. Move to the Inlets list by clicking on the « **Stream»** cell, or by pressing **ALT + L**.
7. Open the drop-down list of available feed streams.
8. Select the stream **CoolOut** by doing one of the following:
   – Click on the stream name in the drop-down list.
   – Press the **DOWN** arrow key to highlight the stream name and press **ENTER**.
9. Move to the **Vapour Outlet** cell by pressing **ALT + V**.
10. Create the vapor outlet stream by typing **OFFGAS** and pressing **ENTER**.

11. Click on the **Liquid Outlet** cell, type the name **LIQ_NH3** and press **ENTER**. The completed Connections page appears as shown in the following figure.

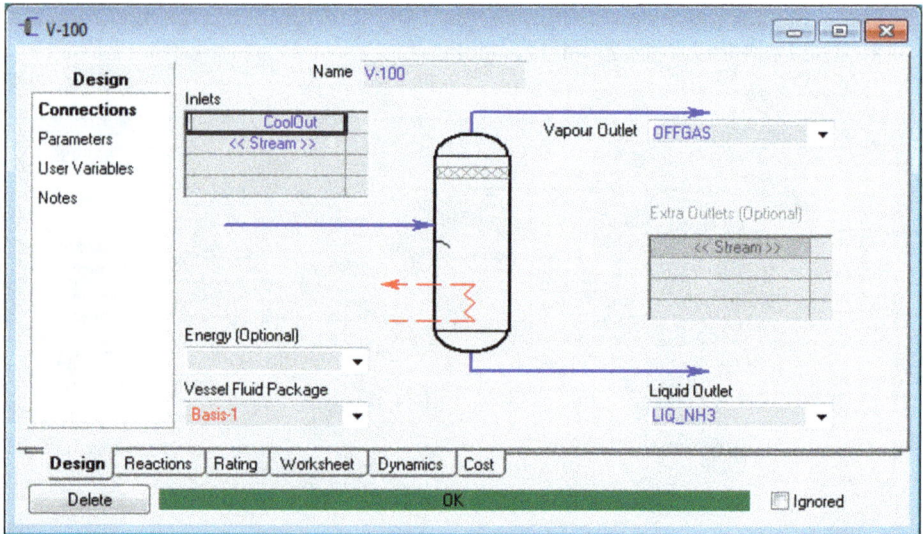

12. Select the **Parameters** page. The current default values for Delta P, Volume, Liquid Volume and Liquid Level are acceptable.

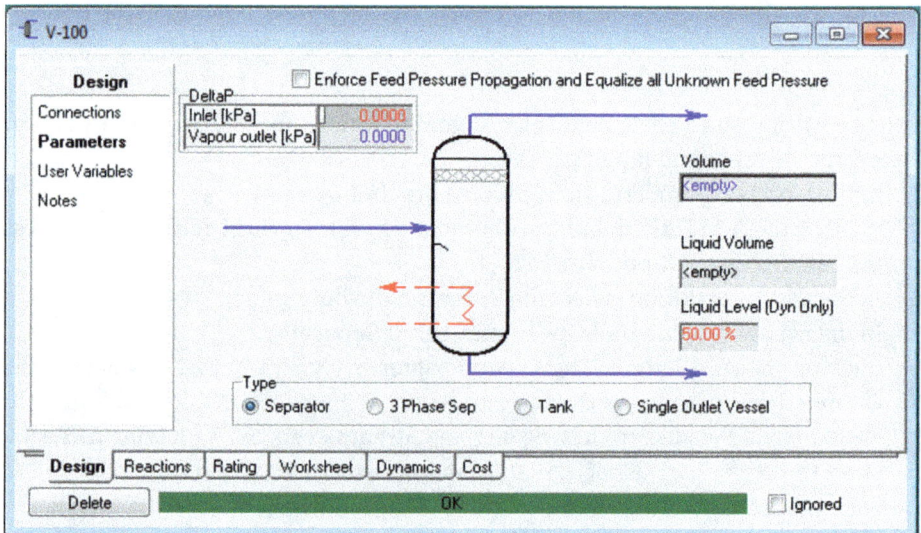

13. To view the calculated outlet stream data, click on the **Worksheet** tab and select the **Conditions** page. The table appearing on this page is shown below.

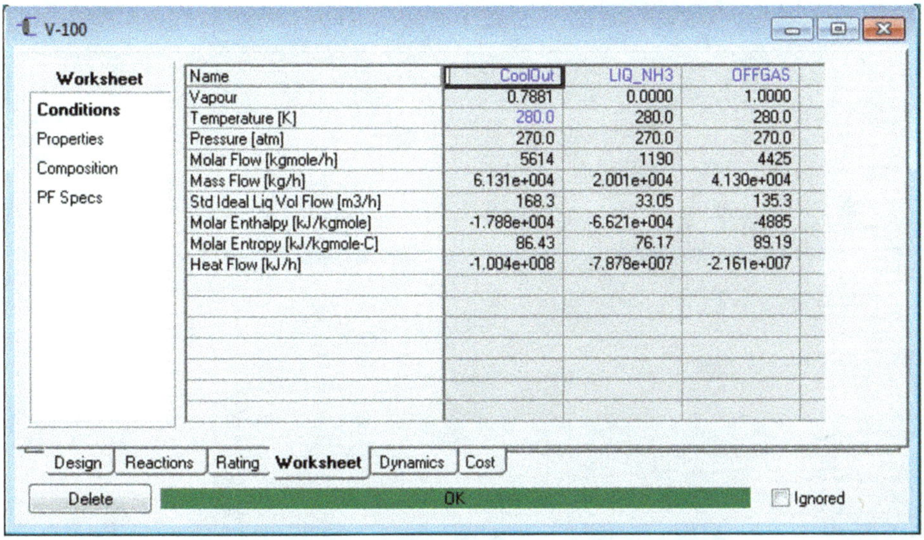
V-100

Worksheet: Conditions | Properties | Composition | PF Specs

Name	CoolOut	LIQ_NH3	OFFGAS
Vapour	0.7881	0.0000	1.0000
Temperature [K]	280.0	280.0	280.0
Pressure [atm]	270.0	270.0	270.0
Molar Flow [kgmole/h]	5614	1190	4425
Mass Flow [kg/h]	6.131e+004	2.001e+004	4.130e+004
Std Ideal Liq Vol Flow [m3/h]	168.3	33.05	135.3
Molar Enthalpy [kJ/kgmole]	-1.788e+004	-6.621e+004	-4885
Molar Entropy [kJ/kgmole-C]	86.43	76.17	89.19
Heat Flow [kJ/h]	-1.004e+008	-7.878e+007	-2.161e+007

Design | Reactions | Rating | Worksheet | Dynamics | Cost

Delete | OK | Ignored

14. When finished, click on the **Close** icon to close the Separator property view. The whole process is shown in Figure 11.8.

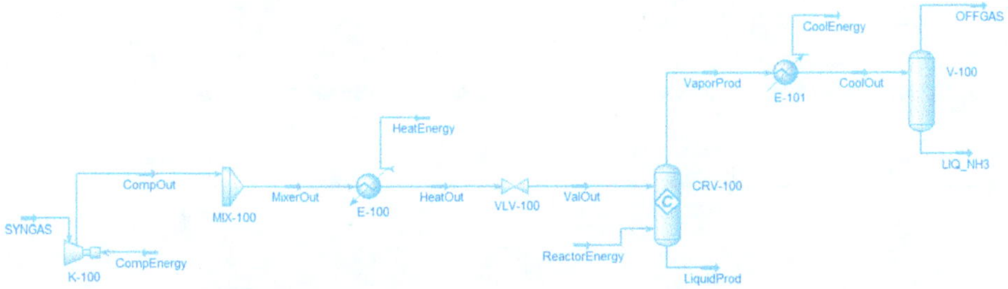

Figure 11.8: Open-loop ammonia synthesis process.

11.9 Review simulation results

1. Open the **Workbook** to access the calculated results for the entire flowsheet. The Material Streams and Compositions tabs of the Workbook appear below.

Name	SYNGAS	CompOut	MixerOut	HeatOut	ValOut	VaporProd	LiquidProd
Vapour Fraction	1.0000	1.0000	1.0000	1.0000	1.0000	1.0000	0.0000
Temperature [K]	553.1	1208	1208	755.0	755.1	755.0	755.0
Pressure [atm]	26.17	271.4	271.4	271.4	270.0	270.0	270.0
Molar Flow [kgmole/h]	7000	7000	7000	7000	7000	5614	0.0000
Mass Flow [kg/h]	6.131e+004	6.131e+004	6.131e+004	6.131e+004	6.131e+004	6.131e+004	0.0000
Liquid Volume Flow [m3/h]	214.1	214.1	214.1	214.1	214.1	168.3	0.0000
Heat Flow [kJ/h]	4.482e+007	1.894e+008	1.894e+008	8.941e+007	8.941e+007	1.358e+007	0.0000
Name	CoolOut	OFFGAS	LIQ_NH3	** New **			
Vapour Fraction	0.7881	1.0000	0.0000				
Temperature [K]	280.0	280.0	280.0				
Pressure [atm]	270.0	270.0	270.0				
Molar Flow [kgmole/h]	5614	4425	1190				
Mass Flow [kg/h]	6.131e+004	4.130e+004	2.001e+004				
Liquid Volume Flow [m3/h]	168.3	135.3	33.05				
Heat Flow [kJ/h]	-1.004e+008	-2.161e+007	-7.878e+007				

Material Streams | Compositions | Energy Streams | Unit Ops

FeederBlock SYNGAS
K-100

☑ Horizontal Matrix

Fluid Pkg All
☐ Include Sub-Flowsheets
☐ Show Name Only
Number of Hidden Objects: 0

Name	SYNGAS	CompOut	MixerOut	HeatOut	ValOut	VaporProd	LiquidProd
Comp Mole Frac (Hydrogen)	0.737143	0.737143	0.737143	0.737143	0.737143	0.548874	0.547290
Comp Mole Frac (Nitrogen)	0.247429	0.247429	0.247429	0.247429	0.247429	0.185095	0.184723
Comp Mole Frac (Methane)	0.010286	0.010286	0.010286	0.010286	0.010286	0.012824	0.012876
Comp Mole Frac (Argon)	0.002714	0.002714	0.002714	0.002714	0.002714	0.003384	0.003391
Comp Mole Frac (CO)	0.002429	0.002429	0.002429	0.002429	0.002429	0.003028	0.003022
Comp Mole Frac (Ammonia)	0.000000	0.000000	0.000000	0.000000	0.000000	0.246794	0.248699
Name	CoolOut	OFFGAS	LIQ_NH3	** New **			
Comp Mole Frac (Hydrogen)	0.548874	0.691448	0.018688				
Comp Mole Frac (Nitrogen)	0.185095	0.233811	0.003936				
Comp Mole Frac (Methane)	0.012824	0.015449	0.003064				
Comp Mole Frac (Argon)	0.003384	0.003967	0.001215				
Comp Mole Frac (CO)	0.003028	0.003800	0.000156				
Comp Mole Frac (Ammonia)	0.246794	0.051524	0.972941				

Material Streams | **Compositions** | Energy Streams | Unit Ops

FeederBlock SYNGAS
K-100

☑ Horizontal Matrix

Fluid Pkg All
☐ Include Sub-Flowsheets
☐ Show Name Only
Number of Hidden Objects: 0

11.10 Saving

After completing this simulation, you should save the file as a.usc file. It is also good practice to save periodically as you create a simulation so you do not risk losing any work. The open-loop simulation is now ready to add a recycle stream, which we will then call a closed-loop simulation. See Module Design-002 for the closed-loop design.

11.11 Closed-loop simulation of an ammonia synthesis plant in UniSim

In the second part of this tutorial we are going to add the recycle to the ammonia synthesis plant. Adding recycles generally leads to convergence issues. We will insert a purge stream, learn how to closed recycle loops and explore closed-loop convergence methods.

Next we will also optimize the process operating conditions to maximize the product composition and flow rates of the process, in other words, how to utilize and build the model analysis tools into a UniSim Design. We will find the optimal purge fraction to meet the desired product specifications and determine the decreasing effect of product composition on cooling efficiency of the preflash cooling unit.

11.11.1 Review of UniSim Design convergence methods

There are several methods UniSim Design can utilize to converge recycle loops. Convergence in UniSim Design is an iterative process consisting of making guesses for tear streams and then comparing the calculated stream values with the guessed values. If these values are equivalent within a certain tolerance, then the simulation has successfully converged. Consider the example shown in the following flowsheet.

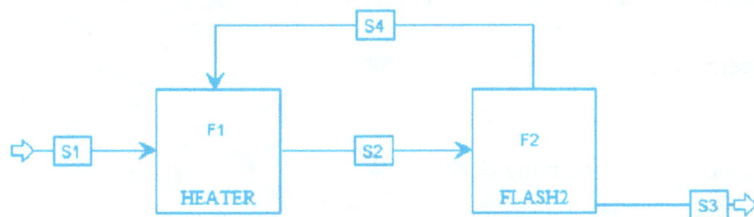

To calculate the properties of stream S4, the properties of stream S2 must be known or calculated. To calculate the properties of stream S2, the properties of streams S1 and S4 must be known or calculated. In mathematical terms we have

$$S4 = F2(S2),$$

$$S2 = F1(S1, S4).$$

The mutual dependency of streams S4 and S2 creates an algebraic loop. This loop can be removed by "tearing" a stream apart. For example, we can choose to hypothetically tear stream S4 into two separate streams. This would result in the following flowsheet:

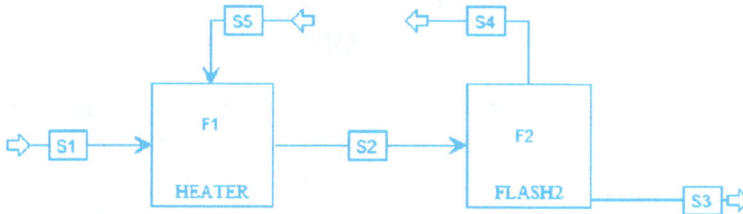

Mathematically we now have the following:

$$S4 = F2(S2),$$

$$S2 = F1(S1, S5).$$

There is no longer mutual dependency between streams S4 and S2. The issue now relies on finding a solution that results in stream S5 being equal to stream S4. This is accomplished by utilizing iterative convergence methods which are briefly described below. Based on the flowsheet one creates, UniSim Design will automatically define tear streams to converge, or alternatively you can input user-defined tear streams.

The following methods are available in UniSim Design:
– Wegstein;
– Dominant Eigenvalue;
– Secant;
– Broyden;
– Sequential quadratic programming (SQP).

The Wegstein method is an extrapolation of the direct substitution method used to accelerate convergence. It attempts to estimate what the final solution will be based on the difference between successive iteration values. This is the default convergence method for system-generated tear convergence blocks and is usually the quickest and most reliable method for tear stream convergence.

The Dominant Eigenvalue method includes interactions between variables being accelerated. Further, the Dominant Eigenvalue option is superior when dealing with non-ideal systems or systems with strong interactions between components.

The secant method uses a succession of roots of secant lines to approximate the root of a function. Compared with Newton's method, the secant method does not require the evaluation of the function's derivative. This enables this method to converge for systems involving non-elementary functions. The secant method can be used for converging single design specifications and is the default method in UniSim Design for design specification convergence.

Broyden's method is a modification of the Newton and secant methods that uses approximate linearization which can be extended to higher dimensions. This method is faster than Newton's method but is often not as reliable. Broyden's method should be used to converge multiple tear streams or design specifications, and is particularly useful when converging tear streams and design specifications simultaneously.

Sequential quadratic programming is an iterative method for flowsheet optimization. This method is useful for simultaneous convergence of optimization problems with constraints and tear streams.

11.11.2 UniSim Design solution

In the first part of this tutorial, the following flowsheet was developed for an open-loop ammonia synthesis process.

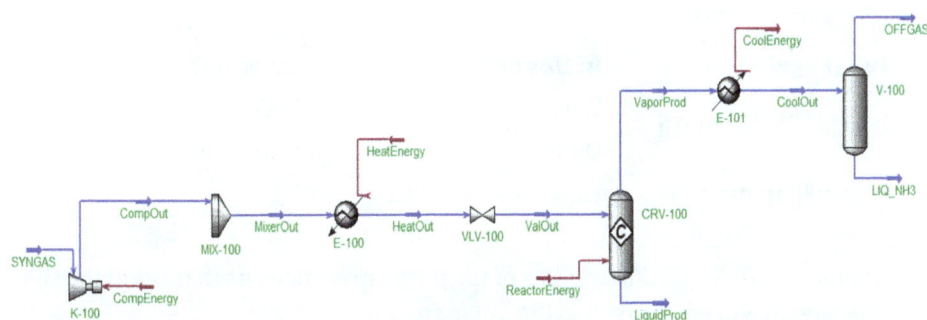

This process produces two outlet streams: a liquid stream containing the ammonia product and a vapor stream containing mostly unreacted hydrogen and nitrogen. It is desired to capture and recycle these unreacted materials to minimize costs and maximize product yield.

11.12 Add recycle loop to ammonia synthesis process

Beginning with the open-loop flowsheet constructed in Part 1 of this series, a recycle loop will be constructed to recover unreacted hydrogen and nitrogen contained in the vapor stream named OFFGAS.

11.12.1 Adding a Tee

The first step will be to add a **Tee** to separate the **OFFGAS** stream into two streams; a purge stream and a recycle stream. As a rule of thumb, whenever a recycle stream exists, there must be an associated purge stream to create an exit route for impurities or by-products contained in the process. Often if an exit route does not exist, impurities will build up in the process and the simulation will fail to converge due to a mass balance error.

1. On the **Object Palette**, right-click and hold the **Tee** icon.
2. Drag the cursor over the PFD. The mouse pointer becomes a bullseye.
3. Position the bullseye pointer above the Separator V-100 and release the mouse button.
4. A Tee icon named TEE-100 appears.
5. Double-click the TEE-100 icon on the PFD to open its property view.
6. In the Tee property view, specify the following connections:

Tab [Page]	In this cell...	In this cell...
Design [Connections]	Inlet	OFFGAS
	Outlet	PURGE, TeeOut
Design [Parameters]	Flow Ratios (PURGE)	0.01

   A value of 0.01 for the Flow Ratios of the purge stream means that 1 % of the OFF-GAS stream will be diverged to the purge stream.
7. Click on the **Close** icon to close the Valve property view.

11.12.2 Adding a Compressor

Next, we must add a Compressor to bring the pressure of the recycle stream back up to the feed conditions.

1. Double click on the **Compressor icon** ▷ in the operations palette. The **Compressor property view** appears. Remember that you can rotate the block icons by right clicking the block and selecting "Transform function."

2. Click on the **Inlet** field, choose **TeeOut** from the list. The status indicator now displays "Requires a product stream."
3. Click in the **Outlet** field. Type **CompOut1** in the cell and press **ENTER**. The status indicator now displays "Requires an energy stream."
4. Click in the **Energy** field. Type **CompEnergy1** in the cell and press **ENTER**. The status indicator now displays "Unknown Duty."
5. Click on the **Worksheet** tab. Choose the **Conditions** page. Set the pressure of the **CompOut1** stream to **271.4 atm**. Now, the status indicator displays a green **OK**, indicating that the operation and attached streams are completely calculated.

	K-101				▢ ▢ ▣
Worksheet	Name		TeeOut	CompOut1	CompEnergy1
	Vapour		1.0000	1.0000	<empty>
Conditions	Temperature [K]		280.0	280.5	<empty>
Properties	Pressure [atm]		270.0	271.4	<empty>
Composition	Molar Flow [kgmole/h]		4380	4380	<empty>
	Mass Flow [kg/h]		4.089e+004	4.089e+004	<empty>
PF Specs	LiqVol Flow [m3/h]		133.9	133.9	<empty>
	Molar Enthalpy [kJ/kgmole]		-4885	-4867	<empty>
	Molar Entropy [kJ/kgmole-C]		89.19	89.20	<empty>
	Heat Flow [kJ/h]		-2.140e+007	-2.132e+007	7.658e+004

| Design | Rating | **Worksheet** | Performance | Dynamics | Cost |

| Delete | | OK | | ☑ On | ☐ Ignored |

6. Now that the Compressor is completely known, close the view to return to the FPD.

11.12.3 Adding a Recycle

The Recycle installs a theoretical block in the process stream. The stream conditions can be transferred in either a forward or a backward direction between the inlet and outlet streams of this block. In terms of the solution, there are assumed values and calculated values for each of the variables in the inlet and outlet streams. Depending on the direction of transfer, the assumed value can exist in either the inlet or the outlet stream. For example, if the user selects Backward for the transfer direction of the Temperature variable, the assumed value is the Inlet stream temperature and the calculated value is the Outlet stream temperature.

There are two ways that you can add a Recycle to your simulation. The first is the following:

1. In the **Flowsheet** menu, click the Add Operation command. The UnitOps view appears.
2. Click on the **Logicals** radio button.
3. From the list of available unit operations, select Recycle.
4. Click on the **Add** button.

The second way is the following.
1. In the **Flowsheet** menu, click the Palette command. The Object Palette appears.
2. Double-click the **Recycle** icon. The Recycle property view appears.
3. In the recycle property view, specify the following connections:

Tab [Page]	In this cell...	In this cell...
Design [Connections]	Inlet	CompOut1
	Outlet	Recycle
Parameters [Variables]	Flow	1.0
	Composition	0.1

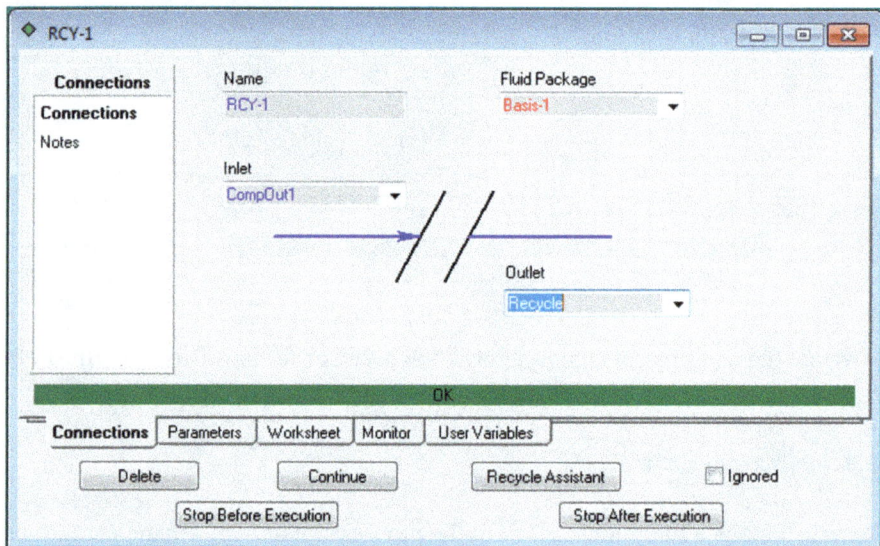

The CompOut1 stream will be the tear stream in this simulation. UniSim Design automatically recognizes and assigns tear streams; however, you can also specify which stream you would like to be tear streams.
4. The Recycle stream is now ready to be connected back to the mixer block to close the loop. Click on the mixer MIX-100, and select Recycle as another feed stream to the mixer in the Inlet cell.

You should see an error stating that block RCY-1 is not in mass balance and that the simulation failed to converge after 10 iterations.

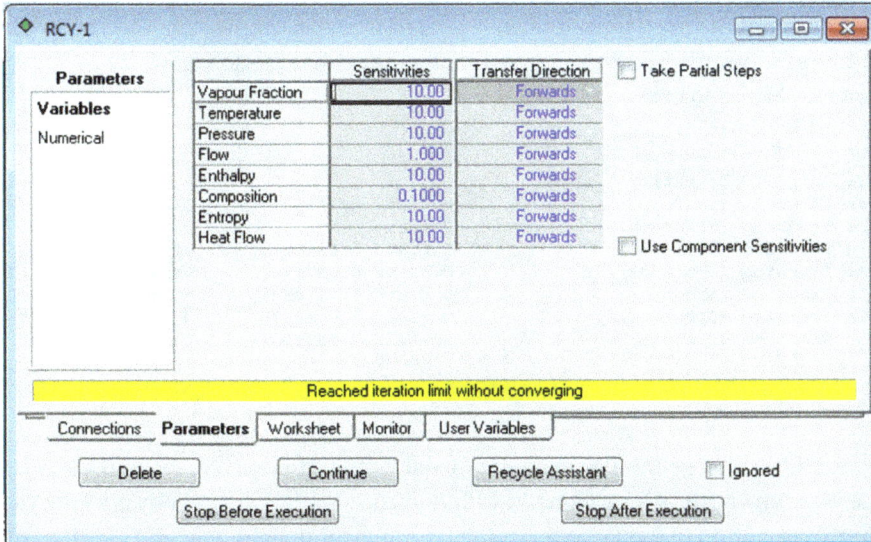

There are several steps you can take to overcome this issue. First, check to see which convergence method is being used and check which stream is being converged.

5. Click in **Parameters** tab, choose the **Numerical** page. UniSim Design is using the **Wegstein** method to converge recycle stream CompOut1. The Wegstein method is a good method to use when trying to converge a single recycle stream, and stream CompOut1 is an appropriate stream to attempt to converge. The next thing you can do is to check the maximum error per iteration to see whether the solver is heading towards convergence or not.

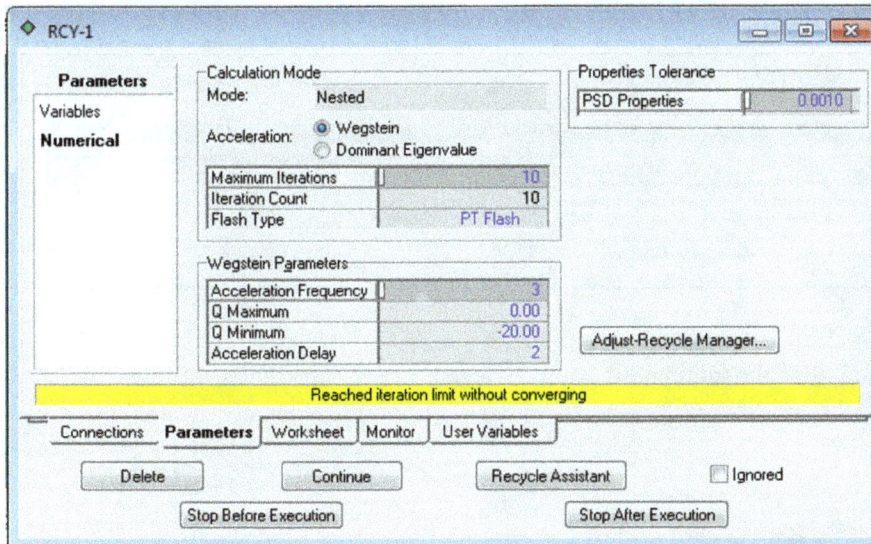

6. By looking at the message window, it is clear that the Wegstein method is on the right track towards finding the solution. It may be that the solver just needs a few more iterations to converge.

```
Recycle: RCY-1    Iteration   10
  Mass Flow                          Outlet: 130553      Inlet: 130822
  Mole Fraction (Hydrogen)           Outlet: 0.543249    Inlet: 0.542764
  Mole Fraction (Nitrogen)           Outlet: 0.231828    Inlet: 0.232617
  Mole Fraction (Methane)            Outlet: 0.0971569   Inlet: 0.0966114
  Mole Fraction (Argon)              Outlet: 0.0168153   Inlet: 0.0167232
  Mole Fraction (CO)                 Outlet: 0.0485375   Inlet: 0.0488934
  Mole Fraction (Ammonia)            Outlet: 0.0624132   Inlet: 0.062391
Maximum Iterations Reached
```

7. Go to RCY-1, choose the Parameters tab, select the Numerical page and increase the **Maximum Iterations** value to 100. Run the simulation again by clicking the **Continue** button of the Recycle. In the Iteration count field, you will see that the solver has converged after 49 iterations.

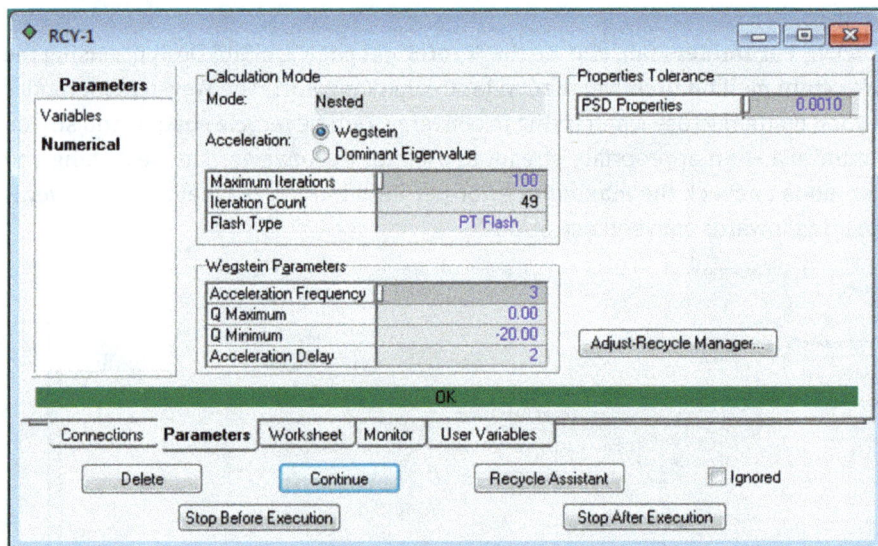

8. Now that the simulation has converged, check the results through the Workbook.

11.13 Optimize the purge rate to deliver desired product

We now wish to determine the purge rate required to deliver a product with an ammonia mole fraction of 0.96. First, check the composition of the current ammonia stream by clicking on the product stream (LIQ_NH3), choosing Worksheet and selecting Composition.

From the composition analysis results, the mole fraction of ammonia in the product stream is only 0.954, which is below the specification of 0.96. We need to determine the purge rate required to reach this product specification.

In this tutorial, we will use the Adjust function to determine the required flow rate ratios of the splitter, which gives a specified ammonia concentration.

11.13.1 Installing, connecting and defining the Adjust

1. Click on the **PFD** icon to display the PFD and access the Object Palette by pressing **F4**.
2. Click on the **Adjust** icon on the Object Palette.
3. Position the cursor on the PFD to the right of the **LIQ_NH3** stream icon.
4. Click to **drop** the Adjust icon onto the PFD. A new Adjust object appears with the default name **ADJ-1**.
5. Click on the **Attach Mode** icon on the PFD toolbar to enter Attach mode.
6. Position the cursor over the left end of the **ADJ-1** icon. The connection point and pop-up Adjusted Object appears.
7. With the pop-up visible, left-click and drag toward the **TEE-100** icon.
8. When the solid white box appears on the **TEE-100**, release the mouse button. The Select Adjusted Variable view appears.

   At this point, UniSim Design knows that the TEE-100 should be adjusted in some way to meet the required target. An adjustable variable for the TEE-100 must now be selected from the Select Adjusted Variable view.

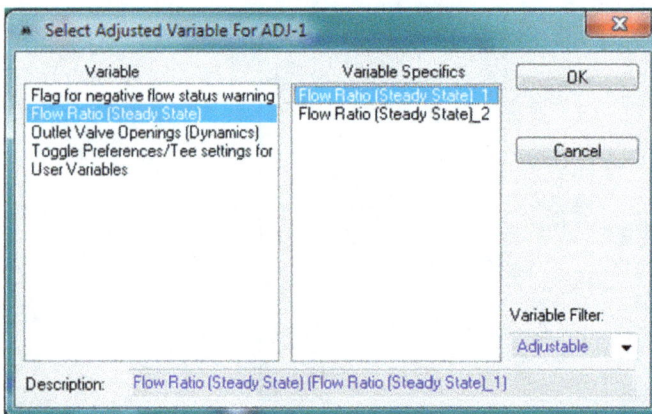

9. From the Variable list, select **Flow Ratio (Steady State)**, and choose **Flow Ratio (Steady State)_1** in the Variable Specifics field.
10. Click on the **OK** button.
11. Position the cursor over the right corner of the **ADJ-1** icon. The connection point and pop-up Target Object appears.

12. With the pop-up visible, left-click and drag toward the **LIQ_NH3** stream icon.
13. When the solid white box appears at the cursor tip, release the mouse button. The Select Target Variable view appears.
14. From the Variable list, select **Master Comp Mole Frac**, and choose **Ammonia** from Variable Specifics.
15. Click on the **OK** button.

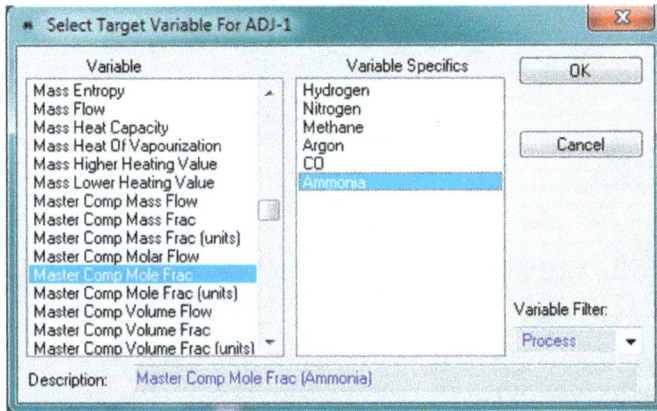

16. Click on the **Attach Mode** icon to leave Attach mode.
17. Double-click the **ADJ-1** icon to open its property view.
   The connections made in the PFD have been transferred to the appropriate cells in the property view.

11.13.2 Adjusting the target variable

The next task is to provide a value for the target variable, in this case the ammonia concentration. A concentration of **0.96** will be used as a desired target.
1. Click on the **Parameters** tab.
2. In the **Initial Step Size** cell, enter **0.01**.
3. In the **Minimum and Maximum** field, enter **0.01** and **0.05**, respectively.
4. In the **Maximum Iterations**, enter **50**.
5. Click on the **Connection** tab and enter **0.96** in the **Specified Target Value** field. Click **OK** when the following message appears.

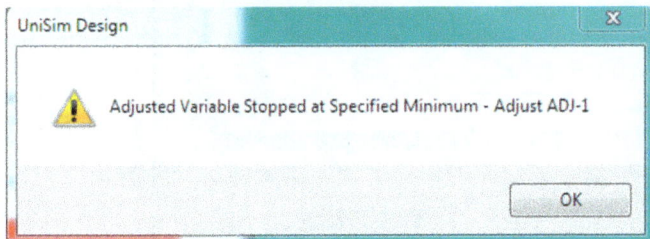

6. Click on the **Reset** button.
   The above message appears again, click **OK** until the status indicator now displays a green **OK,** indicating that the operation and attached streams are completely calculated.

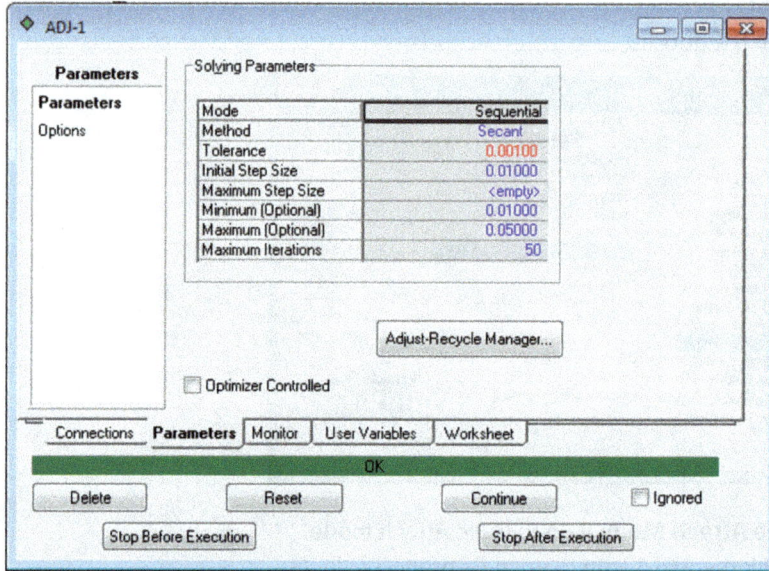

ADJ-1	
Parameters	Solving Parameters
Parameters	Mode — Sequential
Options	Method — Secant
	Tolerance — 0.00100
	Initial Step Size — 0.01000
	Maximum Step Size — <empty>
	Minimum (Optional) — 0.01000
	Maximum (Optional) — 0.05000
	Maximum Iterations — 50

Adjust-Recycle Manager...

☐ Optimizer Controlled

Connections | **Parameters** | Monitor | User Variables | Worksheet

OK

Delete | Reset | Continue | ☐ Ignored
Stop Before Execution | Stop After Execution

7. Click on the **Monitor** tab. This tab enables you to view the calculations. You will see that the mole fraction of ammonia in the product stream has reached **0.96** at a purge fraction of **4.2 %.**

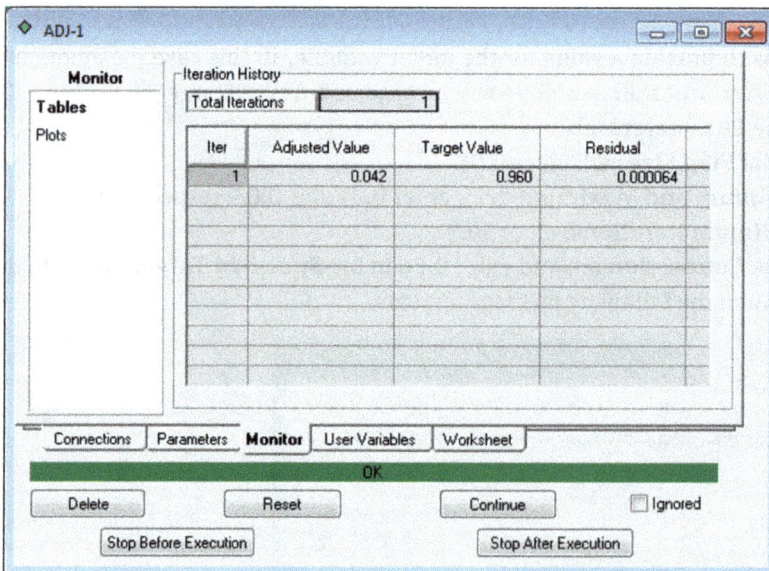

ADJ-1			
Monitor	Iteration History		
Tables	Total Iterations — 1		
Plots	Iter	Adjusted Value	Target Value	Residual
	1	0.042	0.960	0.000064

Connections | Parameters | **Monitor** | User Variables | Worksheet

OK

Delete | Reset | Continue | ☐ Ignored
Stop Before Execution | Stop After Execution

8. Click on the **Close** icon in the Adjust property view.

11.14 Investigate the effects of flash feed temperature on product composition

We would now like to determine how fluctuations in flash feed temperature will affect the product composition and flow rate. Changes in cooling efficiency or utility fluid temperature can change the temperature of the flash feed stream. This change in temperature will change the vapor fraction of the stream, thus changing the composition and flow rate of the product and recycle streams.

In this tutorial, we will use **Databook** to monitor the flash feed temperature variable under a variety of process scenarios, and view the results in a tabular or graphical format.

11.14.1 Defining the key variables

1. To open the Databook, do one of the following:
   - Press **CTRL + D.**
   - Open the **Tools** menu and select **Databook**.
   The Databook appears as shown below.

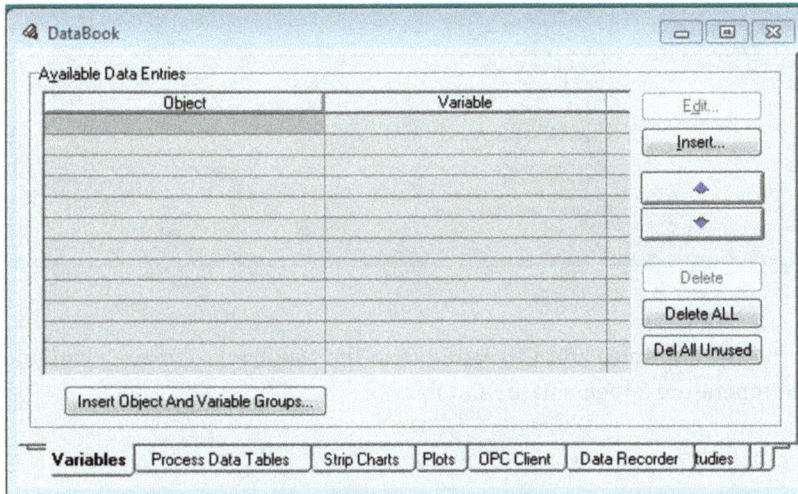

2. Click on the **Variables** tab. Here you will add the key variables to the Databook.
3. Click on the **Insert** button. The Variable Navigator view appears.

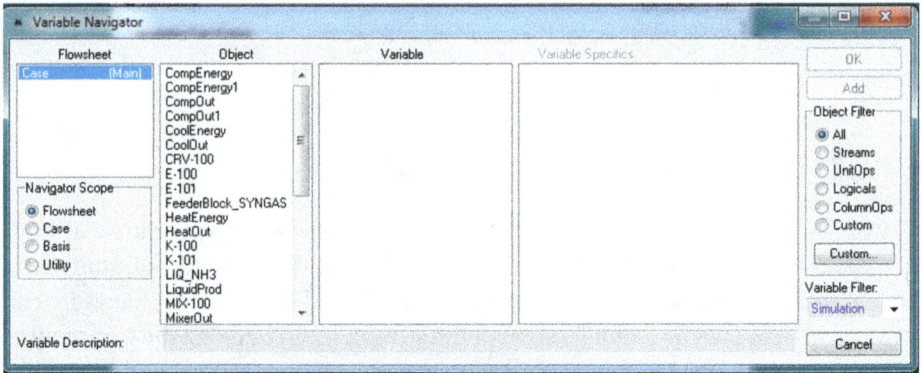

4. In the Object Filter group, select the **UnitOps** radio button. The Object list will be filtered to show unit operations only.
5. In the Object list, select **E-101**. The Variable list available for E-101 appears to the right of the Object list.
6. In the Variable list, select **Product Temperature**. UniSim Design displays this variable name in the **Variable Description** field.

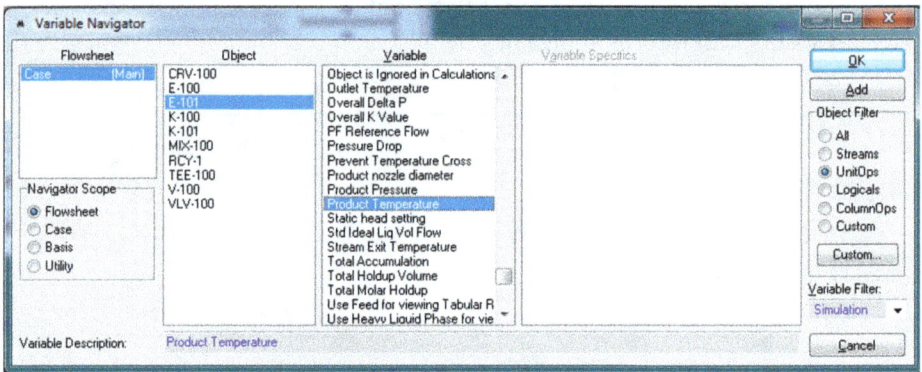

7. Click on the **OK** button to add this variable to the Databook. The new variable Product Temperature appears in the Databook.

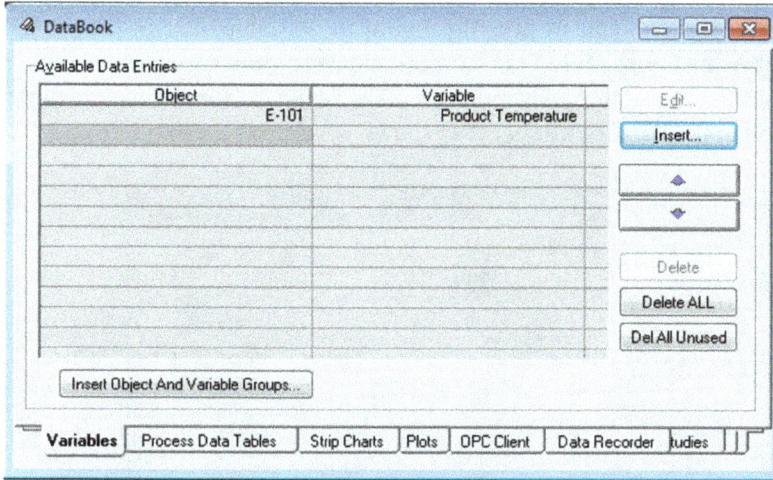

Continue adding variables to the Databook.

8. Click on the **Insert** button. The Variable Navigator reappears.
9. In the Object Filter group, select the **Streams** radio button. The Object list is filtered to show streams only.
10. In the Object list, select **LIQ_NH3**. The Variables list available for material streams appears to the right of the Object list.
11. In the Variable list, select **Master Comp Molar Flow**. Choose **Ammonia** in the Variable Specifics list.
12. Click on the **Add** button. The variable now appears in the Databook, and the Variable Navigator view remains open.

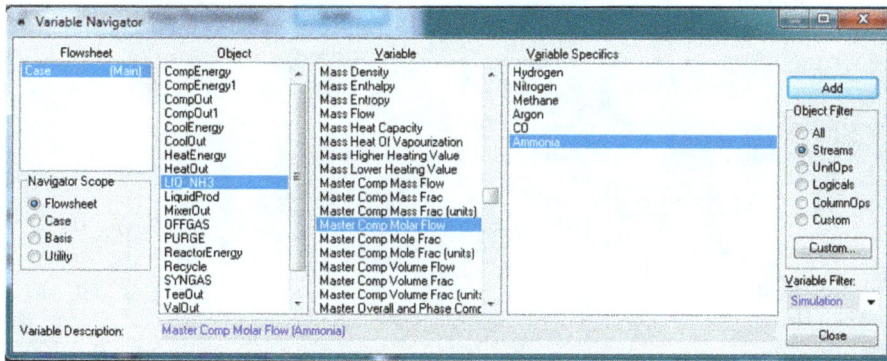

13. Repeat the previous steps to add the following variables to the Databook:
    LIQ_NH3, Master Comp Mole Frac variable and **Ammonia** variable specifics.
14. Click on the **Close** button to close the Variable Navigator view. The completed
    Variables tab of the Databook appears as shown below.

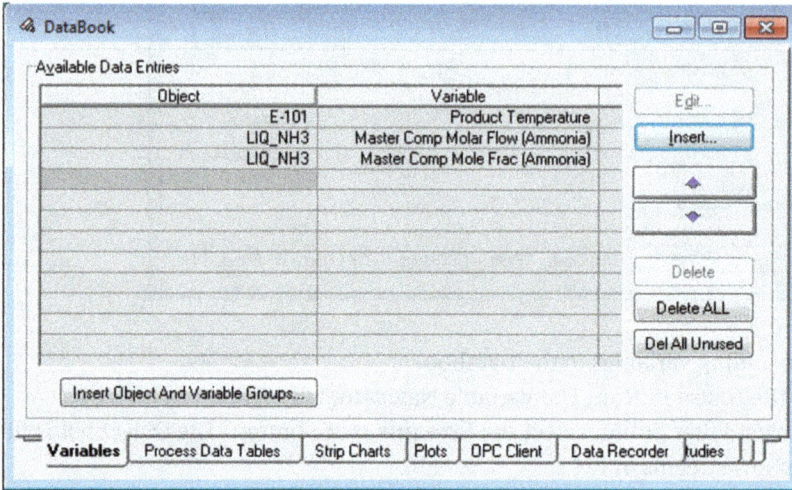

11.14.2 Creating the case study

1. Click on the **Case Studies** tab.
2. In the Available Case Studies group, click on the **Add** button to create Case Study 1.
3. Check the Independent and Dependent Variables as shown below.

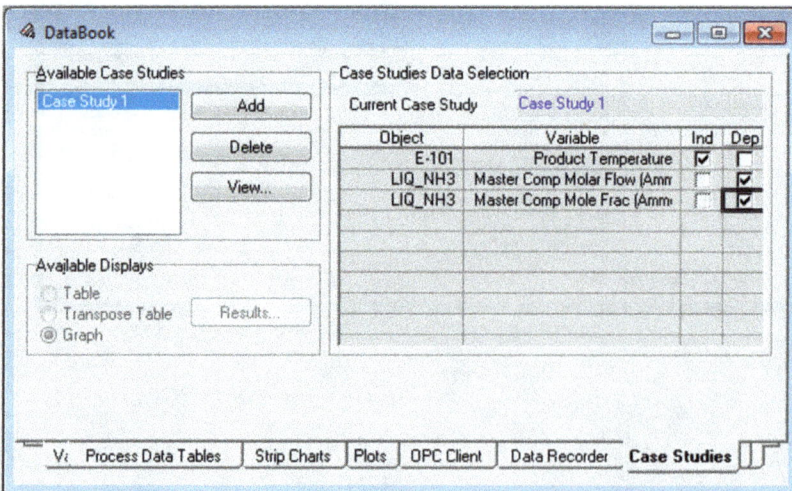

To automate the study, the independent variable range and step size must be given.

4. Click on the **View** button to access the Case Studies Setup view. Define the range and step size for the flash feed temperature as shown below.

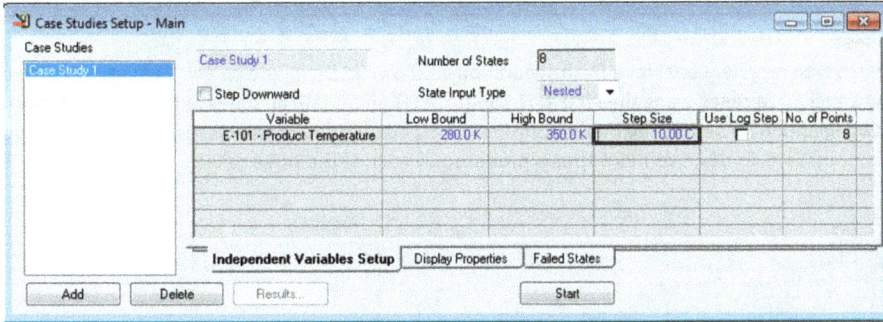

5. To begin the Study, click on the **Start** button.
6. Click on the **Results** button to view the variables. If the results are in graphical form, click on the Table radio button in the Case Studies view.
   The results of this study appear below.

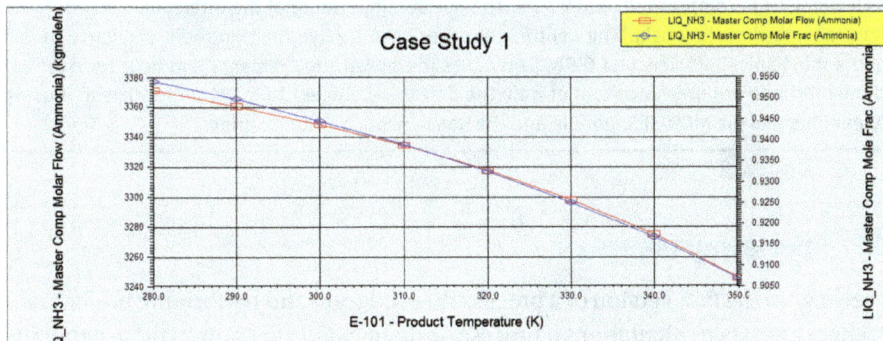

You will see that as the temperature increases, both the product flow rate and product quality decrease, which means that when operating this process it will be very important to monitor the flash feed temperature in order to deliver a high-quality product.

Exercises

1. Flash with recycle: Consider a flash separation as shown in the figure below. If using AspenPlus, solve all three cases using the MIXER, FLASH2, FSPLIT and PUMP modules and the Soave Redlich Kwong option for setting the thermo-physical properties. Compare and discuss the flow rates and compositions for the overhead stream produced by each of the three cases.

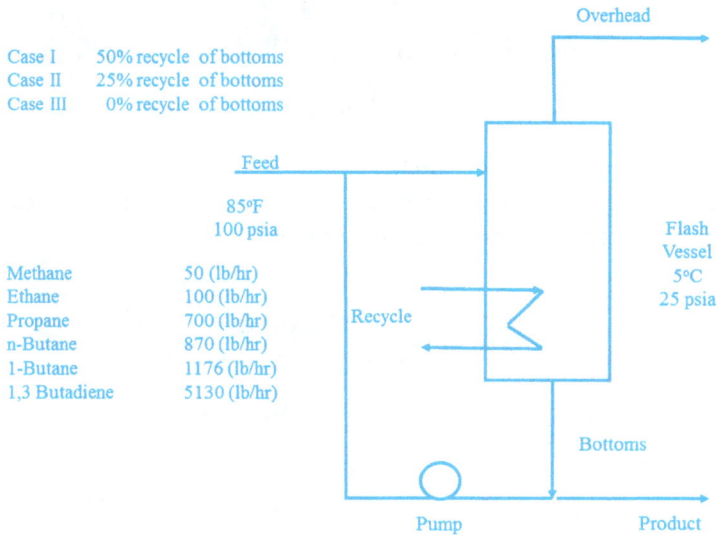

Case I   50% recycle of bottoms
Case II   25% recycle of bottoms
Case III 0% recycle of bottoms

Feed

85°F
100 psia

Methane   50 (lb/hr)
Ethane   100 (lb/hr)
Propane   700 (lb/hr)
n-Butane   870 (lb/hr)
1-Butane   1176 (lb/hr)
1,3 Butadiene 5130 (lb/hr)

Overhead

Flash
Vessel
5°C
25 psia

Recycle

Bottoms

Pump   Product

2. Modify case 3 of exercise 1 to determine the flash temperature necessary to obtain 850 lb/hr of overhead vapor. If using AspenPlus, a design specification can be used to adjust the temperature of the flash drum to obtain the desired overhead flow rate.

3. Use a process simulator to model a two-stage compression system with an intercooler. The feed stream consists of 95 mol% of hydrogen and 5 mol% of methane at 100 °F and 30 psia; 440 lbmol/hr is compressed to 569 psia. The outlet temperature of the intercooler is 100 °F and its pressure drop is 2 psia. The centrifugal compressors have an isentropic efficiency of 0.9 and a mechanical efficiency of 0.98. Determine the power requirements and heat removed for three intermediate pressures (outlet from the first three stages) 100, 130 and 160 psia. If using AspenPlus use the MCOMPR module and the Soave Redlich Kwong option.

11.15 Take-away message

After setting up the first version of a process flow diagram and performing preliminary mass/energy balance calculations a process designer wants to know in more detail the operating conditions of the different units, the dimensions of the equipment and the actual conversions and separation efficiencies throughout the process. To find these answers process simulation is used. There exist many different process simulators which can be classified in equation-oriented packages and modular packages. Often modular packages are used that have extended libraries of preprogrammed unit operations. A user can draw the flowsheet in such packages and the software has different solvers to solve the automated system of equations resulting from the balance equations. Recycles in flowsheets might create potential issues with solver convergence. Typically tear variables and tear streams are needed to deal effectively with recycles.

11.16 Further reading

Crow and Nishio (1975). Convergence promotion in the simulation of chemical processes – the general dominant eigen value method. *AiChE J.*, 21(3).

Myers and Seider (1976). *Introduction to chemical engineering and computer calculations*. Englewood Cliffs, New Jersey: Prentice-Hall.

Westerberg, Hutchison, Motard and Winter (1979). *Process flowsheeting*. Cambridge: Cambridge University Press.

Seider, Lewin, Seader, Widagdo, Gani and Ng (2017). *Product and process design principles, synthesis, analysis and evaluation*, 4th ed. Wiley.

12 Reactor design

12.1 Essence of reactors

The conversion of feed to products is the essence of a chemical process and, thus, the reactor is the heart of a chemical plant. When designing a reactor, an engineer must first collect data about the chemical reaction and then select appropriate reaction conditions, which will help determine suitable materials of construction. Next, the designer should determine the rate limiting step and, from this, the critical sizing parameter. Next, preliminary sizing, layout and costing can be conducted for the reactor. At this point, simulations and experiments can be conducted to verify that the proposed reactor will meet the desired specifications. The design is optimized until these targets are met. Throughout the design process, it is important for the engineer to consider the most appropriate type of reactor to use, any mixing or heat transfer equipment that must be added and safety considerations.

This chapter follows Towler and Sinnott (2013) and the open process design textbook to a large extent.

12.2 Ideal reactors

12.2.1 Batch reactors

In a batch reactor, the reagents are added together and allowed to react for a given amount of time. The compositions change with time, but there is no flow through the process. Additional reagents may be added as the reaction proceeds, and changes in temperature may also be made. Products are removed from the reactor after the reaction has proceeded to completion.

Batch processes are suitable for small-scale production (less than 1,000,000 lb/yr) and for processes where several different products or grades are to be produced in the same equipment (Douglas, 1988). When production volumes are relatively small and/or the chemistry is relatively complex, batch processing provides an important means of quality control.

12.2.2 Plug flow reactor (PFR)

A plug flow reactor (PFR) with tubular geometry has perfect radial mixing but no axial mixing. All materials have the same residence time, τ, and experience the same temperature and concentration profiles along the reactor. The equation for PFRs is given by

$$dM = r\,dV, \tag{12.1}$$

https://doi.org/10.1515/9783110570137-012

where M is the molar flow rate, dV is the incremental volume and r is the rate of reaction per unit volume.

This equation can be integrated along the length of the reactor to yield relationships between reactor resident time and concentration or conversion.

12.2.3 Continuously stirred tank reactor (CSTR)

The stirred tank reactor models a large scale conventional laboratory flask and can be considered to be the basic chemical reactor. In a continuously stirred tank reactor (CSTR), shown in Figure 12.1, there is no spatial variation – the entire vessel contents are at the same temperature, pressure and concentration. Therefore the fluid leaving the reactor is at the same temperature and concentration as the fluid inside the reactor.

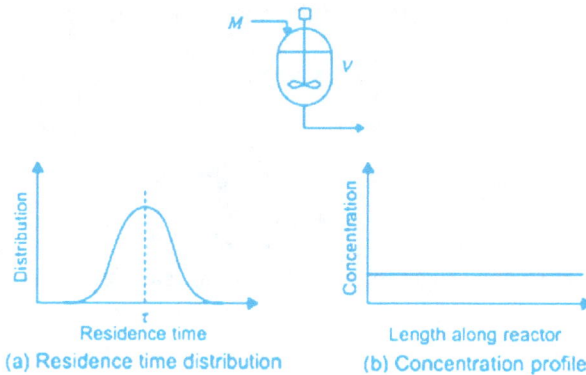

(a) Residence time distribution   (b) Concentration profile

Figure 12.1: Continuously stirred tank reactor.

The material balance across the CSTR is given by

$$M_{in} - M_{out} = rV. \tag{12.2}$$

Some of the material that enters the reactor can leave immediately, while some leaves much later, so there is a broad distribution in residence time, as shown in Figure 12.1.

12.3 General reactor design

The design of the reactor should not be carried out separately from the overall process design due to the significant impact on capital and operating costs on other parts of the process.

Step 1: Collect required data

Out of all process equipment, reactor design requires the most process input data: reaction enthalpies, phase-equilibrium constants, heat and mass transfer coefficients and reaction rate constants. All of the aforementioned parameters can be estimated using simulation models or literature correlations except for reaction rate constants, which need to be determined experimentally (Towler and Sinnott, 2013).

The heat given out in a chemical reaction is based on the enthalpies of the component chemical reactions, which are given for standard temperature and pressure (1 atm, 25 °C). Values for standard heats of reaction can be found tabulated in literature, or can be calculated from heats of formation or combustion. Care must be taken to quote the basis for the heat of reaction and the states of reactants and products. The following equation is used to convert enthalpies from standard conditions to the process conditions:

$$\Delta H_{rPT} = \Delta H_r^o + \int_1^P \left[\left(\frac{\partial H_{\text{prod}}}{\partial P} \right)_T - \left(\frac{\partial H_{\text{react}}}{\partial P} \right)_T \right] dP + \int_{298}^T \left[\left(\frac{\partial H_{\text{prod}}}{\partial T} \right)_P - \left(\frac{\partial H_{\text{react}}}{\partial T} \right)_P \right] dT.$$

$$(12.3)$$

If the effect from pressure is not significant and only temperature needs to be accounted for, the following equation should be used:

$$\Delta H_{rT} = \Delta H_r^o + \Delta H_{\text{prod}} + \Delta H_{\text{react}}. \tag{12.4}$$

The equilibrium constant and Gibbs free energy are related by

$$\Delta G = -RT \ln K, \tag{12.5}$$

where ΔG is the change in Gibbs free energy from the reaction at temperature T, R is the ideal gas constant and K is the reaction equilibrium constant, given by

$$K = \prod_{i=1}^n a_i^{\alpha_i}, \tag{12.6}$$

where a_i is the activity of component i, α_i is the stoichiometric coefficient of component i and n is the total number of components.

Equilibrium constants can be found in the literature and are useful for evaluating the rates of forward and reverse reactions. Care must be taken to the experimental design used for the literature equilibrium constants to make sure they are consistent with the conditions of the actual process reactor. For more complicated reactions consisting of several sequential or simultaneous reactions, the equilibrium is found by minimizing the Gibbs free energy. Commercial process simulation programs use the Gibbs reactor model in this way.

Reaction mechanisms, rate equations and rate constants

In most cases the main process reaction rate equations and rate constants cannot be predicted from first principles and must be approximated (Towler and Sinnott, 2013). This is due to the following:

- use of heterogeneous catalysis or enzymes which lead to Langmuir–Hinshelwood–Hougen–Watson or Michaelis–Menten kinetics;
- mass transfer between vapor and liquid or two liquid phases;
- multi-step mechanisms whose rate expressions do not follow overall reaction stoichiometry;
- competing side reactions.

As a result the main process reaction is usually approximated as first- or second-order over a narrow range of process conditions (temperature, pressure, species concentrations) to estimate the residence time required for a target conversion. Rate equations are always a fit for experimental data and should thus be used for interpolation within the data. It is important to collect more data when extrapolating, especially for exothermic reactions which have the potential for runaway.

Heat and mass transfer properties

Correlations for tube-side heat transfer coefficients for catalyst-packed tubes of a heat exchanger are given below.

For heating

$$\frac{h_i d_t}{\lambda_f} = 0.813 \left(\frac{\rho_f u d_p}{\mu} \right)^{0.9} \exp\left(-\frac{6 d_p}{d_t} \right) \tag{12.7}$$

and for cooling

$$\frac{h_i d_t}{\lambda_f} = 3.50 \left(\frac{\rho_f u d_p}{\mu} \right)^{0.7} \exp\left(-\frac{4.6 d_p}{d_t} \right), \tag{12.8}$$

where h_i is the tube-side heat transfer coefficient for a packed tube, d_t is the tube diameter, λ_f is the fluid thermal conductivity, ρ_f is the fluid density, u is the superficial velocity, d_p is the effective particle diameter and μ is the fluid viscosity.

Diffusion coefficients

Diffusion coefficients are necessary when mass transfer can limit the rate of reaction, such as in catalytic reactions or reactions involving mass transfer processes such as gas absorption, distillation and liquid–liquid extraction. The diffusivity for gases can be estimated by the following correlation (Fuller–Schettler–Giddings):

$$D_v = \frac{1.013 \times 10^{-7} T^{1.75} \left(\frac{1}{M_a} + \frac{1}{M_b} \right)^{\frac{1}{2}}}{P [(\sum_a v_i)^{\frac{1}{3}} + (\sum_b v_i)^{\frac{1}{3}}]^2}, \tag{12.9}$$

where D_v is the diffusivity, T is temperature, M_a, M_b are the molecular masses of components a and b, P is the total pressure and the summations are special diffusion volume coefficients.

Wilke and Chang developed a correlation for estimating the diffusivity of components in the liquid phase:

$$D_L = \frac{1.173 \times 10^{-13}(\phi M_w)^{1/2}T}{\mu V_m^{0.6}}, \tag{12.10}$$

where D_L is the liquid diffusivity, ϕ is an association factor for the solvent, M_w is the molecular mass of the solvent, μ is the solvent viscosity, T is the temperature and V_m is the molar volume of the solute at its boiling point. This correlation holds for organic compounds in water but not for water in organic solvents.

Mass transfer

For multi-phase reactors it is necessary to estimate the mass transfer coefficient. The Gupta–Thodos equation predicts the mass transfer coefficient for a packed bed of particles:

$$\frac{kd_p}{D} = 2.06\frac{1}{\epsilon}\,\mathrm{Re}^{0.425}\,\mathrm{Sc}^{0.33}, \tag{12.11}$$

where k is the mass transfer coefficient, d_p is the particle diameter, D is the diffusivity, Re is the Reynolds number calculated using the superficial velocity through the bed, Sc is the Schmidt number and ϵ is the bed void fraction.

Mass transfer between vapor and liquid in an agitated vessel can be described by the Van't Riet equations.

For air–water

$$k_L a = 0.026\left(\frac{P_a}{V_{\mathrm{liq}}}\right)^{0.4} Q^{\frac{1}{2}} \tag{12.12}$$

and for air–water-electrolyte

$$k_L a = 0.002\left(\frac{P_a}{V_{\mathrm{liq}}}\right)^{0.7} Q^2, \tag{12.13}$$

where k_L is the mass transfer coefficient, a is the interfacial area per unit volume, Q is the gas volumetric flow rate, V_{liq} is the liquid volume and P_a is the agitator power input.

Fair's method for calculating the mass transfer coefficient for low-viscosity systems is given by

$$\frac{(k_L a)_{\mathrm{system}}}{(k_L a)_{\mathrm{air-water}}} = \left(\frac{D_{L,\mathrm{system}}}{D_{L,\mathrm{air-water}}}\right)^{\frac{1}{2}}, \tag{12.14}$$

where D_L is the liquid phase diffusivity. Mass transfer correlations for vapor–liquid systems should be used with caution when there are surfactants.

Step 2: Select reaction conditions

A major determining factor in reactor type selection is the choice of operating conditions. Optimal process operation usually involves optimizing process yield and not necessarily reactor yield. Based on the preliminary economical analysis a target range of yields and selectivities can be chosen. The final reaction conditions must be verified experimentally to ensure target yields and selectivities are realized (Towler and Sinnott, 2013).

Chemical or biochemical reaction

If the desired product is to be produced by a biochemical reaction the chosen conditions must maintain the viability of the biological agent (*e. g.*, microorganisms or enzymes). Proteins denature outside of their specific temperature and pH ranges, while living organisms require specific concentrations of oxygen and other solutes to survive and cannot withstand high shear rates.

Catalyst

A catalyst is used to increase the reaction rate by lowering the activation energy without being consumed in the reaction. The use of a catalyst imposes operating condition constraints as the catalyst must maintain activity for a period of time between catalyst regenerations. Catalyst deactivation can be accelerated by high temperatures as well as contaminants in the feed or recycle streams.

Temperature

Increasing the reaction temperature will increase the reaction rate, diffusivities and mass transfer rates. Temperature also affects the equilibrium constant: higher temperatures increase the equilibrium constant for endothermic reactions and decrease it for exothermic reactions; see Figure 12.2.

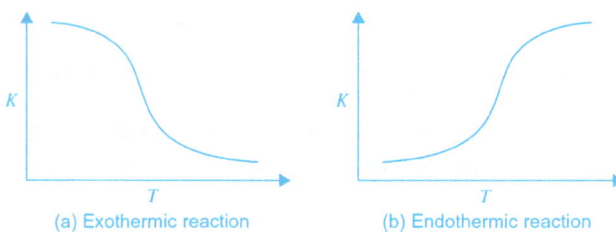

(a) Exothermic reaction   (b) Endothermic reaction

Figure 12.2: Effect of temperature on equilibrium constant.

Increased reaction temperatures will reduce the cost of reactor design except for the following scenarios/considerations:
– biochemical reactions where living organisms could die at high temperatures;
– presence of organic compounds that undergo thermal degradation;

- unwanted side reactions that accelerate with higher temperature, such as polymerization or auto-oxidation;
- oxidation reactions where selectivity decreases at higher temperatures as product
  oxidation tends to increase;
- exothermic reactions as it is more difficult to control the temperature and there is
  risk of reaction runaway;
- construction cost of the reactor can become prohibitive at extremely high temperatures.

Pressure
The main consideration when choosing the reactor pressure is to maintain the reaction at the desired phase for the selected temperature. The pressure can also be chosen to allow for vaporization of a component, making separation of a product easier,
shifting the reaction equilibrium, or removing heat from the reactor. Increasing pressure for reactions that take place in the gas phase increases reactant activity and thus
the reaction rate. Reactor yields follow Le Chatelier's principle: for reactions in which
the number of moles increases, lower pressure will increase equilibrium conversion;
for reactions in which the number of moles decreases, lower pressure will decrease
equilibrium conversion. Increasing the pressure in gas–liquid reactions increases the
solubility of the gas in the liquid, which increases the reaction rate.

Reaction phase
Reactions are usually carried out in liquid or gas phases as fluids are easier to handle, heat, cool and transport than solids. For reagents or products in the solid phase
a suspension in liquid or gas is usually used. The phase of the reaction is usually determined by reactor temperature and pressure. Liquid-phase operation is usually preferred due to the highest concentrations and greatest compactness. However, at temperatures above the critical temperature there cannot be a liquid phase. The pressure
can sometimes be adjusted to keep all reagents in the liquid phase, however when this
is not possible a multi-phase reactor will be necessary. If mass transfer limitations become too significant it can be beneficial to reduce the pressure such that the reaction
temperature is above the dew point and the reaction is carried out in the vapor phase.

Solvent
Solvents are used for liquid-phase reactions and can be used for the following:
- dilution of feed to improve selectivity;
- increasing solubility of gas-phase components;
- dissolving solids in the reacting phase;
- increasing thermal mass which lowers temperature change per unit volume from
  reaction;
- improving miscibility of mutually insoluble components.

Solvents should be inert in the main reaction and should not react with products or feed contaminants. Solvents should also be inexpensive and easily separated from the reaction products. Some widely used process solvents and their properties are given in Table 12.1.

Table 12.1: Summary of suggested flow rates for gas flow as agitation.

Degree of agitation	Liquid depth 9 ft	Liquid depth 3 ft
Moderate	0.65	1.3
Complete	1.3	2.6
Violent	3.1	6.2

Concentrations

Higher concentrations of feed can lead to higher reaction rates; however, for exothermic reactions high feed concentrations should be avoided. Feed compounds are usually not supplied in stoichiometric ratio as using a higher concentration of one feed can lead to increased selectivity towards the desired product.

Understanding the effect of feed contaminants and by-products is essential to reactor design; they can play significant roles in reactor selectivity and performance. When recycling attention must be paid to by-products; those formed through reversible reactions can be recycled, leading to improved overall selectivity. Feed contaminants generally pose a greater issue than by-products due to their ability to poison catalysts or kill biological organisms. If a feed contaminant is particularly detrimental to reactor performance it should be removed upstream of the reactor.

Inert compounds will usually increase reactor cost due to the larger volume required, as well as increased downstream separation costs; they can still be advantageous under the following circumstances:
- inerts in gas-phase reactions reduce partial pressure of reagents, which can increase equilibrium conversion in reactions that lead to an increase in number of moles;
- feed compound reacting with itself or products can be reduced by dilution using inerts;
- inerts can allow operation outside of the flammability envelope;
- reaction solutions can be buffered to control pH.

Step 3: Determine materials of construction

A preliminary analysis of the materials of construction for the reactor can be conducted after the reaction conditions have been specified. Particularly important in this analysis are the temperatures and pressures the process will run at. At extreme conditions, costly alloys may need to be used. In addition, the designer must ensure that process streams will not react with materials used in process equipment.

Step 4: Determine rate limiting step and critical sizing parameters

The key parameters that determine the extent of reaction must be identified by carrying out an experiment plan with a broad range of conditions. In general, the rate of reaction is usually limited by the following fundamental processes. The first three have been discussed in previous sections. Mixing will be developed in more detail in Section 12.4.

- **Intrinsic kinetics:** There will usually be one slowest step that limits the overall rate.
- **Mass transfer rate:** In multi-phase reactions and processes that use porous heterogeneous catalysis, mass transfer can be particularly important. Often, careful experimentation will be needed to separate the effects of mass transfer and the rate of reaction to determine which is the rate limiting step.
- **Heat transfer rate:** The rate of heat addition can become the limiting parameter for endothermic reactions. Heat transfer devices such as heat exchangers or fired heaters may need to be used.
- **Mixing:** The time taken to mix the reagents can be the limiting step for very fast reactions.

Once rate data have been collected, the designer can fit a suitable model of reaction kinetics. Next, a critical sizing parameter can be specified for the reactor.

Step 5: Preliminary sizing, layout and costing of reactor

The designer can estimate the reactor and catalyst volume from the sizing parameter. This calculation will yield a value for the active reacting volume necessary. Clearly, the actual reactor will need additional space. The geometry of the reactor will depend on the desired flow pattern and mixing requirements. The cost of most reactors can be estimated by determining the cost of a pressure vessel with the same dimensions and adding in the cost of the internals.

Step 6: Estimate reactor performance

At this point in the design process, it is important to verify that the proposed reactor will achieve the target conversions and selectivities. A combination of experimental methods, such as pilot plants, and computer simulations can be used to predict the full-scale reactor performance.

Step 7: Optimize the design

The reactor is typically a relatively small fraction of the total capital cost, so minimal time should be devoted to optimization to reduce the reactor cost. However, if the target conversion, yields and selectivities are not met, the process economics could be significantly impacted. Therefore, steps 2 to 6 should be repeated at least until the minimum specifications are met.

12.4 Mixing in industrial reactors

Mixing plays an important role in many processing stages, including reactor performance. It is critical to select the appropriate method of mixing in order to ensure the process produces the desired process yields, product purity and cost effectiveness.

Correlations such as the Reynolds number can be used to determine the extent of mixing and correlate power consumption and heat transfer to the reactor shell (Towler, 2012). In some cases, simple correlations may not be adequate; for example,
-   if dead zones cannot be tolerated for reasons of product purity, safety, etc.;
-   if reactor internals are complex;
-   if reaction selectivity is very sensitive to mixing.

In these cases, it is usually necessary to carry out a more sophisticated analysis of mixing, for example:
-   by using computational fluid dynamics to model the reactor;
-   by using physical modeling ("cold flow") experiments;
-   by using tomography methods to look at performance of the real reactor.

12.4.1 Gas mixing

Gases mix easily because of their low viscosities. The mixing given by turbulent flow in a length of pipe is usually sufficient for most purposes (Towler and Sinnott, 2013). Orifices, vanes and baffles can be used to increase turbulence.

12.4.2 Liquid mixing

In-line mixers can be used for the continuous mixing of low-viscosity fluids. One inexpensive method involves the use of static devices that promote turbulent mixing in pipelines.

When one flow is much lower than the other, an injection mixer should be used. A satisfactory blend will be achieved in about 80 pipe diameters. Baffles or other flow restrictions can be used to reduce the mixing length required. These mixers work by introducing one fluid into the flowing stream of the other through a concentric pipe or an annular array of jets.

Mixing in stirred tanks is conducted by an impeller mounted on a shaft driven by a motor. The reactor usually contains baffles or other internals to induce turbulence and prevent the contents from swirling and creating a vortex. Typically, baffles are 1/10 of the diameter and located 1/20 of the diameter from the wall. Typical arrangements of agitator and baffles in a stirred tank, and the flow patterns generated, are shown in Figure 12.3. Mixing occurs through the bulk flow of the liquid and by the motion of the turbulent eddies created by the agitator. Bulk flow is the predominant mixing mechanism required for the blending of miscible liquids and for solids suspension. Turbulent mixing is important in operations involving mass and heat transfer, which can be considered as shear-controlled processes.

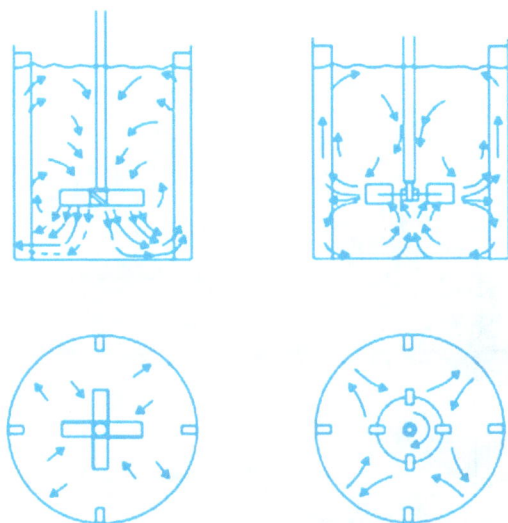

Figure 12.3: Agitator arrangements and flow patterns.

At high Reynolds numbers (low viscosity), one of the three basic types of impeller shown in Figure 12.4 should be used. For processes controlled by turbulent mixing, flat-bladed (Rushton) turbines are appropriate. For bulk mixing, propeller and pitched-bladed turbines are appropriate.

For more viscous fluids, paddle, anchor and helical ribbon agitators (Figures 12.5(a)–(c)) are used. The selection chart given in Figure 12.6 can be used to make a preliminary selection of the agitator type, based on the liquid viscosity and tank volume.

12.4.3 Gas–liquid mixing

Gases can be mixed into liquids using the in-line mixing or stirred tank methods discussed previously. A special type of gas injector, called a sparger (shown in Figure 12.7) can also be used. This is a long injection tube with multiple holes drilled in it.

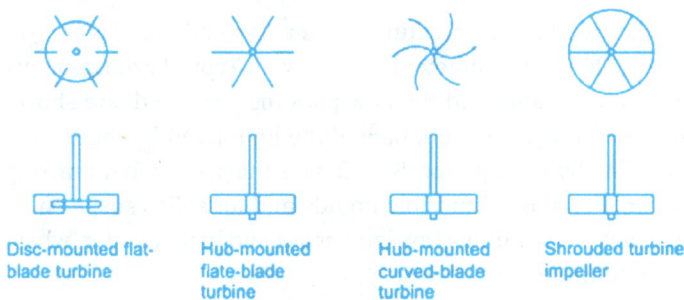

Disc-mounted flat-blade turbine   Hub-mounted fiate-blade turbine   Hub-mounted curved-blade turbine   Shrouded turbine impeller

(a)

(b)   (c)

Figure 12.4: Basic impeller types.

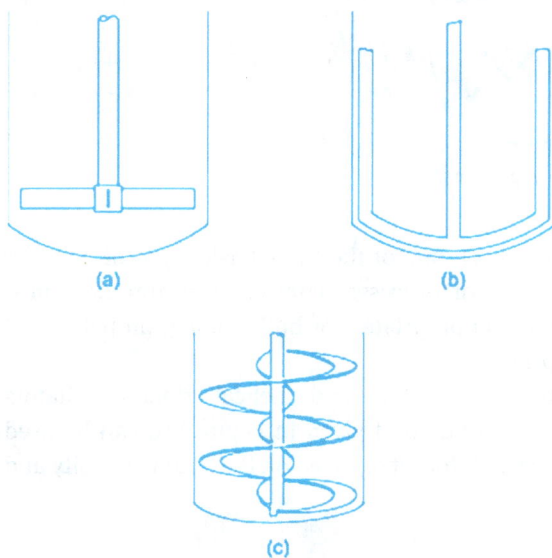

(a)   (b)

(c)

Figure 12.5: Low-speed agitators (Towler and Sinnott, 2013).

A small flow of liquid can be dispersed into a gas stream using a spray nozzle (Figure 12.8).

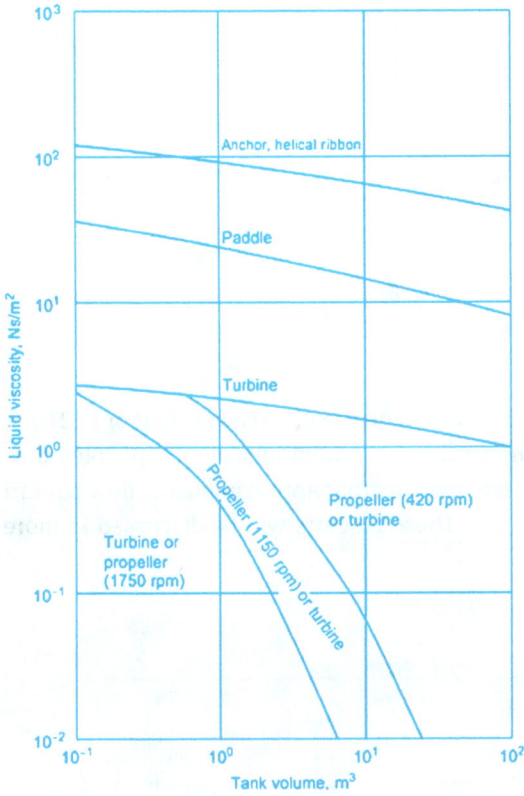

Figure 12.6: Agitator selection guide (Towler and Sinnott, 2013).

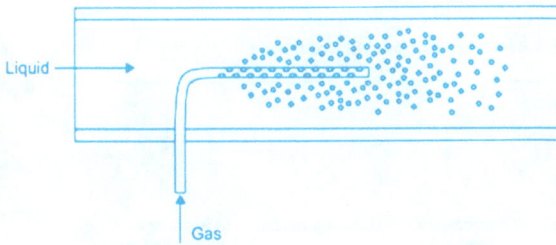

Figure 12.7: Gas sparger.

12.4.4 Solid–liquid mixing

Solids are usually added to a liquid in a stirred tank at atmospheric pressure. In order to allow more accurate control of dissolved solid concentration, mixing of solids and liquids is often carried out as a batch operation.

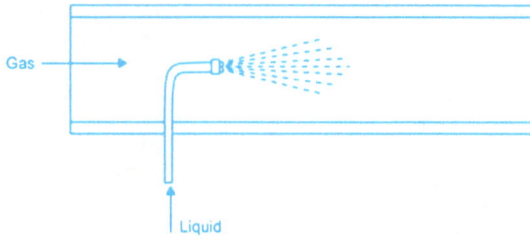

Figure 12.8: Liquid injection into gas.

12.5 Types of reactors

Most reactors used in industry approximate the ideal batch reactor, PFR or CSTR. In fact, real reactors can be modeled as networks or combinations of multiple plug-flow and stirred tank reactors. Examples of real reactors that approximate the flow pattern of ideal reactors are shown in Figure 12.9. These reactors will be discussed in more detail in the following sections.

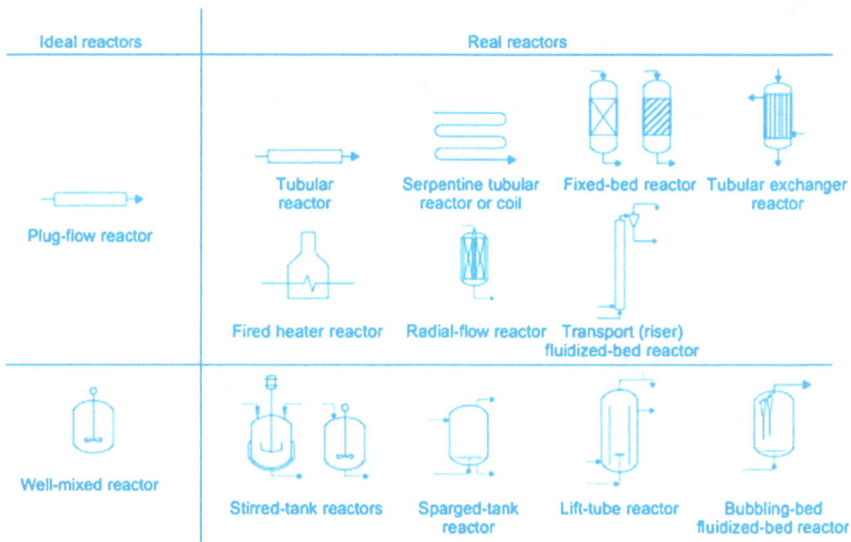

Figure 12.9: Ideal reactors and some real reactors that approximate the same flow pattern.

12.5.1 Vapor–liquid reactors

Vapor–liquid reactions are important in many chemical processes. For example, oxygenation and hydrogenation reactions are usually carried out with the organic component in the liquid phase.

If the residence time requirements are short enough, vapor–liquid contacting columns are preferred because of the large area for mass transfer. Trayed or packed columns can be used to contact vapor and liquid for reaction. The column packing may be catalytically active or could be inert packing.

Stirred tanks or tubular reactors are used when long residence time is needed for the liquid phase.

The basic concept of a sparged reactor is shown in Figure 12.10.

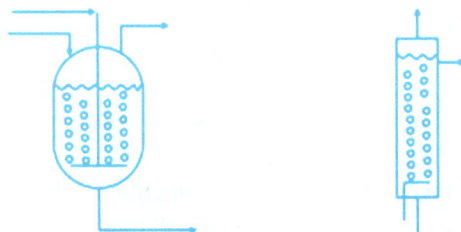

Figure 12.10: Sparged stirred tank and tubular reactors.

The gas is bubbled up through the liquid in a sparged reactor. For smaller bubbles, a porous pipe diffuser can be used instead. The designer must allow some disengaging space at the top of the reactor, or entrainment will be excessive. If the gas flow rate is large, then the gas flow can be used as the primary means of agitation. Perry's Handbook suggests the following air rates (ft^3/ft^2 min) for agitating an open tank full of water at 1 atm:

12.5.2 Catalytic processes

A catalyst increases the rate of a chemical reaction without itself becoming permanently changed by the reaction. Catalysts allow reactions to be run in smaller reactors and operated at lower temperatures and improve selectivity. Therefore, catalysts will almost always lead to a more economically attractive process than a non-catalytic route. Catalysts are normally selected based on performance rather than price, since increases in catalyst selectivity will almost always quickly pay back any price premium expected by the manufacturer. It is important to test the catalysts under conditions that are representative of process conditions.

Catalyst activity often deteriorates over time. Common causes of deactivation include the following:
- poisoning by components in feed (*e. g.*, base destroys acid catalyst);
- blockage of pores or active sites by by-products such as coke;
- thermal or hydrothermal modification of catalyst structure.

Slow activity loss can be compensated by:
- putting in more catalyst (lower space velocity);
- slowly raising reactor temperature.

Rapid activity loss may require moving the catalyst to a continuous regeneration zone.

Catalytic reactions can be either homogenous (catalyst is in the same phase as the reagents) or heterogeneous (catalyst is not in the same phase as the reagents).

Homogeneous catalysis

Homogeneous catalysis can be conducted in the basic batch reactors, PFRs or CSTRs, which have already been discussed. However, when the catalyst is in the same phase as the reagent, recovering this catalyst after the reaction can be difficult and expensive, particularly if the catalyst is sensitive to high temperatures. Providing adequate interfacial area is also a challenge of homogeneous catalysis. A reaction often only occurs at the interface or in the boundary layer between the catalyst and the reagents. Increased mixing can increase the rate and selectivity of the reaction, but this can require detailed and expensive mixing equipment. For these reasons, reactions requiring homogenous catalysts are not usually used unless an easy separation can be found to recover the catalyst.

Heterogeneous catalysis

Catalyst recovery in processes involving heterogeneous catalysis is much easier. However, the rate of reaction is limited by the available inter-phase surface area and the mass transfer of reagents and products to and from the interface. Therefore, reactors for these processes are designed to reduce these limitations.

Fixed bed reactors: In a fixed bed reactor, the reagent flows over a stationary bed of packed catalyst. This is the most common type of reactor used for heterogeneous catalysis as long as the catalyst does not require continuous regeneration and the reaction mixture does not require high agitation.

Radial-flow reactors: When there is very little pressure drop available, the L/D ratio must be much less that one. A common solution to this is to use a radial flow reactor with the catalyst contained in an annulus between vertical perforated or slotted screens. The fluid flows radially through the bed and the direction of flow can be either inwards or outwards.

Moving bed reactors: A moving bed reactor is similar to a radial flow reactor, but the catalyst is moved through the annular space.

Fluidized bed reactors: If the fluid flow is up through the catalyst bed, then the bed can become fluidized if the pressure drop is high enough to support the weight of the catalyst. Fluidized beds usually have a lower pressure drop than downflow at high flow rates. In addition, fluidizing the catalyst eases the transition from one reaction zone to another.

The catalyst bed is fluidized using a distributor to inject fluidization fluid, which is not necessarily the feed. Fluidization occurs when the bed pressure drop balances the weight of the particles.

Fluidization can only be used with relatively small-sized particles (<300 μm with gases). The solid material must be strong enough to withstand attrition in the fluidized bed and cheap enough to allow for make-up to replace attrition losses. Fluidized bed reactors must also allow for separating the fluid-phase product from entrained solids so that solids are not carried out of the reactor.

Trickle bed reactors: Trickle bed reactors are used when all three phases are involved in the reaction. They must ensure good distribution of both the vapor and the liquid, without channeling of either phase. In a trickle bed reactor, the liquid flows down over the surface of a stationary bed of solids. The gas phase usually also flows downwards with the liquid, but countercurrent flow is feasible as long as flooding conditions are avoided. This requires a more sophisticated distributor like those used for packed distillation columns.

Slurry reactors: Liquid is mixed up in the liquid in slurry-phase reactions. Slurry reactors are prone to attrition of the solids, caused by pumping or agitation of the liquid. Slurry-phase operation is usually not preferred for processes that use heterogeneous catalysts because the catalyst tends to become eroded and can be difficult to recover from the liquid.

12.5.3 Bioreactors

Bioreactors have requirements that add complexity compared to simpler chemical reactors. These reactions often are three-phase (cells, water and air), need sterile operation and require heat removal. However, biological systems have the following advantages:
- Some products can only be made by biological routes.
- Large molecules such as proteins can be made.
- Selectivity for desired product can be very high.
- Products are often very valuable.

Enzyme catalysis
Enzymes are the biological equivalent of catalysts. They can sometimes be isolated from host cells. They are usually proteins and, therefore, most are thermally unstable above ~ 60 °C and active only in water at a restricted pH. Enzymes can sometimes be absorbed onto a solid or encapsulated in a gel without losing their structure. In this

case, they can be used in a conventional fixed bed reactor. Typically, homogenous reactions are carried out in batch reactors.

Microorganism design and selection

As an alternative to an enzyme catalyst, engineered microorganisms can be used to produce chemicals of interest. These products could be complex biological compounds, therapeutic proteins or commodity plastics and fuels. Host cells as a platform for modification have so far included bacteria, yeast and mammalian cells. The efficiency of a bioreactor is heavily dependent on the efficiency of the microorganism used. An inefficient cell host that does a poor job of producing the desired product will always result in a poorly designed bioreactor, regardless of the equipment or conditions used. Furthermore, the design of a bioreactor is largely based around the ideal growth conditions of the microorganism. As shown in this section, the design and/or selection of a microbial host is closely related to the design of the bioreactor. Choice of a host demands particular reactor conditions, and in the case of genetically engineered microbes, the cells must be designed to operate in conditions that are feasible and affordable with modern bioreactor technology. This process can involve the rigorous engineering of a novel microorganism, a large screening for high-producing strains or, most likely, a combination of the two. This step of the bioreactor design process requires close collaboration between process engineers and microbiologists.

Fermentation goals

Fermentation as a general practice is carried out with the following goals, many of which are affected directly by microorganism choice.

Cost, yield and productivity

The goal of an efficient microbial host results in four parameters that relate microbe performance to the overall reactor performance. Overall fermentation performance for batch and fed batch processes can be evaluated as follows:

$$\frac{dP}{dt} = q_p X, \tag{12.15}$$

where P represents the concentration of desired product, X the concentration of cells and q_p the specific productivity in mass of product per mass of cells per time. Cell growth can be modeled by

$$\frac{dX}{dt} = uX, \tag{12.16}$$

where u is the specific growth rate per time. Desirable values for these parameters for a scaled bacterial process are a productivity of 0.1 g/l/hr and a growth rate of 0.2–0.7 l/hr. These parameters are specific to cell lines and are difficult to engineer orthogonally.

In addition to growth and product formation, it is important to consider substrate consumption in selecting an efficient microorganism. Often high product titers can be obtained with excessive waste of substrate, leading to high costs and unrealistic reactor sizes.

While many elements are required, it is not necessary to model all of them. For evaluating the consumption of feed, it is useful to model the organism's chemical consumption on an elemental level for only the first four. For example,

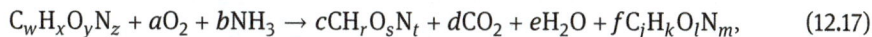

$$C_w H_x O_y N_z + aO_2 + bNH_3 \rightarrow cCH_r O_s N_t + dCO_2 + eH_2O + fC_j H_k O_l N_m, \qquad (12.17)$$

where w, x, y, z indicate substrate composition, r, s, t indicate the relative cell composition and j, k, l, m indicate the composition of the product. However, this design equation cannot be solved for a single solution. Instead, two additional parameters are required that are specific to the cell host. $Y(X/Sp)$ and $Y(P/S)$ represent the yield of cell mass and product mass per mass of fed substrate. These parameters characterize how the cell host utilizes its feed, and again are difficult to orthogonally engineer. Using this stoichiometric design equation and a desired product formation rate, the rate of substrate utilization can be calculated. In this way, the overall fermentation yield relies heavily on the four organism design parameters q_p, $uY(X/S)$ and $Y(P/S)$. For continuous fermentation, the process holds the same dependence on these parameters, which are found in the following design mass balance:

$$FS_0 - FS - \frac{VyX}{Y_{X/S}} - \frac{Vq_pX}{Y_{P/S}} = V\frac{dS}{dt}. \qquad (12.18)$$

Product isolation and purification

Downstream processing of the product is primarily dependent on the nature and chemical properties of the product itself. For example, a particular intracellular protein may be very difficult to separate from other cell internals. However, the choice of microorganism can have a significant impact on early separation steps, specifically the separation of the product from the cell mass. Paramount is whether or not the product is excreted from cells. Bacteria like *E. coli* lack many of the mechanisms required to excrete a desired product into the fermentation broth. This requires the lysing of cells in early downstream processing and separation of the product from cell internals. This process would be executed in batches, which can be timed optimally to maximize use of fermentation and separation equipment. On the other hand, mammalian cells and yeast can be engineered or screened to secrete the product of interest into the fermentation broth. This process removes the requirement of lysis step, and greatly simplifies the purification of product. This also makes the reactor particularly amenable to continuous fermentation. Additionally, eukaryotic cells can produce more complex products, such as glycosylated proteins. The glycosylation of proteins is a mechanism only recently achieved in bacteria.

Operation conditions, equipment and scale-up

It is important for the process engineer to select a microorganism that can operate within reasonable reactor conditions.

- Feed: The microbe must exhibit desirable design parameters when grown on a feed that is not commercially or cost-restrictive.
- Heat: The microbe must exhibit desirable design parameters at a temperature that is reasonable to maintain in a bioreactor. Because fermentation generates excessive heat from substrate breakdown, this generally involves cooling the reactor to between ambient temperature and 37 °C. A microbial host that requires temperatures too high or low is not amenable to controlled fermentation. This is especially salient with extremophiles – microbes that live in extreme conditions that often exhibit naturally high titers of high-value products. In this case, it would be necessary to engineer the extremophile, or choose a more reasonable cell host.
- Oxygen: Microbes can generally grow in aerobic or anearobic conditions. Often, product formation and growth will be favorable in aerobic conditions. If this is the case, it is important to consider the oxygen requirement to maintain aerobic conditions and ensure that the bioreactor designed can meet the requirements of the organism at the desired growth rates and concentrations. When designing a microorganism, it is important to not require an oxygen usage rate that is above what a reasonable bioreactor can provide.

Challenges in microorganism design

Cells must be engineered to produce a heterologous product through recombinant DNA. For a therapeutic protein, this includes identifying the DNA sequence coding for the protein and expressing that DNA in a cell host. For a commodity molecule, enzymes that catalyze the synthesis of that molecule must be identified and expressed in the host cell. The engineering of microorganisms presents a number of formidable challenges. Many companies avoid this issue by screening known microorganisms for strains that naturally produce high titers of product, or close precursors. Expressing heterologous genes in cells causes high stress and disrupts natural metabolic balance.

Many techniques are used for the engineering of microorganisms. These involve mostly the manipulation and delivery of heterologous DNA to the host cell line, to genetically manipulate its phenotype. However, because this chapter focuses on design for the process engineer, those techniques are omitted from this discussion. Instead, the aspects of organism design that impact the process parameters exhibited by the organism will be elucidated. This mainly involves balancing the observed yields to maintain high productivity and growth rate.

Metabolic engineers study ways to relieve these stresses by "rewiring" synthesis networks within cells. This includes largely two parts. The first involves constructing non-native biochemical pathways in cells. This is necessary if the host does not already produce the desired product. Enzymes are expressed in the host that catalyze the correct reactions to synthesize the product. This often puts stress on cells, as it diverts resources in the cell that are typically utilized elsewhere, such as for growth, towards the product. The second aspect of metabolic engineering involves the manipulation and balancing of metabolic fluxes within the cell. This involves controlling the expression of enzymes so that the cell makes enough product, but still has enough resources to grow to an acceptable level. Sometimes, it can be advantageous to only induce production of the product after cells have grown to a high concentration. This requires the heterologous DNA to be expressed with an inducible promoter. For example, production of a product could be induced when the feedstock is switched to methanol.

A parallel strategy to metabolic engineering is protein engineering. This simply involves the design or random testing of proteins, usually enzymes, to either enhance or alter their function. This is used in conjunction with metabolic engineering to either create novel pathways or balance existing pathways.

Cell growth
Cell growth goes through several phases during a batch, shown in Figure 12.11.
- I: Innoculation: slow growth while cells adapt to the new environment.
- II: Exponential growth: growth rate proportional to cell mass.
- III: Slow growth as substrate or other factors begin to limit the growth rate.
- IV: Stationary phase: cell growth rate and death rate are equal.
- V: Decline phase: cells die or sporulate, often caused by product buildup.

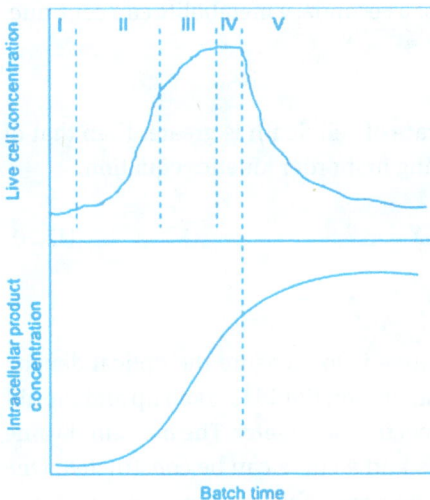

Figure 12.11: Cell growth and product formation in batch fermentation.

Innoculation

The innoculation or lag phase is the first step of cell growth during a batch fermentation process. There is a minimal increase in cell density. This phase is least understood by scientists but has been noticed since the end of the 19th century. There is a lack of data that can adequately explain the physiological and molecular processes that take place during this phase.

Exponential growth

The exponential phase, also known as the logarithmic growth phase, occurs when cells have adjusted to their new conditions. They are dividing at a constant rate, resulting in an exponential increase in cells following first-order kinetics. The following equation illustrates this process:

$$\frac{dX}{dt} = X(\mu - K_d).$$

(12.19)

Cell growth is often substrate-limited, meaning growth will slow down once substrates become less available. Cell growth rate can be measured by different forms of inhibition. These forms include substrate inhibition, product inhibition and toxic compounds inhibition.

Stationary phase

The stationary phase occurs when the number of cells dying and dividing reaches an equilibrium. This can be caused by the depletion of one or more nutrients, the accumulation of toxic by-products or the induction of a gene. Induction causes a stressful environment for cells and increases the death rate. In this phase, production of the primary metabolite stops, but the production of a secondary metabolite can continue.

Decline phase

The decline or death phase occurs when the rate of cell death is greater than that of cell generation. It is represented by the following first-order kinetics equation:

$$\frac{dX}{dt} = -K_d X.$$

(12.20)

Measuring growth

One easy way to quickly establish a growth curve is to measure the optical density with a spectrophotometer. A sample of the fermentation liquid is taken up and the absorbance of the sample is measured with the spectrophotometer. The measured value is then combined with previous measurements and a curve can be constructed. One drawback of this method is that both viable and non-viable cells are measured and taken into account.

Intracellular product accumulation is slow at first because there are a limited number of cells. However, it is important to note that product accumulation continues even after the live cell count falls, since dead cells still contain product.

The growth rate of cells can be limited by many factors, including:
- the availability of the primary subtrate (typically glucose, fructose, sucrose or another carbohydrate);
- the availability of other metabolites (vitamins, minerals, hormones or enzyme cofactors);
- the availability of oxygen;
- mass transfer properties of the reaction system;
- inhibition or poisoning by products or by-products;
- high temperature caused by inadequate heat removal.

All of these factors are exacerbated at higher cell concentrations. Clearly, biological reactions must be carefully controlled. An additional complication in dealing with biological reactions is that the product formation is often not closely tied to the rate of consumption of the substrate, because the product may be made by the cells at a relatively low concentration and some cell metabolic processes may not be involved in formation of the desired product.

Batch or continuous

Batch bioreactors represent the majority of industrial processes. This requires a sterilization phase and inoculation of the culture medium with microorganisms before the reaction can occur. Thee advantages of batch systems include:
- a reduced risk of contamination due to a short growth time;
- lower capital investment;
- more flexibility for biological systems.

The disadvantages include:
- intermediate steps that cause decreased productivity levels;
- high expense when preparing culture for inoculation;
- higher hygiene risks due to close contact with microorganisms.

For a continuous process, medium that is either sterile or contains bacteria is continuously fed into a bioreactor in order to maintain the steady state. The advantages of continuous systems include:
- potential for automation;
- lower costs of labor;
- less time spent sterilizing and preparing;
- consistent product quality.

The disadvantages include:
- only small changes in process are allowed;
- feed quality needs to be specified and maintained;
- higher investment costs;
- risk of cell mutation due to short cultivation.

Mass transfer for bioreactors

Mass transfer is important to keep in mind because it often becomes the limiting step of the overall process. The volumetric oxygen transfer coefficient must be known to accurately design and scale up bioreactors. The following equation shows the mass balance for dissolved oxygen in a well-mixed reactor in the absence of biomass:

$$\frac{dC}{dt} = k_L a(C^* - C). \tag{12.21}$$

The variables that affect the $k_L a$ values are mostly affected by impeller configuration, speed and aeration. An increase in the gas flow rate increases the mass transfer coefficient values.

Types of bioreactors

Stirred tank fermenter: The stirred tank fermenter is the most common reactor used for biological reactions and is similar to the stirred tanks discussed previously. It can be used in both batch and continuous mode. Figure 12.12 shows a stirred tank fermenter.

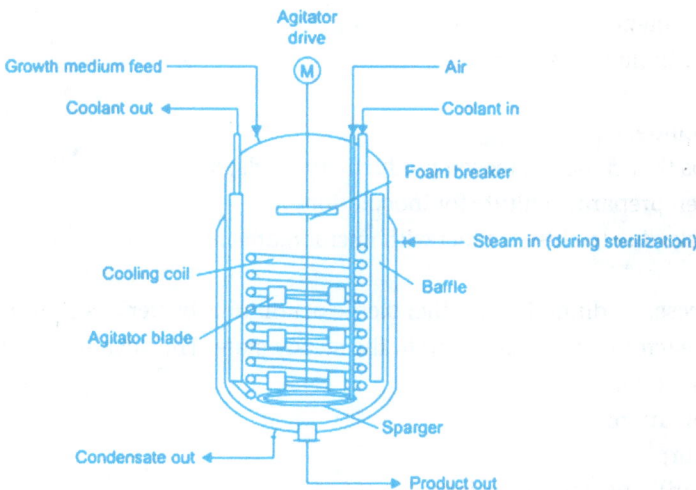

Figure 12.12: Fermentation reactor.

Shaftless bioreactors: Shaftless bioreactors are used when the pump shaft seal is considered a non-permissible source of contamination. These reactors use gas flow to provide agitation of the liquid. The design requires careful attention to hydraulics.

WAVE bioreactors: WAVE bioreactors represent an alternative to standard stainless steel bioreactors. These reactors are flexible and single-use, cutting down time between batches and allowing for a more sterile environment. These disposable reactors are mostly used in mammalian cell culture. Three layers of plastic are the minimum necessary for construction. The first is a structural layer, followed by a barrier layer that allows for permeability. The last layer, the fluid contact layer, is designed to take into account inertness and maintain a good seal.

Packed bed bioreactors: Packed bed bioreactors are structured so that the cells are immobilized and placed on large particles. Although they are relatively simple to construct, they can have blockage issues or poor oxygen transfer. There are three types of flow: downward flow, upward flow and the recycling method. In industry, upward flow is preferred, especially when there is gas production during the fermentation.

Anaerobic bioreactors: Anaerobic reactions are used in ethanol production, winemaking, beer brewing and wastewater treatment. Due to their longstanding history, these processes have become well established and improvements include decreasing cost of production due to new technology. Although continuous production for beer has been patented on a large scale, most investment is still focused on batch production.

Preventing contamination

Since cells are easily affected by both unwanted chemicals and other species in the reactor, bioreactors must be designed in order to avoid contamination. Bacterial spores are the most demanding sterilization challenge in a bioreactor. Bacterial spores are dormant and non-reproductive structures produced by a small number of bacteria. Since they are meant to ensure survival of bacteria in times of environmental stress, they are heat-resistant. In order to ensure that all spores within the medium are killed, the medium must be sterilized. There are many methods of sterilization, including filtering and chemical, thermal and radiation treatment. Thermal sterilization using steam is the most common method, as it is the most economical method for large-scale reactors. Chemical agents cannot be toxic to the product, and UV radiation cannot penetrate fluids easily. Steam sterilizations either occur in the fermentation vessel as a batch sterilization or in a continuous apparatus upstream of the fermentation vessel.

Sterilization

Sterilization is done to prevent issues rising from biological contamination. Problems involved in sterilization increase greatly with scale-up, as sterilization methods used for lab-scale reactions are not acceptable for industry-scale reactions. Given the same medium, a sterilization temperature and time may be enough for a small-scale reactor but not for a larger scale (Figure 12.13).

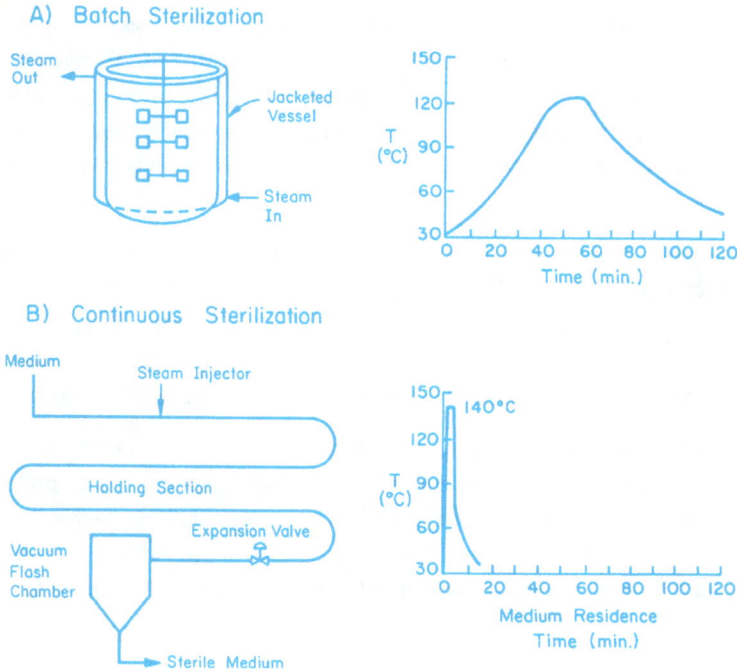

Figure 12.13: Diagrams and temperature profiles of batch and continuous sterilization.

Other considerations

Since spores will germinate in a moist environment, making them easier to kill as vegetative cells, moist heat is preferred for sterilization. Connections in sterilization equipment must be steam-sterilized and trapped air pockets should be avoided. Pipes should be sloped in order to avoid condensate pools and equipment should be pressure-tested for leaks. It should be ensured that no viable host cells are in the waste streams and escape to the environment, and exit gas streams also need to be filtered to prevent the escape of microbes to the environment.

If the medium being sterilized contains heat-sensitive materials, filter sterilization must be used instead of steam. Filtration is also used to sterilize the process air used in the system. Microporous filters are used, so the medium must be prefiltered for larger particulates so that the microporous filter does not get clogged. However, filtration is not as reliable as steam sterilization, as any defects in the membrane can lead to contamination and viruses can often pass through the filter.

Sterile sampling

Sampling the medium in the reactor is necessary to ensure product quality, but it also carries the risk of introducing contamination into the medium. Sampling is usually done about five times a day for a bioreactor.

Cleaning

Cleaning the fermentation vessel at the end of the production run is necessary in order to remove residual substrates that could lead to contamination of future batches. Cleaning consists of the following wash steps:

1. wash with high-pressure water jets;
2. wash with an alkaline cleaning solution, usually 1 M NaOH;
3. rinse with tap water;
4. wash with an acidic cleaning solution, usually 1 M nitric or phosphoric acid;
5. rinse with tap water;
6. rinse with deionized water.

The vessel is drained after each of these steps. For this reason, the vessel should have no internal dead spots where material could accumulate. Also, due to the repeated emptying and filling of the vessel, cleaning leads to significant downtime between batches (Towler 2013). A clean-in-place (CIP) system with a transfer flow plate can make cleaning easier, as it connects all the bioreactors and transfer pipes in a plant to one cleaning system. The transfer plate has removable pipe sections, providing assurance against mixing of different bioreactor contents.

12.6 Heating and cooling of reacting systems

Exothermic and endothermic reactions will require reactors with heat control systems to prevent operating conditions from falling out of the desired range. Reactor performance is often limited by the ability to add or remove heat. Insufficient heat removal can cause runaway reactions, particularly dangerous situations in chemical processing. Before considering the design of a heating or cooling system to couple with a reactor, a few important questions should be asked:

- Can the reaction be carried out adiabatically?
- Can the feeds provide the required heating or cooling? Staged addition of feed can help alleviate the cost of adding a heat exchange network or heat transfer jacket. Also consider adding an inert diluent or hot/cold shots (Seider *et al.*, 2004).
- Would it be more cost-effective to carry out the heat exchange outside of the reactor?
- Would it be more effective to carry out the reaction inside of a heat transfer device? If a reaction requires only a small volume or small quantities of catalyst, it may be possible to utilize a heat exchanger as a temperature controller and as a reaction location.

– Does the proposed design allow the process to be started up and shut down smoothly?
– Are there safety concerns with heating or cooling the reactor?

After considering these aspects of the design, commercial design software such as HYSYS or UniSim can be utilized to estimate heating/cooling requirements. Once this is done, design of the heat exchange system can begin, with different reactor types and reactions requiring different design approaches.

12.6.1 Stirred tank reactors

Heating and cooling of a stirred tank reactor is carried out to ensure a uniform reaction temperature, so that there do not exist hot or cold spots within the reactor that can negatively affect selectivity.

For indirect heat transfer, there are three main alternatives: a heat transfer jacket, an internal coil and an external heat transfer circuit. A jacket is utilized as long as there is sufficient heat transfer area for the heat exchange to take place. If this is not the case, coils are used, although the inclusion of a heating coil will significantly increase reactor volume and utility requirements, leading to a large increase in price for the reactor. External circuits contain a heat exchanger that will heat or cool the product stream as required and recycle this material to the reactor to control temperature. External circuits are useful because they can be designed independently of the reactor; sizing the required pumps and heat exchangers will not fundamentally change the activity of the reactor. For any of these choices, it should be ensured that no corrosion of the involved piping will occur, as utility streams bleeding into the reactor can have a very negative impact on the selectivity of the reaction and on the operation of the reactor on a whole.

Some direct heat transfer alternatives also exist, as long as the reaction in question is compatible with the addition of extra water. Steam can be pumped into the reactor to maintain temperature, which will eliminate the need to design heat transfer surfaces. However, steam injected into the system cannot be recovered, so this will lead to an increase in annual utility costs. Additionally, vapor will be produced if it did not exist previously, so reactors will need to be redesigned to accommodate a vapor removal system.

12.6.2 Catalytic reactors

Slurry reactors: Since slurry reactors already use a mixture of solid catalyst and liquid reactants, any of the methods described in the section on stirred tank reactors can be applied to slurry reactors. It is not recommended to use internal coils in such a design, as reactor slurry will often corrode heat exchange material very easily.

Fixed bed reactors: Indirect heat transfer is not often utilized to control the temperature in fixed bed reactors, as it is hard to maintain uniform temperature across the radial section of the catalyst bed. In cases where temperature control is required, the reactor will be split into smaller sections. After each bed, there will be a heat transfer stage, where the product stream is heated or cooled as necessary and returned to the next catalytic segment.

Fluidized bed reactors: Fluidized bed reactors have high heat transfer coefficients, so indirect heat transfer is highly effective. The solid catalyst particles can be used as a heat transfer medium themselves; heated catalyst contains a reaction location and the necessary heat to maintain the required temperature. Deactivated catalyst is heated during reactivation and recycle.

12.7 Heat exchangers as reactors

It is sometimes necessary to design a reactor as a heat transfer device, like when it is necessary to operate a reactor isothermally and there is a large heat of reaction. Some common situations include high-temperature endothermic reactions that quickly quench without continuous heat input and low-temperature exothermic reactions that must be kept at constant temperature to maintain selectivity. The most common heat transfer equipment used for reactions are shell and tube heat exchangers and fired heaters.

12.7.1 Homogenous reactions

If the reaction does not require a catalyst, then the heat transfer design is the same as for a conventional heat transfer device, with some important changes in the thermal design. The usual heat exchanger equations will not apply to the design of a heat exchanger reactor due to the non-linear behavior of the reaction rate with regard to temperature. In these cases, the usual practice of conservative temperature estimations will not aid in heat transfer design, as greater detail will be required to ensure the proper operation of the reactor. Detailed kinetic models should be developed before designing the internals of the heat transfer device.

12.7.2 Heterogenous reactions

The problems of designing for homogenous reactions still hold for heterogenous ones, with the added complication of solid catalyst beds. Catalyst can be loaded into the tubes of a shell and tube exchanger if the exchanger is mounted vertically and a suitable retaining screen is included at either end of the design. In this instance, hot catalyst can be reliably recycled and heat-treated to reactivate the catalysts and reduce the presence of reactor hot spots. High-temperature endothermic reactions will be even more difficult to design for, as their heat requirements often exceed the amount provided by a heated catalyst. In these cases, a "tube-in-tube" design is utilized, where feed and catalyst are heated simultaneously by an external fired heater. This can be done as long as thermal expansion does not cause damage to the tubes, or else significant catalyst poisoning can occur. The same concerns as detailed in the section on homogenous reactions will still apply for any design utilized for heterogenous ones, so it is again recommended to develop a detailed kinetic model before determining the amount of heat transfer required to maintain proper selectivity.

12.8 Safety considerations in reactor design

Reactors require much attention to safety details in the design process due to the hazards they impose. The highest temperature point in the process is often reached in reactors, reaction heat may be released, and residence times can be long, leading to a large inventory of chemicals. Guidelines exist for inherently safer design principles which seek to remove or reduce process hazards, limiting the impact of unforeseen events. These design methods should be applied throughout the design process as part of good engineering practice; they cannot be retroactively added by a process safety specialist.

Exothermic reactions require special consideration due to their potential to runaway (temperature rises from heat of reaction being released, increasing reaction rate, releasing more heat, etc.). The reactor must be designed such that temperature can be precisely controlled and the reaction shut down if temperature control is lost. The use of solvents or inert species also allows for temperature control by adjusting the heat capacity flow rate relative to the rate of heat release from the reaction. An additional safety feature would allow the reactor to be flooded with cold solvent or diluent.

If there is a cooling system it should be designed to return the process to a desired temperature if the maximum temperature is reached.

Venting and relief of reactors is complicated by the potential to keep reacting if containment is lost or material is discharged into the pressure relief system. The relief system should be designed according to guidelines outlined in the Design Institute for Emergency Relief Systems (DIERS) methodology. The reactor design team must understand the reaction mechanism and kinetics, including the role of any compounds which may accelerate the reaction.

12.9 Capital cost of reactors

Reactors are classified as pressure vessels, and as such the pressure vessel design methods can be used to estimate wall thickness and thus determine capital cost. Additional costs come from reactor internals or other equipment. Jacketed stirred tank reactors require more in-depth analysis than that provided by pressure vessel design. The wall of the reaction vessel may be in compression due to the jacket. For preliminary cost estimation a correlation for jacketed stirred tank reactors operating at pressures below 20 bar can be used:

$$C_e = a + bS^n, \tag{12.22}$$

where C_e is the purchased equipment cost on a US Gulf Coast Basis, a, b are cost constants, S is the size parameter and n is the exponent for that type of equipment.

Exercises

1. Consider the liquid-phase organic esterification reaction taking place in a CSTR. Two streams, an acid stream containing no base and a base stream containing no acid, are fed into the CSTR. The esterification reaction and its rate are given by

$$A + B \rightarrow H_2O + Ester, r = kc_A c_B,$$

in which A is the organic acid and B is the organic base. The acid and base are dissolved in an organic solvent and the acid and base feed streams have feed concentrations $c_{A,f}$ and $c_{B,f}$, respectively.

You may assume that the density of the fluid is independent of concentration over the concentration range of interest here. The reactor volume is constant during the entire operation.

(a) What are the units of k?

(b) What is the volumetric flow rate Q of the effluent stream in terms of the feedstream flowrates Q_A and Q_B? Show your reasoning.

(c) Write out the transient material balances for components A and B. You should have differential equations for dc_A/dt, dc_B/dt when you are finished. What initial conditions do you require for these two differential equations?

(d) Now consider the steady-state problem. Write the steady-state balances for the acid and base concentrations in the reactor, $c_{A,s}$, $c_{B,s}$.

(e) Can you solve these two equations for $c_{A,s}$, $c_{B,s}$ in the general form for all values of the parameters? Is this steady-state solution unique?

(f) Whether or not you were able to solve the equations in the general form, given the following specific parameter values, find the steady-state reactor concentrations $c_{A,s}$, $c_{B,s}$: $\tau = 10$ min, $Q_A = Q_B$, $c_{A,f} = 8$ mol/l, $c_{B,f} = 4$ mol/l, $k = 0.1$ (in units of min, mol, l).

(g) What are the steady-state molar conversions of A and B for these specific parameter values?

12.10 Take-away message

In a reactor the main product of a process is produced. Although the reactor forms the core of a process design, in size and energy costs it only comprises a small part of the overall flowsheet. In continuous reactor design there are two basic options: the continuously stirred tank reactor (CSTR) and the tubular or plug flow reactor (PFR). In determining which reactor is most feasible, an analysis of reaction mechanisms, reaction equations and rate constants has to be carried out. Mass and heat transfer characteristics also immensely influence the overall design of a reactor.

12.11 Further reading

Towler, G., Sinnott, R. (2013). *Chemical Engineering Design: Principles, Practice and Economics of Plant and Process Design*, 2nd ed. Boston: Elsevier.

Douglas, J. M. (1988), *Conceptual Design of Chemical Processes*. New York: McGraw-Hill.

Turton, R., Bailie, R. C., Whiting, W. B., Shaewitz, J. A. and Bhattacharyya, D. (2012). *Analysis, Synthesis, and Design of Chemical Processes*, 4th ed. Upper Saddle River: Prentice-Hall.

Shuler, M. L. and Kargi, F. (2002). *Bioprocess engineering*. New York: Prentice Hall.

Seider, Lewin, Seader, Widagdo, Gani and Ng (2017), *Product and process design principles, synthesis, analysis and evaluation*, 4th ed. Wiley.

13 Batch process design

13.1 Continuous versus batch-wise

Continuous processes are dominant in the process industries. One can think of commodities, plastics, petroleum-based products, paper, etc. Continuous processes are typically used to produce products in large quantities at low production costs.

However, there is also a large range of products, such as specialty chemicals, pharmaceuticals, food products and electronics materials, that have highly specialized properties and are demanded in small quantities/vessels, while the production costs are typically very high. For such products (semi-)batch processing is a much better approach.

The main challenges in batch production are to determine the best size of the vessels (equipment) and the required processing times or batch times. In general this is a bit more complicated for semi-batch (also called fed batch) processes, where the flow rates and concentrations have a strong influence on the overall process performance.

Batch processing is useful if we encounter low flow rates, large residence times and an intermittent product demand. Batch production is also preferred when chemicals are hazardous or toxic, in other words, when safety aspects are of great concern.

Most important is to determine the batch time and the size factor (volume/unit mass transport) in such a way that an objective (for example, the amount of product) is maximized or minimized. Often we need dynamic process models and degrees of freedom that can be adjusted. This leads to a so-called optimal control problem.

Example 1 (The exothermic batch reactor). In a batch reactor the following reaction takes place: $n_1 A \rightleftharpoons n_2 B$. The reaction rate is known:

$$r = c_A^{n1} k_1 \exp\left(-\frac{E_1}{RT}\right) - c_B^{n2} k_2 \exp\left(-\frac{E_2}{RT}\right).$$

For an exothermic reaction, $E_1 < E_2$. The conversion is further defined as

$$X = \frac{C_{A0} - C_A}{C_{A0}}$$

What is the operating time that minimizes batch time while achieving a specified fractional conversion of A? We can set up the component balances for A and B and then find a profile for the temperature with time that maximizes the reaction rate, which leads to

$$T(t) = \frac{E_2 - E_1}{R * \ln\left(\frac{c_B^{n2} k_2 E_2}{c_A^{n1} k_1 E_1}\right)}.$$

A plot of the temperature with time will show a decreasing exponential behavior. As conversion increases, T goes down until equilibrium conversion, and from such plot the optimal batch time can be read.

---

https://doi.org/10.1515/9783110570137-013

13.2 Batch scheduling

In its most general form, scheduling is a *decision making* activity within the manufacturing and service industries. A scheduler tries to allocate *resources* efficiently to *processing tasks* over a given time. Each task requires a certain amount of resources at a specified time instant, often denoted as *processing time*. The resources may be processing equipment in a chemical plant, but they could be just as well runways at airports or crews at construction sites. Tasks may be the operations performed in the chemical facility, but they could also be take-offs and landings at an airport or the activities in a construction project.

Generally, the efficient allocation of resources to tasks involves a certain objective, for example, to minimize the production *make span* (time required to finish all tasks), to maximize the plant throughput, to maximize profit or to minimize costs. The allocation of resources to tasks is usually limited by the availability of the resources. For example, a reactor can only produce one product at the time. Overall, scheduling decisions may include the allocation of resources to tasks, the sequencing of tasks allocated to the same resource and the overall timing (Mendez *et al.*, 2006).

Scheduling is similar to *planning*. However, scheduling normally involves short-term decisions (from days to weeks) while planning activities entail a longer time horizon (months to even years).

Once decisions have been made at the planning level, the optimized planning is sent to the scheduling level, which normally fills in the shorter time. In turn scheduling will optimize the short-term decisions, which ultimately will be passed on to the so-called *control level*, operating at the minutes to hours time scale.

Nevertheless, the focus in this chapter is on scheduling of (chemical) processing plants. While many chemicals are produced in large-scale *continuous processes*, many products are still produced *batch-wise*. This is especially common in facilities that produce low volumes of products. These plants are often made flexible to changing consumption markets and can produce different products in the same equipment (*multi-product*). These multi-product production facilities can be found especially in the production of fine chemicals, pharmaceuticals, foods and to a certain extent polymers (Rippin, 1983).

For these types of facilities production needs to be scheduled, *i. e.*, the order in which the products will be produced and the time allocation to each of them must be decided. It is finally noted that the possibilities in the production scheduling strongly depend on the design of the plant and that this can have major economic consequences (Biegler *et al.*, 1997).

In the following sections we will introduce the reader to the basics of batch scheduling and the formulation of so-called state-task networks. In addition, we will provide a flavor of the mathematical formulation of scheduling problems and with a simple example we will familiarize the reader with the implementation of such problems to an appropriate software program.

13.3 Basics of batch scheduling

Single product versus multi-product

Pharmaceuticals and specialty chemicals are often produced with dedicated equipment that is operated batch-wise. In Figure 13.1 an example of a batch process for the production of a *single product* C is given. There are four processing units that are operated in batch mode; a reactor, a static mixer, a belt filter and a dryer.

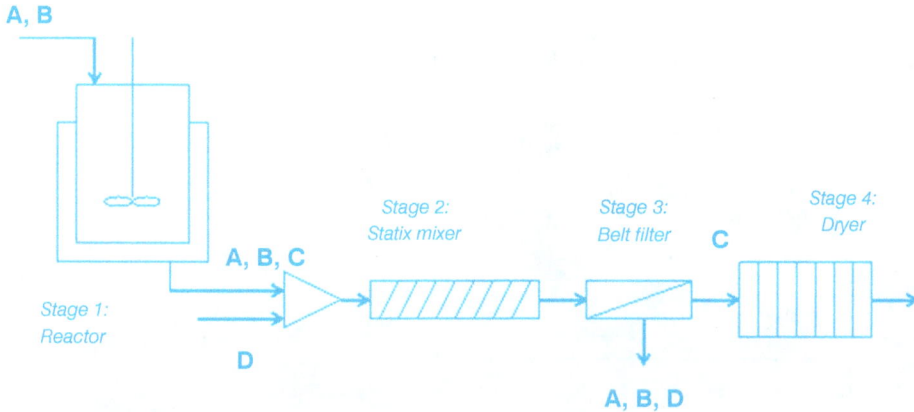

Figure 13.1: Example of a batch process.

In this set-up a single product C is produced according to the following *recipe*:
1. Mix the raw materials A and B. Heat the reactor to 75 °C and react for 3 hours to form product C.
2. Mix the reaction mixture with solvent D for 2 hours at room temperature.
3. Filtrate to recover product C for one hour.
4. Dry the product in a dryer for 3 hours at 50 °C.

The activities at each processing step can be represented by a *Gantt chart*, as shown in Figure 13.2. It is noted that the Gantt chart could also include the times required for filling and emptying the equipment.

Figure 13.2 further shows us that in this case the batch or *lot* is produced in a *non-overlapping* way; a batch is produced after the previous one has completed; two batches cannot be manufactured in parallel. It is of course also possible to simultaneously process the batches. The *idle times* are in such case eliminated as much as possible. Figure 13.3 shows the Gantt chart of an *overlapping* mode of operation.

As can be clearly seen from Figures 13.2 and 13.3, the overlapping mode is more efficient because the idle times are greatly reduced. We can now define the so-called cycle time, CT, for non-overlapping operation, *i. e.*,

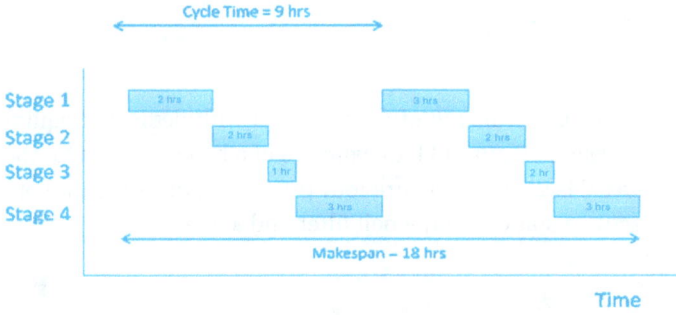

Figure 13.2: Gantt chart for the plant of Figure 13.1 (without transfer times).

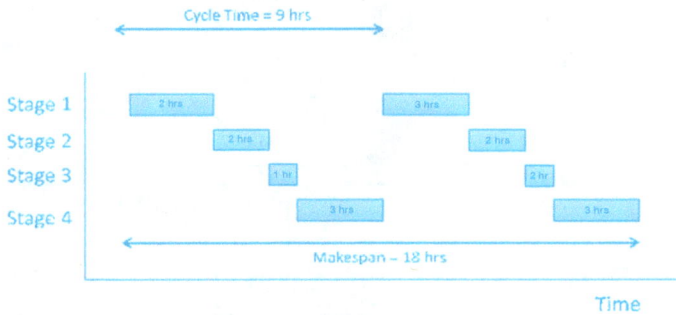

Figure 13.3: Gantt chart with overlapping mode of operation.

$$CT = \sum_{j=1}^{M} T_j, \tag{13.1}$$

and for overlapping operation, *i. e.*,

$$CT = \max\{T_j\}, \tag{13.2}$$

where T_j is the processing time in stage j. For the given example, the cycle time for non-overlapping operation is 9 hours, while the cycle time for overlapping operation is only 3 hours. It is further noted that the *make span* corresponds to the total time needed to produce a given number of batches. For the non-overlapping operation the make span is 18 hours, while for the overlapping operation the make span is only 12 hours.

In a *multi-product plant*, production campaigns are set up to manufacture a fixed number of batches for the various products. Take as an example the production of three batches of products A and B, produced in a plant with three stages. The processing times are given in Table 13.1.

The three batches for product A and B can basically be produced in two different ways, namely in so-called *single-product campaigns* (SPCs) – where all batches of

Table 13.1: Processing times for a two-product plant.

	Stage 1	Stage 2	Stage 3
A	5 hrs	1 hr	1 hr
B	1 hr	2 hrs	2 hrs

one product are produced before switching to the second one – or in *mixed-product campaigns* (MPCs) – where batches are produced according to a chosen switching sequence. In Figure 13.4 the two production campaigns are compared.

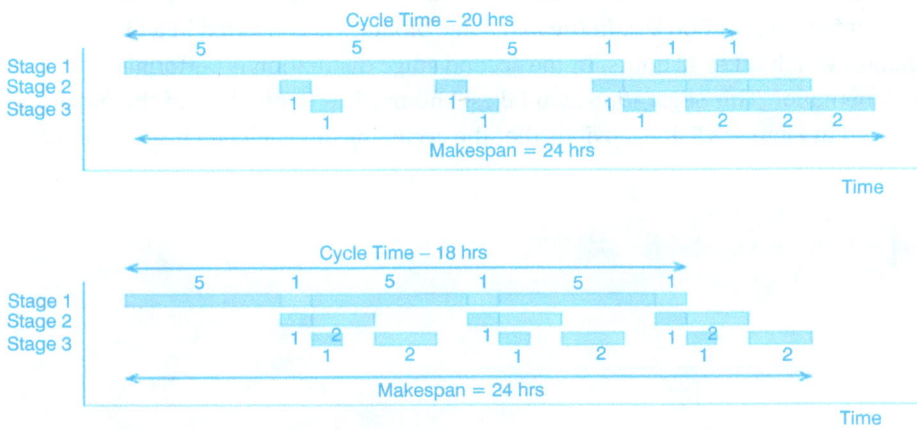

Figure 13.4: Schedules for single and mixed-product campaigns. Light bars for product A, dark bars for product B.

For the example sketched above the MPC seems to be more efficient, as the cycle time is only 18 hours, as compared to the single product campaign, which has a cycle time of 20 hours. It is noted here that MPCs are not necessarily more efficient than SPCs as clean-up time or product changeovers could influence the cycle time significantly. What would be the cycle times if the equipment after each batch requires a cleaning procedure of one hour?

Flowshop versus jobshop

The multi-product plant example given in Section 13.3 is sometimes referred to as a *flowshop plant*; all products go through all stages, following the same order of operations. If a plant produces very similar products, for example different polymer grades, the plant adopts a flowshop approach. However, there are also plants where not all products require all stages or follow the same processing sequence; these plants are called *jobshop plants* or *multi-purpose plants*.

Transfer policies, parallel units, storage and inventory

The so-called *zero wait transfer policy* is used when a batch at any stage is immediately transferred to the next stage because there is no intermediate storage vessel available or it cannot be kept in the current vessel. This policy is extremely restrictive. The other extremum is the *unlimited intermediate storage* policy, where a batch can be stored without any capacity limit in a storage vessel. Lastly there is a transfer option called *no intermediate storage*, which allows the batch to be kept inside the vessel. Normally the zero wait transfer requires the longest cycle time. In practice plants normally have a mixture of the three transfer policies.

The addition of parallel units can increase the efficiency of the plant considerably. Let us consider a simple biorefinery that is operated in zero wait mode. Each batch is the same (1,000 kg). In the first stage biomass is converted in a bioreactor to ethanol, which takes 12 hours. In the second stage distillation is performed to purify the bioethanol. This separation step takes 3 hours. The Gantt chart of the process is depicted in Figure 13.5. It is obvious that the cycle time for each batch is 12 hours.

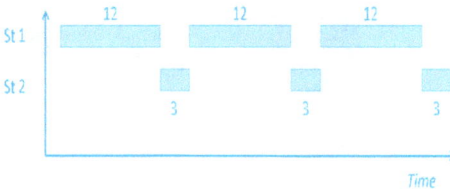

Figure 13.5: Gantt chart for the biorefinery.

Adding a parallel unit to stage 1, the plant can be operated as shown in Figure 13.6. The cycle time has been reduced to 6 hours. The cycle time for a zero wait transfer policy plant can be calculated by

$$CT = \max\{T_{ij}/NP_j\}, \tag{13.3}$$

where NP_j is the number of parallel units with $j = 1 \ldots M$.

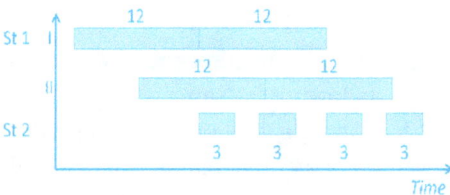

Figure 13.6: Plant with parallel units in the bioreactor.

If a large number of batches has to be processed, the batch size could be reduced to 500 kg because the cycle time has been halved. Another option to increase the efficiency of the plant is by introducing intermediate storage between the stages. This basically allows the two stages to operate independently from each other with their own cycle times and batch sizes.

Inventory plays an important role in the selection of the production cycle. The main trade-off is between inventory and clean-ups. The shorter the production cycle, the less inventory we need since the products are available more frequently. However, the number of transitions will increase. If the production cycle is longer, the number of transitions decreases. However, the inventories will increase, as the products are produced less frequently.

13.4 State-task networks

In the beginning of the 1990s the so-called *recipe networks* were introduced (Reklaitis, 1991). These networks look very similar to process flowsheets of continuous plants. However, these networks are intended to describe the process and not so much a plant. In Figure 13.7 an example of a recipe network is given. Basically, each block in the sequence represents a task that should be executed.

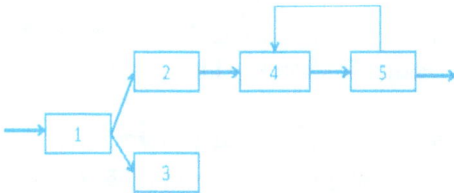

Figure 13.7: Recipe network representation of a chemical process.

Recipe networks are certainly useful for serial processing structures, but they can also lead to ambiguities when they are applied to more complex networks (Kondili *et al.*, 1993). From Figure 13.7, it is not clear whether task 1 will produce *two* different products or whether it has only *one* type of product, which will be shared between task 2 and 3. As a second example, it is not clear whether task 4 requires *two* different feedstocks (from task 2 and 5) or whether it only needs one type of feedstock by either task 2 *or* task 5. Both interpretations are plausible.

To deal with the ambiguities in recipe networks Kondili *et al.* (1993) introduced the concept of the *state-task network* (STN) representation. The main difference between the STN and the recipe network is that it contains two types of nodes; *tasks* and *states*. The state nodes represent the feeds and intermediate and final products, while the task nodes represent the processing operation that transforms materials in one form to another. The state nodes are represented by circles, the tasks by rectangles.

In Figures 13.8 and 13.9 two STNs are shown which both correspond to the recipe network of Figure 13.7.

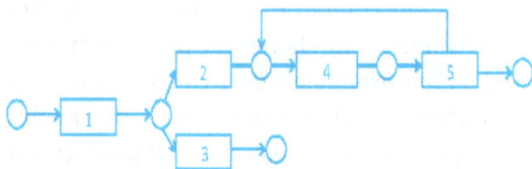

Figure 13.8: STN representation of a chemical process.

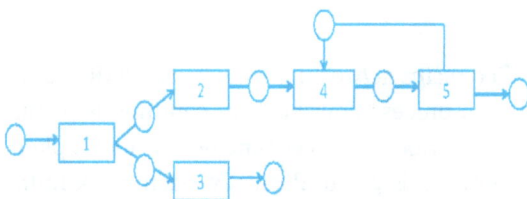

Figure 13.9: STN representation of a chemical process.

Figure 13.8 shows clearly that task 1 has only one product, which is then shared by tasks 2 and 3. In addition, task 4 only requires one feedstock, which is produced by tasks 2 and 5.

Figure 13.9 shows that task 1 actually has two different products, forming the inputs to tasks 2 and 3, respectively. Furthermore it shows that task 4 also has two different feedstocks, produced by task 2 and task 5.

13.5 Mathematical formulations of scheduling problems

From the STN representation we can go to a mathematical formulation of the batch scheduling problem. This class of optimization models is called *mixed-integer linear programs (MILPs)*.

The MILP model that we will describe in this chapter is based on the one by Kondili *et al.* (1993). For this model, the assignment of equipment to tasks does not need to be fixed, the batch size may be variable and can be handled with the possibility of mixing and splitting and different storage and transfer policies are allowed, as well as the limitation in the resources.

The MILP model will be formulated according to a provided STN, given one or more feeds, demands and deadlines. Now we will have to decide the timing of the operations, the assignments of the equipment to the operations and the flows of the materials through the network. The objective is to maximize a given profit function. We will further work with a uniform, discretized time domain, where h time periods of equal size are considered. Task i is defined as follows:

S_i = Set of states inputs to task i
$S*_i$ = Set of states outputs of task i
$\rho_{i,s}$ = Proportion input to task i from state s
$\rho*_{i,s}$ = Proportion output of task i for state s
P_i = Processing time for task i
K_i = Set of units j capable of processing task i

The last parameter K relates the process equipment units to the STN. The state s is defined as follows:

T_s = Set of tasks receiving material from state s
$T*_s$ = Set of tasks producing material for state s
IP = Set of states s corresponding to products
IF = Set of states s corresponding to feeds
II = Set of states s corresponding to intermediates
$d_{s,t}$ = Minimum demand for state s of IP at the beginning of period t
$r_{s,t}$ = Maximum purchase for state s of IF at the beginning of period t
C_s = Maximum storage of state s

The unit j may be capable of performing one or more tasks and is characterized as follows:

V_j = Maximum capacity
I_j = Set of tasks i for which equipment j can be used

As for the variables, we will require binary and continuous variables:

$W_{i,j,t}$ = 1 if unit j starts processing task i at the beginning of period t
$B_{i,j,t}$ = Amount of material starts task i in unit j at the beginning of period t
$S_{s,t}$ = Amount of material stored in state s at the beginning of period t
$U_{u,t}$ = Demand of utility u over time interval t
$R_{s,t}$ = Purchases of state s at the beginning of period t
$D_{s,t}$ = Sales of state s at the beginning of period t

With all variables and parameters in place we can start with the formulation of the constraints of the MILP model. We will start with constraining the assignment of equipment j to tasks i over the various time periods t. We have

$$\sum_{i \in I_j} \sum_{t*=t}^{t-p_i-1} W_{i,j,t*} \leq 1, \quad \forall j, t. \tag{13.4}$$

If $W_{i,j,t} = 1$, unit j cannot be assigned to tasks other than i during the interval $[t-p_i+1, t]$. The capacity limits for the units and storage tanks can be expressed as

$$0 \leq B_{i,j,t} \leq V_j W_{i,j,t} \quad \forall i, t, j \in K_i, \tag{13.5}$$

$$0 \leq S_{s,t} \leq C_s, \quad \forall s, t. \tag{13.6}$$

The mass balances for each state and time are given as

$$S_{s,t-1} + \sum_{i \in T *_s} \rho *_{i,s} \sum_{j \in K_i} B_{i,j,t-pi} + R_{s,t} = S_{s,t} - 1 + \sum_{i \in T_s} \rho_{i,s} \sum_{j \in K_i} B_{i,j,t} + D_{s,t} \quad \forall s, t, \tag{13.7}$$

i. e., the initial amount plus the amounts produced and purchased must be equal to the holdup plus the amounts consumed and sold. Furthermore, the following bounds on the purchases and sales apply:

$$D_{s,t} \geq d_{s,t}, \quad s \in \text{IP}, \tag{13.8}$$

$$R_{s,t} \leq r_{s,t}, \quad s \in \text{IF}. \tag{13.9}$$

For the maximum amount of utility that is available we can introduce the following constraint, where α and β are the appropriate cost factors:

$$U_{u,t} = \sum_i \sum_{j \in K_i} \sum_{\theta=1}^{p_i-1} (\alpha_{u,i} W_{i,j,t-\theta} + \beta_{u,i} B_{i,j,t-\theta}), \quad \forall u, t, \tag{13.10}$$

$$0 \leq U_{u,t} \leq U_u^{\max}. \tag{13.11}$$

Lastly, the objective function can be defined as "sales − purchases + final inventory − utilities":

$$Z = \sum_s \sum_{t=1}^h C_{s,t}^D D_{s,t} - \sum_s \sum_{t=1}^h C_{s,t}^R R_{s,t} + \sum_s C_{s,h=1} S_{s,h+1} - \sum_u \sum_{t=1}^h C_{u,t} U_{u,t}, \tag{13.12}$$

where $C_{s,t}^D$, $C_{s,t}^R$, $C_{s,h+1}$ and $C_{u,t}$ are the appropriate cost coefficients. It is possible to include a zero wait policy by adding constraints that specify that task $i*$ follows task i:

$$\sum_{j \in K_i} W_{i,j,t} = \sum_{j \in K *_i} W_{i*,j,t+pi}. \tag{13.13}$$

The objective function in equation (13.12) subject to constraints (13.4)–(13.11) and (13.13) is an MILP problem that can be solved within reasonable computational time, if the number of time intervals is not too large.

13.6 Example: scheduling of an ice cream factory

In the following section, a small scheduling problem is formulated and solved for the so-called fast-moving consumer good (FMCG) industry. Although the example is not a typical chemical process, the food industry is of great interest to the process engineering community. The example concerns a small ice cream factory. The problem is formulated in a way very similar to the MILP formulation introduced in the previous section. The interested reader is referred to (van Elzakker *et al.*, 2012), in which a scheduling problem at industrial scale for the FMCG industry is solved.

For this example, consider a small ice cream factory containing two production stages. In the first stage the ingredients are mixed and the ice cream is frozen. In the second stage the ice cream is packed. Between mixing and packing the products are stored in an intermediate storage area with unlimited capacity. The factory contains one mixing line and two packing lines, as shown in Figure 13.10.

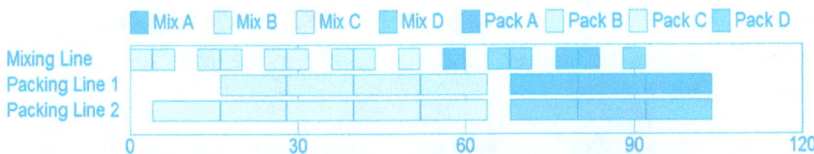

Figure 13.10: Schematic overview of the ice cream production process.

Four varieties of ice cream are produced in this factory. The mixing line can produce all four products. Products A and B are packed on packing line 1 and products C and D are packed on packing line 2. The mixing rate is 3,000 kg/hr and the packing rate is 1,000 kg/hr. When switching from one product to another, a 4-hour changeover is required to clean the mixing/packing line. The ice cream is produced in batches of 12,000 kg. The demand, which is given in Table 13.2, must be met at the end of the week. The objective is to minimize the make span.

Table 13.2: Demand for the different products.

	Demand
Product A	36,000 kg
Product B	48,000 kg
Product C	60,000 kg
Product D	36,000 kg

We can use the following model to optimize the production schedule for this ice cream factory. First, we ensure that at most one task can be active on any unit at any time, *i. e.*,

$$\sum_{i \in I_j} \sum_{t*=t}^{t+p_i-1} W_{i,j,t*} \leq 1, \quad \forall j,t. \tag{13.14}$$

Second, different tasks on the same unit cannot start before the end of the current task plus the changeover time, *i. e.*,

$$\sum_{(i' \neq i) \in I_j} \sum_{t'=t}^{t+P_i+\tau_j-1} W_{i',j,t'} \leq M(1 - W_{i,j,t}) \quad \forall i \in I_j, j, t. \tag{13.15}$$

If task i is active on unit j in period t, then the amount of material undergoing this task is equal to the batch size, *i. e.*,

$$B_{i,j,t} = \text{BatchSize} \cdot W_{i,j,t} \quad \forall i,j,t. \tag{13.16}$$

The amount of material in state s at the end of period t is equal to the amount in the previous period plus the amount produced by the tasks that are completed at the end of the current period minus the amount that is consumed in the tasks starting in the current period. We have

$$S_{s,t} = S_{s,t-1} + \sum_{i \in T_s^+}\left(\rho_{i,s} \cdot \sum_{j \in K_i} B_{i,j,t-P_i}\right) - \sum_{i \in T_s^-}\left(\rho_{i,s}^* \cdot \sum_{j \in K_i} B_{i,j,t}\right) \quad \forall s,t. \tag{13.17}$$

The amount of material in state s should be equal to the demand at the end of the scheduling horizon, so

$$S_{s,\text{last}(t)} = D_s \quad \forall s. \tag{13.18}$$

The make span is equal to the latest end time of all tasks, *i. e.*,

$$MS \geq t \cdot W_{i,j,t} + P_i - 1 \quad \forall i \in I_j, j, t. \tag{13.19}$$

This MILP model can be used to optimize the weekly production schedule. The resulting schedule will have a make span of 104 hours. It should be noted that several different schedules with a 104-hour make span could be obtained. One example of such a schedule is shown in Figure 13.11.

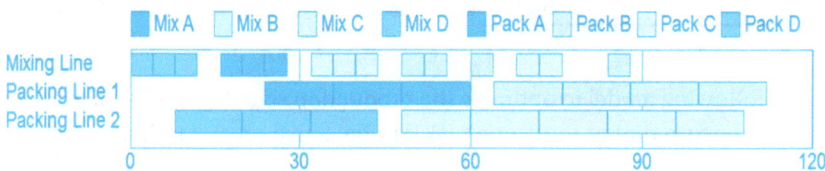

Figure 13.11: Ice cream production Gantt chart.

However, in practice the schedule should satisfy another constraint. After the mixing stage, the ice cream must remain in the intermediate storage room for several hours before it can be packed to allow the product to age. The minimum required ageing time depends on the product. Products A and D have an ageing time of 4 hours, product B of 8 hours and product C of 12 hours.

We can consider this ageing time by modifying equation (13.17). To ensure that an intermediate product cannot be packed before it is aged, the amount of material in state s is only increased after the production and ageing time of this material in state s, i. e.,

$$S_{s,t} = S_{s,t-1} + \sum_{i \in T_s^+} \left(\rho_{i,s} \cdot \sum_{j \in K_i} B_{i,j,t-P_i-\mathrm{Age}T_i} \right) - \sum_{i \in T_s^-} \left(\rho_{i,s}^* \cdot \sum_{j \in K_i} B_{i,j,t} \right) \quad \forall s, t. \qquad (13.20)$$

Using the modified model, a schedule with a make span of 112 hours is obtained. A Gantt chart of this schedule is given below. This time products A and D are produced first because their shorter ageing time allows the packing to start earlier. The Gantt chart of the modified model is shown in Figure 13.12.

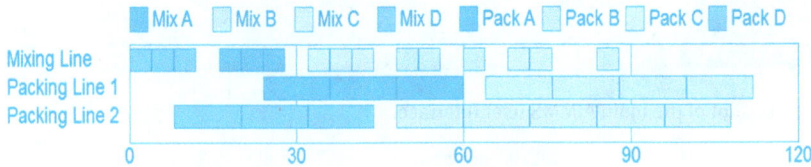

Figure 13.12: Ice cream production Gantt chart for the modified model.

13.7 Implementation

The above formulated MILP model was coded in the software package AIMMS. The model contains 1058 constraints and 721 variables of which 240 are binary. In Section 13.8 the modeling code can be found.

AIMMS offers an all-round development environment for the creation of high-performance decision support and advanced planning applications to optimize strategic operations. It allows organizations to rapidly improve the quality, service, profitability and responsiveness of their operations. The AIMMS development environment possesses a unique combination of advanced features and design tools, such as the graphical model explorer, which allow the user to build and maintain complex decision support applications and advanced planning systems in a fraction of the time required by conventional programming tools.

As a beginning user of AIMMS it is advised to visit the website of AIMMS (http://aimms.com/downloads/aimms). The reader can find the software package for which a free trial license can be obtained.

A 1-hour instruction to the basics of AIMMS can be obtained at http://aimms.com/aimms/download/manuals/aimms_tutorial_beginner.pdf in different languages.

After installation and start-up of AIMMS a *new project* can be created. The Main AIMMS interface opens; see Figure 13.13. In the so-called *Model explorer* a small *declaration icon* appears that can be used to insert new *types of identifiers*, such as constraints, parameters and data.

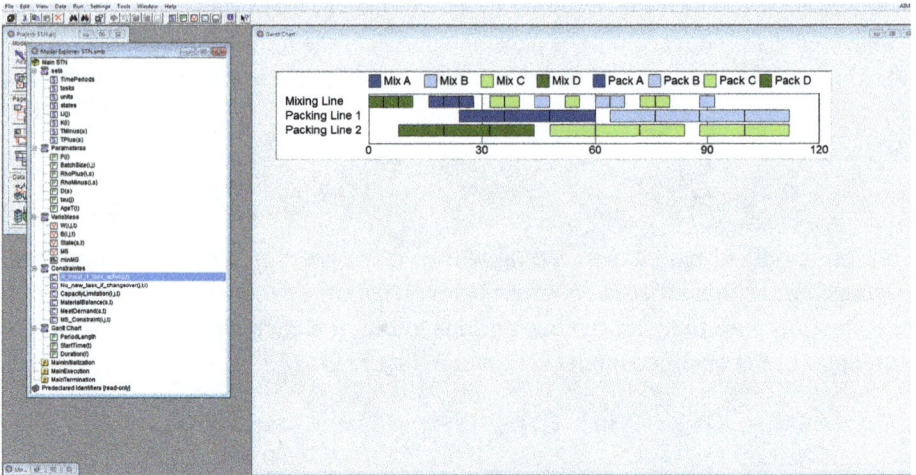

Figure 13.13: Screenshot of the main AIMMS user interface.

In Figure 13.14 an example of a constraint type of identifier is shown. Under the header *Definition* the constraint can be inserted.

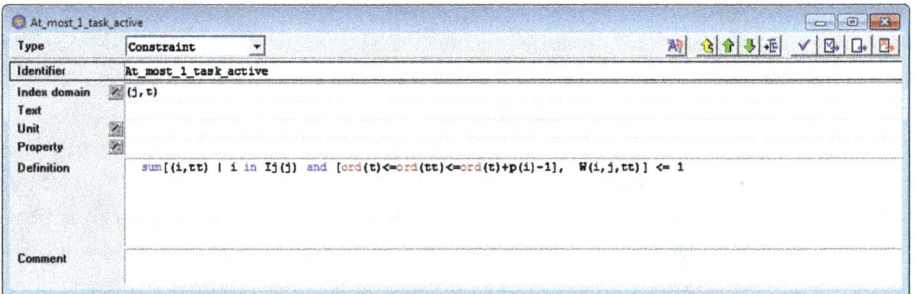

Figure 13.14: Screenshot of an identifier block for a constraint.

AIMMS uses a predefined solver configuration. However, if the user wishes to change the settings under the folder *Settings\Project Options* specific preferences for solver and options can be modified. Figure 13.15 shows a screenshot of the AIMMS options folder.

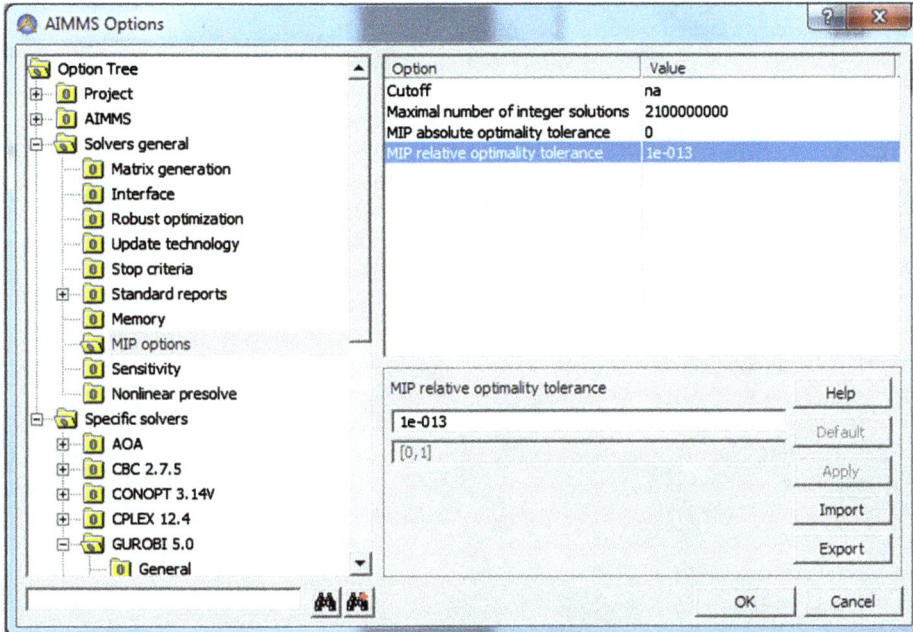

Figure 13.15: Screenshot of the AIMMS Options with the solver settings.

In the model explorer it is also possible to define the preferred output, for example as *Gantt chart* or *Table*. Finally, to run the model the user selects from the Model explorer the *Main Execution* folder and selects *Run Procedure* (see Figure 13.16).

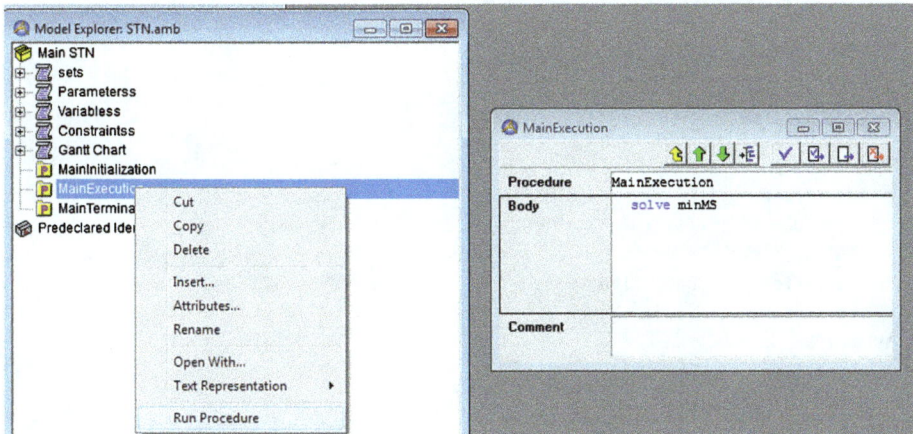

Figure 13.16: Screenshot for running the AIMMS model.

13.8 AIMMS code for the scheduling model

```
MAIN MODEL Main\_STN
 DECLARATION SECTION sets
   SET:
       identifier   :  TimePeriods
       indices      :  t, tt
       definition   :  elementrange(001,030,1,"Period ") ;
   SET:
       identifier   :  tasks
       indices      :  i, ii ;
   SET:
       identifier   :  units
       index        :  j ;
   SET:
       identifier   :  states
       index        :  s ;
   SET:
       identifier   :  IJ
       index domain :  (j)
       subset of    :  tasks ;
   SET:
       identifier   :  K
       index domain :  (i)
       subset of    :  units ;
   SET:
       identifier   :  TMinus
       index domain :  s
       subset of    :  tasks ;
   SET:
       identifier   :  TPlus
       index domain :  s
       subset of    :  tasks ;
 ENDSECTION  ;
 DECLARATION SECTION Parameterss
   PARAMETER:
       identifier   :  P
       index domain :  (i) ;
   PARAMETER:
       identifier   :  BatchSize
       index domain :  (i,j) ;
   PARAMETER:
       identifier   :  RhoPlus
       index domain :  (i,s) ;
   PARAMETER:
       identifier   :  RhoMinus
       index domain :  (i,s) ;
   PARAMETER:
       identifier   :  D
```

```
      index domain :  (s) ;
   PARAMETER:
      identifier   :  tau
      index domain :  (j) ;
   PARAMETER:
      identifier   :  AgeT
      index domain :  (i) ;
ENDSECTION  ;
DECLARATION SECTION Variabless
   VARIABLE:
      identifier   :  W
      index domain :  (i,j,t) | i in IJ(j)
      range        :  binary ;
   VARIABLE:
      identifier   :  B
      index domain :  (i,j,t)| i in IJ(j)
      range        :  nonnegative ;
   VARIABLE:
      identifier   :  State
      index domain :  (s,t)
      range        :  nonnegative ;
   VARIABLE:
      identifier   :  MS
      range        :  free ;
   MATHEMATICAL PROGRAM:
      identifier   :  minMS
      objective    :  MS
      direction    :  minimize
      constraints  :  AllConstraints
      variables    :  AllVariables
      type         :  Automatic ;
ENDSECTION  ;
DECLARATION SECTION Constraintss
   CONSTRAINT:
      identifier   :  At\_most\_1\_task\_active
      index domain :  (j,t)
      definition   :  sum[(i,tt) | i in Ij(j) and [ord(t)<=ord(tt)<=ord(t)+p(i)-1],
W(i,j,tt)] <= 1 ;
   CONSTRAINT:
      identifier   :  No\_new\_task\_if\_changeover
      index domain :  (j,t,i) | i in Ij(j)
      definition   :  sum[ii | ii in Ij(j) and ord(ii)<>ord(i), sum[tt | (ord(tt)>=
ord(t)) and (ord(tt) <= ord(t) + P(i)+tau(j)-1), W(ii,j,tt)]] <= 20*(1 - W(i,j,t)) ;
   CONSTRAINT:
      identifier   :  CapacityLimitation
      index domain :  (i,j,t)
      definition   :  B(i,j,t) = BatchSize(i,j) * W(i,j,t) ;
   CONSTRAINT:
      identifier   :  MaterialBalance
```

```
     index domain :  (s,t)
     definition   :  State(s,t) = State(s,t-1)
                     + sum[i | i in TPlus(s), RhoPlus(i,s) *
sum[j | j in K(i), B(i,j,t- P(i)-AgeT(i))]]
                     - sum[i | i in TMinus(s), RhoMinus(i,s) *
sum[j | j in K(i), B(i,j,t)]] ;
  CONSTRAINT:
     identifier   :  MeetDemand
     index domain :  (s,t) | ord(t)=30
     definition   :  State(s,t) = D(s) ;
  CONSTRAINT:
     identifier   :  MS\_Constraint
     index domain :  (i,j,t) | i in IJ(j)
     definition   :  MS >= 4*(ord(t)* W(i,j,t) + P(i)-1) ;
ENDSECTION  ;
DECLARATION SECTION Gantt\_Chart
  PARAMETER:
     identifier   :  PeriodLength ;
  PARAMETER:
     identifier   :  StartTime
     index domain :  (t)
     definition   :  Periodlength*(ord(t)-1) ;
  PARAMETER:
     identifier   :  Duration
     index domain :  i
     definition   :  Periodlength*P(i) ;
ENDSECTION  ;
PROCEDURE
  identifier :  MainInitialization
ENDPROCEDURE  ;
PROCEDURE
  identifier :  MainExecution
  body       :
    solve minMS
ENDPROCEDURE  ;
PROCEDURE
  identifier :  MainTermination
  body       :
    return DataManagementExit();
ENDPROCEDURE  ;
ENDMODEL Main\_STN ;
```

Exercises

1. A batch process requires the following operations to be completed in a sequence: 3 hrs of mixing, 5 hrs of heating, 4 hrs of reaction, 7 hrs of purification and 2 hrs of transfer.

   (a) When the five operations are carried out in vessels U1, U2, U3, U4 and U5, respectively, determine the cycle times and construct a Gantt chart corresponding to zero wait, intermediate storage and unlimited intermediate storage inventory strategies.

(b) When a new purification vessel U4A is purchased so that now two 7-hour purification steps can take place in parallel, determine the system bottleneck using the intermediate storage inventory strategy.

13.9 Take-away message

One of the first decisions a process designer has to make is whether the process will be run batch-wise or in a continuous way. Often batch-wise processing is selected when specialty, high-value chemicals are produced in small quantities. The same batch process equipment can be used for different tasks; for example, a vessel could be used as a reactor or as a storage tank. Batch processes are often designed for multi-product and multi-purpose use. The two most important batch design parameters are the batch time and the size factor. One important outcome of batch process design is the Gantt chart, which shows in which order which batches should be produced.

13.10 Further reading

Mendez, C. A., Cerda, J., Grossmann, I. E., Harjunkoski, I. and Fahl, M. (2006). State-of-the-art review of optimization methods for short term scheduling of batch processes. *Comp. & Chem. Eng.*, 30, pp. 913–946.

Rippin, D. W. T. (1983). Simulation of single- and multiproduct batch chemical plants for optimal design and operation. *Comp. & Chem. Eng.*, 7(3), pp. 137–156.

Biegler, L. T., Grossmann, I. E. and Westerberg, A. W. (1997). *Systematic methods of chemical process design*. Prentice Hall.

Reklaitis, G. V. (1991), *Perspectives of scheduling and planning of process operations*. PSE'91, Montebello, Canada.

Kondili, E., Pantelides, C. C. and Sargent, R. W. H. (1993). A general algorithm for short-term scheduling of batch operations-I. MILP formulation. *Comp. & Chem. Eng.*, 17(2), pp. 211–227.

van Elzakker, M. A. H., Zondervan, E., Raikar, N. B., Grossmann, I. E. and Bongers, P. M. M. (2012). Scheduling in the FMCG Industry: An industrial case study. *Industrial and Engineering Chemistry Research*, 51(22), pp. 7800–7815.

Ierapetritou, M. G. and Floudas, C. A. (1998). Effective continuous-time formulation for short-term scheduling. 1. Multipurpose batch processes. *Ind. Eng. Chem. Res.*, 37, pp. 4341–4359.

Floudas, A. F. and Lin, X. (2004). Continuous-time versus discrete-time approaches for scheduling of chemical processes: a review. *Comp. & Chem. Eng.*, 28, pp. 2109–2129.

14 Separation train design

14.1 Separations in process development

One might argue that the reaction sections in a process design are the most important. Indeed, the products are formed in a reaction set-up, but a great number of separations are required to bring the product to its required purity. In addition, the unreacted raw materials need separations before they can be recycled back to the reactor and waste and utility components need treatment before they can be released to the environment. Figure 14.1 shows where in a process separations are required: basically everywhere.

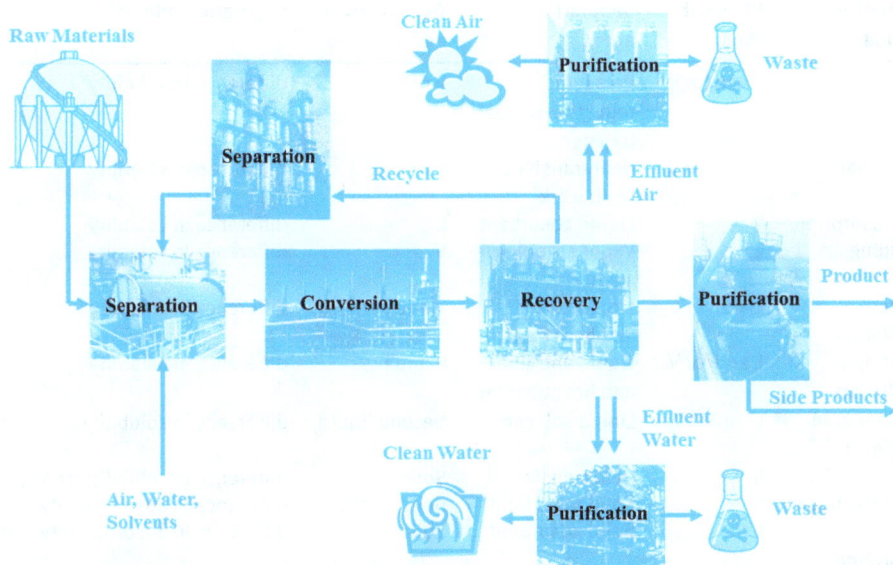

Figure 14.1: Separations throughout the flowsheet.

It turns out that the major investment and operating costs of a process are the ones related to the separation equipment.

For two-component mixtures it is relatively easy to identify a separation method that can accomplish the separation task in one piece of equipment. But if mixtures are more complex, more complex separation systems have to be designed.

14.2 Energy and separations

Separation of mixtures of chemicals generally need some form of energy. Separation can be achieved by forcing different species into different spatial locations by combination of four well-known industrial techniques:

https://doi.org/10.1515/9783110570137-014

1. by heat transfer, shaft work or pressure reduction of a second phase that is immiscible with the feed phase (energy separation agent);
2. introduction of a second fluid phase in the system which subsequently is removed (mass separation agent);
3. addition of a solid phase at which adsorption can occur;
4. the placement of a membrane barrier.

Table 14.1 shows the common industrial separation methods for different feeds, separation agents and separation principles.

Table 14.1: Industrial separation methods (Seider *et al.*).

Separation method	Phase of the feed	Separation agent	Developed or added phase	Separation principle
Equilibrium flash	L and/or V	Pressure reduction or heat transfer	V or L	difference in volatility
Distillation	L and/or V	Heat transfer or shaft work	V or L	difference in volatility
Gas absorption	V	Liquid absorbent	L	difference in volatility
Stripping	L	Vapor stripping agent	V	difference in volatility
Extractive distillation	L and/or V	Liquid solvent and heat transfer	V and L	difference in volatility
Azeotropic distillation	L and/or V	Liquid entrainer and heat transfer	V and L	difference in volatility
Liquid–liquid extraction	L	Liquid solvent	Second liquid	difference in solubility
Crystallization	L	Heat transfer	Solid	difference in solubility or m. p.
Gas adsorption	V	Solid adsorbent	Solid	difference in adsorbabililty
Liquid adsorption	L	Solid adsorbent	Solid	difference in adsorbabililty
Membranes	L or V	Membrane	Membrane	difference in permeability and/or solubility

14.3 Selection of a suitable separation method

The development of an appropriate separation train needs the selection of (1) suitable separation methods, (2) energy separation agents or mass separation agents, (3) separation equipment, (4) optimal arrangement or sequencing of the equipment and (5) optimal operating conditions (*e. g.*, temperature, pressure) of the equipment.

The selection of the separation method depends of course on the feed condition. If the feed comes as a vapor, partial condensation, distillation, absorption, adsorption or permeation (membranes) is common. If the feed comes as a liquid, distillation, stripping, extraction, crystallization, adsorption or membranes (via dialysis or reverse osmosis) is common. If the feed is solid, typically drying (wet feed) or leaching (dry feed) are selected as separation technique.

An important factor that defines the degree of separation achievable between two components is the separation factor (SF). This factor, for the separation of component 1 from component 2 between phases I and II, is defined as

$$SF = \frac{C_1^I / C_2^I}{C_1^{II} / C_2^{II}}, \tag{14.1}$$

where C is the composition variable and I and II are phases rich in components 1 and 2. The separation factor is in general bounded by thermodynamic equilibrium. For example, in the case of distillation, while using mole fractions as the composition variables and letting phase I be the vapor phase and II be the liquid phase, the limiting value of the separation factor is given in terms of the vapor–liquid equilibrium ratios as

$$SF = \frac{y_1 / x_1}{y_2 / x_2} = \frac{K_1}{K_2} = \alpha_{12}, \tag{14.2}$$

which actually equals P_1^S / P_2^S for ideal liquids and vapors.

For vapor–liquid separations that use a mass separation agent that leads to a non-ideal liquid solution (for example in extractive distillation), the separation factor is given as

$$SF = \alpha_{12} = \frac{\gamma_1 P_1^S}{\gamma_2 P_2^S}. \tag{14.3}$$

If the mass separation agent is used to create two liquid phases (for example in liquid–liquid extraction) the separation factor is modeled as the relative selectivity β, i. e.,

$$SF = \beta_{12} = \frac{\gamma_1^{II} / \gamma_2^{II}}{\gamma_1^I / \gamma_2^I}. \tag{14.4}$$

Generally, mass separating agents for extractive distillation and liquid–liquid extraction are selected on the basis of their ease to recover for recycles and to achieve large values of the separation factor.

Figure 14.2 shows how relative volatilities for distillation columns relate to relative volatilities and selectivities needed for extractive distillation or liquid–liquid extraction systems with the same performance.

14.4 The sequencing of separations

The number of possible sequences of distillation columns depends on the number of components one wants to separate. The number of sequences increases factorially with the number of components according to

$$N_S = \frac{[2(P - 1)]!}{P!(P - 1)!}, \tag{14.5}$$

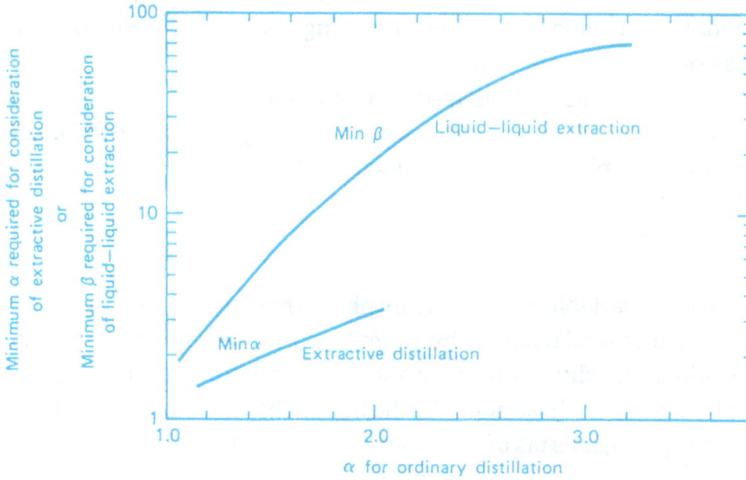

Figure 14.2: Relative volatilities for equal cost separators (Souders, 1964).

where N_S is the number of columns and P the number of produced products. A mixture of three components to be separated requires two distillation columns which can be sequenced in two different ways. A mixture of four components can be separated by three separators which can be sequenced in five different ways, while a mixture of eight components needs seven separations which can be sequenced in 429 different ways. Figure 14.3 shows the five different sequences for the separation of a four-component mixture.

The best sequencing might be derived on the basis of the following six heuristics (by Seider *et al.*):

1. remove thermally unstable, corrosive or chemically reactive components early in the sequence;
2. remove final products one-by-one as distillates;
3. remove components with greatest molar percentage in the feed as early as possible in the sequence;
4. sequence the separations in order of decreasing relative volatilities such that most difficult splits are made in the absence of the other components;
5. sequence the separations in such way that the last separations give the highest purity of the products;
6. sequence separations in such way that equimolar amounts of distillate and bottoms are favored (the reboiler duty is not excessive).

14.5 The sequencing of ordinary distillation columns

Multi-component mixtures can be separated by sequences of ordinary distillation columns, provided that (Seider *et al.*):

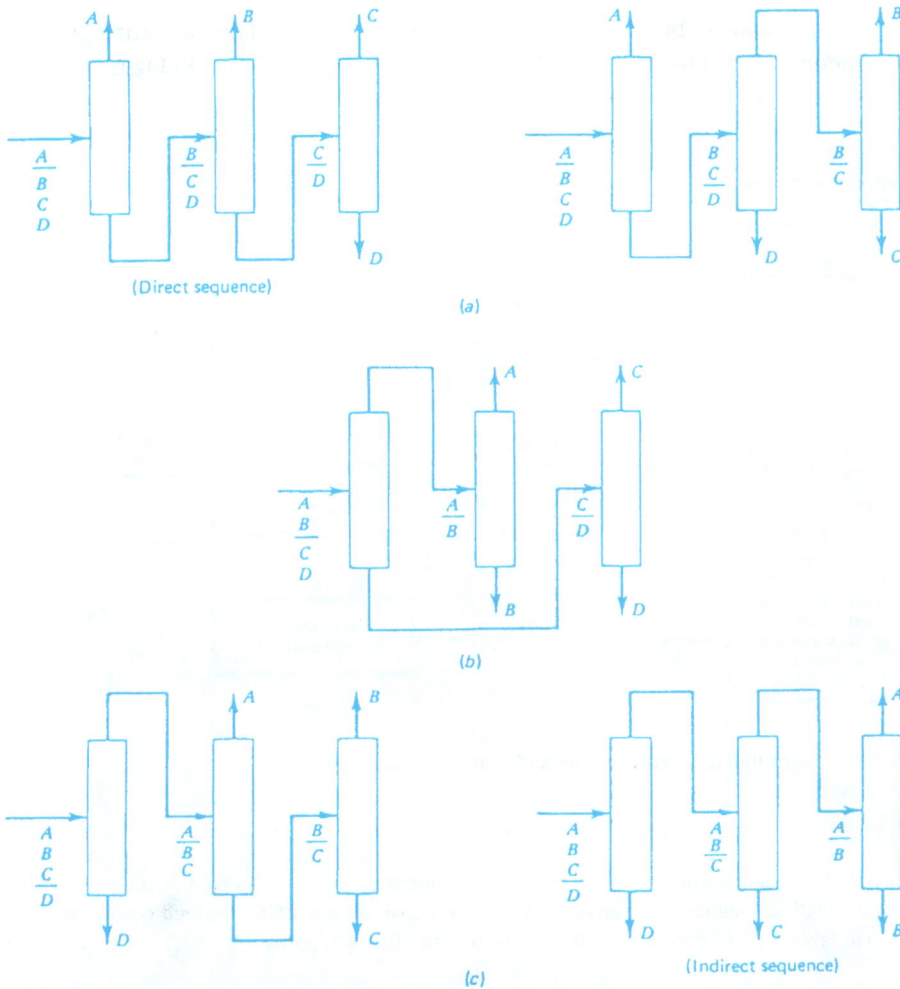

Figure 14.3: Five different separation sequences for a four-component mixture.

1. *a* in each column is larger than 1.05;
2. the reboiler duty is not excessive;
3. the tower pressure does not cause the mixture to approach the critical temperature of the mixture;
4. the column pressure drop is acceptable, particularly if the operation takes place under vacuum;
5. the overhead vapor can be at least partially condensed at the column pressure to provide reflux without excessive refrigeration requirements;
6. the bottom temperature for the tower pressure is not so high that chemical decomposition occurs;
7. azeotropes do not prevent the desired separation.

Seider *et al.* propose a basic algorithm for the determination of the operating pressure and condenser types for ordinary distillation sequences, as shown in Figure 14.4.

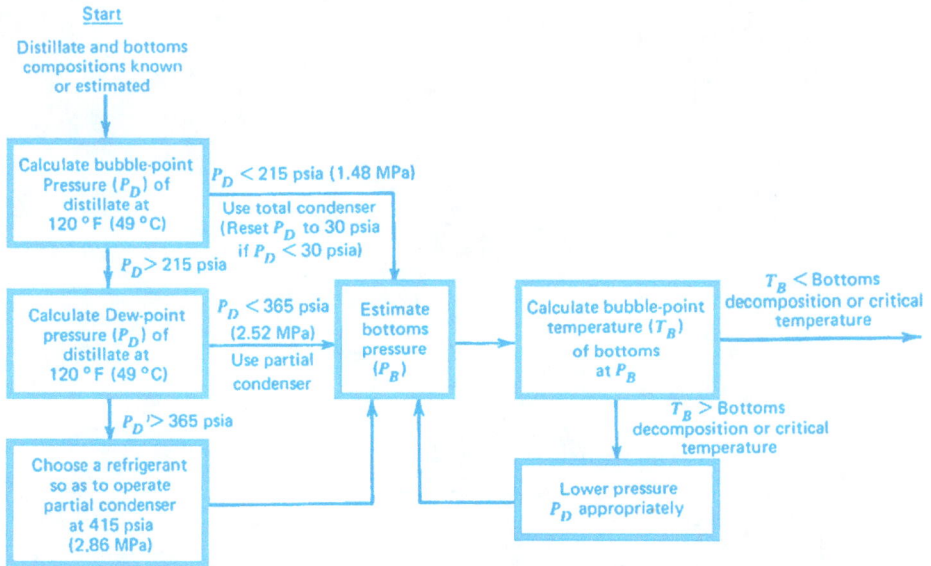

Figure 14.4: Algorithm to select pressure and condenser type.

Example 1 (Sequencing of distillation columns). Suppose we want to design a sequence of ordinary distillation columns that separates a mixture of five components. The feed composition is given in Table 14.2. All chemicals should be recovered with 98 % purity.

Table 14.2: Feed composition of a mixture (left) and relative volatilities of component pairs (right).

Species	Kmoles/hr	Component pair	Approximate α at 1 atm
Propane (C3)	45.4	C3/C4	3.6
Isobutane (iC4)	136.1	iC4/nC4	1.5
n-Butane (nC4)	226.8	nC4/iC5	2.8
Isopentane (iC5)	181.4	iC5/nC5	1.35
n-Pentane (nC5)	317.5		
	907.2		
Feed	37.8 °C and 1.72 MPha		

A possible distillation sequence is shown in Figure 14.5.

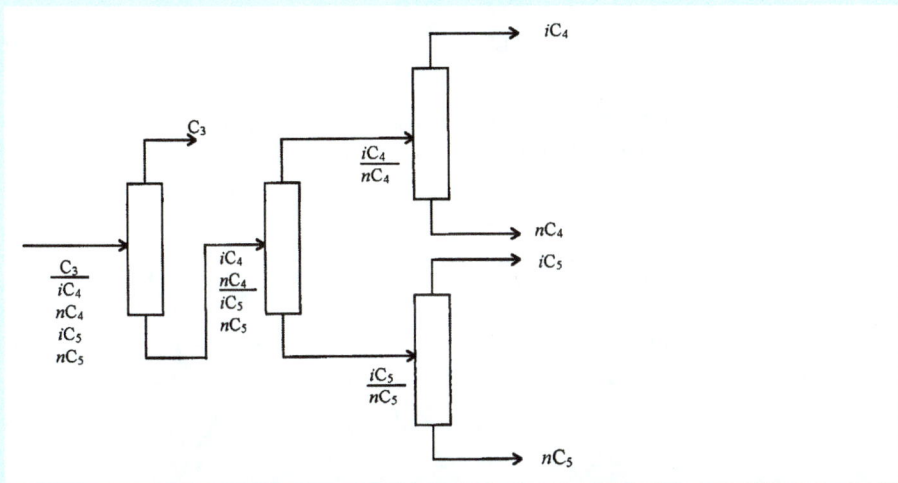

Figure 14.5: Possible sequence of ordinary distillation columns for the separation of the five-component mixture (Hendry and Hughes, 1972).

14.6 Complex column configurations for ternary mixtures

In some cases rather complex distillations should be considered when developing a separation sequence. In Figure 14.6 seven advanced separation sequences are sketched.

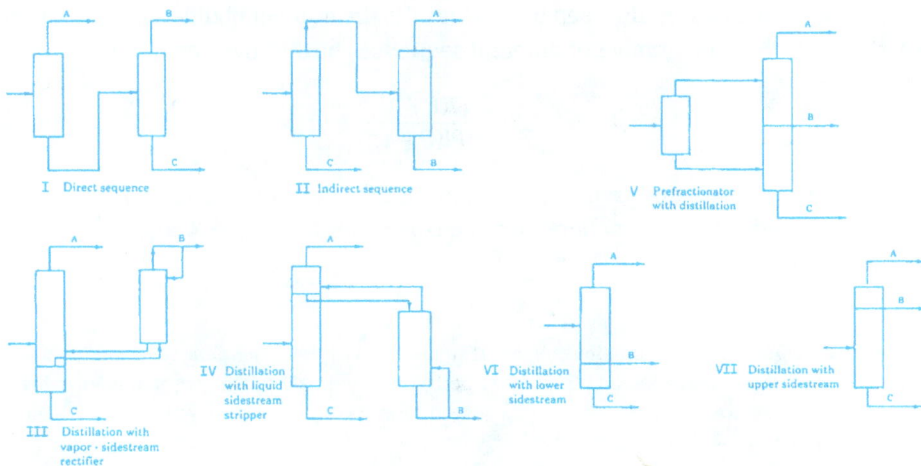

Figure 14.6: Complex columns for ternary mixtures (Tedder and Rudd, 1978).

The configurations sketched in Figure 14.6 have optimal operating regions depending on feed composition and the ease-of-separation index, which is given as

$$\text{ESI} = \frac{\alpha_{AB}}{\alpha_{BC}}. \tag{14.6}$$

Figure 14.7 shows the ternary diagrams and the optimal regions of operation of the seven complex sequences for separating ternary mixtures.

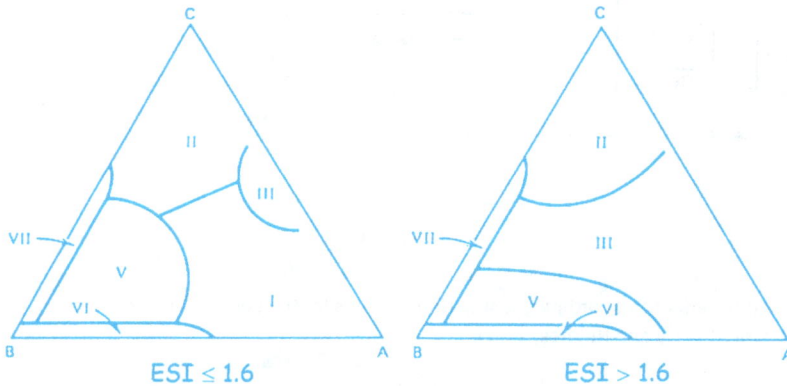

Figure 14.7: Regions of optimality for complex sequences of distillation columns for separating a three-component mixture (Tedder and Rudd, 1978).

When simple distillation is not practical for all separators in a multi-component mixture other types of separators must be used and the order of volatilities or the separation index may be different for each type.

If all units are two-product separators and T is the number of different separation technologies, then the number of different sequences now is given as

$$N_S^T = T^{P-1} \frac{[2(P-1)]!}{P!(P-1)!}. \tag{14.7}$$

For example, if $P = 3$ and ordinary distillation, extractive distillation with two solvents and liquid–liquid extraction are to be considered, then $T = 4$ and 32 different sequences exist.

Example 2 (Sequencing of distillation columns (Example 1 revisited)). The data collected for this problem are summarized in Table 14.3. Because both *trans*- and *cis*-2-butene are contained in the butenes product and are adjacent when species are ordered by relative volatilities they do not need to be separated.

Table 14.3: Data for six species.

Species		b. pt. (°)	Tc (°)	Pc (MPa)
Propane	A	−42.1	97.7	4.17
1-Butene	B	−6.3	146.4	3.94
n-Butane	C	−0.5	152.0	3.73
trans-2-Butene	D	0.9	155.4	4.12
cis-2-Butene	E	3.7	161.4	4.02
n-Pentane	F	36.1	196.3	3.31

We now have five components that need to be separated. If we considered two separation techniques (ordinary distillation and extractive distillation), $T = 2$ and $P = 5$, we could have 224 different sequences.

It would be very useful to reduce the number of sequences to a number that can be analyzed in more detail. In other words, we need to eliminate infeasible separations and enforce ordinary distillation for separations with acceptable volatilities.

Table 14.4: Approximate relative volatilities under ideal conditions for adjacent binary pairs.

Adjacent binary pair	a_{ij} at 65.5 °C
Propane/1-Butene (A/B)	2.45
1-Butene/n-Butane (B/C)	1.18
n-Butane/trans-2-Butene (C/D)	1.03
cis-2-Butene/n-Pentane (E/F)	2.50

From Table 14.4 the following may be concluded.
- Splits A/B and E/F should be performed by ordinary distillation *only* ($\alpha \approx 2.5$).
- Split C/D is infeasible by ordinary distillation ($\alpha = 1.03$). Split B/C is feasible, but an alternative method may be more attractive.
- Use of 96 % furfural as a solvent for extractive distillation increases volatilities of paraffins to olefins, causing a reversal in volatility between 1-Butene and n-Butane, altering the separation order to ACBDEF, and giving $\alpha_{C/B} = 1.17$. Also, split $(C/D)_{II}$ with $\alpha = 1.7$ should be used instead of ordinary distillation.
- Thus, the splits to be considered, with all others forbidden, are $(A/B\ldots)_I$, $(\ldots E/F)_I$, $(\ldots B/C\ldots)_I$, $(A/C\ldots)_I$, $(\ldots C/B\ldots)_{II}$, and $(\ldots C/D\ldots)_{II}$.

14.7 Estimating the annualized costs for separation sequences

For each separation, the annualized cost can be estimated assuming 99 mol% recovery of light key in distillate and 99 mol% recovery of heavy key in bottom residue. The following steps should be taken:

- Set distillate and bottom column pressures using the algorithm presented in Figure 14.4.
- Estimate the number of stages and reflux ratio with a simulation package such as AspenPLus or UniSim.
- Select tray spacing (typically 0.5 m) and calculate the column height H.
- Compute the tower diameter D (simulation packages like AspenPLus or UniSim might be employed using short-cut methods).
- Estimate installed cost of tower (scales of economy might be used).
- Estimate size and cost of ancillary equipment (condenser, reboiler, reflux drum) and the sum of total capital investment, C_{TCI}.
- Compute annual cost of heating and cooling utilities (COS).
- Compute the annualized costs C_A assuming ROI (typically $r = 0.2$) and $C_A = \text{COS} + rC_{TCI}$.

Example 3 (Separation train sequencing with a superstructure). Once all technically feasible separation sequences are established, a superstructure as shown in Figure 14.8 can be obtained. Each arc in this superstructure represents a separation step with an associated annualized cost (in $/yr).

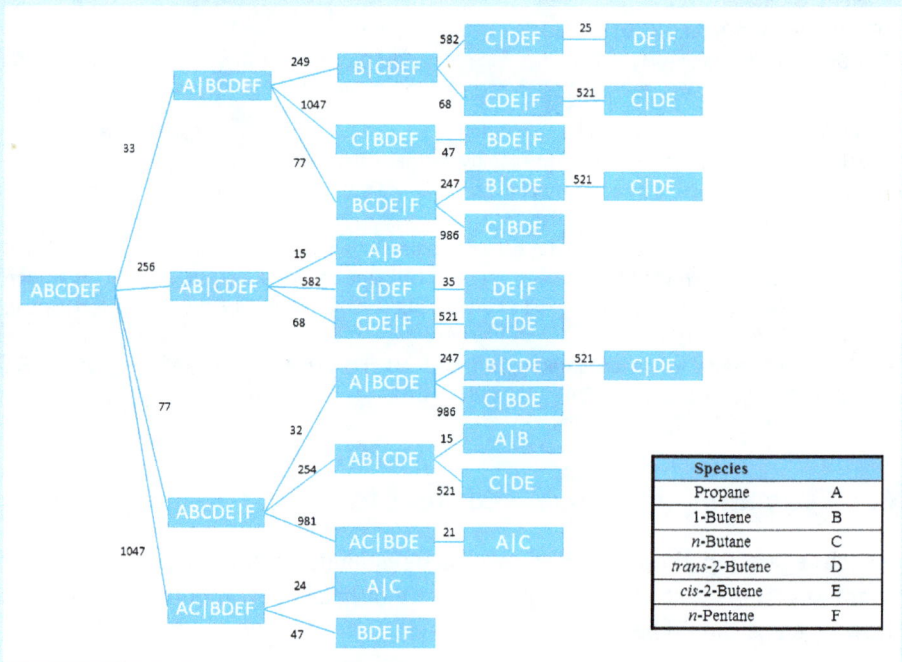

Figure 14.8: Superstructure showing all feasible separation sequences.

The superstructure can be used to identify the most promising route (in terms of annualized costs). The most cost effective separation route is shown in Figure 14.9, with an annualized cost of US$840,000 per year.

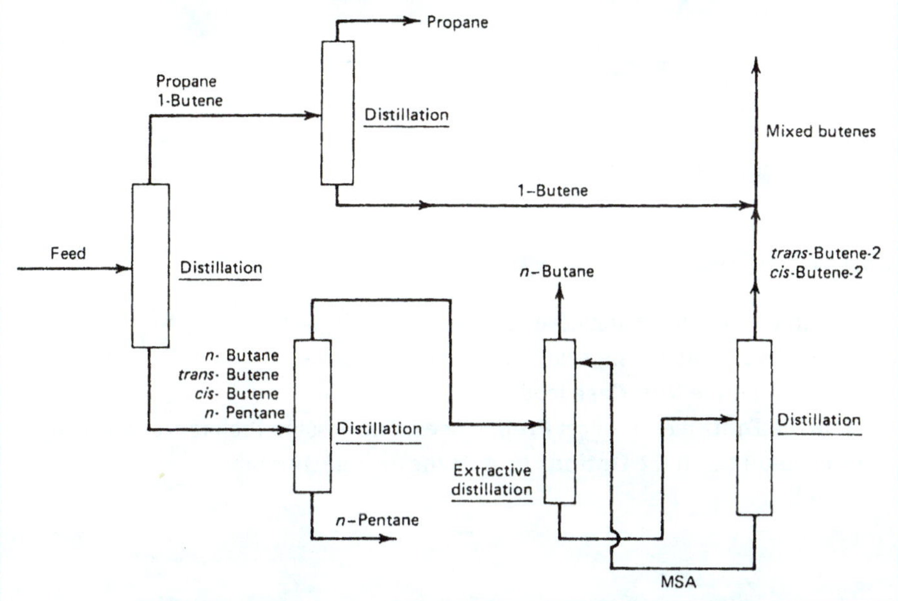

Figure 14.9: Optimized separation sequence.

14.8 Distillation column design with a process simulator

In this section we introduce the process simulator UniSim from Honeywell. Chapter 10 gives a much more detailed overview of process simulation and the reader is referred to this chapter for more information regarding licenses of UniSim.

In this section we will learn how to utilize a distillation column to recover methylcyclohexane (MCH). The process flow diagram and operating conditions are shown in Figure 14.10.

MCH and toluene form a close-boiling system that is difficult to separate by simple binary distillation. In the recovery column, phenol is used to extract toluene, allowing relatively pure methylcyclohexane to be recovered in the overhead. The purity of the recovered methylcylohexane depends on the phenol input flow rate. In this section we create a UniSim Design simulation that allows us to investigate the performance of the column.

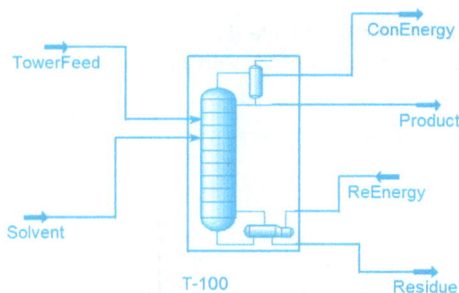

Figure 14.10: Flowsheet of the MCH recovery process.

14.8.1 Setting your session preferences

1. To start a new simulation case, do one of the following:
   - From the **File** menu, select **New** and then **Case**.
   - Click on the **New Case** icon.
2. From the **Tools** menu, select **Preferences**. The Session Preferences view appears. You should be on the **Options** page of the **Simulation** tab.

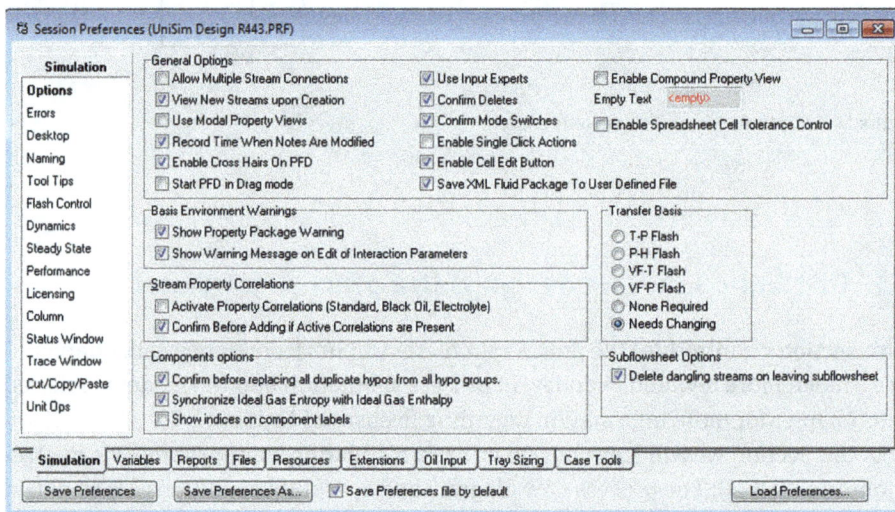

The UniSim Design default session settings are stored in a Preference file called **Unisimdesign Rxxx.prf**. When you modify any of the preferences, you can save the changes in a new Preference file by clicking either **Save Preferences**, which will overwrite the default file, or **Save Preferences As...** and UniSim Design will prompt you to provide a name for the new Preference file. Click on the **Load Preferences** button to load any simulation case. Check the **Save Preferences File by**

default to automatically save the preference file when exiting UniSim Design. In the General Options group, ensure the **Use Modal Property Views** checkbox is unchecked.

14.8.2 Building the simulation

Creating a component list

Now that you have created a new case, your next task is to select the components.

1. In the **Components** tab of the Simulation Basis Manager view, click on the **Add** button in the Component Lists group. The Component List View appears.

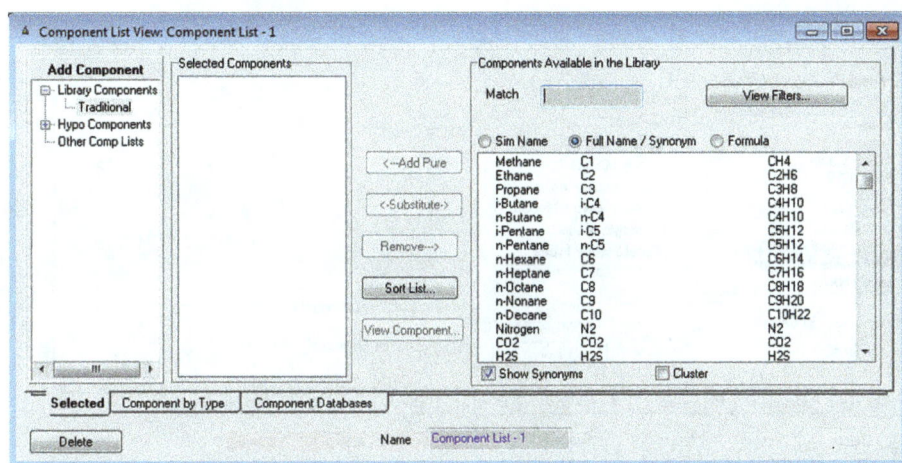

Here you will use **Toluene, Phenol and Methylcyclohexane**. The completed component list appears below.

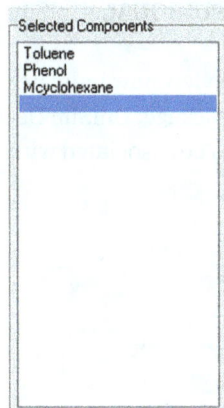

Creating a Fluid Package

The next step is to add a Fluid Package. As a minimum, a Fluid Package contains the components and a property method (for example an equation of state) that UniSim Design uses in its calculations for a particular flowsheet. Depending on what is required in a specific flowsheet, a Fluid Package may also contain other information, such as reactions and interaction parameters.

1. In the Simulation Basis Manager view, click on the **Fluid Pkgs** tab.
2. Click on the **Add** button, and the property view for your new Fluid Package appears.

    The property view is divided into a number of tabs. Each tab contains the options that enable you to completely define the Fluid Package.

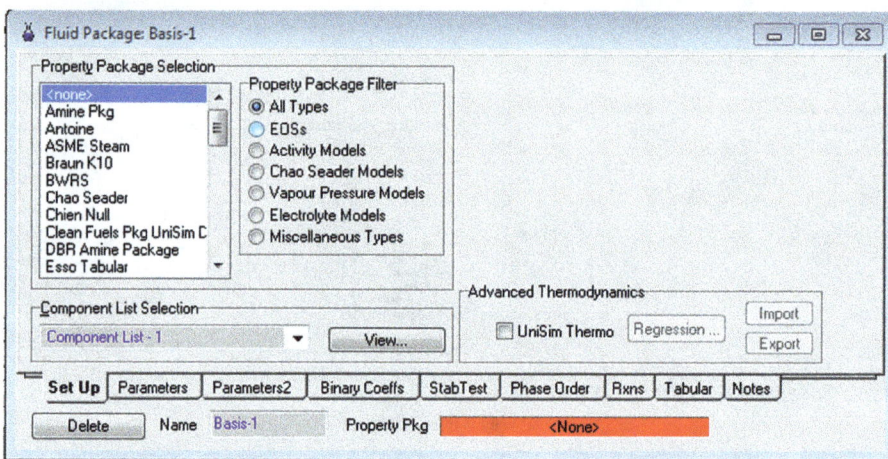

The first step in configuring a Fluid Package is to choose a Property Package on the **Set Up** tab. The current selection is **<none>**. Here, you will select the **UNIQUAC** property package.

Select **<none>** in the Property Package Selection list and type **UNIQUAC**. UniSim Design automatically finds the match to your input.

The **Property Pkg** indicator at the bottom of the Fluid Package view now indicates that **UNIQUAC** is the current property package for this Fluid Package. UniSim Design has also automatically created an empty component list to be associated with the Fluid Package.

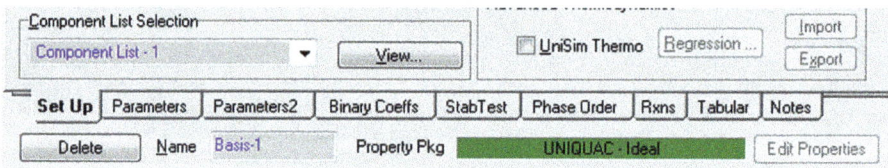

Providing binary coefficients

The next task in defining the Fluid Package is providing the binary interaction parameters.

1. Click on the **Binary Coeffs** tab of the Fluid Package view.

In the Activity Model Interaction Parameters group, the Aij interaction table appears by default. UniSim Design automatically inserts the coefficients for any component pairs for which library data are available. You can change any of the values provided by UniSim Design if you have data of your own. However, for this example, you will use one of UniSim Design's built-in estimation methods.

Next, you will use the **UNIFAC VLE** estimation method to estimate all pairs.

2. In the Coeff Estimation group, ensure the **UNIFAC VLE** radio button is selected.
3. Click on the **ALL Binaries** button. UniSim Design provides values for all pairs. The final Activity Model Interaction Parameters table for the **Aij** coefficients appears, as shown below.

4. To view the **Bij** coefficient table, select the **Bij** radio button. For this example, all the **Bij** coefficients will be left at the default value of zero.
5. Go to the simulation environment by clicking the **Enter Simulation Environ-ment** button on the Simulation Basis Manager.

Installing unit operations

Adding a RadFrac column: UniSim Design has a number of prebuilt column templates that you can install and customize by changing attached stream names, the number of stages and default specifications.

In this section, you will install a Distillation Column.
1. From the **Tools** menu, select **Preferences**.
2. On the **Simulation** tab, **Options** page, ensure that the **Use Input Experts** check-box is selected (checked), then close the view.
3. Press **F4** to access the Object Palette.
4. Double-click on the **Distillation Column** icon on the Object Palette. The first page of the Input Expert view appears.

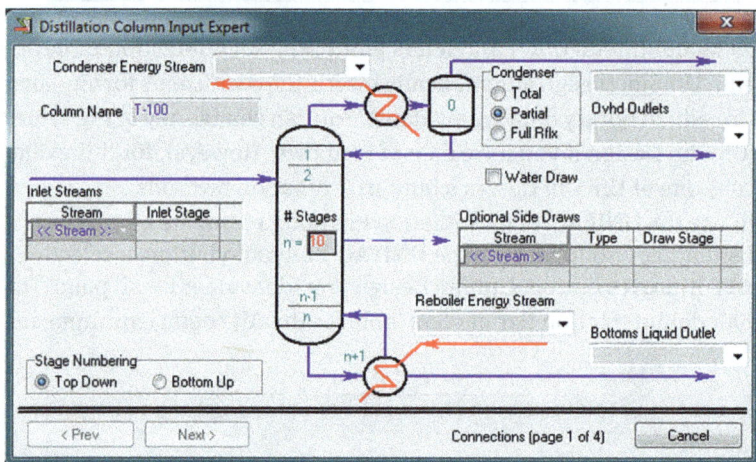

When you install a column using a prebuilt template, UniSim Design supplies certain default information, such as the number of stages. The current active cell is # Stages (number of stages), indicated by the thick border around this cell and the presence of 10 (default number of stages).

The following points are worth noting:
– These are theoretical stages, as the UniSim Design default stage efficiency is one. If you want to specify real stages, you can change the efficiency of any or all stages later.

- The Condenser and Reboiler are considered separate from the other stages, and are not included in the Numb Stages field.

For this example, **22** theoretical stages will be used, so fill 22 into the # Stages box.

5. Click on the **«Stream»** cell in the Inlet Streams table. Type **TowerFeed** in the cell and press **ENTER**.

UniSim Design will supply a default feed location in the middle of the Tray Section (TS), in this case stage 11 (indicated by 11_Main TS). However, in this example, the TowerFeed will enter above **stage 14**.

6. Click on the **Inlet stage** cell, and choose 14_Main TS.
7. Repeat steps 5 and 6 with a new stream as **Solvent** and **stage 7**.

This column has Overhead Liquid (distillate) and Bottoms Liquid products, but no Overhead Vapour product.

8. In the Condenser group, select the **Total** radio button.

The Vapour stream disappears. This is the same as leaving the Condenser as Partial and later specifying a zero vapor rate.

9. Enter the streams and Column names as shown in the figure below.

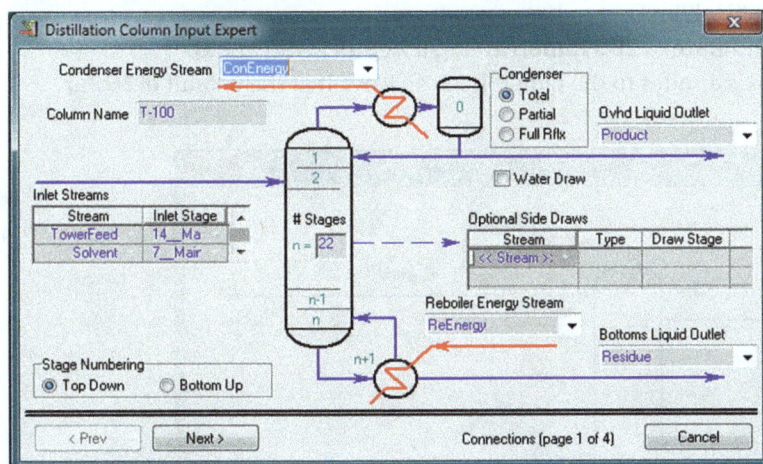

When you are finished, the **Next** button becomes active, indicating sufficient information has been supplied to advance to the next page of the Input Expert.

10. Click on the **Next** button to advance to the **Pressure Profile** page.
11. In the **Condenser Pressure** field, enter **16 psia**.
12. In the **Reboiler Pressure** field, enter **20.2 psia**.

The Condenser Pressure Drop can be left at its default value of zero.

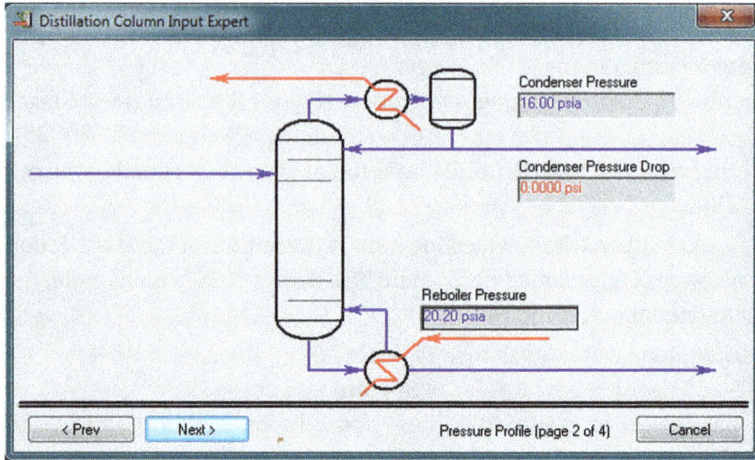

13. Click on the **Next** button to advance to the **Optional Estimates** page.
14. Click on the **Next** button to advance to the fourth and final page of the Input Expert.

    The Specifications page allows you to supply values for the default column specifications that UniSim Design has created.
15. Enter a **Liquid Rate** of **200 lbmol/hr** and a **Reflux Ratio** of **8.0**.

    The Flow Basis applies to the Liquid Rate, so leave it at the default of **Molar**.

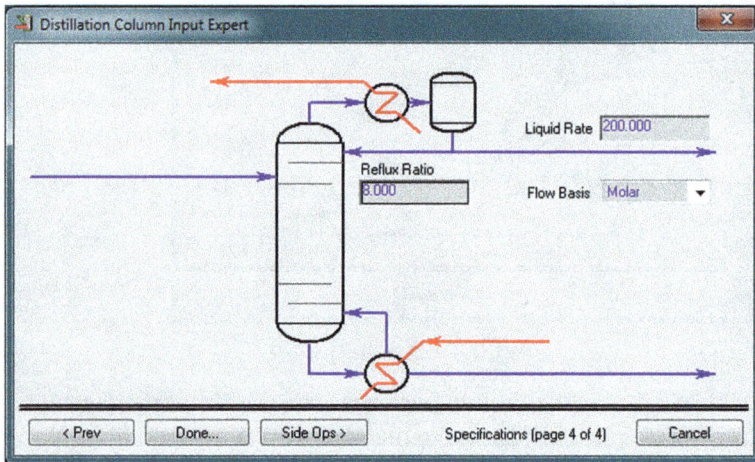

16. Click on the **Done** button.

    The Distillation Column property view appears.

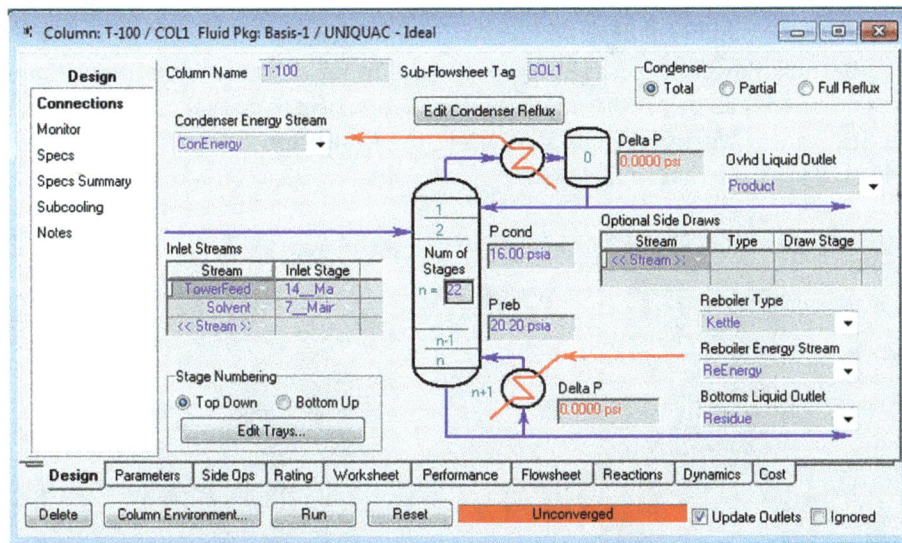

17. Select the **Monitor** page.

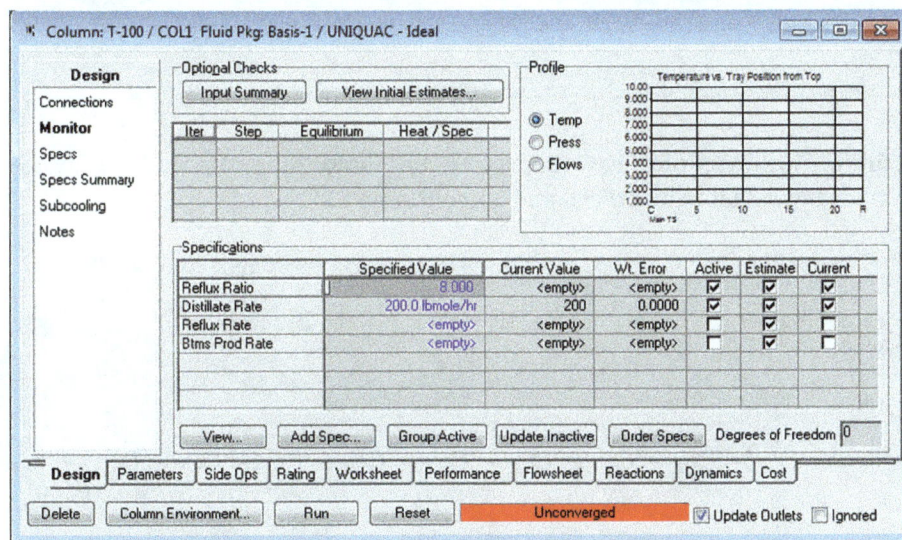

The Monitor page displays the status of your column as it is being calculated, updating information with each iteration. You can also change specification values and activate or deactivate specifications used by the Column solver directly from this page.

The current number of degrees of freedom is zero, indicating the column is ready to be run. However, we need to specify the input streams before the column calculations can be performed. We will use Worksheet to do that.

18. Click on the **Worksheet** tab, and select the **Condition** page. Your Worksheet should appear as shown below.

Worksheet	Name		TowerFeed	Solvent	Residue	Product	
Conditions	Vapour		<empty>	<empty>	<empty>	<empty>	
Properties	Temperature [F]		<empty>	<empty>	<empty>	<empty>	
	Pressure [psia]		<empty>	<empty>	20.20	16.00	
Composition	Molar Flow [lbmole/hr]		<empty>	<empty>	<empty>	<empty>	
PF Specs	Mass Flow [lb/hr]		<empty>	<empty>	<empty>	<empty>	
	Std Ideal Liq Vol Flow [barrel/day]		<empty>	<empty>	<empty>	<empty>	
	Molar Enthalpy [Btu/lbmole]		<empty>	<empty>	<empty>	<empty>	
	Molar Entropy [Btu/lbmole-F]		<empty>	<empty>	<empty>	<empty>	
	Heat Flow [Btu/hr]		<empty>	<empty>	<empty>	<empty>	

Column: T-100 / COL1 Fluid Pkg: Basis-1 / UNIQUAC - Ideal

Design | Parameters | Side Ops | Rating | **Worksheet** | Performance | Flowsheet | Reactions | Dynamics | Cost

Delete | Column Environment... | Run | Reset | Unconverged | ☑ Update Outlets ☐ Ignored

19. In the Worksheet property view, choose the **Conditions** page and specify the following data for **TowerFeed** and **Solvent** streams.

Worksheet	Name		TowerFeed	Solvent	Residue	Product	
Conditions	Vapour		<empty>	<empty>	<empty>	<empty>	
Properties	Temperature [F]		220.0	220.0	<empty>	<empty>	
	Pressure [psia]		20.00	20.00	20.20	16.00	
Composition	Molar Flow [lbmole/hr]		<empty>	<empty>	<empty>	<empty>	
PF Specs	Mass Flow [lb/hr]		<empty>	<empty>	<empty>	<empty>	
	Std Ideal Liq Vol Flow [barrel/day]		<empty>	<empty>	<empty>	<empty>	
	Molar Enthalpy [Btu/lbmole]		<empty>	<empty>	<empty>	<empty>	
	Molar Entropy [Btu/lbmole-F]		<empty>	<empty>	<empty>	<empty>	
	Heat Flow [Btu/hr]		<empty>	<empty>	<empty>	<empty>	

Column: T-100 / COL1 Fluid Pkg: Basis-1 / UNIQUAC - Ideal

Design | Parameters | Side Ops | Rating | **Worksheet** | Performance | Flowsheet | Reactions | Dynamics | Cost

Delete | Column Environment... | Run | Reset | Unconverged | ☑ Update Outlets ☐ Ignored

20. Select the **Composition** page. By default, the components are listed by **Mole Fractions**.

21. Click on the field next to the Toluene cell. The **FeedTower stream property view** appears.

22. Click on the **Composition** page, click on the **Edit** button and choose Mole Flows in **Composition Basic**. Click on the **CompMoleFlow** cell for each component, and fill in as follows: **Toluene = 200 lbmol/hr, MCH = 200 lbmol/hr**.

23. Click on the **OK** button, and UniSim Design accepts the composition. The stream is now completely defined, so UniSim Design flashes it at the conditions given to determine its remaining properties.
24. Close the TowerFeed stream property view.
25. Repeat from step 20 to 25 to specify the **Solvent** stream as follows: **Phenol = 1200 lbmol/hr**.

Now, all streams and the column are specified, so the column is ready to be calculated.
26. Click on the **Design** tab and choose the **Monitor** page.
27. Click on the **Run** button to begin calculations.
    The information displayed on the **Monitor** page is updated with each iteration. The column converges quickly, in six iterations.

Column: T-100 / COL1 Fluid Pkg: Basis-1 / UNIQUAC - Ideal

Design — Connections — **Monitor** — Specs — Specs Summary — Subcooling — Notes

Optional Checks: Input Summary | View Initial Estimates...

Iter	Step	Equilibrium	Heat / Spec
2	1.0000	0.046657	0.037346
3	0.0001	0.006996	0.018442
4	0.0001	0.000137	0.002126
5	1.0000	0.000005	0.000512
6	1.0000	0.000000	0.000066

Profile: ◉ Temp ○ Press ○ Flows — Temperature vs. Tray Position from Top

Specifications:

	Specified Value	Current Value	Wt. Error	Active	Estimate	Current
Reflux Ratio	8.000	8.00	0.0000	✓	✓	✓
Distillate Rate	200.0 lbmole/hr	200	0.0000	✓	✓	✓
Reflux Rate	<empty>	1.60e+003	<empty>	□	✓	□
Btms Prod Rate	<empty>	1.40e+003	<empty>	□	✓	□

View... | Add Spec... | Group Active | Update Inactive | Order Specs — Degrees of Freedom 0

Design | Parameters | Side Ops | Rating | Worksheet | Performance | Flowsheet | Reactions | Dynamics | Cost

Delete | Column Environment... | Run | Reset | Converged | ☑ Update Outlets □ Ignored

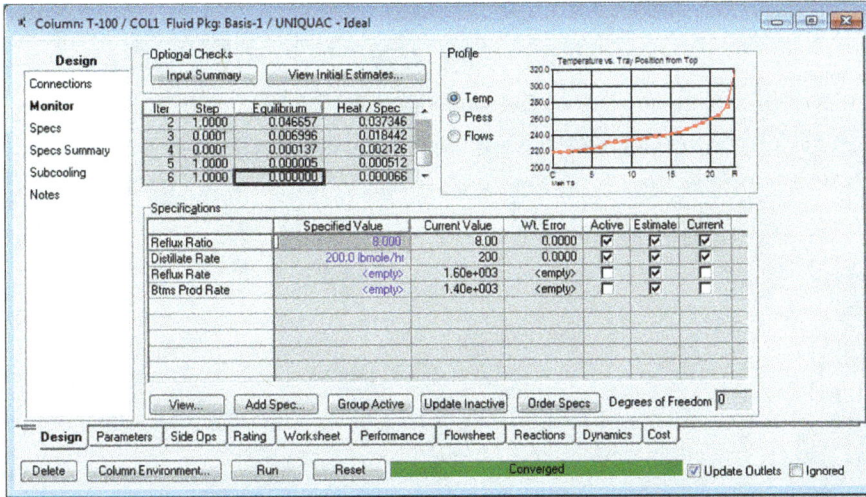

The table in the Optional Checks group displays the Iteration number, Step size and Equilibrium error and Heat/Spec error.

The column temperature profile appears in the Profile group. You can view the pressure or flow profiles by selecting the appropriate radio button.

The status indicator has changed from Unconverged to Converged.

28. Click on the **Performance** tab and select the **Column Profiles** page to access a more detailed stage summary.

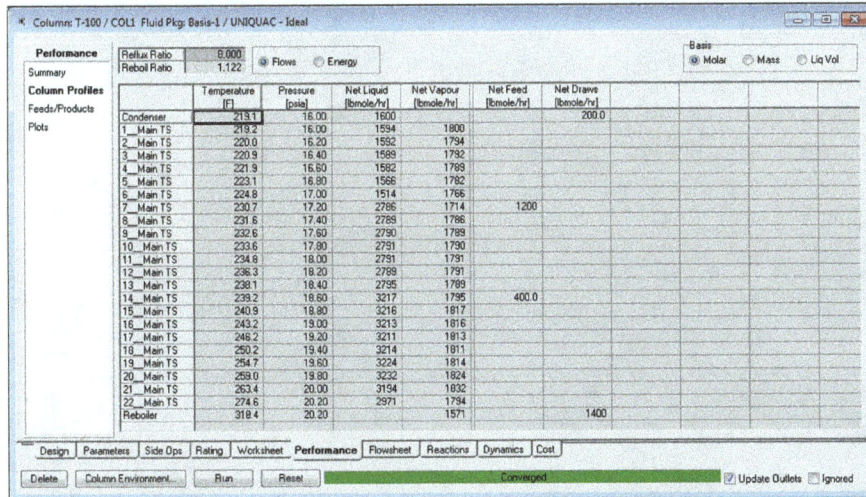

Column: T-100 / COL1 Fluid Pkg: Basis-1 / UNIQUAC - Ideal

Performance — Summary — **Column Profiles** — Feeds/Products — Plots

Reflux Ratio 8.000 | Reboil Ratio 1.122 | ◉ Flows ○ Energy | Basis: ◉ Molar ○ Mass ○ Liq Vol

	Temperature [F]	Pressure [psia]	Net Liquid [lbmole/hr]	Net Vapour [lbmole/hr]	Net Feed [lbmole/hr]	Net Draws [lbmole/hr]
Condenser	219.1	16.00	1600			200.0
1_Main TS	219.2	16.00	1594	1800		
2_Main TS	220.0	16.20	1592	1794		
3_Main TS	220.9	16.40	1589	1792		
4_Main TS	221.9	16.60	1582	1789		
5_Main TS	223.1	16.80	1566	1782		
6_Main TS	224.8	17.00	1514	1766		
7_Main TS	230.7	17.20	2786	1714	1200	
8_Main TS	231.6	17.40	2789	1786		
9_Main TS	232.6	17.60	2790	1789		
10_Main TS	233.6	17.80	2791	1790		
11_Main TS	234.8	18.00	2791	1791		
12_Main TS	236.3	18.20	2789	1791		
13_Main TS	238.1	18.40	2795	1789		
14_Main TS	239.2	18.60	3217	1795	400.0	
15_Main TS	240.9	18.80	3216	1817		
16_Main TS	243.2	19.00	3213	1816		
17_Main TS	246.2	19.20	3211	1813		
18_Main TS	250.2	19.40	3214	1811		
19_Main TS	254.7	19.60	3224	1814		
20_Main TS	259.0	19.80	3232	1824		
21_Main TS	263.4	20.00	3194	1832		
22_Main TS	274.6	20.20	2971	1794		
Reboiler	318.4	20.20		1571		1400

Design | Parameters | Side Ops | Rating | Worksheet | **Performance** | Flowsheet | Reactions | Dynamics | Cost

Delete | Column Environment... | Run | Reset | Converged | ☑ Update Outlets □ Ignored

29. When considering the column, you might want to focus only on the column sub-flowsheet. You can do this by entering the column environment.

30. Click on the **Column Environment** button at the bottom of the column property view. UniSim Design desktop now displays the Column Subflowsheet environment.

File Edit Simulation Flowsheet Column Tools Window Help

Environment: T-100 (COL1)
Mode: Steady State

31. In this environment you can do the following:
    – Click on the **PFD** icon to view the column subflowsheet PFD.

    – Click on the **Workbook** icon to view a Workbook for the column subflow-
       sheet objects.

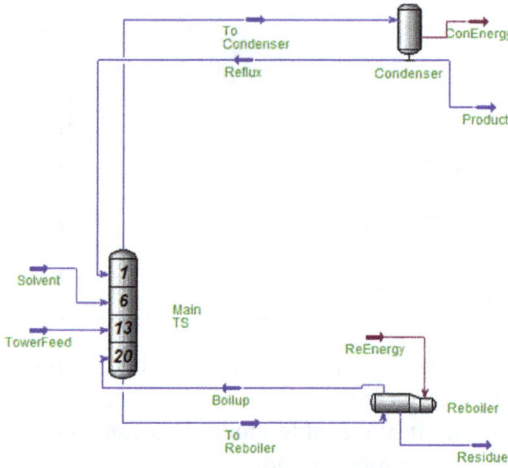

Workbook - T-100 (COL1)

Name	Reflux	To Condenser	Boilup	To Reboiler	Product	Residue
Vapour Fraction	0.0000	1.0000	1.0000	0.0000	0.0000	0.0000
Temperature [F]	219.1	219.2	318.4	274.6	219.1	318.4
Pressure [psia]	16.00	16.00	20.20	20.20	16.00	20.20
Molar Flow [lbmole/hr]	1600	1800	1571	2971	200.0	1400
Mass Flow [lb/hr]	1.570e+005	1.766e+005	1.461e+005	2.774e+005	1.962e+004	1.314e+005
Liquid Volume Flow [barrel/day]	1.389e+004	1.562e+004	1.087e+004	1.965e+004	1736	8779
Heat Flow [Btu/hr]	-1.175e+008	-1.081e+008	9.486e+006	-8.371e+007	-1.468e+007	-6.253e+007
Name	TowerFeed	Solvent	** New **			
Vapour Fraction	0.0000	0.0000				
Temperature [F]	220.0	220.0				
Pressure [psia]	20.00	20.00				
Molar Flow [lbmole/hr]	400.0	1200				
Mass Flow [lb/hr]	3.807e+004	1.129e+005				
Liquid Volume Flow [barrel/day]	3191	7323				
Heat Flow [Btu/hr]	-1.270e+007	-7.112e+007				

Material Streams   Compositions   Energy Streams   Unit Ops

Condenser
Main TS

Fluid Pkg All

☑ Horizontal Matrix

☐ Show Name Only
Number of Hidden Objects: 0

    – Click on the **Column Runner** icon to access the inside column property
       view.
    This property view is essentially the same as the outside, or main flowsheet,
    property view.

32. When you are finished in the column environment, return to the main flowsheet by clicking the **Enter Parent Simulation Environment** icon ⬆ in the tool bar or the **Parent Environment** button on the column Worksheet view.

14.8.3 Review simulation results

1. Open the Workbook for the main case to access the calculated results for all streams and operations.
2. Click on the **Material Streams** tab.

Workbook - Case (Main)					
Name	TowerFeed	Solvent	Residue	Product	** New **
Vapour Fraction	0.0000	0.0000	0.0000	0.0000	
Temperature [F]	220.0	220.0	318.4	219.1	
Pressure [psia]	20.00	20.00	20.20	16.00	
Molar Flow [lbmole/hr]	400.0	1200	1400	200.0	
Mass Flow [lb/hr]	3.807e+004	1.129e+005	1.314e+005	1.962e+004	
Liquid Volume Flow [barrel/day]	3191	7323	8779	1736	
Heat Flow [Btu/hr]	-1.270e+007	-7.112e+007	-6.253e+007	-1.468e+007	

Material Streams | Compositions | Energy Streams | Unit Ops

FeederBlock_TowerFeed
T-100

Fluid Pkg All

☐ Include Sub-Flowsheets
☐ Show Name Only
☑ Horizontal Matrix   Number of Hidden Objects: 0

3. Click on the **Compositions** tab.

Workbook - Case (Main)					
Name	TowerFeed	Solvent	Residue	Product	** New **
Comp Mole Frac (Toluene)	0.500000	0.000000	0.140765	0.014649	
Comp Mole Frac (Phenol)	0.000000	1.000000	0.856928	0.001511	
Comp Mole Frac (Mcyclohexane)	0.500000	0.000000	0.002308	0.983841	

Material Streams | **Compositions** | Energy Streams | Unit Ops

FeederBlock_TowerFeed
T-100

Fluid Pkg All

☐ Include Sub-Flowsheets
☐ Show Name Only
☑ Horizontal Matrix   Number of Hidden Objects: 0

Exercises

1. Stabilized effluent from a hydrogenation unit is separated by ordinary distillation into five rela- tively pure products. Four distillation columns are required. How many sequences are possible? Draw the different sequences.

Component	Feed flow rate (lbmol/hr)	Relative volatility compared with C5
Propane (C3)	10	8.1
1-Butane (B1)	100	3.7
n-Butan (NB)	341	3.1
2-Butane isomers (B2)	187	2.7
n-Pentane (C5)	40	1.0

2. Researchers have studies all 14 sequences for separating the following mixture at a flow rate of 200 lbmol/hr into its five components with about 98 % purity for each.

Species	Feed mole fraction	Approximate volatility relative to n-Pentane
Propane (A)	0.05	8.1
Isobutane (B)	0.15	4.3
n-Butane (C)	0.25	3.1
Isopentane (D)	0.20	1.25
n-Pentane (E)	0.35	1.0

For each sequence they determined the annual operating costs including depreciation of the cap- ital investment. Cost data for the best three and the worst sequence are given in the figure below. Explain why the best sequences are best and the worst are worst. Which heuristics are the most important?

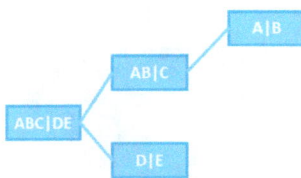

Best sequence Cost $858.780/yr

Third best sequence Cost $871.460/yr

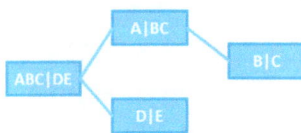

Second best sequence Cost $863.500/yr

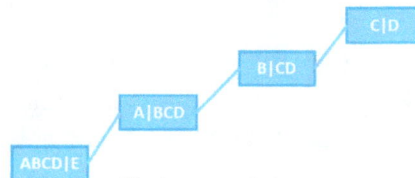

Worst sequence Cost $939.400/yr

14.9 Take-away message

Separations are needed throughout the entire process, for purifying the final product, for recovering unreacted raw materials or solvents, for cleaning gaseous and aqueous waste streams, for utilities generation, etc. Separations take up until 80 % of the overall energy requirement of a process. The design of a separation train can lead to many different routes, depending on the number of components in the mixture and the alternative separation technologies. Typically the number of routes is factorial to components and technologies. Process knowledge and heuristics can be used to reduce the number of alternative routes and select the most promising ones.

14.10 Further reading

Hendry and Hughes (1972). Generating separation process flowsheets. *Chem. Eng. Prog.*, 68(6), p. 69.

Souders (1964). The countercurrent separation process. *Chem. Eng. Prog.*, 60(2), pp. 75–82.

Tedder and Rud (1978). Parametric studies in industrial distillation: Design comparison. *AIChEJ*, 24, pp. 303.

Tedder and Rudd (1978). Parametric studies in industrial distillation II: heuristic optimization. *AIChEJ*, 24, pp. 316.

De Haan and Bosch (2015). *Industrial separation processes, Fundamentals*. de Gruyter.

Seider, Lewin, Seader, Widagdo, Gani and Ng (2018). *Product and process design principles, synthesis, analysis and evaluation*, 4th ed. Wiley.

15 Plant-wide control

15.1 Process control

The field of process control is concerned with the development of dynamic models for control, optimization and prediction. Especially in suppressing unexpected disturbances and optimizing start-ups, shut-downs and switchover of operations, process control is very valuable. Control can take place at different temporal and spatial scales, as shown in Figure 15.1.

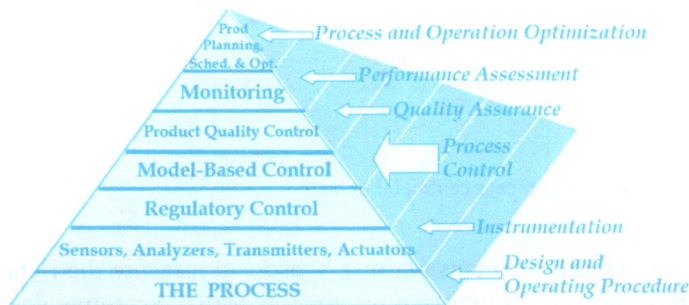

Figure 15.1: Process control hierarchy.

In the 1940s, process control in plants was mainly done manually. The operators relied on information flows and manually executed the control actions that had to be taken. In the 1950s, pneumatic and mechanical devices (bellows and springs) were developed and the basics of feedback control were implemented. In the 1960s electrical, analog control came into use and feedback, feedforward and cascaded control systems were set up. When computers arose, electrical, digital control systems evolved and model predictive controllers enabled (petro)chemical companies to control their processes very efficiently.

15.2 Incentives for process control

Process control is needed for a process to guarantee safety, to satisfy product specifications (quality), to obey environmental regulations, to satisfy operational or economic constraints and very often to account for errors that were made during the process design stages.

To accomplish good process operation, continuos monitoring and control should be done to suppress the influences of external disturbances, to ensure the stability of the process and to ensure optimum performance and controllability of the process.

In process design the fixed costs are evaluated *versus* the variable costs, while in process control the focus is only on the variable costs. Process control has gained more and more importance over the last years as plants have more mass/energy integration

https://doi.org/10.1515/9783110570137-015

and smaller or no buffer tanks and because the modern plant is highly flexible with respect to feed (flow and composition).

Control is only possible if the engineer provides the required equipment during the process design stages. In Figure 15.2 the main hardware for proper process control is depicted.

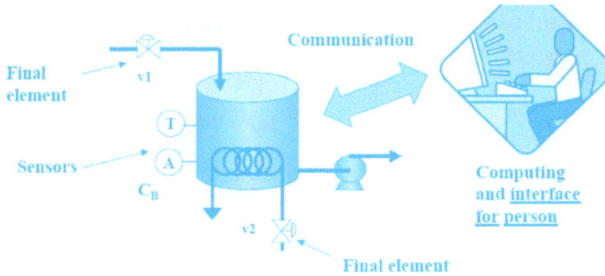

Figure 15.2: Required elements for proper process control (after Marlin, 2000).

For proper control one needs final elements (valves), sensors to measure the right properties and software/computers to interface the process with a user/operator. These three items form the critical elements for a (feedback) control system.

Figure 15.3 shows how these elements are integrated in a block diagram. A process has outputs (*e. g.*, the tank level in a vessel), inputs (*e. g.*, the outflow out of the tank) and disturbances (*e. g.*, the inflow in a tank). Sensors are used to measure the outputs (*e. g.*, tank level), while final elements can be used to manipulate the inputs (*e. g.*, the outflow could be influenced via a valve). The controller uses the information of the sensor (how far is the tank level from its setpoint, or desired value) to calculate a control action (how much should the valve be opened or closed).

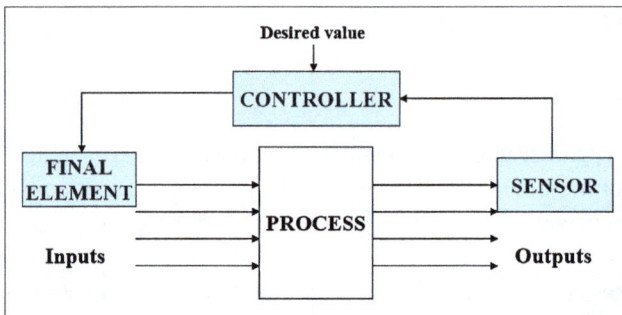

Figure 15.3: Block diagram of a (feedback) control loop (after Marlin, 2000).

Nowadays process control is done from the central control room in a plant, which registers all measurements from sensors and coordinates all actions for control.

15.3 Importance of modeling

All modern control is "model-based," *i. e.*, controllers use process models to calculate the control actions for the plant. Such dynamic models can be first principle models (mass, energy, component, impulse) or black box models (regressed structures). The dynamic model can be used for prediction when measurements are lacking. Dynamic process models are also very useful in operator training simulations (OTSs).

15.4 Block diagrams

After establishing the goals for control, sketching the process (process flow diagram) and setting the boundaries and assumptions for models, the first stage of modeling is often to set up block diagrams.

A block diagram indicates what the relationships between process inputs and outputs are. Inputs are usually flows, but could also be inlet process conditions such as inlet temperature, concentration or pressure. Outputs are generally state variables such as level, pressure, concentration and temperature. Before any modeling can take place the relationships between inputs and outputs have to be understood.

It is obvious that a block model can look different when different goals are set, or different assumptions are made. Figure 15.4 illustrates this by presenting two alternative block diagrams for a boiler controller that has different goals.

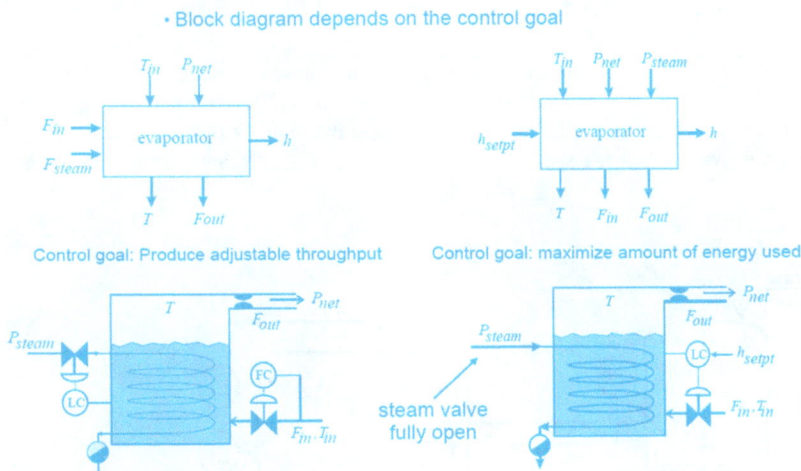

Figure 15.4: Two block diagram for boilers.

For a proper design of the block diagram one has to ensure that the process is well known, that one has sufficient physical understanding of the system (what depends on what) and that one analyzes the control strategy by determining the goal of control, the constraints, the external process variables (disturbances), the controlled variables (inputs) and the control variables (outputs).

Typical disturbances are variations in utilities (steam, cooling water), changes in the process load (*e. g.*, feed), changes in raw material quality, atmospheric disturbances (rain or snow on a distillation column), tear, wear and deterioration.

Typical controlled process variables are the quantity and quality of the product, non-self regulating mass accumulations (liquid level, gas pressure, etc.) and self-regulating process conditions such as temperature, concentration and pressure.

Self-regulation means that a process variable stays within certain limits without direct control. Most process variables return to a new steady state after being perturbed and sometimes a variable stays within a certain limit because another variable is being controlled, for example steam (P–T relation) or concentration (x–T relation).

The selection of control variables is usually the process flows, for example valve position or pumping frequency.

15.5 Control schemes

After setting up the block diagram a control scheme can be devised. The control scheme combines the controlled variables with the control variables. A control scheme is intended to maximize the power and speed of control and minimize control loop interaction. In setting up a control scheme, common sense is usually a good indicator! When setting up a control scheme, standard conventions for drawing sensors, final elements and controllers are used.

Example 1 (Setting up a control scheme). Aniline ($x_{A,in}$), which is present as pollution in water, can be removed with benzene (F_B). Steam raises the temperature to 50 °C. The aniline concentration in waste water should be lowered to x_A (ppm), and in benzene it should be increased to x_B (ppm) (y = weight fraction water). A phase separator separates the phases. A scheme of the process is shown below.

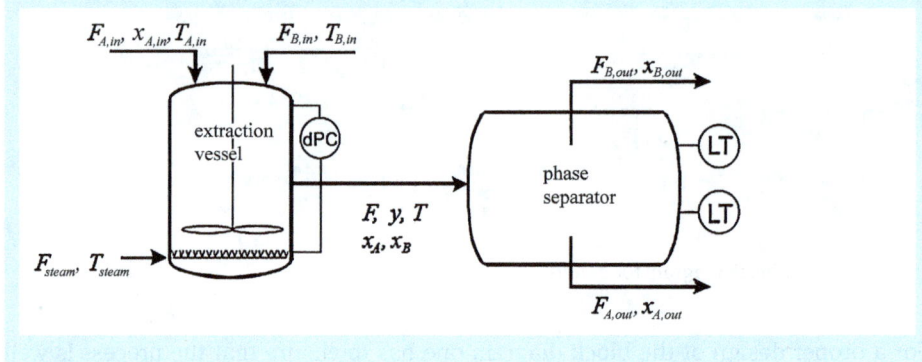

Exercise: Design a block diagram for the goal mentioned, and indicate disturbances and controlled variables. Design a control scheme to achieve the goal(s).

A suitable block diagram is shown below. The control goals are to lower the aniline concentration in the waste water and the control output variables can be controlled by manipulating the flows.

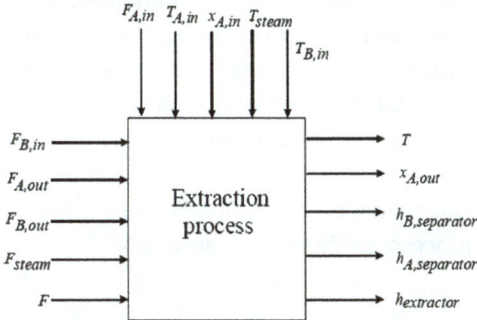

Control goals:
- lower aniline in waste water to x_A
- separate the phases

Control output variables by manipulating $F_{B,in}$, $F_{A,out}$, $F_{B,out}$, F_{steam} and F

Information flow diagram extraction process

A control scheme can be set up to achieve the control goals by using temperature and pressure measurements for controlling the input and outputs of the mixing unit, while using level controllers for the extraction unit, as shown in the picture below.

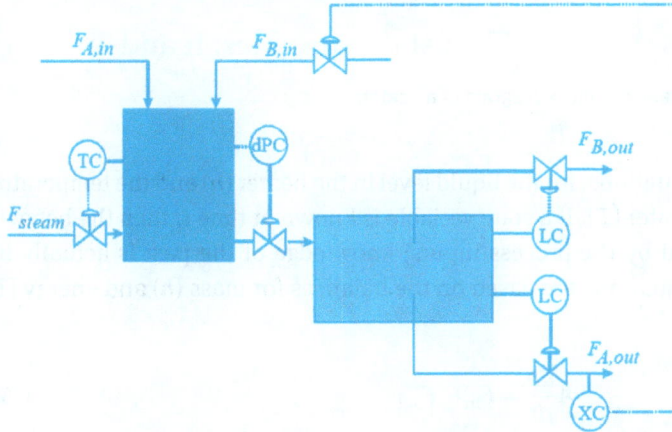

Control scheme to achieve control goals

15.6 Dynamic model development and behavioral diagrams

Dynamic models are very useful descriptions of a process in mathematical equations. First principle models are very powerful as they can be employed over a large area of operation and they can be used for various processes. Black box models have limited use, but they are often more compact and simpler in structure.

A first principles model is generally based on laws of conservation (mass, components, energy or impulse). These quantities can be expressed in a balance equation: accumulation = supply – discharge + production – consumption. Often additional equations are needed that connect transfer relationships for heat, mass and impulse, kinetic equations (for reactions), phase equilibrium relationships and/or thermodynamic equations of state.

Suppose we want to set up the dynamic model for a heater. We have a sketch of the process and a block diagram of the process, as shown in Figure 15.5.

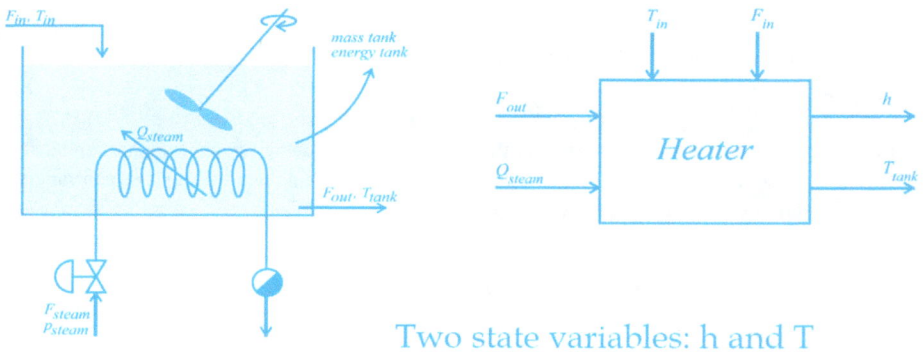

Figure 15.5: Process diagram and block diagram of a heater.

There are two state equations, for the liquid level in the heater (h) and the temperature of the liquid in the heater (T). If a state variable is known at time t, then the behavior is entirely determined by the process inputs; knowledge of the past is actually not required. The state equations are based on the balances for mass (h) and energy (T), i.e.,

$$A\frac{dh}{dt} = F_{in} - F_{out},\tag{15.1}$$

$$Ah\frac{dT}{dt} = F_{in}(T_{in} - T) - \frac{Q}{\rho C_P}.\tag{15.2}$$

The model outputs are a function of the states, the control inputs and the disturbances. If the derivatives are zero an equilibrium (steady state) is reached. If a small change in a process input is made, a new equilibrium or steady state can be reached. For example, by increasing the inflow of the boiler a bit, the level and the temperature will move to new stationary values.

Figure 15.6 shows how all inputs and outputs can be connected via arcs and circles into a so-called behavioral diagram.

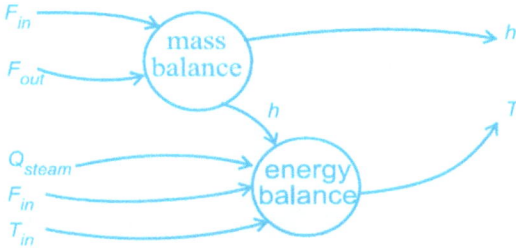

Figure 15.6: Behavioral diagram of a boiler.

In the current version of the dynamic boiler model there are two equations (mass and energy balance), two unknowns (h and T), two disturbances (F_{in} and T_{in}), three constants (ρ, A and C_p) and two degrees of freedom (F_{out} and Q).

Of course this model can be expanded (more detail), for example by assuming that there is a static relationship between the steam flow and the heating duty or by assuming that the steam flow is affected by the valve stem position. Then the model would include a steam balance and a valve equation and if we also ignore heat transfer through the wall, we have

$$C\frac{dP}{dt} = F_{steam} - \frac{Q}{\Delta H},\tag{15.3}$$

$$F_{steam} = \frac{c_v(x)\sqrt{P_{net} - P_{steam}}}{\rho_{steam}},\tag{15.4}$$

$$Q = UA(T_{in} - T).\tag{15.5}$$

In this case the behavioral diagram also has to be modified, as shown in Figure 15.7.

The starting point for a behavioral model is the block diagram (zoom in). The next step is to write down the conservation balances and add the additional equations.

One should always try to keep the model as simple as possible, by making the right assumptions. Accumulations should be taken into account when they are important for control, and accumulations should be ignored when they are small as compared to other accumulations. The constants for the operating region, e.g., physical properties, should be determined, and the model should only be expanded if required. Lumped parameters should be used as much as possible to keep the model compact.

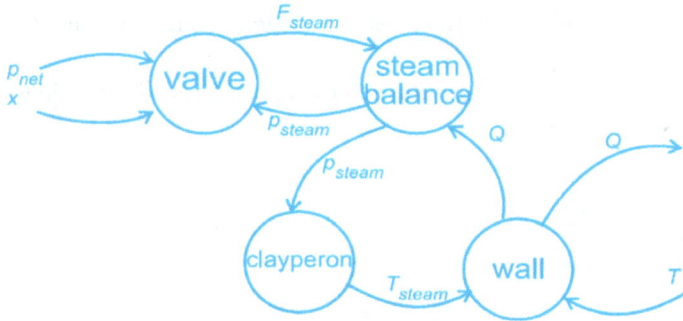

Figure 15.7: Modified behavioral diagram for the boiler.

15.7 Linearizations and Laplace transforms

For a control systems engineer, the representation of a model via (ordinary) differential equations is not very convenient. Often these equations are translated to the Laplace domain. The main idea is that Laplace transforms of differential equations lead to algebraic equations which can be handled and solved much easier. But Laplace transforms only work when the differential equations are linear. If the differential equations are non-linear, first a linearization via a Taylor series has to be carried out.

Suppose we want to describe the tank level in a tank as a function of time, inflow and outflow. We could setup a mass balance, *i. e.*,

$$A\frac{dh}{dt} = F_{\text{in}} - \beta\sqrt{h}. \tag{15.6}$$

Equation (15.6) assumes that the tank level outflow is related to the tank level via Torricelli's law. The process control engineer wants to convert this equation to the Laplace domain. Equation (15.6) is a non-linear differential equation, so the first thing to be done is making the differential equation linear. We have

$$-A\frac{dh}{dt} + F_i = \beta\sqrt{h} \tag{15.7}$$

$$\approx \beta\sqrt{h_0} + \left[\frac{d\beta\sqrt{h}}{dh}\right]_{h=h_0}(h - h_0) + \left[\frac{d^2\beta\sqrt{h}}{dh^2}\right]_{h=h_0}(h - h_0)^2 \tag{15.8}$$

$$\approx \beta\sqrt{h_0} + \frac{\beta}{2\sqrt{h_0}}(h - h_0). \tag{15.9}$$

In other words,

$$A\frac{d(h_0 + \delta h)}{dt} = F_{i0} + \delta F_i - \beta\sqrt{h_0} + \frac{\beta}{2\sqrt{h_0}}(h_0 + \delta h - h_0), \tag{15.10}$$

or

$$A\frac{d(\delta h)}{dt} = \delta F_i - \frac{\beta}{2\sqrt{h_0}}(h - h_0), \tag{15.11}$$

which is equal to

$$A\frac{d(\delta h)}{dt} = \delta F_i - \left[\frac{d\beta\sqrt{h}}{dh}\right]_{h=h_0}\delta h. \tag{15.12}$$

A Taylor series expansion (linearization) only works if we linearize a small area around a working point (denoted with h_0). The result of equation (15.12) is now linear (and depends on what are called difference variables δh and δF_i) and can be converted to the Laplace domain (using the Laplace transform of a derivative):

$$As\delta h = \delta F_i - \frac{\beta}{2\sqrt{h_0}}\delta h, \tag{15.13}$$

where s is the Laplace operator (or, quick and dirty, the d/dt). Equation (15.13) can be rearranged into

$$\left(As + -\frac{\beta}{2\sqrt{h_0}}\right)\delta h = \delta F_i \rightarrow \frac{\delta h}{\delta F_i} = \frac{1}{\left(As + -\frac{\beta}{2\sqrt{h_0}}\right)} = \frac{\left[\frac{1}{\frac{\beta}{2\sqrt{h_0}}}\right]}{\left[\frac{A}{\frac{\beta}{2\sqrt{h_0}}}s + 1\right]} = \frac{K}{\tau s + 1}. \tag{15.14}$$

The final result, equation (15.14), is called a transfer equation and shows how a small change in the inflow affects the tank level. The lumped parameters K and τ are called the process gain and time constant of the process. The process gain gives information on how large the change caused by a disturbance of the inflow on the tank level is, and the time constant gives information on how fast the change of the disturbance is.

An operator in a plant can conduct a simple step change experiment that will enable the engineer to quickly determine the process gain and time constant of the process. Such information can be used to set up straightforward process models for control and optimization.

It is noted that many processes have more complex transfer functions that exhibit "higher-order" dynamics.

15.8 Basic control loops

With dynamic process models in place the control systems engineer can set up control loops for the process. The most common control loop is called the feedback control loop. Feedback control is as old as civilization, or at least 1000 years old. Figure 15.8 shows a klepshydra, which is ancient Greek for "waterthief"!

The klepshydra is a device to record time on the basis of water drops falling in a controlled way into a reservoir. The picture of Figure 15.8 is a copy of a medieval Chinese manual of constructing a water clock. The klepshydra is one of the first recorded studies on feedback control.

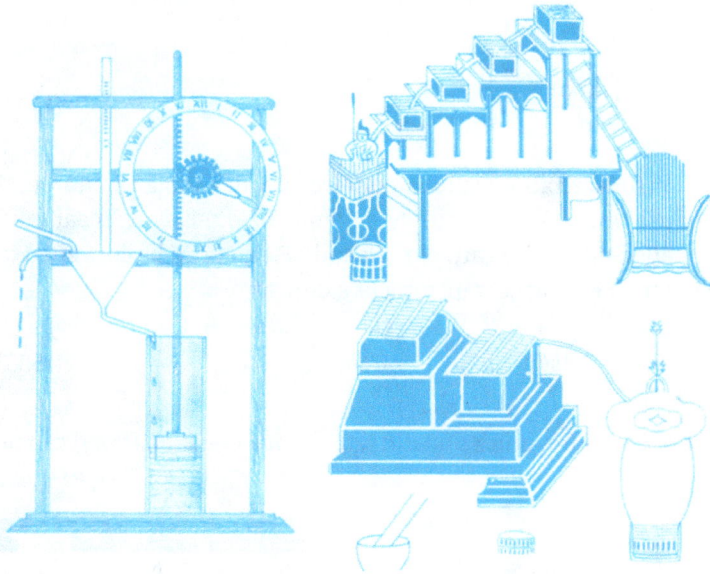

Figure 15.8: An ancient water clock.

The most common concept for control is the feedback controller, as shown in Figure 15.9 (controlling the level in the tank). Most processes can be controlled with feedback controllers. The corrective action for a feedback controller is taken after a few deviates from the setpoint, which is normally not a problem.

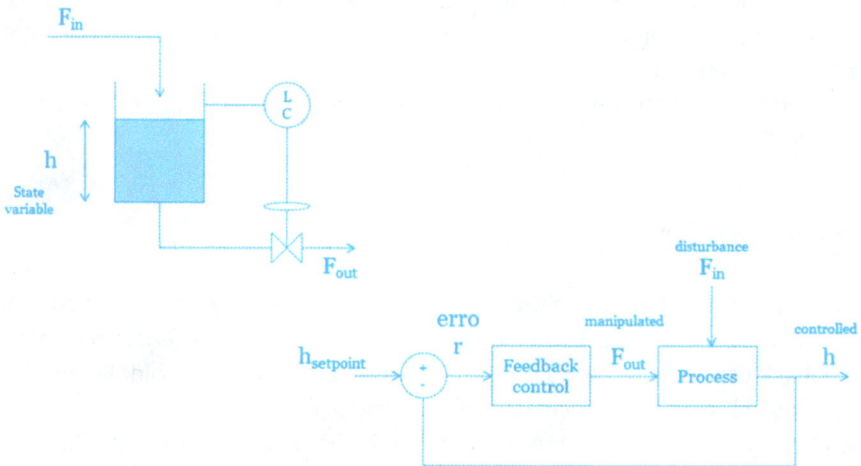

Figure 15.9: The feedback control concept: control the level in a tank by manipulating the outflow of the tank. The inflow is the disturbance here.

The control action can be calculated with the PID controller equation

$$\delta u = K_p\left(1 + \frac{1}{\tau\_S} + \tau_d s\right)\delta\epsilon. \tag{15.15}$$

The control action (δu) depends on the mismatch between the actual process value and the set point or error ($\delta\epsilon$). There are three constants in equation (15.15), the proportional gain (K_p), the integral gain (τ_i) and the derivative gain (τ_d), which lead to the proportional integral derivative (PID) controller. The appropriate values for these constants can be found by trial-and-error or by using so-called tuning protocols, like the Ziegler–Nichols method.

Contrary to feedback control there exists also a feedforward controller. The feedforward controller measures the disturbance and immediately takes action. This makes feedforward control very fast, but differences between setpoint and the actual (steady-state) output signal might be large (this is called an offset).

Feedforward control is used frequently in cases of slow changing processes where the state variable responds slowly to a disturbance (e. g., in distillation columns). Figure 15.10 show the feedforward control loop.

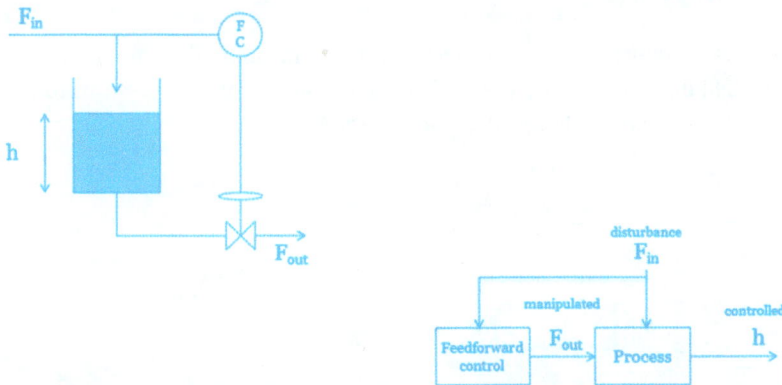

Figure 15.10: The feedforward control loop, where disturbances in the inflow are measured and used to calculate corrective actions for the outflow, in such way that a desired tank level can be maintained.

The feedforward controller depends on the actual process model of the disturbance:

$$G_{FF} = -\frac{G_1}{G_2},$$

where G_1 is the transfer function between the tank level and the inflow and G_2 is the transfer function between the tank level and the outflow.

Of course it is possible to combine feedback control with feedforward control, as shown in Figure 15.11.

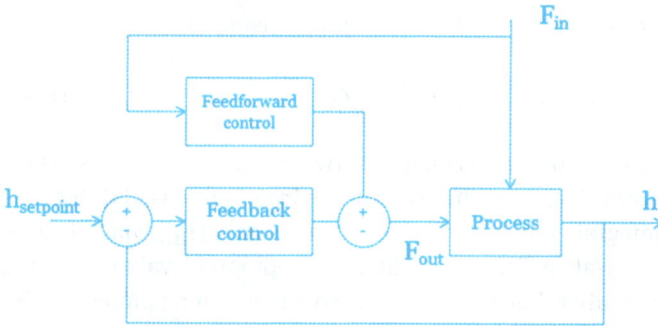

Figure 15.11: Combined feedback and feedforward control.

There are many more advanced control concepts that can be used when the process has specific issues. For example, the cascade control loop can be of use if there are more outputs to control than there are inputs. And the Smith predictor is a very useful model predictive controller that filters out the effect of dead times in the process.

15.9 Sensors and valves

The big four sensors for measuring quantities in a process measure flow, temperature, pressure and level. Figure 15.12 shows the basic measurements for flow (venturi tube), temperature (thermo couple) and pressure (membrane).

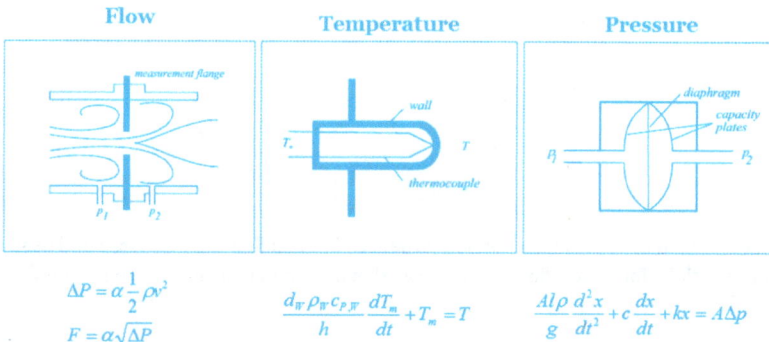

Flow	Temperature	Pressure

$$\Delta P = \alpha \frac{1}{2} \rho v^2$$
$$F = \alpha \sqrt{\Delta P}$$

$$\frac{d_w \rho_w c_{P,w}}{h} \frac{dT_m}{dt} + T_m = T$$

$$\frac{Al\rho}{g} \frac{d^2 x}{dt^2} + c \frac{dx}{dt} + kx = A\Delta p$$

Figure 15.12: Main measurement devices in the process industry.

Control valves can be categorized basically into air-to-open (Fail close) and air-to-close (Fail open) valves. These two valves are controlled pneumatically or electrically and open, or close, when air pressure or electricity is cut off (in case of emergency). This will ensure that components are confided or released in the case of an accident. The stem position of the valve could be linear or exponentially related to the flow through the valve. The main control valve concepts are shown in Figure 15.13.

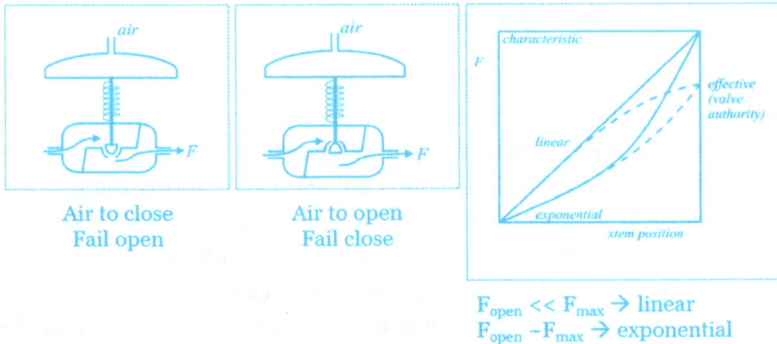

$$F_{open} << F_{max} \rightarrow linear$$
$$F_{open} - F_{max} \rightarrow exponential$$

Figure 15.13: Main control valve concepts.

15.10 Process interlocks

Interlocks prevent operators from departing from the set procedure. These interlocks may be pneumatic, electric, mechanical or programmed. For example, interlocks prevent operators from entering values higher or lower than a set, reasonable value. For example, interlocks may prevent an operator from entering a value for reactor temperature beyond the capabilities of the associated heating unit or beyond safe operating limits for the reactor.

Alarms and safety trips

Alarms are intended to alert operators to deviations in process conditions that have the potential to severely impact product quality or safety. Alarms can create audio or visual cues to alert the operator about the situation. Alternatively, alarm systems can be programmed as automatic trip systems that do not need operator input, especially in situations where lack of response from the operator could lead to serious issues. An automatic trip system is composed of a sensor, a link that transfers the signal and an actuator to perform the desired action.

Safety instrumented system (SIS)

A safety instrumented system (SIS) is initiated when a critical process variable exceeds allowable limits to try to get the process back to a safe condition. The SIS involves extreme action such as starting or stopping a pump or completely shutting down a process unit. Because such actions may have a significant negative impact on the process equipment and/or the product quality, the SIS is used as a last resort for safety. In order for the SIS to effectively serve as a back-up to BPCS, it must be able to function independently of the BPCS. Thus, it is recommended that the SIS have separate sensors and actuators. This is essential in cases such as a power outage, which would take out the BPCS entirely. Because the SIS involves more drastic measures, there tends to be redundant instrumentation to prevent an errant signal from kicking in the SIS.

15.11 Process control over the entire process

Often process design and process control are seen as activities independent of the engineer, but this leads to mismatches and errors. That is why process control is seen as the panacea for mistakes that were made in the design process.

It is important that process control questions are addressed during the process design stages. As we move from conceptual design towards actual production, the impact of the decisions we make decreases, while the actual expenditure is increasing. Four basic stages of process design can be recognized: (1) the conceptual design stage, where a selection of alternative material pathways and flowsheets have to be made (typically steady-state calculations); (2) the preliminary steady-state design, where feasibility studies are done, the unit operations are selected and the heat integration is realized (steady state); (3) the detailed steady-state design, which leads to optimization of key process variables and addresses the sensitivity of disturbances and uncertainties (steady-state but moving towards dynamics); and (4) the controllability analysis and control systems design, where the flowsheet controllability is evaluated. Transient responses have to be generated and the control system needs to be configured (dynamics).

The design of a control system for a plant is guided by the objective to maximize profits by transforming raw materials into useful products while satisfying product specifications, safety, operational constraints and environmental regulations. The mapping of inputs, outputs, disturbances and the degrees of freedom can be done via a set of selection rules.

Selection of controlled variables:

Rule 1: *Select variables that are not self-regulating.*

Rule 2: *Select output variables that would exceed the equipment and operating constraints without control.*

Rule 3: *Select output variables that are a direct measure of the product quality or that strongly affect it.*

Rule 4: *Choose output variables that seriously interact with other controlled variables.*

Rule 5: *Choose output variables that have favorable static and dynamic responses to the available control variables.*

Selection of manipulated variables:

Rule 6: *Select inputs that significantly affect the controlled variables.*

Rule 7: *Select inputs that rapidly affect the controlled variables.*

Rule 8: *The manipulated variables should affect the controlled variables directly rather than indirectly.*

Rule 9: *Avoid recycling disturbances.*

Selection of measured variables:

Rule 10: *Reliable, accurate measurements are essential for good control.*
Rule 11: *Select measurement points that are sufficiently sensitive.*
Rule 12: *Select measurement points that minimize time delays and time constants.*

After the identification and selection of the appropriate variables in the process a plant-wide control design can be made. Luyben *et al.* (1999) suggest a method for the conceptual design of plant-wide control systems, which consists of the following steps:

Step 1: *Establish the control objectives.*

Step 2: *Determine the control degrees of freedom,* or simply stated, the number of control valves – with additions if necessary.

Step 3: *Establish the energy management system.* Regulation of exothermic or endothermic reactors and placement of controllers to attenuate temperature disturbances.

Step 4: *Set the production rate.*

Step 5: *Control the product quality and handle safety, environmental and operational constraints.*

Step 6: *Fix a flow rate in every recycle loop and control vapor and liquid inventories (vessel pressures and levels).*

Step 7: *Check component balances. Establish control to prevent the accumulation of individual chemical species in the process.*

Step 8: *Control the individual process units. Use remaining degrees of freedom to improve local control, but only after resolving more important plant-wide issues.*

Step 9: *Optimize economics and improve dynamic controllability. Add nice-to-have options with any remaining degrees of freedom.*

As an example we can look at an acyclic process.

Example 2 (Plant-wide control of an acyclic process).

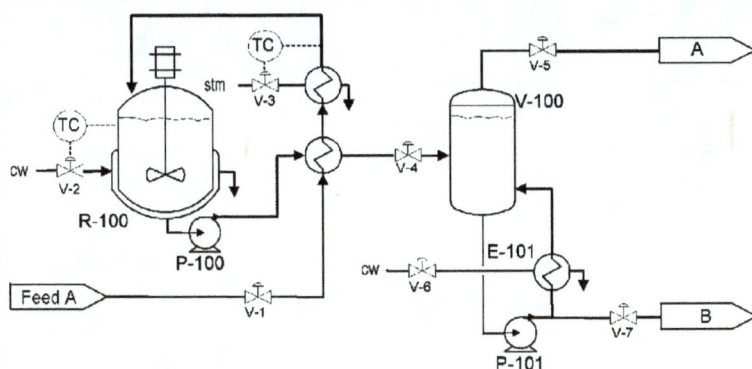

Steps 1 & 2: *Establish the control objectives and degrees of freedom*
1. Maintain a constant production rate.
2. Achieve constant composition in the liquid effluent from the flash drum.
3. Keep the conversion of the plant at its highest permissible value.

Step 3: *Establish energy management system*
1. Need to control reactor temperature: Use V-2.
2. Need to control reactor feed temperature: Use V-3.

Step 4: *Set the production rate*
1. For on-demand product: Use V-7.

Step 5: *Control product quality and meet safety, environmental and operational constraints*
☆ To regulate V-100 pressure: Use V-5.
☆ To regulate V-100 temperature: Use V-6.

Step 6: *Fix recycle flow rates and vapor and liquid inventories*
2. Need to control vapor inventory in V-100: Use V-5 (already installed).
3. Need to control liquid inventory in V-100: Use V-4.
4. Need to control liquid inventory in R-100: Use V-1.

Step 7: *Check component balances (N/A)*

Step 8: *Control the individual process units (N/A)*

Step 9: *Optimization*
1. Install composition controller, cascaded with TC of reactor.

Exercises

1. Consider the mixing vessel shown in the figure below. The feed stream flow rate, F1, and composition, C1, are considered to be disturbance variables. The feed is mixed with a control stream of flow rate F2 and constant known composition C2. To ensure a product of constant composition, it is also possible to manipulate the flow rate F3 of the product stream. Perform a degrees-of-freedom analysis and suggest alternative control system configurations. Note that unsteady-state balances are required.

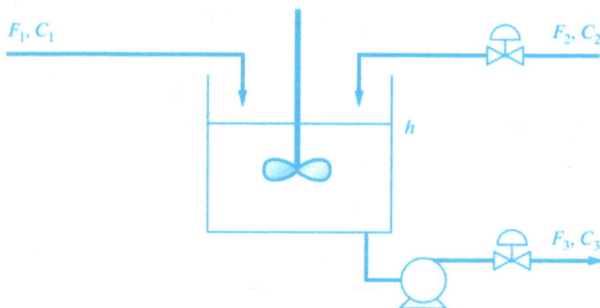

2. A control system is suggested for the exothermic reactor in the figure below. Suggest alternative configurations and compare them with the original configuration.

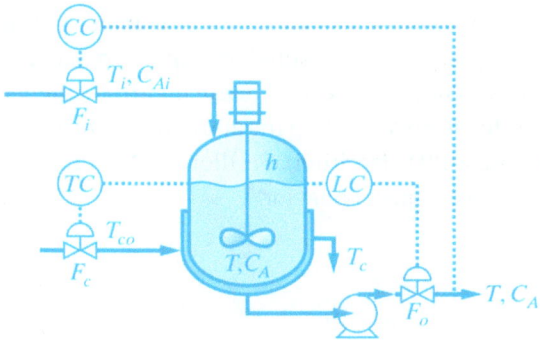

3. The figure below shows a process for the isothermal production of C from A and B (A + B → C). The two reagents are fed to a CSTR R-100, where complete conversion of B is assumed. The reactor effluent stream, consisting of C and unreacted A, is separated in a distillation column, T-100, where the more volatile A is withdrawn in the distillate and recycled, and product C is withdrawn in the bottom stream. Your task is to devise a conceptual plant-wide control system for the process. Hint: It may be helpful to reposition the feed stream of A.

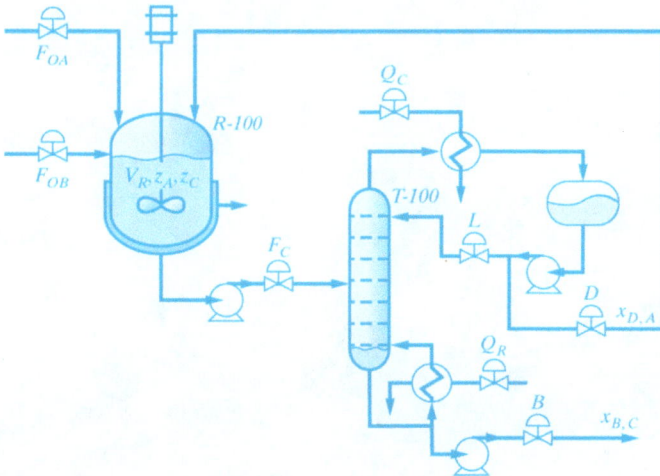

15.12 Take-away message

Plant-wide control concerns the extension of the flowsheet with sensoring and actuation. Control is used to deal with variations of important process variables over time. Control is needed throughout a process to guarantee safety, to reach product specifications, to deal with fluctuations in inlet feed conditions, to protect equipment from damage and to deal with flaws during the process design stages. It is important to establish first the goal for control. Subsequently the measured, manipulated and disturbance variables have to be determined. These can be used to calculate the number of degrees of freedom, which allows us to effectively set up the allowed control loops. Often feedback control is employed (with a PID algorithm). Feedforward control and model-based controllers are used for more complex behavior.

15.13 Further reading

Luyben (1990). *Process modeling, simulation and control for chemical engineers*, 2nd ed. McGraw-Hill, New York.

Roffel & Betlem (2003). *Advanced practical process control*. Springer.

Roffel & Betlem (2006). *Process dynamics and control – modeling for control and prediction*. Wiley.

Marlin (2000). *Process control – designing processes and control systems for dynamic performance*, 2nd ed. McGraw-Hill, Europe.

16 Heat integration

16.1 Pinch analysis

Heat and power integration is concerned with using the energy in the high-temperature streams that need to be cooled or condensed to heat and/or vaporize the cold streams and provide power to compressors for turbines and heat engines where possible. In most designs it is common to disregard power demands favoring the designing of an effective network by heat exchangers without using the energy of the high-temperature streams to produce power. In this chapter we will discuss how pinch analysis can be used as a tool to achieve heat integration in the process. This chapter is based on the extensive work of Linhoff and Townsend and *The chemical engineers resource*.

The term "pinch analysis" was first coined in the 1970s by Linnhoff to represent a new set of thermodynamically based methods that guarantee minimum energy levels in the design of heat exchanger networks (HENs). Since the 1970s it has become a standard method in process design. The term "pinch analysis" is often used to represent the application of the tools and algorithms for studying industrial processes.

Pinch technology presents a simple methodology for systematically analyzing chemical processes and the surrounding utility systems with the help of the First and Second Laws of Thermodynamics.

The First Law of Thermodynamics provides the energy equation for calculating the enthalpy changes (H) in the streams passing through a heat exchanger. The Second Law determines the direction of heat flow. That is, heat energy may only flow in the direction of hot to cold. This prohibits "temperature cross-overs" of the hot and cold stream profiles through the exchanger unit.

In a heat exchanger a hot stream cannot be cooled below the cold stream supply temperature, nor can a coldstream be heated to a temperature higher than the supply temperature of the hot stream. In practice the hot stream can only be cooled to a temperature defined by the "temperature approach" of the heat exchanger. The temperature approach is the minimum allowable temperature difference (ΔT_{min}) in the stream temperature profiles for the heat exchanger unit. The temperature level at which ΔT_{min} is observed in the process is referred to as "pinch point." The pinch defines the minimum driving force allowed in the exchanger unit.

16.2 Motivation for heat integration by pinch analysis

Consider the simple process in Figure 16.1, where the feed stream to a reactor is heated before the inlet to a reactor and the product stream has to be cooled. The heating and cooling are done by use of steam (Heat Exchanger-1) and cooling water (Heat Exchanger-2), respectively.

https://doi.org/10.1515/9783110570137-016

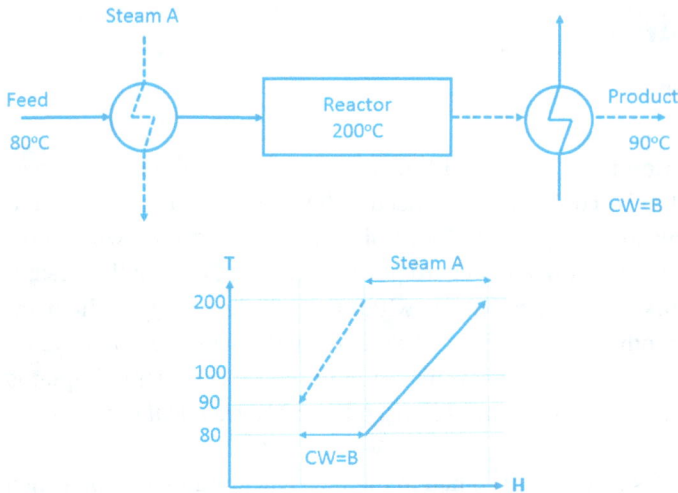

Figure 16.1: Heat integration example of a reactor. Dashed lines, cold streams; continuous lines, hot streams.

The temperature (T) *versus* enthalpy (H) plot for the feed and product streams depicts the hot (Steam) and cold (CW) utility loads when there is no vertical overlap of the hot and cold stream profiles.

An alternative, improved scheme is shown in Figure 16.2. The addition of a new "Heat Exchanger-3" recovers product heat (X) to preheat the feed. The steam and cooling water requirements are also reduced by the same amount (X). The amount of heat recovered (X) depends on the "minimum approach temperature" allowed for the new exchanger. The minimum temperature approach between the two curves on the vertical axis is ΔT_{min} and the point where this occurs is defined as the "pinch." From the $T-H$ plot, the X amount corresponds to a ΔT_{min} value of 20 °C. Increasing the ΔT_{min} value leads to higher utility requirements and lower area requirements.

When the process involves single hot and cold streams (as in the above example) it is easy to design an optimum heat recovery exchanger network intuitively by heuristic methods. In any industrial set-up the number of streams is so large that the traditional design approach has been found to be limiting in the design of a good network. With the development of pinch technology in the late 1980s, not only optimal network design was made possible, but also considerable process improvements could be discovered.

16.3 The pinch analysis approach

Process integration using pinch technology offers a novel approach to generate targets for minimum energy consumption before heat recovery network design. Heat recovery and utility system constraints are then considered in the design of the core process.

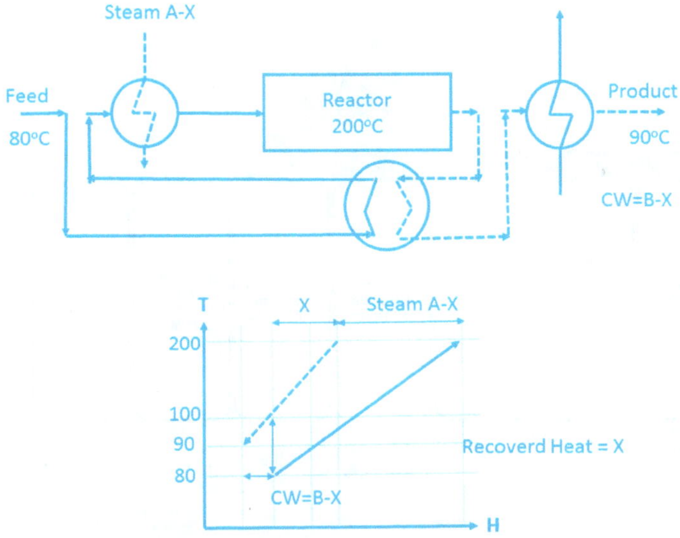

Figure 16.2: Improved heat integration example of a reactor. Dashed lines, cold streams; continuous lines, hot streams.

Interactions between the heat recovery and utility systems are also considered. The pinch design can reveal opportunities to modify the core process to improve heat integration. The pinch approach is unique because it treats all processes with multiple streams as a single, integrated system. This method helps to optimize the heat transfer equipment during the design of the equipment.

Most industrial processes involve transfer of heat either from one process stream to another process stream (interchanging) or from a utility stream to a process stream. In the present energy crisis scenario all over the world, the target in any industrial process design is to maximize the process-to-process heat recovery and to minimize the utility (energy) requirements. To meet the goal of maximum energy recovery or minimum energy requirement (MER) an appropriate HEN is required. The design of such a network is not an easy task considering the fact that most processes involve a large number of process and utility streams. With the advent of pinch analysis concepts, the network design has become very systematic and methodical.

Pinch analysis follows a well-defined procedure (Figure 16.3). It should be noted that these steps are not necessarily performed on a once-through basis, independent of one another. Additional activities such as resimulation and data modification occur as the analysis proceeds and some iteration between the various steps is always required.

Step 1: identification of the hot, cold and utility streams in the process

"Hot Streams" are those that must be cooled or are available to be cooled, *e. g.*, product cooling before storage. "Cold Streams" are those that must be heated, *e. g.*, feed pre-

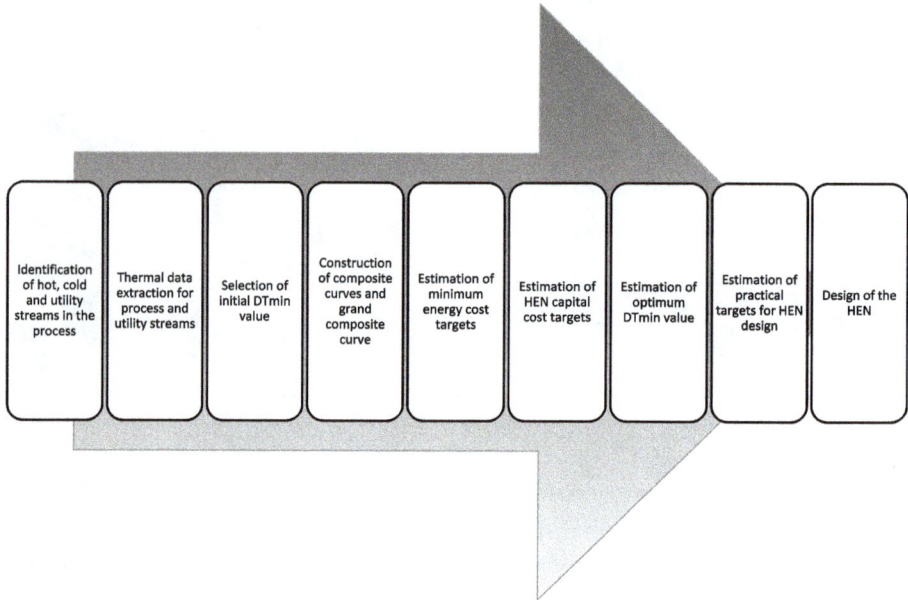

Figure 16.3: Steps of pinch analysis.

heat before a reactor. "Utility Streams" are used to heat or cool process streams when heat exchange between process streams is not practical or economic. A number of different hot utilities (steam, hot water, flue gas, etc.) and cold utilities (cooling water, air, refrigerant, etc.) are used in industry.

The identification of streams needs to be done with care as sometimes, despite undergoing changes in temperature, the stream is not available for heat exchange. For example, when a gas stream is compressed, the stream temperature rises because of the conversion of mechanical energy into heat and not by any fluid-to-fluid heat exchange. Hence such a stream may not be available to take part in any heat exchange. Such a stream may or may not be considered to be a process stream.

Step 2: thermal data extraction for process and utility streams
For each hot, cold and utility stream identified, the following thermal data are extracted from the process material and heat balance flow sheet:
- supply temperature (T_S, °C): the temperature at which the stream is available;
- target temperature (T_T, °C): the temperature the stream must be taken to;
- heat capacity flow rate (c_P, kW/°C): the product of flow rate (m) in kg/sec and specific heat (C_P, kJ/kg °C), where we have

$$c_P = mC_p; \tag{16.1}$$

- enthalpy change (H) associated with a stream passing through the exchanger is given by the First Law of Thermodynamics:

$$H = Q \pm W. \tag{16.2}$$

In a heat exchanger, no mechanical work is being performed, so

$$W = 0. \tag{16.3}$$

The above equation simplifies to

$$H = Q, \tag{16.4}$$

where Q represents the heat supply or demand associated with the stream. It is given by the relationship

$$Q = C_P(T_S - T_T) \tag{16.5}$$

and the enthalpy change

$$H = C_P(T_S - T_T). \tag{16.6}$$

Here the specific heat values have been assumed to be temperature-independent within the operating range. The stream data and their potential effect on the conclusions of a pinch analysis should be considered during all steps of the analysis.

Any erroneous or incorrect data can lead to false conclusions. To avoid mistakes, the data extraction is based on certain qualified principles.

Step 3: selection of initial ΔT_{min} value

The design of any heat transfer equipment must always adhere to the Second Law of Thermodynamics, which prohibits any temperature cross-over between the hot and the cold stream, *i. e.*, a minimum heat transfer driving force must always be allowed for a feasible heat transfer design. Thus the temperature of the hot and cold streams at any point in the exchanger must always have a minimum temperature difference (ΔT_{min}). This ΔT_{min} value represents the bottleneck in the heat recovery.

The value of ΔT_{min} is determined by the overall heat transfer coefficient (U) and the geometry of the heat exchanger. In a network design, the type of heat exchanger to be used at the pinch will determine the practical ΔT_{min} for the network. For example, an initial selection for the ΔT_{min} value for shells and tubes may be 3–5 °C (at best) while compact exchangers such as plate and frame often allow for an initial selection of 2–3 °C. The heat transfer equation, which relates Q, U, A and the log mean temperature difference (LMTD), is

$$\text{LMTD} = \frac{(T_{SH} - T_{TC})(T_{TH} - T_{SC})}{\ln(\frac{T_{SH}-T_{TC}}{T_{TH}-T_{SC}})}, \tag{16.7}$$

with

$$LQ = U * A * \text{LMTD}. \tag{16.8}$$

The temperatures under consideration are given in Figure 16.4.

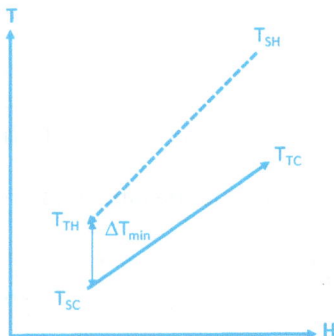

Figure 16.4: Heat transfer relations.

For a given value of heat transfer load (Q), if smaller values of $\Delta T_{\min}$ are chosen, the area requirements rise. If a higher value of $\Delta T_{\min}$ is selected the heat recovery in the exchanger decreases and demand for external utilities increases. Thus, the selection of the $\Delta T_{\min}$ value has implications for both capital and energy costs.

This concept will become clearer with the help of composite curves and total cost targeting discussed later. Just as for a single heat exchanger, the choice of $\Delta T_{\min}$ (or approach temperature) is vital in the design of a HEN. To begin the process an initial $\Delta T_{\min}$ value is chosen and pinch analysis is carried out. Typical $\Delta T_{\min}$ values based on experience are available in literature for reference.

Step 4: construction of composite curves and Grand Composite Curve
Temperature-enthalpy $(T–H)$ plots, known as "composite curves," have been used for many years to set energy targets ahead of design. Composite curves consist of temperature (T)-enthalpy (H) profiles of heat availability in the process (the hot composite curve) and heat demands in the process (the cold composite curve) together in a graphical representation.

In general any stream with a constant heat capacity (CP) value is represented on a $T–H$ diagram by a straight line running from stream supply temperature to stream target temperature.

When there are a number of hot and cold streams, the construction of hot and cold composite curves simply involves the addition of the enthalpy changes of the streams in the respective temperature intervals. An example of hot composite curve construction is shown in Figure 16.5. A complete hot or cold composite curve consists of a series of connected straight lines; each change in slope represents a change in overall hot stream heat capacity flow rate (C_p).

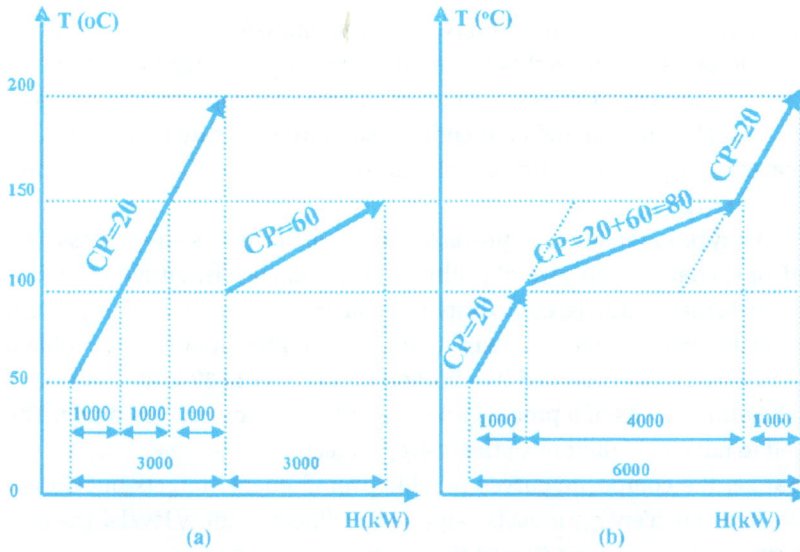

Figure 16.5: Temperature-enthalpy relations used to construct composite curves.

For heat exchange to occur from the hot stream to the cold stream, the hot stream cooling curve must lie above the cold stream heating curve. Because of the "kinked" nature of the composite curves (Figure 16.6), they approach each other most closely at one point defined as the minimum approach temperature (ΔT_{min}). ΔT_{min} can be measured directly from the $T-H$ profiles as being the minimum vertical difference between the hot and cold curves. This point of minimum temperature difference represents a bottleneck in heat recovery and is commonly referred to as the "pinch."

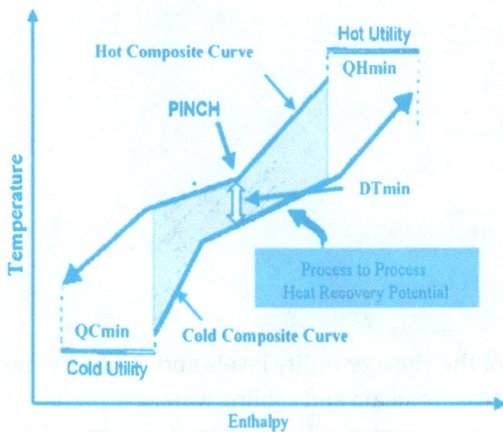

Figure 16.6: Combined composite curves.

Increasing the ΔT_{min} value results in shifting of the curves horizontally apart, resulting in lower process-to-process heat exchange and higher utility requirements. At a particular ΔT_{min} value, the overlap shows the maximum possible scope for heat recovery within the process. The hot end and cold end overshoot the indicated minimum hot utility requirement (QH_{min}) and minimum cold utility requirement (QC_{min}) of the process for the chosen ΔT_{min}.

Thus, the energy requirement for a process is supplied via process-to-process heat exchange and/or exchange with several utility levels (steam levels, refrigeration levels, hot oil circuit, furnace flue gas, etc.). Graphical constructions are not the most convenient means of determining energy needs. A numerical approach called the Problem Table Algorithm (PTA) was developed by Linnhoff and Flower in 1978 as a means of determining the utility needs of a process and the location of the process pinch. The PTA lends itself to hand calculations of the energy targets.

To summarize, the composite curves provide overall energy targets but do not clearly indicate how much energy must be supplied by different utility levels. The utility mix is determined by the Grand Composite Curve.

In selecting utilities to be used, determining utility temperatures and deciding on utility requirements, the composite curves and PTA are not particularly useful. A new tool, the Grand Composite Curve (GCC), was introduced in 1982 by Itoh, Shiroko and Umeda. The GCC (Figure 16.7) shows the variation of heat supply and demand within the process. Using this diagram the designer can find which utilities are to be used.

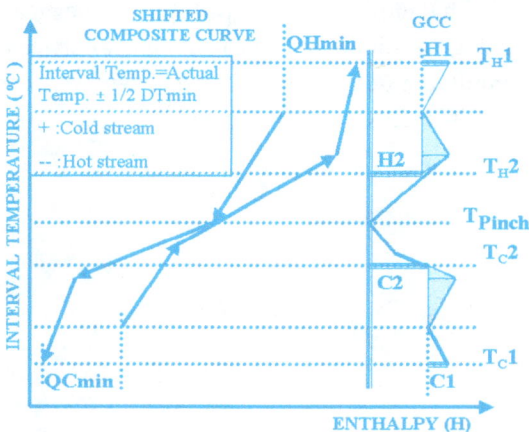

Figure 16.7: The Grand Composite Curve.

The designer aims to maximize the use of the cheaper utility levels and minimize the use of the expensive utility levels. Low-pressure steam and cooling water are preferred over high-pressure steam and refrigeration, respectively. The information required for the construction of the GCC comes directly from the PTA developed by Linnhoff and

Flower. The method involves shifting (along the temperature [vertical] axis) the hot composite curve down by $\frac{1}{2}\Delta T_{min}$ and that of the cold composite curve up by $\frac{1}{2}\Delta T_{min}$. The vertical axis on the shifted composite curves shows the process interval temperature.

The curves are shifted by subtracting part of the allowable temperature approach from the hot stream temperatures and adding the remaining part of the allowable temperature approach to the cold stream temperatures. The result is a scale based on process temperature having an allowance for temperature approach (ΔT_{min}).

The GCC is then constructed from the enthalpy (horizontal) differences between the shifted composite curves at different temperatures. On the GCC, the horizontal distance separating the curve from the vertical axis at the top of the temperature scale shows the overall hot utility consumption of the process.

Figure 16.7 shows that it is not necessary to supply the hot utility at the top temperature level. The GCC indicates that we can supply the hot utility over two temperature levels, TH1 (high-pressure steam) and TH2 (low-pressure steam). Recall that, when placing utilities in the GCC, intervals, and not actual utility temperatures, should be used.

The total minimum hot utility requirement remains the same, i. e., QH_{min} = H1 (HP steam) + H2 (LP steam). Similarly, QC_{min} = C1 (refrigerant) + C2 (CW). The points TH2 and TC2, where the H2 and C2 levels touch the GCC, are called the "utility pinches." The shaded pockets represent the process-to-process heat exchange.

The GCC is one of the most basic tools used in pinch analysis for the selection of the appropriate utility levels and for targeting of a given set of multiple utility levels. The targeting involves setting appropriate loads for the various utility levels by maximizing the least expensive utility loads and minimizing the loads on the most expensive utilities.

Step 5: estimation of minimum energy cost targets
Once the ΔT_{min} value is chosen, minimum hot and cold utility requirements can be evaluated from the composite curves. The GCC provides information regarding the utility levels selected to meet QH_{min} and QC_{min} requirements. If the unit cost of each utility is known, the total energy cost can be calculated from

$$\text{TOTAL ENERGY COST} = \sum_{U=1}^{U} Q_U C_U, \tag{16.9}$$

where Q_U is the duty of utility U (kW), C_U are the unit cost of utility (U) in ($/kW yr) and U is the total number of utilities used.

Step 6: estimation of heat exchanger network capital cost targets
The capital cost of a HEN is dependent on three factors:

1. the number of exchangers;
2. the overall network area;
3. the distribution of area between the exchangers.

Pinch analysis enables targets for the overall heat transfer area and minimum number of units of a HEN to be predicted prior to detailed design. It is assumed that the area is evenly distributed between the units; the area distribution cannot be predicted ahead of design.

The calculation of surface area for a single countercurrent heat exchanger requires the knowledge of the temperatures of streams in and out (TLM, $i. e.$, LMTD), the overall heat transfer coefficient (U) and total heat transferred (Q). The area is given by the relation

$$A = \frac{Q}{UT_{LM}}. \tag{16.10}$$

The composite curves can be divided into a set of adjoining enthalpy intervals such that within each interval, the hot and cold composite curves do not change slope. Here the heat exchange is assumed to be "vertical" (pure countercurrent heat exchange). The hot streams in any enthalpy interval, at any point, exchange heat with the cold streams at the temperature vertically below it.

The total area of the HEN (A_{min}) is given by

$$\text{HEN } AREA_{min} = A_1 + A_2 + \cdots + A_i = \sum_i \left[\frac{1}{\Delta T_{LM}} \sum_j \frac{q_j}{h_j} \right]. \tag{16.11}$$

The actual HEN total area required is generally within 10 % of the area target as calculated above. With inclusion of temperature correction factors area targeting can be extended to non-countercurrent heat exchange as well.

For the minimum number of heat exchanger units (N_{min}) required for MER, the HEN can be evaluated prior to HEN design by using a simplified form of Euler's graph theorem. In designing for the MER, no heat transfer is allowed across the pinch and so a realistic target for the minimum number of units (N_{min}MER) is the sum of the targets evaluated both above and below the pinch separately, $i. e.$,

$$N_{min}\text{MER} = [N_h + N_c + N_u - 1]_{AP} + [N_h + N_c + N_u - 1]_{BP}, \tag{16.12}$$

where N_h is the number of hot streams, N_c is the number of cold streams, N_u is the number of utility streams and AP and BP denotes above and below pinch.

The targets for the minimum surface area (A_{min}) and the number of units (N_{min}) can be combined together with the heat exchanger cost law to determine the targets for HEN capital cost (CHEN). The capital cost is annualized using an annualization factor that takes into account interest payments on borrowed capital. The equation used for calculating the total capital cost and exchanger cost law is

$$C(\$)_{HEN} = \left[N_{min}\left(a + b\left(\frac{A_{min}}{N_{min}} \right)^c \right) \right]_{AP} + \left[N_{min}\left(a + b\left(\frac{A_{min}}{N_{min}} \right)^c \right) \right]_{BP}, \qquad (16.13)$$

where a, b and c are constants in exchanger cost laws. For the exchanger cost equation shown above, typical values for a carbon steel shell and tube exchnager would be $a = 16,000$, $b = 3,200$ and $c = 0.7$. The installed cost can be considered to be 3.5 times the purchased cost given by the exchanger cost equation.

Step 7: estimation of optimum ΔT_{min} value by energy-capital trade-off

To arrive at an optimum ΔT_{min} value, the total annual cost (the sum of total annual energy and capital cost) is plotted for varying ΔT_{min} values (Figure 16.8). Three key observations can be made from Figure 16.8. An increase in ΔT_{min} values results in higher energy costs and lower capital costs, and a decrease in ΔT_{min} values results in lower energy costs and higher capital costs. An optimum ΔT_{min} exists where the total annual cost of energy and capital costs is minimized. Thus, by systematically varying the temperature we can determine the optimum heat recovery level or the optimum ΔT_{min} for the process.

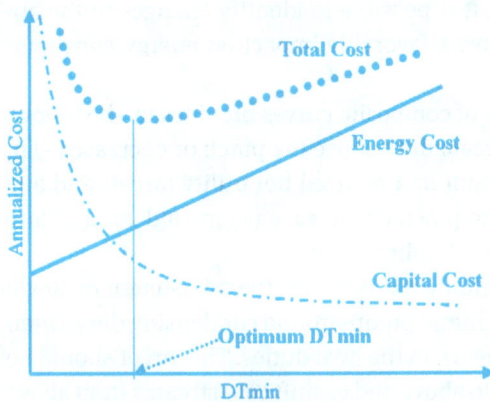

Figure 16.8: Energy-Capital cost trade-off.

Step 8: estimation of practical targets for HEN design

The HEN designed on the basis of the estimated optimum ΔT_{min} value is not always the most appropriate design. A very small ΔT_{min} value, perhaps 8 °C, can lead to a very complicated network design with a large total area due to low driving forces.

The designer, in practice, selects a higher value (15 °C) and calculates the marginal increases in utility duties and area requirements. If the marginal cost increase is small, the higher value of ΔT_{min} is selected as the practical pinch point for the HEN design.

Recognizing the significance of the pinch temperature allows energy targets to be realized by design of appropriate heat recovery networks. The pinch divides the process into two separate systems, each of which is in enthalpy balance with the utility. The pinch point is unique for each process. Above the pinch, only the hot utility is required. Below the pinch, only the cold utility is required. Hence, for an optimum design, no heat should be transferred across the pinch. This is known as the key concept in pinch technology.

Pinch technology gives three rules that form the basis for practical network design:
1. no external heating below the pinch;
2. no external cooling above the pinch;
3. no heat transfer across the pinch.

Violation of any of the above rules results in higher energy requirements than the minimum requirements theoretically possible. The Plus/Minus Principle states the following: The overall energy needs of a process can be further reduced by introducing process changes (changes in the process heat and material balances). There are several parameters that could be changed, such as reactor conversions, distillation column operating pressures and reflux ratios, feed vaporization pressures and pump-around flow rates. The number of possible process changes is nearly infinite. By applying the pinch rules as discussed in this chapter, it is possible to identify changes in the appropriate process parameter that will have a favorable impact on energy consumption.

Applying the pinch rules to the study of composite curves provides us the following guidelines: any increase (+) in hot stream duty above the pinch or decrease (–) in cold stream duty above the pinch will result in a reduced hot utility target, and any decrease (–) in hot stream duty below the pinch or increase (+) in cold stream duty below the pinch will result in a reduced cold utility target.

These simple guidelines provide a definite reference for the adjustment of single heat duties such as vaporization of a recycle and pump-around condensing duty. Often it is possible to change temperatures rather than the heat duties. The target should be to shift hot streams from below the pinch to above and to shift cold streams from above the pinch to below.

The process changes that can help achieve such stream shifts essentially involve changes in the following operating parameters:
1. reactor pressure/temperatures;
2. distillation column temperatures, reflux ratios, feed conditions, pump-around conditions, intermediate condensers;
3. evaporator pressures;
4. storage vessel temperatures.

For example, if the pressure for a feed vaporizer is lowered, vaporization duty can shift from above to below the pinch. The leads to a reduction in both hot and cold utilities.

Apart from the changes in process parameters, proper integration of key equipment in the process with respect to the pinch point should also be considered. The pinch concept of "appropriate placement" (integration of operations in such a way that there is a reduction in the utility requirement of the combined system) is used for this purpose.

Appropriate placement principles have been developed for distillation columns, evaporators, heat engines, furnaces and heat pumps. For example, a single-effect evaporator having equal vaporization and condensation loads should be placed such that both loads balance each other and the evaporator can be operated without any utility costs.

This means that appropriate placement of the evaporator is on either side of the pinch and not across the pinch. In addition to the above pinch rules and principles, a large number of factors must also be considered during the design of heat recovery networks.

The most important are operating cost, capital cost, safety, operability, future requirements and plant operating integrity. Operating costs are dependent on hot and cold utility requirements as well as pumping and compressor costs. The capital cost of a network is dependent on a number of factors, including the number of heat exchangers, heat transfer areas, materials of construction, piping and the cost of supporting foundations and structures.

With a little practice, the above principles enable the designer to quickly plan through 40–50 possible modifications and choose three or four that will lead to the best overall cost effects. The essence of the pinch approach is to explore the options of modifying the core process design, heat exchangers and utility systems with the ultimate goal of reducing the energy and/or capital cost.

Step 9: design of a heat exchanger network

The design of a new HEN is best executed using the "Pinch Design Method" (PDM). The systematic application of the PDM allows the design of a good network that achieves the energy targets within practical limits. The method incorporates two fundamentally important features:

(1) it recognizes that the pinch region is the most constrained part of the problem (consequently it starts the design at the pinch and develops by moving away); and

(2) it allows the designer to choose between match options.

In effect, the design of network examines which "hot" streams can be matched to "cold" streams via heat recovery. This can be achieved by employing "tick off" heuristics to identify the heat loads on the pinch exchanger. Every match brings one stream to its target temperature. As the pinch divides the heat exchange system into two thermally independent regions, HENs of above and below pinch regions are designed separately. When the heat recovery is maximized the remaining thermal needs must be supplied by hot utility. A graphical method of representing flow streams and heat recovery matches is called a "grid diagram" (Figure 16.9).

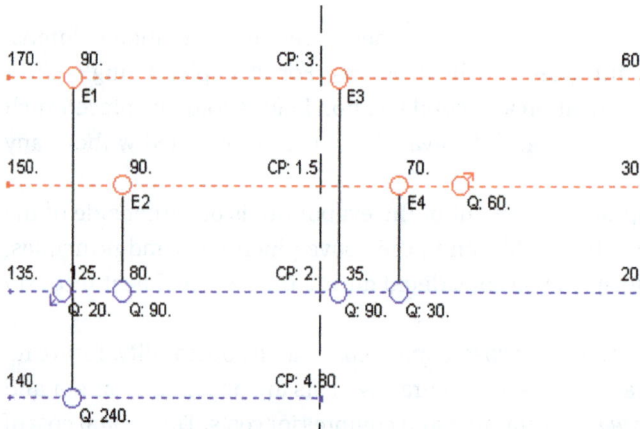

170.   90.   CP: 3.   60.
       E1   E3

150.   90.   CP: 1.5   70.   30.
                E2   E4   Q: 60.

135.   125. 80.   CP: 2.   35.   20.
       Q: 20. Q: 90.   Q: 90.   Q: 30.

140.   CP: 4.30.
       Q: 240.

Figure 16.9: The cold (blue lines) and hot (red lines) streams are represented by horizontal lines. The entrance and exit temperatures are shown at either end. The vertical line in the middle represents the pinch temperature. The circles represent heat exchangers. Unconnected circles represent exchangers using utility heating and cooling. The design of a network is based on certain guidelines, like the "CP Inequality Rule," "StreamSplitting," "Driving Force Plot" and "Remaining Problem Analysis".

Having made all the possible matches, the two designs above and below the pinch are then brought together and usually refined to further minimize the capital cost. After the network has been designed according to the pinch rules, it can be further subjected to energy optimization. Optimizing the network involves both topological and parametric changes of the initial design to minimize the total cost.

Exercises

1. Four streams are to be cooled or heated:

Stream	T_S (°C)	T_T (°C)	C (kW/°C)
H1	180	60	3
H2	150	30	1
C1	30	135	2
C2	80	140	5

(a) For $\Delta T_{min} = 10\,°C$ find the minimum heating and cooling utilities. What are the pinch temperatures?

(b) Design a HEN for MER on both the hot and the cold side of the pinch.

2. Consider the design of a network of heat exchangers that requires the minimum utilities for heating and cooling. It is true that a pinch temperature can occur *only* at the inlet temperature of a hot or a cold stream? Hint: Sketch a typical composite hot and cold curve for two hot and two cold streams.

16.4 Take-away message

The different units in a process all require or deliver heat. The systematic matching of heat requiring and delivering heat streams via heat exchangers is called heat integration. From the First Law of Thermodynamics a very practical strategy for the pairing of heat streams can be proposed: pinch analysis. Heat integration using pinch technology offers a novel approach to generate targets for minimum energy consumption before heat recovery network design. Heat recovery and utility system constraints are considered in the design of the core process. Interactions between the heat recovery and utility systems are also considered. The pinch design can reveal opportunities to modify the core process to improve heat integration.

16.5 Further reading

Linnhoff and Flower (1978). Synthesis of heat exchanger networks: I: systematic generation of energy optimal networks. *AIChE. J.*, 24, p. 633.

Townsend and Linnhoff (1983). Heat and power networks in process design II. Design procedures for equipments selection and process matching. *AIChE. J.*, 29, p. 748.

Linnhoff, Townsend, Boland, Hewitt, Thomas, Guy and Marsland (1994). *A user guide on process integration for the efficient use of energy, Revised*, 1st ed. Rugby, England: The IChemE.

Smith (2005). *Chemical process design and integration*. Southern Gate, Chichester, England: Wiley the Atrium.

17 Process economics and safety

17.1 Process safety

In designing new products and processes, one of the main criteria for decision making is whether or not this process or product can be designed and operated in a safe fashion. Process safety is concerned with preventing fire, explosions and accidental chemical releases in chemical processing facilities or other auxiliary processes dealing with hazardous materials, such as refineries and oil and gas production installations. Safety directly influences the costs of capital and operation.

A common tool to explain and assess the risk for hazards is the Hazard and Operability Analysis (HAZOP). The objective of a HAZOP study is to offer a structured approach to identify deviations from design intent and normal process conditions in a process installation. Process characteristics such as temperature, pressure and composition are combined with deviations such as more, less and reverse in order to identify operational and safety-related problems in a structured way. If necessary, actions with responsible action holders can be defined.

Over the years there have been numerous examples of disasters that were the result of mistakes during the process design stage where safety issues were neglected.

One illustrative example is the Flixborough disaster, which occurred in June 1974. Fifty tons of cyclohexane were released from Nypro's KA plant (oxidation of cyclohexane), leading to the release of a vapor cloud and its detonation. The plant was completely destroyed and there were 28 deadly casualties. The Nypor's KA plant involved a highly reactive system with low conversions and large inventory in the plant. The process involved six 20-ton stirred tank reactors.

The discharge was caused by the failure of a temporary pipe that was installed to replace a cracked reactor; this so-called "dog-leg" was not able to withstand the operating conditions of the process (10 bars, 150 °C). The dog-leg construction is shown in Figure 17.1.

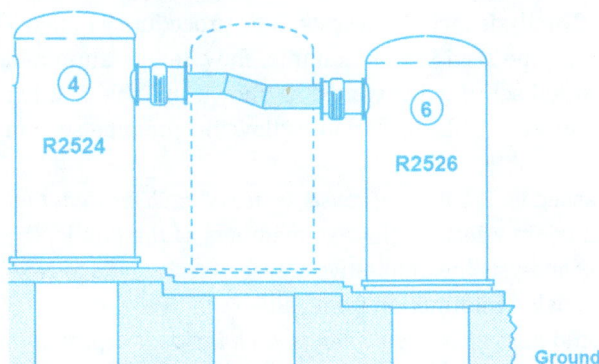

Figure 17.1: Dog-leg pipe replacing a cracked reactor.

https://doi.org/10.1515/9783110570137-017

From the Flixborough disaster it can be learned that processes should be designed with low inventory, especially the process involves storage of flashing fluids or other hazardous substances. Furthermore, it is always worthwile to perform a systematic search for the possible causes of a problem before a process is modified. A HAZOP analysis should be carried out.

17.1.1 HAZOP analysis of a typical pipe section

Normally a HAZOP study is executed for each pipe section in a flowsheet. Figure 17.2 shows a feed section of a process, consisting of a storage tank (1) and a pump (2) which supply organic feed to a distillation unit via a valve (4) as well as to an upstream plant on start-up (5). There is a kick-back line on the pump (6) that guarantees that there is flow through the pump, even when the upstream valve is closed. There is also a check valve (3) on the pump discharging line.

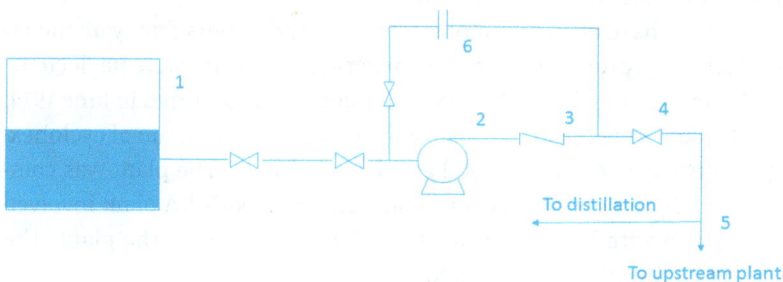

Figure 17.2: Flowsheet of the feed section of a process.

The first question to be asked is: *What could go wrong?* The most important stage in a hazard study is to determine the principal causes of failure that lead to accidents or operating problems. A HAZOP analysis includes a systematic procedure where relevant variables associated with a pipe section (for example, flow, temperature and pressure) are queried with a limited set of guide words, as shown in Table 17.1. For example, a possible concern connected with the variable "flow" in the pipe section would be "reverse flow."

The second question to be asked is: *What is the consequence of each incident that can occur?* These consequences might affect employees, members of the public, the plant and its profits. In a HAZOP analysis the qualitative assessment is made to categorize the severity of the hazard, using information from Table 17.2.

The third question one should answer is: *How often is each incident expected to occur?* In a HAZOP study a qualitative estimate of the expected likelihood of the occurrence is made, using a categorization as listed in Table 17.3.

Table 17.1: Guide words for a HAZOP analysis.

Guide word	Deviations
NONE	No forward flow when flow is required.
MORE OF	More of any relevant physical property than required, for example higher flow, higher temperature.
LESS OF	Less than any relevant physical property than required, for example lower flow, lower temperature.
REVERSE	Flow in reverse direction.
PART OF	Composition of system different than required, for example ratio different or species missing.
MORE THAN	More species present than required, for example extra phases or impurities present.
OTHER THAN	Other occurrences than normal departures from operating conditions, for example in start-up, shut-down or failure of services.

Table 17.2: Degree of severity of hazard.

Severity	Significance
1	High fatality or serious injury hazard or hazard leading to loss of greater than six months of production or €10 million.
2	Medium-high injury hazard or hazard leading to loss of between 1 and 6 months of production or between €1–10 million.
3	Low-medium injury hazard or hazard leading to loss of 1 to 4 weeks of production or €0.1–1 million.
4	Low or no injury hazard or hazard leading to loss of less than one week of production or €100,000

Table 17.3: Degree of likelihood of hazard occurrence.

Likelihood	Significance
1	High hazard, expected more than once per year.
2	Medium-high hazard, expected several times in the plant lifetime.
3	Low-medium hazard, not expected more than once in the plant lifetime.
4	Low hazard, not expected at all in the plant lifetime.

Lastly, one should ask: *How can each incident be prevented?* The risk of the occurrence is determined by its severity and likelihood using a chart given in Figure 17.3. Each risk calls for a decision on the changes of the piping section to prevent or to reduce the probability of incidents to acceptable levels. A hazard that has been classified as having a strong possibility of leading to fatalities (Severity rank 1) or as being likely to occur more than once in the plant's lifetime (Likelihood rank 3) is classified in the chart as having an undesirable risk level C. It requires a reduction of risk to level B at the very worst.

Ranking	Significance
A	Acceptable risk level.
B	Almost acceptable risk level. Acceptable if suitably controlled by management. Should check that suitable procedures and/or control systems are in place.
C	Undesirable risk level. Must be reduced to level B at the most by engineering or management control.
D	Unacceptable risk level. Must be reduced to level B at the most by engineering or management control.

		Severity			
		1	2	3	4
	1	D	D	C	A
Likelihood	2	D	C	B	A
	3	C	B	A	A
	4	B	A	A	A

Figure 17.3: Risk ranking in HAZOP analysis.

17.2 Equipment sizing

In equipment sizing two important parameters (and their minimum and maximum values) play an important role: the temperature and the pressure. The temperature and pressure are important because the strength of the metal changes when for example the temperature increases. By calculating the design pressures and temperatures we can account for uncertainties in the process and this will ensure safety when the plant comes in operation.

Common margins used for pressure vessel design temperatures are as follows.

The maximum design temperature is the highest mean metal temperature expected in operation plus a margin (typically 50 °F). The minimum design temperature is the lowest mean metal temperature expected in operation minus a margin (typically 25 °F).

The design pressure is the maximum operating pressure plus a margin. The margin is typically the greater value between 10 % of the maximum operating pressure or 25 psi. The maximum operating pressure is the highest expected pressure, potentially during start-up, shut-down or emergencies. The specified pressure is usually near the relief valve at the top of the vessel.

Vessels under external pressure (jacketed or those under vacuum) need to be able to resist the maximum differential pressure that can occur in the process. Often, vessels under external pressure are also fitted with internal stiffening rings. Additionally, it is recommended that vessels under vacuum be designed at a pressure of –1 bar unless fitted with an effective vacuum breaker.

17.2.1 Vessel geometry

Typically pressure vessels are cylinders with at least a 2:1 ratio of height to width; 3:1 and 4:1 ratios are most common. Pressure vessels can be oriented either vertically or horizontally. Vertical vessels are more common because they use less land space and the smaller cross-sectional area of the vessel allows for easier mixing. Horizontal vessels are used when more phase separation is required because larger cross-sectional areas allow for less vertical velocities and therefore less entrainment. Settling tanks and flash vessels are typically horizontal for this reason. Horizontal vessels also allow easier cleaning, so heat exchangers are primarily horizontal.

There are three different designs for the ends of the pressure vessel: hemispherical, ellipsoidal and torispherical (Figure 17.4).

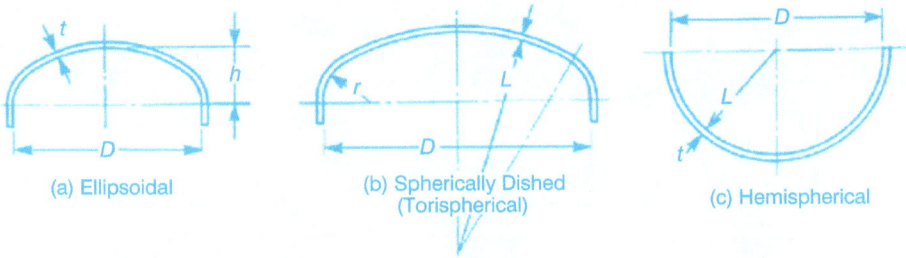

(a) Ellipsoidal (b) Spherically Dished (Torispherical) (c) Hemispherical

Figure 17.4: Types of pressure vessel heads.

Hemispherical heads are best for high-pressure systems; they provide the largest internal volume of the three options, they are half the thickness of the shell and they are the most expensive to make and combine with the shell.

Ellipsoidal heads are cheaper than hemispherical heads and provide less internal volume, they are the same thickness as the shell and they are most common for systems with more than 15 bar.

Torispherical heads are the cheapest of the three options, and are most commonly used when pressures do not exceed 15 bar. The two junctions in a torispherical end closure are between the head and the cylinder, and at the junction of the crown and knuckle radii. Bending and shear stresses can occur at these points and needs to be accounted for. The crown radius should not be larger than the diameter of the cylinder section, and the ratio of the knuckle to the crown radii should not be less than 0.06.

Openings and branches in pressure vessels are used to connections, manways and instrument fittings. Having an opening inherently weakens the shell, making stress concentrations likely. In order to compensate for this, wall thickness is increased in the area around an opening. It is necessary to provide support without changing the existing dilation pattern around the opening of the vessel. If the wall is overreinforced, it can cause a "hard spot" which makes the wall less flexible and can cause additional secondary stresses. To calculate the minimum amount of reinforcement needed for an opening or branch, we refer to ASMA BPV Code, Section VIII D.1, Part UG-37.

The most common method to reinforce openings is by welding a collar or pad around the opening, typically with an outer diameter about 1.5 to 2 times the opening's diameter (Figure 17.5(a)). This method, while common, does not provide the best support around the opening because thermal stresses can arise due to poor thermal conductivity at the junction of the shell and the pad.

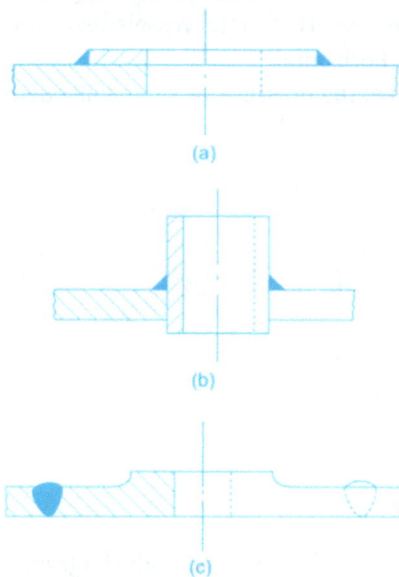

Figure 17.5: Compensation methods for openings. (a) Welded pad. (b) Inset nozzle. (c) Forged ring.

Branches are commonly reinforced by allowing the branch to protrude into the vessel (Figure 17.5(b)). It is important to use caution with this method, as the protruding branch can trap particulates and corrosion can occur in the created crevices.

Enforcing rings (Figure 17.5(c)) are the most reliable method of reinforcement, but are also the most expensive. Because of this, they are typically only used for large openings in vessels under severe operating conditions.

17.2.2 Stresses and strains

There are a variety of potential stresses on a pressure vessel that must be accounted for during design and construction: internal and external pressure, weight of vessel, weight of contents, weight of internals (distillation trays, heating/cooling coils, packing supports), weight of attached equipment, thermal expansion, cyclic loads caused by condition changes, friction loads and environmental loads (wind/snow/seismic).

17.2.3 Wall thickness

There are two main stresses that can occur on the shell portion of the pressure vessel; hoop stress and longitudinal stress. For both, a different correlation can be used to estimate the required wall thickness.

For hoop stress one can use

$$\text{Wall thickness} = \frac{PD}{2SE - 1.2P}, \tag{17.1}$$

And for longitudinal stress one can use

$$\text{Wall thickness} = \frac{PD}{4SE + 0.8P}, \tag{17.2}$$

where P is the pressure, D is the diameter, S is the maximum allowable stress and E is the welded joint efficiency. The thicker of the two is chosen as the wall thickness. The minimum wall thickness (without considering corrosion allowances) is 1/16 inch.

Typically walls are much thicker. In high-pressure vessels, internal pressure has the largest magnitude. In low-pressure vessels, wall thickness is designed to resist vacuum.

17.2.4 Head thickness

Different equations can be used to calculate the appropriate head thickness.

For hemispherical heads,

$$\text{Thickness} = \frac{PD}{4SE - 0.4P}, \tag{17.3}$$

For ellipsoidal heads,

$$\text{Thickness} = \frac{PD}{2SE - 0.2P} \tag{17.4}$$

And for torispherical heads,

$$\text{Thickness} = \frac{0.885PR_c}{SE - 0.1P} \tag{17.5}$$

where R_c is the crown radius.

17.2.5 Corrosion allowance

A margin of wall thickness must be added to account for corrosion of the vessel over time. This margin is usually between 1/16″ and 3/16″. In heat exchangers where wall thickness can affect heat transfer, smaller margins are used.

Example 1 (Determine wall thickness). What is the wall thickness of a 304 stainless steel pressure vessel with a 5-ft diameter, 400-psi design pressure and 500 °F design temperature? Assume double-welded butt joints were used ($E = 0.85$).

First, the max allowable stress must be calculated. Using a table found in ASME BPV Code, Section VIII D.1, Section II, Part D[4], the maximum allowable stress under these conditions is 12,900 psi. The hoop stress is calculated as follows:

$$t = \frac{400(5)(12)}{2(12,9000)(0.85) - 1.2(400)} \approx 1.12 \text{ inch.}$$

The longitudinal stress is then calculated as follows:

$$t = \frac{400(5)(12)}{4(12,900)(0,85) + 0.8(400)} \approx 0.54 \text{ inch.}$$

The larger of the two is hoop stress. Adding the corrosion allowance and rounding off to the nearest quarter inch gives a wall thickness of 1.25 inches.

Example 2 (Nozzle design). Find the wall thickness required for the following nozzle. Given: interior pressure (P) of 201.4 psi, allowable stress (S) of 17,100 psi. Assume a nozzle efficiency (E) of 1, a nozzle corrosion allowance (nca) of 0.01 inches and an undertolerance allowance (UTP) of 12.5 %.

First, we need to determine R for the nozzle, i. e.,

$$R = \frac{D}{2} - N_{wall} + nca + (N_{wall})(UTP) = \frac{4.5}{2} - 0.237 + 0.01 + (0.237)(0.125) = 2.05 \text{ inch.}$$

Next, we can use the same equations used for determining stresses in pressure vessel heads. We have

$$t = \frac{(P)(R)}{(E)(S) - 0.6(P)} + nca$$

$$t = \frac{(201.4)(2.05)}{(17,100)(1) = 0.6(201.4)} + 0.01 = 0.034 \text{ inch.}$$

Finally, we check if N_{wall} is large enough, i. e.,

$$0.237 > 0.034.$$

The wall thickness is large enough.

17.3 Estimation of capital

One of the most important aspects of determining the overall economic viability of a chemical process is determining the capital cost. In addition to the purchase price of the equipment, capital costs include delivery and installation of equipment, preparation of land for construction, salaries of contractors and construction workers and any other costs associated with building a chemical plant. For this reason, the cost associated with process equipment is not as straightforward as the sticker price.

17.3.1 Fixed capital investment

The fixed capital investment is the total cost associated with constructing the plant. This cost includes design, site remediation, purchasing process equipment, developing infrastructure and contingency charges, and includes the raw material costs as well as labor. It is divided into the following four categories.

Inside Battery Limits (ISBL): Plant costs including the procurement, installation of all equipment, purchasing, shipping, land costs, infrastructure, piping, catalysts and construction. ISBL are the "inner costs of the plant."

Outside Battery Limits (OSBL): Plant costs including the offsite costs, water or electricity used from the main grid, fencing security, wastewater treatment, fire fighting facilities, etc. The OSBL plant costs are often estimated as 40 % of the ISBL plant costs.

Contingency charges, which are basically a correction for when prices change during the project or when unanticipated costs arise. The contingency charges are typically around 10 % to 40 % of the ISBL plant costs.

The working capital is basically the money needed to address irregularities in process operation that may or may not be spent; the value of inventory, products, by-products, process equipment and spare parts. Sometimes the working capital is estimated as 7 weeks of production costs minus 2 weeks of feedstock costs. Another estimate is that the working capital is around 10 % to 20 % of the annual operating costs.

If the working capital is too low, the costs for operation might not be covered, while if the working capital is too high the cash is not increasing its value.

The working capital turnover is defined as the ratio of the annual revenues to the working capital. Companies try to maximize their working capital turnover. Companies like Dow and Dupont had a working capital turnover of around 4 in 2014, while companies like Praxair reached working capital turnovers of more than 30.

17.3.2 Project financing

Because of the magnitude of costs associated with the start-up and maintenance of a chemical plant, there are often different project financing methods required to cover the capital needs. The two main methods of project financing are debt and equity financing. Debt financing usually involves the issuing of bonds. Equity financing involves the issuing of common stock. However, most companies utilize a combination of these two methods to successfully finance a project.

Debt financing basically means that the company issues bonds to banks and/or investors. After selling a bond the company is in dept to the buyer (also called a creditor).

In equity financing the company will sell ownership, often by means of stocks (*i. e.*, public ownership).

There are different ways of measuring growth and profitability of a company. One indicator is the return on investment (ROI), which is the ratio of the net profit a company makes to the cost of investment.

Example 3 (Calculating an ROI). An investor buys US$1,000. worth of stock and sells them two years later for US$1,200. Calculate the ROI.

$$\text{ROI} = \frac{\text{Net profit}}{\text{Cost of investment}} * 100\,\% = \frac{200}{1,000} * 100\,\% = 20\,\%$$

Another quantitative indicator of profitability is the payback period, which compares the initial investment with the cash inflow per period.

Example 4 (Calculating a PB). A company may invest US$ 150 million in a new project. This project runs over a 10-year period. The project is expected to generate US$ 25 million profit. Calculate the PB.

$$\text{PB} = \frac{\text{Initial investment}}{\text{Cash inflow per period}} = \frac{105}{25} = 4.2 \text{ years}$$

17.3.3 Capital cost estimates

At the initial phase of a project we can only use very rough measures to estimate the capital costs of equipment or even of a complete plant. A common way of getting these rough estimates is by the so-called economy of scale equation:

$$C_E = C_B * \left(\frac{Q}{Q_B}\right)^n, \tag{17.6}$$

where C_E is the costs estimate for an equipment with capacity Q and C_B is the reference cost of the equipment with capacity Q_B; n is a constant that depends on the equipment type. In many cases, the exponent is not equal to 1, *i. e.*, the costs do not linearly increase with the equipment size. In Table 17.4 a few equipment capacities and their associated capital cost correlation parameters are presented. Note that the scaling factors have been lumped. A much more detailed overview can be found in Seider *et al.*

Table 17.4: Purchase costs of processing equipment.

Equipment	Size factor (S)	Range of S	Cost equation
Adsorber (with activated carbon)	Bulk volume, S, ft^3		$CE = 41S$
Crystalizer (continuous cooling)	Length, L, ft	15–200 ft	$CE = 1644L^{0.67}$
Dryer (batch tray)	Tray area, A, ft^2	20–200 ft^2	$CE = 4400A^{0.35}$
Evaporator (horizontal tube)	Heat transfer area, A, ft^2	100–8,000 ft^2	$CE = 4604A^{0.53}$
Heat exchanger (plate and frame)	Heat transfer area, A, ft^2	150–15,000 ft^2	$CE = 10070A^{0.42}$
Liquid-liquid extractor	$S = $ height (ft) × diameter (ft)$^{1.5}$	3–2,000 ft$^{2.5}$	$CE = 363S^{0.84}$
Membranes (ultrafiltration)	Membrane surface, A, ft^2		$CE = 10-25A$
Mixer (tumbler)	Volume, V, ft^3	50–270 ft^3	$CE = 3856V^{0.42}$
Hydrocyclone	Liquid feed rate, Q, gal/min	8–1,200 gal/min	$CE = 275Q^{0.5}$
Filter (belt)	Volumetric flow rate, ft^3/hr	120–500 ft^3/hr	$CE = 813S^{0.38}$
Storage tank (spherical)	Volume, V, gal	10,000–1,000,000 gal	$CE = 68V^{0.78}$

Before the chemical plants can be built, capital cost estimates must be made. This is done by using the factorial method. Accuracy and reliability of the estimate will heavily depend on the availability of the data and the level of the design at the time. Lang proposes capital cost equipment by the following equation:

$$\text{Total capital costs} = F * \sum_i C_i, \tag{17.7}$$

where C_i are the costs of equipment I and F is an installation factor, or Lang factor. This factor is a kind of correction depending on the type of process one is designing. For example, $F = 3.1$ if one is designing a process that handles solids, and $F = 4.74$ when processing fluids.

17.3.4 Estimation of production cost and revenues

The overall production costs depend on the costs for raw materials, utilities (steam, electricity, cooling costs), disposal of waste and labor. There is a correlation that can be used to determine the number of operators that one needs for the plant, *i. e.*,

$$N = \left(6.29 + 31.7 * P^2 + 0.23 * N_P\right)^{0.5}, \tag{17.8}$$

where N is the number of operators in an 8-hour shift, P is the number of processing steps in which solids are handled and N_P is the number of other processing steps throughout the process.

In production costs, normally also the maintenance costs for the plant are included, which are typically around 6 % of the capital investment. There are associated costs for research and development, taxes and insurances (around 3 % of the capital investment) and so-called plant overhead (for example, costs for the salary administration department, employee benefits, etc.). Sometimes the costs for licensing and royalties are also added to the overall production cost.

17.3.5 Market effects

In the chemical industry, pricing has a dominant effect on the profits. Most industry analysts rely on the competitive model, according to which capacity and demand determine the price. In fact, a non-competitive model in which supply (not capacity) and demand determine the price may fit the industry better. The difference is that whereas capacity represents physical ability produce, supply represents willingness to produce. The behavior of the chemical industry during two critical periods has encouraged the use of the competitive model; see, for example, the developments in crude oil pricing since 1960 in Figure 17.6.

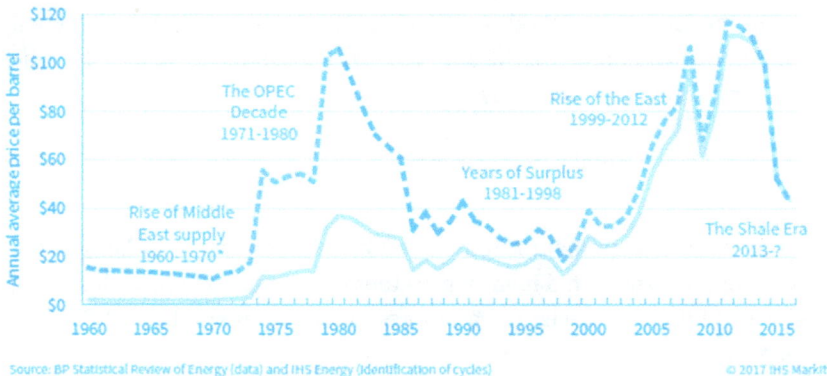

Source: BP Statistical Review of Energy (data) and IHS Energy (identification of cycles)   © 2017 IHS Markit

Figure 17.6: Crude oil pricing cycles since 1960.

Obviously, process design is inherently dependent on global markets and the economy as a whole. Process design can depend on locational markets. For example, if an expected long-run exchange rate between two specific currencies makes production of a specific chemical more profitable in a colder climate such as in Russia as opposed to a warmer climate such as in Mexico, this may have an effect on how the process to produce this specific chemical should be designed.

Relative inflation rates of different countries can have similar effects on process design. Additionally, markets for key production inputs can have an effect on process design. If a company was designing a large-scale production plant which required massive amounts of steel, the global market for steel (or to a lesser extent, iron mines) would impact how/where this company would choose to design and build their process. Also worth noting is the fact that the state of the economy as a whole may impact process design as well. If the economy is in a recession, funds are likely to be more tightly managed within a company. Thus, it may make sense for a process engineer to design a process that operates on a smaller scale. Not only will this decrease the cost of running the process (lower utilities costs, lower costs for process inputs), but since less product will be produced, this will also reduce the risk of having the company running an inventory.

This is a positive, as running an inventory can lead to losing even more money during a recession. Conversely, a healthy economy may encourage the design of facilities with larger production capabilities. This section aims to look at several recent examples of macroeconomic markets affecting process design, similar to the hypothetical situations described above.

17.3.6 The gross (profit) margin

Any economic evaluation starts with calculating the gross (profit) margin, which effectively is the difference between the revenues (sales of products and by-products) and the costs for purchasing raw materials. Many chemical companies only have a 10 % GPM.

17.4 Engineering economic analysis

The purchasing power of our money is changing with time; the things that can be bought with one euro now are slightly less than the year before because of inflation. Investors who have money available expect a return on their funds beyond the rate of inflation. The change of value over time due to these two trends is called the time value of money. Governments are tracking inflation by the consumer price index (CPI) and the producer price index (PPI).

Investors expect that their investments will do more than keep pace with inflation. This expectation can be incorporated in calculations through discounting or counting on future returns, when measured in real terms as worth less than the same amount of money invested in the present.

The simple payback adds all cash flows into and out of the project. This is also called the net present value (NPV). Of course the NPV of a project has to be positive. When a project runs over multiple years, we can consider a break-even point at the year in which the total income (annuities) surpasses the initial costs.

Present, annual and future value are related via the following equation:

$$F = P(1 + i)^N,$$ (17.9)

where F is the future value, P the present value, i the interest and N the horizon in years. The present value can also be related to the annuity (A), i.e.,

$$P = \frac{A(1 + i)^N - 1}{i(1 + i)^N}.$$ (17.10)

Combining equations (17.9) and (17.10) will give

$$F = A\frac{(1 + i)^N - 1}{i}.$$ (17.11)

Example 5 (Calculating the future value). A cashflow stream consists of an annuity of €100,000 earned for 10 years at 7 % interest. What is the future value at the end of this term?

$$F = A\frac{(1 + i)^N - 1}{i} = 100.000\frac{(1 + 0.07)^{10} - 1}{0.07} = €1,381,645$$ (17.12)

It is also possible to discount a non-uniform set of annuities to its equivalent present worth (PW) value by treating each annuity as single payments to be discounted from the future to the present, i.e.,

$$PW = \sum^N \frac{A_n}{(1 + i)^n}.$$ (17.13)

Example 6 (Investment in a new small power plant). An investor considers to invest in the construction of a small power plant. The installation costs are €6,500,000. The power plant will generate €400,000 each year for a period of 25 years. The salvage value at the end of the lifetime is €1,000,000. Would this be a project worth investing in?

In the table below the cash inflows and outflows for two different interest rates are given, tabulated as well as in a graph.

Year	Capital expense or revenue (× $1,000)	Annuity (× $1,000)	Net flow (× $1,000)	Discounted Net Flow @i = 5 % (× $1,000)	Discounted Net Flow @i = 4.2 % (× $1,000)
0	−6,500	0	−6,500	−6,500	−6,500
1	0	400	400	381	384
2	0	400	400	363	369
3	0	400	400	346	354
Etc.	Etc.	Etc.	Etc.	Etc.	Etc.
25	1,000	400	1,400	413	504
Total	n/a	n/a	4,500	−567	0

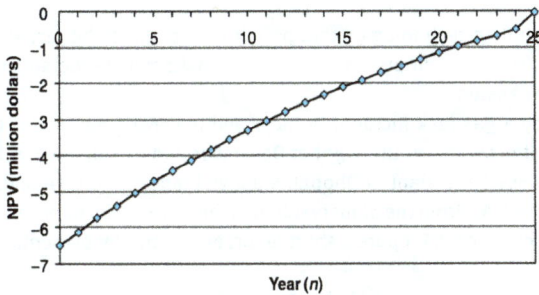

Year (*n*)

For an interest rate of 4.2 % the NPV reaches 0 at the end of the project running period, which motivates an investment. However, generally NPVs should be positive when the project is running for 3–5 years to be attractive for investors.

17.5 Computer tools for cost estimating

It is difficult for smaller companies that do not specialize in process design to maintain accurate data on process costs and perform the necessary analysis for these data to be useful. Instead, most companies use costing software and other computer tools to perform economic analysis.

Several computer tools by Aspen are available for estimating capital costs. Aspen's Economic Evaluation Product Family builds off of its original ICARUS technology. In the aspenONE product suite, the primary capital estimation tool is Aspen Capital Cost Estimator. It couples with Aspen Economic Evaluation to provide capital evaluations during process design and operation.

Some issues that have arisen in the past utilizing ICARUS or Aspen Capital Cost Estimator are as follows:

Mapping equipment from process simulations to ICARUS can simplify design or map dummy equipment that is not real process equipment.

It is good practice to include design factors for safety throughout the process. However, Aspen will map the equipment exactly as specified in HYSYS and therefore will not include any design factors in calculating the capital costs.

Pressure vessels are coasted exactly according to ASME Boiler and Pressure Vessel Code, Section VIII, Division 1. However, in some cases, this may be an inadequate pressure vessel design. In these cases, the design should be manually entered.

Some processes require non-standard components that HYSYS has no way of modeling correctly and for which ICARUS has no appropriate equipment category. Aspen has the capability to include non-standard equipment libraries which often can be obtained by equipment manufacturers. Adding these libraries allows for the use of the costing software for cost estimates.

Exercises

1. Regarding the recent advances in drilling technologies that permit natural gas to be recovered from shale deposits cost-competitively, present arguments from an environmental perspective in favor and opposed to its rapid expansion.

2. The Bhopal disaster was caused by a gas leak incident in India and is considered one of the world's worst industrial disasters. It occurred on the night of December 2nd to 3rd, 1984 at the Union Carbine India Limited (UCIL) pesticide plant in Bhopal, Madhya Pradesh, India. A leak of methyl isocyanate gas and other chemicals from the plant resulted in the exposure of hundreds of thousands of people to highly toxic chemicals. Prepare a short report or powerpoint presentation describing:
   (a) an account of the steps leading up to the reported disaster,
   (b) a list of root causes of the reported disaster,
   (c) the ethical dilemma inherent in the reported disaster because a choice between at least two alternatives had to be made,
   (d) an additional good alternative that could have either prevented the disaster or reduced the severity of its consequences.

3. A concern borrows €50,000 at an annual effective compound interest rate of 10 %. The concern wishes to pay off the debt in 5 years by making equal payments at the end of each year. How much will each payment have to be?

4. A power plant operator is considering an investment in increased efficiency of their plant that costs US$1.2 million and lasts for 10 years, during which time it will save US$250,000/year. The investment has no salvage value. The plant has an MARR of 13 %. Is the investment viable?

5. A municipality is considering an investment in a small renewable energy power plant with the following parameters. The cost is $360,000, and the output averages 50 kW year-round. The price paid for electricity at the plant gate is $0.039/kWh. The investment is to be evaluated over a 25-year time horizon, and the expected salvage value at the end of the project is $20,000. The MARR is 6 %. Calculate the NPV of this investment. Is it financially attractive? Calculate the operating credit per kWh which the government would need to give to the investment in order to make it break even financially. Express your answer to the nearest 1/1,000 dollar.

17.6 Take-away message

In developing new processes, the priority is to set up the design in an inherently safe way, safe for operators, environment and society. A first step in safety analysis is the HAZOP, the Hazard and Operability Analysis, which is a method that uses guided questions to assess the safety of the different processing units. A process should be safe and technically feasible, but also economically attractive. There are different models for calculating the economic feasibility of a new process. The simplest metric is the gross profit margin, which rates the earnings of the product as compared to the costs of the materials and the processing. The profit margin should be positive. Often projects run over multiple years and the time value of money should also be included in the economic analysis (*e. g.*, inflation). A suitable indicator that includes the time value of money is the NPV, which discounts all in- and out-going cash flows over a certain time period.

17.7 Further reading

Lang, H. J. (1948). Simplified approach to preliminary cost estimates, *Chem. Eng.*, 55(6), pp. 112–113.

Peters, Timmelhaus and West (2003). *Plant design and economics for chemical engineers*, 5th ed. New York: McGraw Hill.

American Institute of Chemical Engineers (1990). *Safety, health and loss prevention in chemical processes: problems for undergraduate engineering curricula – student problems*. New York: AIChE.

Green and Perry (2008). *Perry's chemical engineers handbook*, 8th ed. New York: McGraw Hill.

Seider, Lewin, Seader, Widagdo, Gani and Ng (2017). *Product and process design principles, synthesis, analysis and evaluation*, 4th ed. Wiley.

18 Design for sustainability

When we think about process design from a product-driven perspective, it makes sense to find ways to design a sustainable process to pair with a sustainable product. Preserving the plant so that future generations are able to live well is a lofty goal, but a challenging one that humans have never come close to reaching. Overconsumption is a major problem for sustainable development, although it drives demand and thus economic prosperity. As chemical engineers, we must seek techniques and technologies to utilize renewable resources, create biodegradable or recyclable products and find ways to provide the energy needs for an ever-growing population without polluting our planet.

18.1 Taking the long-term view

For chemical engineering processes, sustainability means designing production systems which do not consume non-renewable resources and do not produce waste streams which cannot be reused or biodegraded to renew the resources we need. This is a great challenge, since most of the earth's resources we have been utilizing since the beginning of the industrial revolution have been non-renewable: oil, coal, natural gas and most mineral resources. Furthermore, the consumption of those resources has led to an ever-increasing pollution burden, evidenced most strikingly perhaps by the carbon dioxide-driven global warming. Human society has been very slow to recognize the problem of resource limitation, and unregulated economic activity has encouraged the extirpation of species, the creation of uninhabitable regions due to pollution and the continued destruction of habitats needed both for wildlife and for air-cleansing forests. Efforts by organizations such as the European Union, which has developed the ISO 14000 standards for environmental management by industrial concerns, are hampered by those seeking short-term profits, and thus are usually voluntary. If society and government are not willing or able to drive a push for sustainable development, then we must do so individually, as consumers, small production firms and multinational corporations, so that the earth will be habitable for our descendants.

We must also consider that the economic benefits of uncontrolled resource consumption have not been distributed at all equitably. The earth's population is expected to level off at around 10 billion people, and all of them will eventually want to live at the standard of a current average citizen of the industrialized world. Yet even at our current population, we struggle to provide food, clean water and electricity, let alone the chemical products that make life in the industrialized world pleasant. Ideas of using renewable biomass to make fuels and chemicals sometimes look to improve the sustainability of the chemical industries, but the replacement of food crops by fuel crops endangers the food supply. So the great challenge is to increase the production

https://doi.org/10.1515/9783110570137-018

of the products necessary for life, and improve the standard of living for all while not expending the limited resources we have.

In the chemical industry, the main non-renewable resources utilized are oil, natural gas, coal and mineral resources. Modern petrochemical processing is in general quite wasteful, with millions of tons of natural gas being flared yearly worldwide, simply because it is not cost-effective to capture, transport and sell the gas which is co-produced with oil. Coal and oil resources are being consumed quickly, and many authors believe that oil resources will not last far beyond the year 2100. Therefore, finding novel feedstocks, which will allow the production of current chemical products or perhaps new products serving the same functions but being more biodegradable, will become a critical task for chemical engineers in the coming decades. The burning of fossil fuels to produce energy cannot be continued long-term, as the global warming consequences of this will be catastrophic even before those resources are fully consumed. Many countries are planning now for a complete phase-out of fossil fuel-based energy production, and chemical process designers must take this into account as we build plants intended to operate far into the future.

So how can we apply the principles of sustainable development within the design of chemical production facilities? First, we need metrics to give us a measure of the level of sustainability our design can reach. This allows at least comparative analysis of design alternatives from a sustainability perspective. Secondly, we need to apply those metrics within the process synthesis phase, either heuristically or via true multi-objective optimization. For already existing chemical facilities, we can consider retrofit to improve sustainability, and also a redesign of the product, to make it biodegradable or recyclable. In all cases, we must consider the use of renewable energy, perhaps from non-grid sources, and how the intermittency of those sources affects our production processes. Continuous chemical processing also lends itself very well to heat integration, which can be implemented within new designs, or added to operating chemical facilities via plant retrofits.

18.2 Metrics for sustainability

Many organizations have published guidelines or heuristics to help drive chemical process design in a more sustainable fashion. One of the most important steps in process synthesis is the selection of a process chemistry. The application of *green engineering* adds another dimension to this selection, since now we seek to create sustainable, environmentally friendly processes as well as efficient and cost-effective ones. The American Chemical Society published a list of 12 Principles of Green Chemistry in 1989. These principles, provided below, each help guide the process synthesis procedure in different ways.

1. Prevention of waste: Processes should be designed to avoid the formation of waste streams, rather than treating such streams later. Treatment of waste requires ex-

cess energy, and disposal of non- or slowly degradable waste requires landfill space which is limited.

2. Atom economy: Processes should be designed to incorporate as many of the atoms in the feedstocks as possible into final products, thus minimizing waste production. This can be accomplished by recycling feed materials, by reprocessing unwanted by-products into feed materials, and by designing multi-product plants which convert by-products into other valuable chemical substances.

3. Use of less hazardous chemical synthesis processes: Processes should be designed to avoid the use of highly toxic, reactive or explosive reagents. This can be a significant challenge, since highly reactive reagents often allow for lower-energy reaction processes. But often homogeneous catalysts can be replaced with heterogeneous ones, toxic solvents can be replaced with less toxic ones, and chemicals with a high vapor pressure can be replaced with those less likely to escape our processes.

4. Creation of less toxic products: When a novel chemical product is designed, not only should the primary function be considered, but also the toxicity and the biodegradability should be reviewed. Sometimes, functional groups having little or no effect on the primary function of a molecular product can be altered to change environmental and sustainability properties, when these objectives are considered early in the product design process. The hazards associated with the decomposition products of chemical products must also be taken into account.

5. Use of smaller quantities of safer solvents and auxiliary chemicals: Chemical processes should be designed to minimize the use of solvents, avoid dangerous or toxic solvents and minimize the use of additives whenever possible. The risk inherent in using an auxiliary chemical is the product of the probability of that chemical escaping into the environment times the consequence of that release. So using smaller quantities of solvents, avoiding those which will rapidly spread in the environment and avoiding those for which the adverse effects on humans and the environment are large will greatly enhance the environmental performance of a chemical process.

6. Design for energy efficiency: chemical processes should be designed to be as energy-efficient as possible. This implies operating at lower temperatures and at pressures near atmospheric whenever possible. Vacuum systems and refrigeration expend large amounts of energy and should be avoided when possible. High-pressure streams which need to be lowered in pressure, or higher-pressure gaseous waste streams, should be routed through turbines whenever practical to recover energy. When less overall energy is needed, the use of renewable, variable energy sources is made more feasible, since energy storage systems can make up an energy deficit when the total energy demand is low. Decreasing energy use also decreases production costs in nearly every scenario within chemical processing, so this should be a high priority for any chemical facility being newly designed.

7. Use of renewable feedstocks: The use of renewable or recycled feedstocks is prefer-able in all cases to consumption of limited non-renewable resources. There are cases, such as metals processing, for which recycled materials are in fact easier to process than original raw materials, since the original source materials (ores) have the desired metal present at a much lower concentration than recycled materials.

8. Reduction of the use of derivative compounds: Many chemical synthesis processes employ intermediate compounds, sometimes unnecessarily. The use of such inter-mediates can pose a hazard in terms of safety, can lead to excess energy use and can require increased numbers of waste streams. Direct synthesis of products is always preferable, unless the energy use in separation of our product from by-products is much higher.

9. Catalysis: The use of catalysts, rather than stoichiometric reagents, is preferred for sustainability. Well-designed catalysts can increase the selectivity, conversion and energy consumption of our reactions, and thus decrease overall energy con-sumption. Heterogeneous catalysts are preferred over homogeneous ones, since energy-intensive separation processes can be avoided.

10. Design for degradation: Chemical products should be designed to be biodegrad-able, by including structures accessible by common biological processes. All products have a useful lifetime, and the long-term fate of our products must be considered.

11. Real-time analysis for pollution prevention: Online systems need to be employed to monitor chemical processes and to ensure that operating temperatures and pressures are being kept near design values, to prevent side reactions and in-creased waste generation. Significant pollution occurs during excursions, which are deviations in chemical plant performance far from the desired steady state. Real-time monitoring, well-tuned process controllers and redundant safety and pollution abatement systems need to be in place to avoid or mitigate the effects of excursionary behavior.

12. Inherently safer chemistry for accident prevention: All chemicals present within a plant should be chosen with a view to avoiding substances which create a sub-stantial risk of fire, explosion or toxic release. If a chemical does not exist in a plant, it cannot leak or cause an accident. While risks of accidents can never be completely removed, we can significantly reduce risk through the application of inherent safety principles within chemical engineering synthesis and design.

These green chemistry principles provide an excellent set of heuristics for us to use when we evaluate the sustainability of a given process alternative.

In order to quantify the measurement of sustainability for chemical production facilities, we need to consider not only the plant itself, but also the extraction of the raw materials, the production of the energy required for our process, the fate of the waste streams produced by our process and the eventual discarding and degradation of our product. The most common methodology for consideration of these "cradle-to-grave" factors is called *life cycle assessment* (LCA), and this technique is part of many process design procedures. It should be noted that LCA does not provide a stand-alone measure which can be used to determine if a chemical engineering project should go forward. It should only be used in a comparative sense, to judge between process and product alternatives. Judgements of how sustainable a production process should be, or whether the benefits to society of the production of a given product outweigh the environmental impacts or that production, are beyond the scope of LCA. But LCA does help broaden the viewpoint of environmental impact assessment, by evaluating the impact of processes both internal and external to the process; by compiling an inventory of all chemical substances employed, all energy and material inputs used and all environmental releases by a process; and by providing a system for interpreting the results to help make informed decisions about alternative processes.

An LCA procedure involves four basic steps: (i) the definition of the goals and the scope of the study, (ii) the analysis of the inventories of flows of energy and materials involved with the process, (iii) the assessment of the environmental impacts of flows of energy and materials into and out of the natural world and (iv) the interpretation of the information to provide an overall assessment. The procedure begins by defining the scope of the study. This involves defining the boundaries of the chemical process, determining whether impacts of the use of energy from the power grid will be considered and the impact categories to be considered. The product of interest is called the functional unit, and we must consider all of the impacts of the production, use and eventual disposal of that unit. When defining the impacts of our functional unit, a complication occurs when a chemical plant produces many products. The environmental impact of the production process must be allocated across the product portfolio, which can be a subjective procedure. The impact categories chosen should be those for which we expect our process to be relevant, so that we are not focusing on areas for which no impact will be present. For example, chemical processes which do not involve substances likely to react with ozone in the atmosphere do not need to be evaluated in terms of ozone depletion potential. Remember however that the impacts of those processes to which our process will be compared should also be computed.

Once the system boundaries are clearly defined, a life cycle inventory can be completed. The inventory is usually displayed by drawing a life cycle inventory diagram, which shows all material flows into and out of the system, with both chemical composition and flow rate. If energy usage and production is to be considered, then flows for both energy and materials must be shown. The product is of course a flow leaving the system as well, and evaluating its impact on the environment will allow us to consider possible degradation products.

Once a clear picture of all of the flows in and out of our system is obtained, we can then evaluate each of the impact categories for each flow. This can be challenging when the fate of a given chemical is unknown or the production process for a raw material or for energy varies over time. But normally these effects can be averaged, and the errors in the estimation of these effects are consistent across multiple systems, such that the LCA results for different process alternatives are comparable. The impacts are often subdivided into *first impacts*, which are caused by the extraction of the raw materials, the construction of the facility and the manufacturing of the product; *use impacts*, which consider the maintenance of the facility and the energy and water resources required to continuously operate the facility; and *end-of-life impacts*, which include product disposal, facility decommissioning and processing of waste streams.

The final phase of an LCA assessment is the interpretation of all of the impacts. This begins by considering the accuracy of each impact assessment, since clearly some of the data used to perform the assessment are more reliable that other data. A review of the completeness of the impact assessment and the inventory should be performed to ensure that major possible impacts are not overlooked. It should be noted that many hazardous situations which have occurred within engineering systems were not in any way accounted for or predicted by the designer. If a given impact is not conceptualized, it cannot be evaluated. So broad-based thinking, using scenario analysis, can help to avoid missing potential impacts. A review of accidents and incidents for similar facilities in the past can also help guide the search for relevant impacts. For this phase, it is also important to consider the sensitivity of each impact to the flow rates and compositions denoted within the life cycle inventory. Would a small change in separation efficiency or reaction temperature lead to a highly toxic chemical escaping the plant, leading to a major environmental hazard? Consideration of the sensitivity of each impact assessment also can guide the consideration of alternative designs, since those altering the more sensitive parameters are likely to have a significantly different assessment profile.

18.3 Including sustainability metrics within the design process

Now that we have seen some logical heuristics for developing more sustainable process alternatives and developed the methodology of LCA to be able to evaluate and interpret combinations of environmental and sustainability metrics, the question remains as to how to include these metrics within the process synthesis and design procedure. The heuristics from the American Chemical Society listed above lend themselves to a screening approach in which early-stage process alternatives are considered for their likely sustainability and clearly unsuitable alternatives are removed from further consideration. For example, an option for many catalytic reactions is to use a mercury containing catalyst. Processes using such catalysts may meet economic objectives and even environmental regulations, since the mercury does not leave the system under normal conditions. But the production of such catalysts does release mercury into the environment, and the eventual disposal of the catalyst at the end stage of the plant's lifetime will also be a source of environmental mercury. So a simplified sustainability review would bring up significant concerns about such processes. The heuristics are particularly helpful when we select raw materials, when we choose a process chemistry and when we consider whether producing a specific product is a good choice in the long-term sense.

For a more thorough sustainability review, late-stage process alternatives can be compared via multi-objective optimization. A superstructure formulation can be created, which includes all of the remaining process alternatives, using discrete variables to model the equipment and reaction scheme selection decisions. The various environmental impacts (as considered in the LCA methodology) must then be quantified, such that a numerical score can be generated for a given technology selection and set of continuous variables. The problem can be formulated with an aggregated environmental objective, combining all of the impacts listed, or with separate objectives for each impact. An economic objective (usually minimizing production cost) is always included as well. If an aggregate environmental impact score is used, then visualization of the pareto-optimal set of designs is reasonable. We can immediately discard all dominated solutions, and consider the trade-offs between the pareto-optimal designs. Usually the set is small, so decisions can be made as to whether improvements in one objective justify weaker performance in the other.

18.4 Retrofit for improved sustainability

A common task in the chemical process industries at present is to find and implement strategies to make an existing chemical facility more sustainable. The first and most obvious way to improve sustainability is to consider the raw materials in a process. Could those materials be replaced or augmented with renewably sourced ones? There could be bio-based feedstocks already on the market which could replace some or all of a hydrocarbon feedstock, leading to greater flexibility. In some industries, the use of recycled materials is possible and may even lead to lower energy usage. Such feedstocks may require preprocessing and may not be available at the same price levels of current feedstocks. So the chemical concern must decide how much value they place on more sustainable production.

A second major area in which many chemical facilities can be improved in terms of sustainability is waste generation. Each waste stream should be analyzed carefully, to judge its value from not only an economic, but also an environmental perspective. Smaller subplants may be built alongside a major facility, with the goal of producing higher-value products from streams formerly going to waste. The energy in a waste stream is also potentially useful, and can be recovered through heat integration, or in some cases via a mechanical turbine. In some cases, a separation process to create more pure streams allows a single waste stream to be used as a feedstock for another chemical process. The cost of doing so may be higher than the simple disposal cost, but if a sustainability consideration is used for the plant, such an option may be viable.

Additionally, all auxiliary chemicals used in the process should be carefully considered. Solvents, additional reactants, reaction and separation media and heat transfer fluids all impact the sustainability of a given process, and nearly all of them can be altered while still producing our product. The mining industry, for example, is now seeking ways to eliminate the use of cyanide compounds, which have been used in great quantities to separate ores via reaction. Salts are formed in the cyanide processes, which often are simply disposed of. This leads to a large demand for cyanide-based reagents, making the mining industries some of the least sustainable chemical production processes. Newer materials for separation of ores will be reusable, in that the ions will be recovered in future processing steps and reused again, as is commonly done with solvents and reaction media in the chemical industries.

Separation processes often use numerous solvents, and here we see a strong opportunity for sustainability improvements. Newer solvent categories such as ionic liquids and supercritical carbon dioxide allow for many separations to avoid the use of volatile organic solvents, which endanger the environment from their production, use and eventual disposal. Hybrid separation schemes involving membranes coupled with distillation systems can avoid the use of solvents from some azeotropic separations. And modern methods of molecular design allow for the synthesis of designer solvents, having physical properties tuned specifically for a given separation, thus leading to lower energy requirements. Other ancillary molecules for chemical processing, such as reaction media and blowing agents, can also be replaced with environmentally friendly options. For example, Dow Chemical discovered that supercritical carbon dioxide works equally well as chlorofluorocarbons or simple hydrocarbons as a blowing agent for polystyrene foams. The carbon dioxide can be sourced from other plants' emissions, decreasing the carbon footprint of another process, and the polystyrene thus produced is actually more easily recycled than what was previously produced. In the pharmaceutical industries, both GlaxoSmithKline and Pfizer have produced lists of acceptable solvents to be used by their drug discovery and development teams to ensure that the production of future products uses greener technologies.

Energy efficiency is a third major area where retrofit of an existing process can lead to significant sustainability improvements. Chemical facilities can invest in renewable energy technologies themselves, or seek to join cooperative associations which produce renewable energy. Processes aiming to employ a high percentage of renewable energy must often be redesigned to handle the significant intermittency inherent in these energy production systems. While energy storage systems on-site may be feasible options for smaller facilities which implement chemical processes which are not energy-intensive, larger facilities will need to rely on the traditional power grid in order to avoid repeated shut-downs, which can damage equipment and lead to production shortfalls. In general, more flexible chemical production facilities are required when renewable energy systems are involved. Flexibility can be retrofitted into existing chemical plants through the use of additional equipment, resizing and the implementation of storage systems for energy, excess raw materials and sometimes excess products.

Finally, water reuse is an area for which retrofitting of chemical facilities is well suited. Most chemical plants are not water-efficient, since this was an unnecessary complication when they were initially designed. But water recycle systems are not difficult to add into an existing chemical plant, and water-based separation processes can often be redesigned such that less water is required. This type of redesign simply demonstrates that priorities change over time. The concepts of sustainable chemical production are new for the chemical industry, but this also provides an opportunity to improve the way chemicals are currently produced.

18.5 Take-away message

This chapter has discussed methods by which principles of sustainable engineering can be employed to improve the environmental impact of chemical production. Given the fact that most of the resources currently used to produce chemical products are limited, we need to find ways to employ renewable resources and renewable energy sources to make the products needed by an ever-increasing population. It is not tenable to continue to pollute the planet, especially the atmosphere, at the rate currently occurring. We then considered twelve principles of green chemistry, and how these principles can be implemented within chemical engineering designs. The LCA method was introduced, which provides metrics for the cradle-to-grave environmental impacts caused by a given chemical process and the product being produced. These metrics can then be included within the process synthesis and design procedure, either as heuristic screening rules, or within a multi-objective optimization formulation. In either case, the goal is to derive a small set of alternative designs, which represent options in terms of the trade-offs between environmental and economic performance. We also have many options in terms of the retrofitting of chemical plants, which allows for incremental sustainability improvements, sometimes at very low cost.

Exercises

1. Use the literature to learn about the production and use of methyl *tert*-butyl ether (MTBE) as a gasoline additive. Why was it developed, and then later on removed from some markets? How could sustainability metrics have been used during the development of this product, and what do you think the result would have been if they had?
2. Consider the use of solar power within an electrolysis system to produce hydrogen in a sustainable manner. If such a process is stand-alone (no power is taken from any source except for the solar array), what do you think the ramifications of this will be on the process? How would the design change if more constant sources of energy such as the burning of natural gas are employed?
3. Use literature sources to learn about the many production paths for the commodity chemical 1,3-butadiene. Which of these processes do you think are more sustainable than the others, and why? Roughly, how much of the environmental impact of these processes is due to the process itself, and how much comes from the production of the raw materials?

18.6 Further reading

Allen, D. T. and Shonnard, D. R. (2001). *Green engineering: environmentally conscious design of chemical processes*. Pearson Education.

Anastas, P. T. and Warner, J. C. (1998). *Green Chemistry: Theory and Practice*. Oxford University Press.

Heijungs, R. and Suh, S. (2013). *The computational structure of life cycle assessment*. Springer Science & Business Media.

Hessel, V., Kralisch, D. and Kockmann, N. (2014). *Novel process windows: innovative gates to intensified and sustainable chemical processes*. John Wiley & Sons.

Graedel, T. E. and Allenby, B. R. (2010). *Industrial Ecology and Sustainable Engineering: International Edition*. Pearson Education.

19 Optimization

The application of optimization is ubiquitous throughout the engineering design field. One could argue that design without optimization is simply wasteful, in that solutions could be found which operate at lower energy use, higher profit and decreased environmental impact. Once a chemical production facility is built, there is little scope for significant improvement in terms of the objectives discussed in Chapter 9. Flow rates, temperatures and pressures can be altered within a small range, but significant improvements, like choosing a better process chemistry or a more efficient separation process, require a major retrofit. Thus the time for structural optimization of a chemical engineering design is early, after a base-case design is chosen and simulated. Modern simulators now have optimization routines built-in, but it is important for the student to understand how such routines work, how optimization problems are formulated and, importantly, when a globally optimal solution can be expected and when near-optimal or locally optimal solutions must be accepted.

19.1 Classification of optimization problems

Optimization problems are grouped by the type of variables present, the linearity of the equations and inequalities used to define the problem and the existence of constraints limiting the solution space. All optimization problems involve one or more objective functions, which include variables defining the degrees of freedom of our system and are to be maximized or minimized. For example, our goal might be to minimize the energy usage within a production process. All practical optimization problems arising from chemical engineering systems involve constraints which limit variables to physically reasonable values. We can define a *feasible* solution as one which satisfies all constraints. A *globally optimal* solution satisfies all constraints and maximizes or minimizes a single objective function, and a *locally optimal* solution gives the maximum or minimum objective function value relative to points in a nearby region.

While all engineering systems involve continuous variables such as temperatures, pressures and flow rates, many engineering systems are modeled using integer or binary variables. The order of separations to be performed when a multi-component mixture is to be split into pure components, for example, must be modeled using discrete variables. This adds an extra layer of complexity to the optimization problem, since functions of discrete variables are non-continuous, and thus solution methods based on calculus are not directly applicable. Constraints and objective functions may be linear or non-linear functions of the variables, and of course non-linearities lead to further complications, especially the existence of local solutions. Optimization problems are also classified by considering *convexity*. A convex region is one for which all points on a line segment drawn between any two points in the region are also within

https://doi.org/10.1515/9783110570137-019

the region. So convex regions have a unimodal boundary. Furthermore, if the set of constraints defines a convex feasible region for the optimization problem, then there will be only one locally optimal point within that region, for any objective function, and thus that point will be globally optimal. Table 19.1 provides classifications for commonly encountered optimization problems in plant design.

Table 19.1: Classification of Optimization Problems.

Objective function	Variables	Constraints	Classification
Single, linear	All continuous	All linear	Linear program, or LP
Single, non-linear	All continuous	Non-linear	Non-linear program, or NLP
Single, linear	Some continuous, some integer/binary	All linear	Mixed-integer linear program, or MILP
Single, non-linear	Some continuous, some integer/binary	Non-linear	Mixed-integer non-linear program, or MINLP
Multiple, non-linear	Some continuous, some integer/binary	Non-linear	Multi-objective MINLP
Single, non-linear	Some continuous, some integer/binary	Non-linear with differential equations	Dynamic MINLP

These classifications are helpful to our understanding, since we often select solution methods specific to a given problem type. Also, the existence of local solutions for some classifications is ruled out, and this will be further explained in Section 19.4.

19.2 Objectives and constraints

As mentioned in Chapter 9, our goals for the process design task can be formulated as objective functions. Usually, an economic objective function is used, and environmental and safety considerations are written as constraints to our problem. Since most process design tasks seek to create a plant to make a set capacity of material, and the selling price of that product is determined by market forces beyond our control, the economic objective may be formulated in terms of minimizing production cost. Fixed costs should be annualized, such that they can be combined with operating costs in order to get a full picture of the funds required to produce one unit of product. However, when we consider depreciation of the equipment and the concomitant tax ramifications, we can imagine that simply minimizing the operating costs often leads to an optimal design. If we have already fixed the process chemistry, then the amount of raw material required can only vary by a small percentage, based on the quality of the separation processes. In these cases, minimizing energy consumption per kilogram of product is a fine objective. This also usually corresponds to the minimum carbon dioxide emission solution (including sources caused by energy generated off-site), and so environmental factors are at least partially taken into account as well.

In order to consider both environmental and economic objectives simultaneously, a multi-objective optimization problem must be formulated and solved. In this case, a family of solutions exists for which no objective can be improved without making another objective worse. This so-called *pareto-optimal set* of solutions represents the trade-offs inherent in process design and gives a designer options in terms of balancing the multiple objectives. One can always assign a monetary value to environmental, safety or other objectives, and solve the system as a single-objective optimization problem. This however is equivalent to assigning a single weighting to the different objectives, which then gives a single member of the pareto-optimal set of solutions. For process design, this corresponds to a single set of choices for equipment, flow rates, temperatures and pressures to be used to produce the product. Often, the weighting selected overvalues the cost minimization, leading to designs which pollute the environment right up to the regulatory limit. Consideration of the pareto-optimal set of designs can allow solutions to be considered for which a moderate increase in cost provides a significant reduction in environmental impact.

The constraints of our process design optimization problem come from the basic expressions which model our system: the mass and energy balances for each unit operation, thermodynamic relations and kinetic expressions defining the rates of consumption and generation of species. Limits on process variables based on material and safety considerations also become inequality constraints within our formulation. When the optimization problem is formulated using a superstructure to provide technology choices, then constraints must be written conditionally on the existence of a given unit type, and limits on the number of technology choices must also be provided. For example, it is very unlikely that both distillation and extraction would be selected to perform a single specific separation within a chemical process. Thus binary variables would be assigned to each technology, and the sum of those two variables would be forced to equal one. This forces the optimal solution to include exactly one of the technologies, and the continuous variables (temperatures, pressures) related to that unit operation type would be constrained based on the type of unit chosen.

When considering a plant design optimization problem, we note that non-linearities occur only in specific expressions, and this can be used to help guide solution algorithms. Mass balances are normally formulated as linear equations, except for reaction-based terms. The generation and consumption terms are of course exponential functions of temperature, and the *ad hoc* linearization of such terms leads to highly inaccurate models in most cases. Many thermodynamic models are also nonlinear, and while for some systems simplified, linear thermodynamic models may be good enough to make choices regarding separations, systems which are azeotropic, or exhibit multiple liquid phases, must always be modeled using non-linear expressions, or these important physical phenomena cannot be taken into account. It is important to realize that the non-linear terms which have just been highlighted only involve the continuous variables within a plant design optimization problem. It is usually possible to reformulate such a problem to ensure that the discrete variables, corresponding to design decisions, only occur linearly. Thus we see a specific subclass of optimization problems, which we will define as a *plant design MINLP*, as shown in the following equation:

$$\min \quad \text{Cost} = f(\bar{x}) + \bar{c}^T \bar{y},$$

$$\text{s.t.} \quad h_i(\bar{x}) = 0,$$

$$g_j(\bar{x}) + \overline{\overline{M}}\bar{y} \leq 0,$$

$\bar{x}$ continuous, $\bar{y}$ binary/integer,

where: $f(\bar{x})$ nonlinear but convex,

$h_i(\bar{x})$ linear,

$g_j(\bar{x})$ nonlinear but convex. $\qquad$ (19.1)

While this problem is still very difficult to solve in the general case, it is in fact significantly simpler than the general MINLP problem, which could include non-linear combinations of discrete and continuous variables.

19.3 Specific formulations for process design and operations

As shown in the last section, the optimization of chemical process designs and operation leads to very specific types of formulations. The simplest case is the linear program, which often occurs when we select amounts of different products to produce in batch mode, or when dealing with blends of polymers or additives combined with an active ingredient. The interactions between molecules in such cases tend to be additive, so linear mixing rules are appropriate.

Example 1 (Linear programming formulation of a batch production problem). Consider the production of three types of lawn fertilizer: regular, garden and extra-strength. We will assume that only two ingredients are needed to produce these products, nitrate and phosphate. The current inventory contains 80 metric tons of nitrate and 50 metric tons of phosphate. The ratios of raw materials required to produce each type of fertilizer and the net profit obtained by producing each type are listed in the table below. Assuming no more raw materials can be purchased (but raw materials may be left in inventory) and sufficient demand exists to sell all produced products, formulate an optimization problem to determine how much of each type of fertilizer to produce to maximize the profit.

Raw material requirements and net profit for fertilizer types.

	Nitrate (tons/1,000 bags produced)	Phosphate (tons/1,000 bags produced)	Profit (€/1,000 bags produced)
Regular	4	2	300
Garden	2	2	400
Extra-strength	4	3	500

Since the profit and the consumption of raw materials are linearly proportional to the amount of each type of fertilizer, we can see that this problem can be formulated as a linear program. First of all, we define the variables x_r, x_g and x_e to be the number of 1,000 bag batches of regular, garden and extra-strength fertilizer to be produced, respectively. Then the objective function may be written as

$$Profit = 300x_r + 400x_g + 500x_e.$$

Constraints are necessary to limit the amount of raw materials consumed to those present in inventory and to ensure feasibility, *i. e.*,

$$4x_r + 2x_g + 4x_e \leq 80,$$

$$2x_r + 2x_g + 3x_e \leq 50,$$

$$x_r, x_g, x_e \geq 0.$$

Note the final constraint is very important, since it prevents solutions which produce negative amounts of products. While such solutions are obviously impossible, this makes it easy to forget to exclude them, which can lead to completely incorrect results. In this case, a solution with a negative production amount would actually create raw materials, which would then be consumed to make more profitable products! Careful formulation is necessary to avoid such errors, which can be very hard to find later for a large problem.

A significant number of optimization problem formulations for plant design involve technology selection, and thus must include discrete variables. Another problem involving discrete optimization is batch scheduling, when we must decide how many batches of a set of products to produce, and in which order those products should be produced. Note that in all of these problems, a simplification occurs if we assume all the variables are continuous. If the objective function and constraints are linear and we make this simplification, then a linear program results which is called the *LP-relaxation* of the original problem. While the solution to that problem is usually not feasible to the original MILP, and rounding the solution to the nearest integer can result in a huge error or infeasible schedule, the objective function value obtained from that LP-relaxation contains valuable information. Since the relaxation from discrete to continuous variables only increases the size of the feasible region (the number of possible solutions), the optimal solution to the relaxed problem must be as good as or better than the optimal solution to the original MILP. If the optimal solution to the

LP-relaxation (which is not so hard to find, as we will see in the next section) is not good enough for us to continue with the project, then we have no need to solve the original MILP, since its solution is worse. For example, if the objective is to maximize profit, and the optimal solution to the LP-relaxation is negative, then clearly none of the options defined by the MILP are good enough to be considered. The following example shows an MILP formulation of an optimization problem arising from a chemical engineering task.

Example 2 (Batch production of two products). Consider the production of two products, P1 and P2. Each product is to be made in batches, and all batches are constrained by equipment requirements to create 2,000 lb of product. Due to time and labor limitations, we can only produce 8,000 lb of P1 per day and 10,000 lb of P2 per day. The chemical formulas to produce the two products, based on the three raw materials A, B and C, are as follows:

$$0.4 \text{ lb A} + 0.6 \text{ lb B} = 1 \text{ lb P1},$$

$$0.3 \text{ lb B} + 0.7 \text{ lb C} = 1 \text{ lb P2}.$$

While there are no supply limits on A or C, the amount of B which can be procured is no more than 6,000 lb per day. Profits are \$0.16 per lb of P1 and \$0.20 per lb of P2. Formulate an optimization problem to compute the optimal number of batches.

First of all, we must define variables to store the number of batches. Since partial batches are disallowed, we can simply define an integer variable for the number of each type of batch to be produced: y_{P1} and y_{P2}. Next, we write the objective function to maximize profit, *i. e.*,

$$\text{Profit (\$1,000/day)} = 0.32 y_{P1} + 0.4 y_{P2}.$$

Note the profit per pound of each product must be multiplied by two to get the coefficients of the objective function, since the variables are defined as the number of 2,000 lb batches. We also need constraints on raw material consumption and production of each product, *i. e.*,

$$1.2 y_{P1} + 0.6 y_{P2} \leq 6,$$

$$0 \leq y_{P1} \leq 4,$$

$$0 \leq y_{P2} \leq 5.$$

Note again that the coefficients in the resource limitation must be doubled, since the batches create 2,000 lb of product each. Also, we are careful to avoid negative numbers of batches, as seen in the previous example. Finally, we explicitly state that the values of the variables must be integers, *i. e.*,

$$y_{P1}, y_{P2} \in 0, 1, 2, \ldots.$$

This completes the formulation of the problem.

As stated in the last section, the plant design MINLP is a form which commonly occurs when non-linear mass balances arising from kinetic and/or thermodynamic expressions are combined with linear mass and energy balances and discrete variables involving technology selections. These are particularly challenging problems to formulate and solve, but the reward for doing so can be quite large. Most chemical facilities are designed and operated at a feasible point far from the optimum. This provides a motivation for process improvement, since often lower costs, better environmental performance or enhanced process safety can be obtained, and sometimes all three objectives can be simultaneously achieved. Furthermore, improvements in formulation approaches and solution techniques for MINLPs since the 1990s provide tools to give at least near-optimal solutions to many of these problems.

19.4 Solution algorithms for optimization problems

Great progress has been made on the solution of the types of optimization problems commonly encountered in chemical process design since 1948. George Dantzig's development of the simplex algorithm enabled the solution of linear programs, first by hand and now with high-performance computing. Since the feasible region of a linear program is always convex, we know that there can be at most only one optimal solution, a global one. The simplex method first generates a basic feasible solution, meaning a solution at a corner of the feasible region, and then sequentially generates new basic feasible solutions which improve the objective function. In this way, the algorithm can be shown to always converge to the global optimum. Practically, even large linear programs derived from chemical engineering applications can be solved in just a few seconds, and the simplex algorithm is built into common software systems such as Microsoft Excel, Mathematica and MATLAB®. Thus even when a more complex optimization problem is formulated, creating an LP-relaxation of that problem and solving it provides valuable information (a lower bound when we are minimizing an objective), and is quite easy to obtain.

Example 3 (Solution of a linear program). Let us consider the linear program created in Example 1:

$$\text{Maximize:} \quad \text{Profit} = 300x_r + 400x_g + 500x_e.$$

$$\text{Subject to:} \quad 4x_r + 2x_g + 4x_e \leq 80,$$

$$2x_r + 2x_g + 3x_e \leq 50,$$

$$x_r, x_g, x_e \geq 0.$$

This is a linear program in standard form. Inputting this problem into commonly available mathematical software gives us the following result:

$$x_r = 0, \quad x_g = 25, \quad x_e = 0, \quad \text{Profit} = 10{,}000.$$

We discover that, based on the current price structure, our best strategy is to use all of our resources to create the garden-type fertilizer. This makes sense when we consider that the garden type requires less of the limiting resource (phosphate) than the extra-strength, and provides 1.5 times more profit per bag than the regular type, which consumes the same amount of phosphate. Note not all of the nitrate is consumed, and there is no reason why it should be. The feasible solution $x_r = 15, x_g = 10, x_e = 0$, for example, consumes all of the phosphate and nitrate, but only leads to a profit of 8,500.

In comparison to linear programs, mixed-integer linear programs are much more difficult to handle. The most straightforward approach to their solution is the branch-and-bound method, which seeks to sequentially prune the set of all possible solutions, and use bounding to prove that entire sets of feasible solutions cannot include the global optimum.

At each step, an upper bound and a lower bound on the optimal solution are computed, and when these bounds meet, the optimal solution has been found. Each relaxed subproblem within this tree is a linear program, so we can repeatedly apply the simplex method to compute the value of the objective for each node. Again, the convexity of the feasible region ensures that when an optimal point is found, it is a global one, and no further searching is required.

The branch-and-bound algorithm starts by solving the LP-relaxation of the original problem. This means that we relax the integrality conditions on all discrete variables, treating them as continuous. Assuming we are maximizing a single objective function, the LP-relaxation provides an upper bound on the optimal solution to the original MILP. If the LP-relaxed optimal solution is feasible to the original problem, then it is the global optimum to the MILP. Normally, that solution is infeasible to the MILP, since some of the variable values will be fractional. Assuming that value is good enough so that we might actually employ the optimal solution, we continue the process by *branching* on one of the variables which was found to be fractional in the previous step. We write new constraints, disallowing the fractional region in which that solution lies, and solve two new problems: linear programs in which the branched variable is constrained to be either above or below its previous value. For example, if a variable y was found to be 2.3 at the optimal solution to the LP-relaxation, then two new problems are formed: one for which the constraint $y \leq 2$ is added, and one for which the constraint $y \geq 3$ is added. In this way, the two new problems are more constrained than the LP-relaxation, but still less constrained than the original LP. See Example 4 for a diagram of the branch-and-bound tree. If a problem which has no feasible solutions is encountered during the process, that pathway is discarded.

The main advantage of the branch-and-bound method is known as *implicit fathoming*, which allows us to systematically prove that entire sets of solutions cannot be better than one we already have. This is done by showing that the upper bound for a branch of the solution tree is lower than the lower bound (value of a feasible solution) elsewhere in the tree. Since a lower bound corresponds to a feasible solution to the problem, any branch with an upper bound below the best lower bound must only contain suboptimal solutions. This means that the entire tree of solutions need not be evaluated, which saves orders of magnitude of computation time for large MILP problems.

Example 4 (Branch-and-bound solution of an MILP). Consider the batch processing MILP derived in Example 2

$$\text{Profit (\$1,000/day)} = 0.32y_{P1} + 0.4y_{P2},$$
$$1.2y_{P1} + 0.6y_{P2} \leq 6,$$
$$0 \leq y_{P1} \leq 4,$$
$$0 \leq y_{P2} \leq 5,$$
$$y_{P1}, y_{P2} \in 0, 1, 2, \ldots.$$

This example has few enough feasible solutions (30) that each could be evaluated separately, and the one giving the best objective function is chosen. Nevertheless, let us look at the performance of the branch-and-bound method for this case. First, we evaluate the LP-relaxation of the problem, *i. e.*,

$$\text{Profit} = 0.32y_{P1} + 0.4y_{P2},$$
$$1.2y_{P1} + 0.6y_{P2} \leq 6,$$
$$0 \leq y_{P1} \leq 4,$$
$$0 \leq y_{P2} \leq 5,$$
$$y_{P1}, y_{P2} \text{ continuous.}$$

The optimal solution, obtained by the simplex method, is $y_{P1} = 2.5$, $y_{P2} = 5$, Profit = 2.8. Assuming 2.8 is an acceptable profit in this case (it is the maximum we could ever have), we continue by branching on the fractional variable, y_{P1}. We draw the tree of solutions with the root node at the top, corresponding to the LP-relaxation, and branched problems shown underneath.

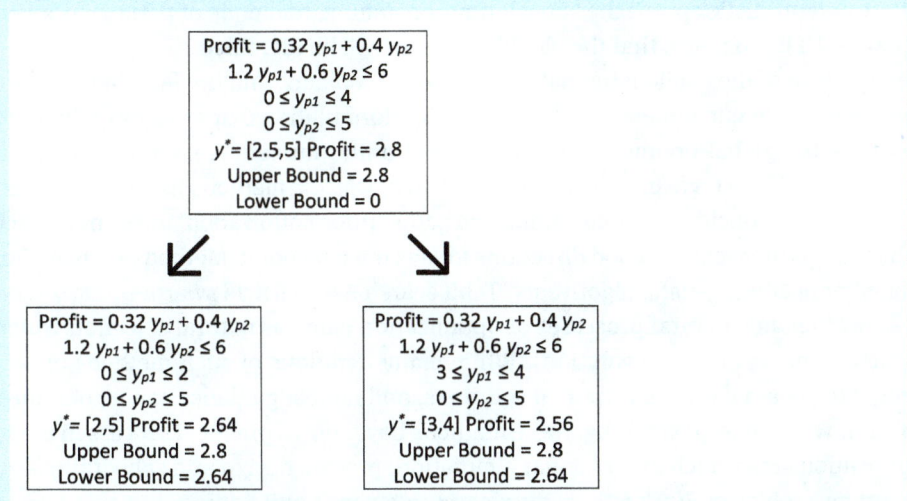

The branch-and-bound tree shows that the final solution has been found by only evaluating three linear programs. The node in the bottom left shows the result of branching y_{P1} below its fractional value of 2.5. When we solve the resulting linear program, we find an integer solution, meaning that this solution is feasible to the original problem. Thus the optimal solution to that LP forms a lower bound on the solution to the original problem, *i. e.*, we know that we can at least reach a profit value of 2.64. Next we branch above the y_{P1} value of 2.5, in the mode on the lower right. We see that another integer solution is found as the solution to that LP. Thus another lower bound is computed, but the value of this lower bound, 2.56, is below that of the lower bound previously determined via the left node. Therefore, we cannot branch any further, and we know that the optimal solution corresponds to the leftmost node, with $y_{P1} = 2$ and $y_{P2} = 5$.

Note that if the rightmost node had given a fractional value for either y_{P1} or y_{P2}, that would correspond to an upper bound on the right half of the tree. No solution found by branching below that node could be better than the upper bound found there. If that lower bound was found to be below 2.64, then no further branching would be required, since any solution feasible to the original problem found below that node would have an objective value below 2.64, and thus would be worse than the one we already have. This demonstrates the implicit fathoming idea.

The solution of MINLP problems is much more difficult than for MILPs. We can apply the branch-and-bound method as just described, but there is a challenging issue which complicates matters. The subproblems to be solved at each node become nonlinear programs in this case, and they are not necessarily convex. This means local and not globally optimal solutions may be obtained at those nodes, and those locally optimal solutions can correspond to false bounds. If the bounds are incorrect, then the implicit fathoming which we demonstrated does not work directly. Thus more sophisticated approaches are needed. These include finding a convex region to approximate the feasible region of the original problem; decomposing the original problem into subproblems which can be solved sequentially, and by which reliable bounds can be derived; or by considering local linearizations for which the error in each step is small. These deterministic approaches are all timeconsuming, and none of them guarantee, for all MINLP problems, that the global optimum will be identified.

Because of the challenging nature of MINLP problems, and the fact that they are very commonly encountered in process design formulations, approaches which do not seek the global optimum, but instead try to find numerous near-optimal solutions, have also been explored. These so-called *stochastic* methods use a guided random search to consider feasible points, and gather information about the search space which helps them choose good directions to seek the next point. Methods such as simulated annealing, genetic algorithms, Tabu search and particle swarm optimization all seek to mimic natural processes of optimization (such as natural selection which optimizes organisms via evolution). While many iterations of such methods are required to generate the near-optimal solutions, and no real guidance can be obtained to know when to stop searching, these methods have shown practical success in many application areas, such as molecular simulation, process design and scheduling and planning problems. Stochastic methods are in a sense quite crude, but this also is advantageous in that they become highly parallelizable. This means that similar computations can be made on many processors of a multi-processor computing platform simultaneously, and the results can simply be compared at the end. The lack of communication between processors required with these methods stands in stark contrast to the deterministic methods such as branch-and-bound, in which each step of the algorithm depends on the previous one, making the algorithms unsuitable for naïve parallel implementation.

The necessity of finding the global optimum should always be considered when selecting an algorithm to solve a large MINLP. If the constraints and objective functions are inexact, such as when they include projected future costs and product demands, or when they use simplified models to represent the complex behavior of multi-component chemical mixtures, then consideration should be given to the use of stochastic optimization techniques. The extra effort and time required to formulate and implement a deterministic algorithm is likely outweighed by the accuracy of the final result. A stochastic method which could provide a number of solutions within a few percent of the global optimum is often more useful to a process designer.

19.5 Take-away message

This chapter has provided an overview of the types of optimization problems commonly encountered in chemical process design. Design and optimization are tightly coupled, in that it makes little sense to implement a suboptimal design when an optimal one can be determined. Optimization problems consist of one or more objective functions, which represent the goals of our design, and a set of constraints which enforce the laws of physics and chemistry, as well as logical limits on variables. The variables present in such problems may be continuous ones like temperature and flow rates, or discrete variables defining decisions to be made as part of the design, such as equipment selections, or the order in which batches are to be produced. A consideration of the linearity of the model expressions, and the types of variables encountered, leads to a classification of optimization problems. This classification system can help us to select appropriate solution methods, or software packages containing these methods. Linear and mixed-integer programs have feasible regions which are convex, meaning a single, globally optimal point exists for a well-formulated problem, and the simplex method (combined with the branch-and-bound method for the mixed-integer case) provides an algorithm which can solve these problems effectively. For the difficult case of MINLPs, we must choose between deterministic algorithms, which will provide a guarantee of a global optimum, but may fail to find any solutions at all, and stochastic algorithms, which use a direct approach to searching the space of feasible solutions, and often find a number of near-optimal solutions quickly.

Exercises

1. Formulate the following problem as a linear program, and use commercial software to find the optimal solution.

   A farmer owns a farm which can produce three crops: corn, oats and soybeans. There are 24 acres of land available for cultivation, and each crop requires different amounts of labor and capital investment. These data, along with the net profit from each crop, are listed in the following table. The farmer has $400 in available capital, and has sufficient laborers to provide 48 hours of work. How much of each crop should the farmer plant to maximize profit?

	Labor (hrs)	Capital ($)	Net profit ($)
Corn (per acre)	6	32	45
Oats (per acre)	3	21	25
Soybeans (per acre)	7	25	32

2. Solve the following MILP problem using the branch-and-bound method. Use a commercial software package to solve the individual linear programs which occur at each node of the branch-and-bound tree.

$$\max \quad z = 86y_1 + 4y_2 + 40y_3,$$
$$\text{s.t.} \quad 774y_1 + 76y_2 + 42y_3 \leq 875,$$
$$67y_1 + 27y_2 + 53y_3 \leq 875,$$
$$y_1, y_2, y_3 \in 0, 1.$$

3. Perform a literature search to familiarize yourself with the simulated annealing algorithm. What are the basic steps involved, and why is it chosen over deterministic algorithms for certain applications? How are random values involved with the computations, and why does the use of random numbers improve the performance of the algorithm?

19.6 Further reading

Deb, K. (2012), *Optimization for engineering design: Algorithms and examples*. PHI Learning Pvt. Ltd.

Edgar, T. F., Himmelblau D. M. and Lasdon, L. S. (1988), *Optimization of chemical processes*. New York: McGraw-Hill.

Kolman, B. and Beck, R. E. (1995), *Elementary linear programming with applications*. Academic Press.

Rao, S. S. (2009), *Engineering optimization: theory and practice*. John Wiley & Sons.

Yang, X.-S. (2010), *Engineering optimization: an introduction with metaheuristic applications*. John Wiley & Sons.

20 Enterprise-wide optimization

20.1 What is Enterprise-wide optimization?

According to Grossmann, enterprise-wide optimization (EWO) is an area at the inter-face of chemical engineering (process systems engineering) and operations research. Optimizing the operations of supply, manufacturing (batch or continuous) and distri-bution in a company are its main features.

The major operational activities include planning, scheduling, real-time opti-mization and inventory control. Supply chain management is similar to EWO. There is a significant overlap between the two terms; an important distinction is that supply chain management is aimed at a broader set of real-world applications with an em-phasis on logistics and distribution, which usually involve linear models, traditionally the domain of operations research. In contrast, in EWO, the emphasis is on the man-ufacturing facilities with a major focus being their planning, scheduling and control which often requires the use of non-linear process models and, hence, knowledge of chemical engineering.

Many process companies are adopting the term EWO to reflect both the impor-tance of manufacturing within their supply chain and the drive to reduce costs through optimization.

20.2 Fast-moving consumer goods supply chains

An important area of EWO is the so-called fast moving consumer good (FMCG) in-dustry. Every day, billions of people worldwide use FMCGs. These FMCGs are typi-cally products one can buy at a supermarket, such as food products, drinks and de-tergents. These products are mostly produced by large multi-national companies such as Unilever, Procter & Gamble and Nestlé, which all have a yearly revenue of over €50 billion. These companies produce a wide range of products to satisfy an increasing demand for product variety. In fact, even a single product category, such as ice cream, can consist of a thousand Stock Keeping Units (SKUs). These SKUs are products that may vary in composition or packaging.

Generally, FMCGs are fully used up or replaced over a short period of time, ranging from days to months depending on the product. In addition, seasonality plays an im-portant role in the FMCG industry. Not only are the products often seasonal, but many of the ingredients are seasonal as well. In addition, many FMCGs have a maximum shelf-life and can, therefore, only be stored for a limited amount of time.

When a FMCG is sold out, a consumer will typically buy a substitute product rather than waiting for the product to become available. This substitute product might very well be a product of one of the competitors. Therefore, it is very important that a suffi-cient amount of product is available at the retailers. However, the costs of maintaining a sufficient inventory throughout the supply chain to ensure a high customer service

https://doi.org/10.1515/9783110570137-020

level are generally significant. Even though FMCGs are profitable because they are produced and sold in large quantities, they generally have a low profit margin. Therefore, it is crucial to obtain the right balance between minimizing the total costs of operating the supply chain, including inventory costs, and ensuring that a sufficient amount of product is available to meet the demand. However, the scale and complexity of the enterprise-wide supply chains in which these products are produced have increased significantly due to globalization.

Therefore, EWO has become a major goal in the process industries. EWO is a relatively new research area that lies at the interface of chemical engineering and operations research. It involves optimizing the procurement, production and distribution functions of a company. Although similar to supply chain management, it typically places a greater emphasis on the manufacturing facilities.

The major activities included in EWO can be divided into three temporal layers, as is shown in Figure 20.1.

Figure 20.1: Temporal layers in enterprise-wide optimization.

20.3 Scheduling

Scheduling plays an important role in most manufacturing and service industries. It involves allocating limited resources to activities such that the objectives of a company are optimized. Various objectives can be used in the scheduling. For example, the objective can be to minimize the time to complete all activities, the total tardiness or the total costs.

The type of scheduling considered in this chapter is short-term production scheduling of a single production facility. The main decisions of production scheduling are to select the tasks to be executed, to decide on which equipment the tasks should be executed and to determine the sequence and the exact timing of the tasks.

Production scheduling has historically been done manually using pen and paper, heuristics, planning cards or spreadsheets. However, especially when the utilization percentage of the equipment is high, the scheduling can become complex, and even obtaining a feasible schedule can become difficult. As a result, optimization support can substantially improve the capacity utilization, leading to significant savings. Typically an increase in effective production capacity of 10 %–30 % can be achieved by using a multi-stage scheduling model rather than manually scheduling the separate stages in the factory.

A variety of approaches have been developed to facilitate the production scheduling and to improve the solutions. These approaches include expert systems, mathematical programming and evolutionary algorithms. In chemical engineering, MILP is one of the most commonly used methods for optimizing scheduling problems.

For example, state-task network (STN) and resource-task network (RTN) MILP formulations have been applied to a large variety of scheduling problems.

One of the most important characteristics of scheduling models is their time representation. A discrete time model divides the scheduling horizon into a finite number of time slots with a prespecified length. However, to accurately represent all activities of various length, a fairly short time slot length must typically be used. As a result, the total number of time slots might be large. Since most variables and constraints are expressed for each time slot, the resulting model can become very large. On the other hand, the advantage of a formulation based on a discrete time horizon is that it is typically tight.

A continuous-time model also divides the scheduling horizon into a finite number of time slots, but the length of these time slots is determined in the optimization. Therefore, the number of time slots can be reduced considerably, and the resulting models are substantially smaller. Nevertheless, they do not necessarily perform better computationally than discrete time formulations. In addition, it can be challenging to determine the optimal number of time slots in a continuous-time model.

In a precedence-based model, the tasks are not directly related to a timeline, but instead the focus is on binary precedence relationships between tasks executed on the same unit. As a result, precedence-based models can handle sequence-dependent changeover times in a straightforward manner. However, the disadvantage of precedence-based models is that the number of sequencing variables scales quadratically with the number of batches to be scheduled. As a result, precedence-based models can become relatively large for real-world applications.

Another important characterization of scheduling models is their scope. A generic model aims at addressing a wide variety of problems. On the other hand, in a problem-specific formulation the computational efficiency can be increased by using the problem characteristics. However, this often means that the problem-specific model relies on these problem characteristics and is, therefore, only applicable to a smaller range of problems.

Nevertheless, problem-specific formulations can often be used to optimize problems that are too large or complex for generic models.

20.4 Planning

Tactical planning seeks to determine the optimal use of the procurement, production, distribution and storage resources in the supply chain. Often an economic objective, such as the maximization of profit or the minimization of costs, is used in the tactical planning.

Tactical planning typically considers a multi-echelon supply chain instead of the single production facility considered in scheduling. In addition, it considers a longer time horizon, which usually covers one year. This one-year time horizon should be divided into periods to account for seasonality and other time-dependent factors. As a result of the larger scope of tactical planning, aggregated information is often used. For example, products could be aggregated into families, resources into capacity groups and time periods into longer periods. Moreover, the level of detail considered in the tactical planning is lower than in scheduling. For example, the sequence dependency of set-up times and costs is typically not considered in a tactical planning problem.

The optimal management of the supply chain can offer a major competitive advantage in the global economy. Therefore, both academia and industry have acknowledged the need to develop quantitative tactical planning models to replace commonly used qualitative approaches. These optimization models can resolve the various complex interactions that make supply chain management difficult. Probably the most commonly used type of optimization models in supply chain management are mathematical programming models.

Originally, these models focused on subsets of the tactical planning decisions. However, optimizing the tactical planning of the various supply chain functions separately leads to suboptimal solutions due to the lack of cross-functional coordination. Therefore, it is desirable to consider the entire supply chain in a tactical planning model, for example, increasing both profit and customer service levels by integrating production and distribution decisions in an MILP model rather than using a decoupled two-phase procedure commonly found in industry.

In addition to mathematical programming, multi-echelon inventory systems are also often used to optimize some of the tactical planning decisions. In particular, these models can be used to determine the optimal safety stock levels across the supply chain considering demand uncertainty. A planner-led effort at Procter & Gamble that was supported by single- and multi-echelon inventory management tools reduced the inventory levels and resulted in US$1.5 billion savings.

20.5 Mixed-integer programming

As mentioned in the previous two sections, mathematical programming is a commonly used method in the optimization of scheduling and planning. MILP models, which are a specific class of mathematical programming models, are suitable for these type of problems because they can accurately capture the important decisions, constraints and objectives in supply chain problems. In addition, demonstrably good solutions to these problems can be obtained with MILP methods. In fact, given a sufficient amount of time, MILP methods can even yield optimal solutions. In this section we will provide a brief introduction to MILP models.

In an MILP model, a linear objective is optimized subject to linear constraints over a set of variables. An MILP model contains both continuous variables, which can assume any non-negative value, and integer variables, which are limited to non-negative integer values.

In the most common case of MILP models, the integer variables are binary variables, which are constrained to values of 0 or 1. This type of MILP model will be considered in the rest of this chapter. These binary variables typically represent yes/no decisions. For example, the decision whether a product should be produced in a certain factory and in a certain week can be modeled with a binary variable. The general form of an MILP model with binary variables is given below:

$$\min Z(a, x, y), \tag{20.1}$$

subject to

$$g(b, x, y) = 0, \tag{20.2}$$

$$h(c, x, y) > 0, \tag{20.3}$$

$$x^L \leq x \leq x^U, \tag{20.4}$$

$$y \in Y, \tag{20.5}$$

where Z is the objective function that is minimized (economic or environmental criteria), x and y are decision variables (continuous or discrete) and a, b, c are parameters in the model. The g are the equality constraints (mass, energy balances) and the h are the inequality constraints (operating or design windows).

The major difficulty in optimizing MILP problems comes from the binary variables. For any choice of 0 or 1 for all of the elements of the vector of binary variables y, a resulting linear programming (LP) model containing the x variables can be optimized. The simplest method of optimizing an MILP problem would then be to optimize these LP models for all possible combinations of the binary variables. However, the number of possible combinations of the binary variables increases exponentially with the number of binary variables. Consequently, the required computational effort of optimizing the MILP model with such a brute force approach would also grow exponentially. Rather than using a pure brute force approach, MILP models have traditionally been solved via branch-and-bound (B&B) techniques. The B&B method is an implicit enumeration approach that theoretically can solve any MILP by solving a series of LP-relaxations. An LP-relaxation of the MILP problem can be obtained by removing the integrality requirements from the integer/binary variables. In other words, binary variables can assume any value from 0 to 1 in the relaxed model.

The first step in the B&B method is to optimize the LP-relaxation of the MILP model. If all relaxed binary variables are exactly 0 or 1 in the obtained solution, the solution of the LP-relaxation is a feasible and optimal solution of the MILP. The feasible region of the original MILP problem is a subset of the feasible region of the LP-relaxation. Therefore, assuming a minimization objective, when non-integer values are obtained for the 0–1 variables, the LP-relaxation provides a lower bound on the objective of the MILP model. As a result, if the optimal solution of the LP model is feasible for the MILP model, it will also be the optimal solution of the MILP model.

However, in most cases the relaxed solution will contain binary variables that have taken a value between 0 and 1, and therefore, the solution will be infeasible for the MILP model.

The B&B method then selects one of the binary variables that violate the integrality requirement and branches on this variable. That is to say, it will create two subproblems, one where the binary variable is fixed at 0 and one where the binary variable is fixed at 1. These subproblems are denoted as nodes. An LP-relaxation of these subproblems can then be solved. The solutions to these LP-relaxations provide lower bounds for their respective nodes.

This procedure is then repeated until a feasible solution is obtained. This feasible solution provides an upper bound. After all, any solution with an objective that is higher than this upper bound will be inferior to the obtained solution. The upper bound is updated each time a new feasible solution with an objective value that is less than the current upper bound is obtained. At any time during the optimization, there will be a set of nodes that still need to be investigated. A search strategy determines the order in which the nodes are selected for branching. Examples of search strategies are depth-first and best-bound.

The advantage of the B&B method compared to a pure brute force approach lies in the pruning of nodes. If the lower bound of a certain node is equal to or higher than the current upper bound, any solution that could be obtained in this node, or any of the successors of this node, will have an objective that in the best case is equal to the best obtained solution. Therefore, the successors of this node do not need to be investigated since they will not provide better solutions. The B&B algorithm will terminate once all nodes have been pruned, since that signifies that either the optimal solution has been obtained or no feasible solution exists.

Often these B&B algorithms are optimized with a specified optimality tolerance. In this case, any node whose lower bound is within, for example, one percent of the best obtained solution will be pruned. The advantage is that this can greatly speed up the optimization. The disadvantage is that the obtained solution might no longer be optimal. The only guarantee is that it is within 1% of the optimal solution. Nevertheless, the available optimization time is usually limited in practice. As a result, it is often preferable to obtain a near-optimal solution in a reasonable amount of time rather than an optimal solution in a considerably longer amount of time. Even with this B&B algorithm and with a small optimality tolerance, MILP problems often become difficult to solve when the number of binary variables increases. In fact, 0–1 MILP problems belong to the class of NP-complete problems.

However, since the 2000s great progress has been made in solution algorithms and computer hardware. Current state-of-the-art commercial solvers, such as CPLEX and GUROBI, incorporate a wide variety of approaches into the B&B algorithm. For example, cutting planes can be used in a B&B based approach to considerably improve the obtained bounds, and thereby greatly reduce the required amount of enumeration.

These algorithms lead to improvements to prepoprcessing, branching and primal heuristics. In addition, branch-and-cut and branch-and price versions have lead to further improvements of the B&B algorithm.

As a result of these and other improvements, a purely algorithmic speed-up of more than 55,000 times has been reported between CPLEX versions 1.2 and 12.2. The performance improvement in CPLEX between 1991 and 2009 is shown in Figure 20.2. Due to the combination of this algorithmic speed-up and the improvement in computer hardware, solving MILP problems has become around 100 million times faster since 2000. Due to this drastic improvement, many MILP problems that were unsolvable a decade ago can be solved within seconds today. This allows for the development of more realistic models that include more details and have a wider scope. Nevertheless, even with the vastly improved optimization capabilities, many realistically sized optimization problems are still challenging from a computational point of view, since MILP problems are NP-complete, which means that in the worst case the computational time increases exponentially with the problem size.

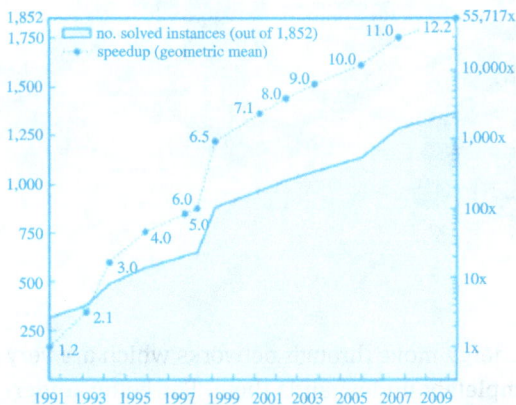

Figure 20.2: Performance improvement of CPLEX versions 1.2 through 12.2. The geometric mean speed-up is shown on the right axis and the number of instances that could be solved out of a test set of 1,852 instances is shown on the left axis. The shown improvements are purely due to algorithmic improvements.

20.6 Optimization challenges

Steve Shashihara, author of *The optimization edge*, asks: "Why do some companies become industry leaders, while others never rise to the top? Who has the competitive advantage and why?" The answer is: leading enterprises have tight focus, they focus attention on making optimal decisions, squeezing every ounce of value from the assets under management.

Optimization is a decision making process and a set of related tools that employ mathematics, algorithms and computer software not only to sort and organize data but also to use these data to make recommendations faster and better than humans can. This definition by Shashihara fits very nicely a definition of Ignacio Grossmann on what is process systems engineering: "PSE concerns systematic analysis and optimization of decision making processes for the discovery, manufacture and distribution of products." In other words: optimization is key!

We have just seen that FMCG networks lead to more and more connectivity (humans, industry, grid), more and more decisions to make (technology, capacity, etc.) and more and more uncertainty (production, storage, conversion, demand). This does not only hold for FMCG networks, but also for agro-food supply chains, water networks and energy systems. See for example Figure 20.3 on complexity in smart grids.

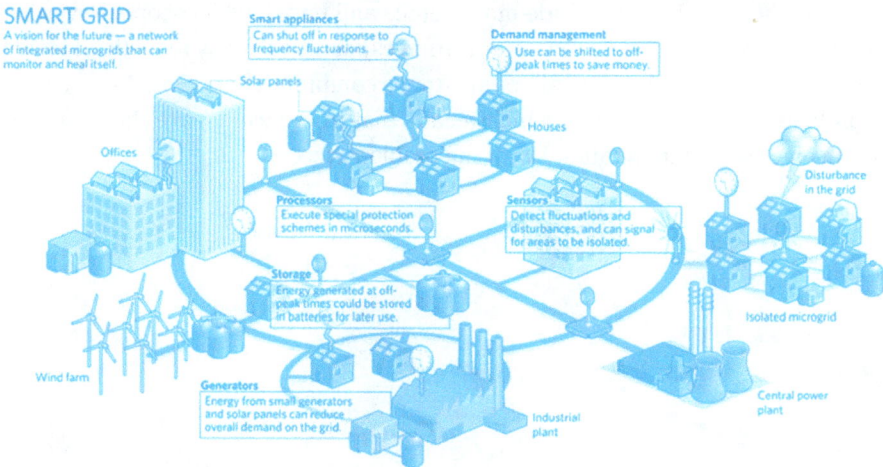

Figure 20.3: Smart grid developments.

Data, people, diseases, products and energy move through networks which are very complex; so complex that nobody completely understands them. For better understanding and controlling such networks several challenges have been identified; for example, the modeling at different spatial and temporal scales of networks is seen as a major challenge.

Three challenges are listed in this chapter:

1.   The challenge of flexibility: when one designs or operates systems/networks; what is the best modus when inlets, outlets and hardware performance vary. In other words, how to deal with uncertainty.
2.   The challenge of sustainability: when one designs or operates systems/networks; what is the best modes or trade-off for multiple (often) conflicting objectives (economics, environment, society)?
3.   The challenge of complexity: when one faces complexity (combinatorial), how does one deal with it in such way that one can still yield accurate information?

20.6.1 The challenge of flexibility

When we incorporate uncertainty in our models, the form (as was given in equations (20.1)–(20.5)) becomes different:

$$\min E[Z(a, x, y)] + R, \tag{20.6}$$

subject to

$$g(\beta, x(\xi), y(\xi)) = 0, \tag{20.7}$$

$$h(\gamma, x(\xi), y(\xi)) > 0, \tag{20.8}$$

$$x^L \le x(\xi) \le x^U, \tag{20.9}$$

$$y(\xi) \in Y, \tag{20.10}$$

$$\xi = \{\xi \mid \alpha \in A, \beta \in B, \gamma \in \Gamma\}. \tag{20.11}$$

The model now has several uncertain parameters, which are replaced by distributions (a, b, g). The objective function is now an expected value and depends on the outcome of uncertainty parameters. And often stochastic models, as given in equations (20.6)–(20.11) contain recourse terms in the objective (a kind of penalty) and non-anticipativity (constraints that ensure that the stochastic model will only generate one outcome). Figure 20.4 shows in a very simplistic example how uncertainty affects the solution.

20.6.2 The challenge of sustainability

Often decision makers want to balance the outcomes that they obtain: good economic performance, good environmental performance and good societal performance (sustainability criteria). This has led to the development of multi-criterion decision making tools. Although there are several of such tools at hand for a decision maker (weighted sum approach, goal programming, preemptive optimization), we will solely look here at the generation of so-called pareto-frontiers.

We have to understand that once more the formulation of our optimization problem changes, especially the objective function, which is now not a single objective, but a set of objectives:

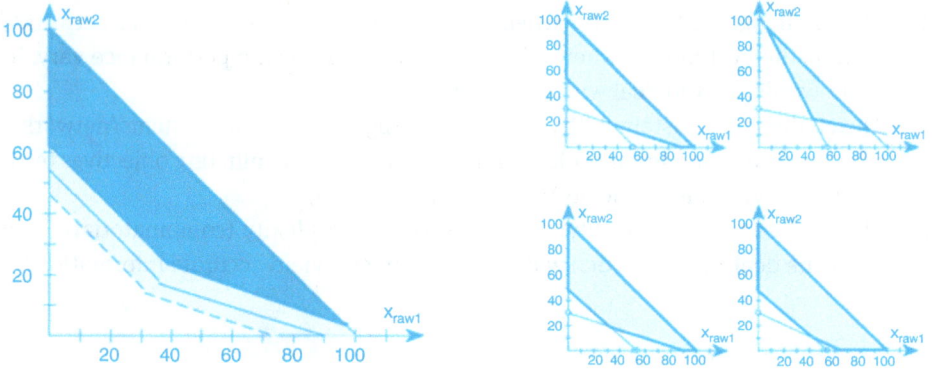

Figure 20.4: The effect of uncertainty parameters on the decision-space. Left: translation, constraints end up higher or lower. Right: rotation, constraints take different slopes. If the feasible space changes, the optimal solution also changes.

$$\min[Z_{1(a_1,x,y)}, Z_2(a_2, x, y), Z_3(a_3, x, y) \dots]. \qquad (20.12)$$

Suppose that there are two criteria to be balanced with each other (economics and environmental impact). For each feasible point, we could actually calculate what would be the objective values. In other the decision-space is translated to the objective space. A simple example is given in Figure 20.5.

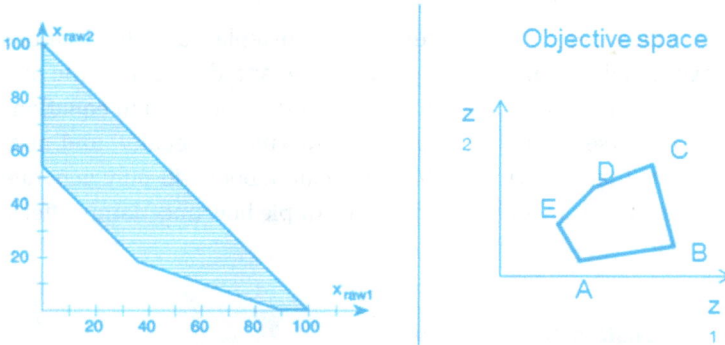

Figure 20.5: Left: decision-space. Right: objective space.

From the objective space a decision maker can take useful information; for example, solution A is the solution with minimum value of Z_2, while solution E is the optimum for minimum Z_1. Actually all points on the line segment E–A are kind of allowed trade-offs. These trade-offs are called non-inferior solutions or pareto-optimal solutions.

Realistic optimization problems with thousands of decisions and constraints cannot be represented geometrically and more advanced tools are needed such as the ϵ-constraint method.

Example 1 (E-constraint method for FMCG supply chains). In this example a model for tactical planning of FMCG supply chains is worked out. The network topology is given in the figure below. The time horizon for planning is 52 weeks and demand, ingredient availability, recipes, production rates, inventory levels and all cost factors are known. In this specific example we optimize an ice cream supply chain.

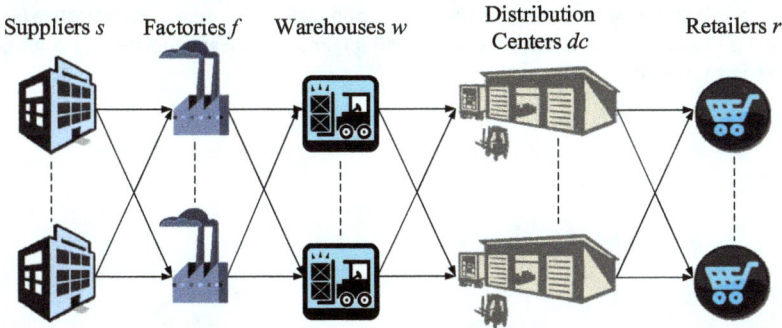

We look at a small European supply chain, with 10 suppliers, 4 factories, 4 warehouses, 10 distribution centers and 20 retailers. Through this supply chain 10 different products (or Stock Keeping Units, SKUs) are moved. The decision maker does not want only to minimize costs, but also to minimize the environmental impact of the supply chain. For that reason he includes data from the life cycle assessment database Eco-99 indicator. Only two main ingredients of ice cream are considered, *i. e.*, milk-based ingredients (milk and cream) and sugar. There are two alternatives for both ingredients, as shown in the table below.

	Non-Organic Milk	Organic Milk	Beet Sugar	Cane Sugar
Environmental impact [ECO-99 units/t ingredient]	24.0	17.9	8.0	9.2

There are also further environmental data for the energy mix used for the production/cooling processes. Energy consumption in the production process, approximately 0.65 MJ/kg ice cream, and environmental impact depend on the energy mix.

Country	Environmental impact per kWh [ECO 99 units/kWh][5]	Environmental impact per ton ice cream [ECO 99 units/t product]
Austria	0.018	3.25
Belgium	0.024	4.33
Greece	0.062	11.19
Portugal	0.047	8.49

Also the environmental impact of the transport is included on a 40-ton truck with a 50 % capacity load and 0.015 ECO 99 units/t km.

The ε-constraint method is now used to create 26 points on a pareto front. First a single objective optimization is run, minimizing the costs. Then a second run is executed, fixing the costs at their optimum value and minimizing the environmental impact.

Then the environmental impact is minimized. This minimum value is now fixed and the costs are once more minimized. From these four optimization runs the overall range of environmental and economic impact is found. Now for a set of 22 additional optimization runs where one objective (say the economics) is fixed at different values between their minimum and maximum value, the graph as given below is generated.

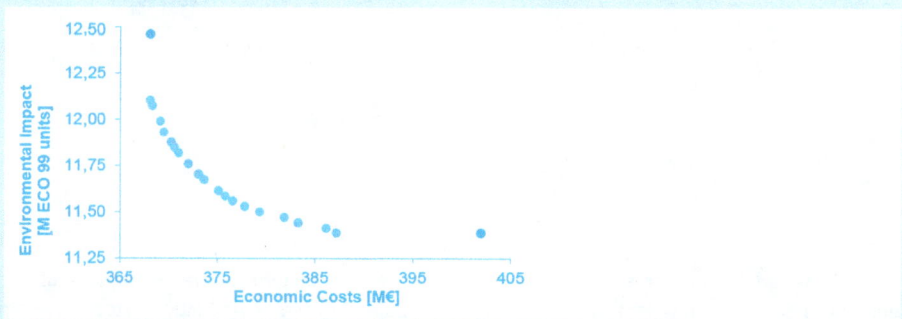

The pareto frontier shows that environmental impact can be reduced significantly without incurring additional costs, and *vice versa*.

20.6.3 The challenge of complexity

The optimization models that are formulated for enterprise networks have a complex nature: large-scale, many equations, variables. They are often discrete in nature (with binary or other integer variables) and the models are highly non-linear and non-convex.

There are several solutions at hand for models. They could for example reduce model sizes by making assumptions or rank effects and eliminate the smallest effects. They could decompose models (for example break them up in smaller parts and solve the smaller parts individually). They could reformulate models: linearizing them or transforming them to other domains. Now with the rise of computer power another way of tackling complexity is by installing more processors/cores or to improve solution algorithms for handling optimization problems.

20.6.4 Perspectives

The combined effort of dealing with flexibility, sustainability and complexity leads to new tools for multi-criterion decision making, uncertainty assessment and sensitivity analysis tools, all with the aim to influence or shift the pareto frontier into a favorable form (this is also called reverse engineering: which properties do I need to set up the best network?).

Exercises

1. Given is the following multi-objective optimization problem:

$$\max Z(x_1, x_2) = \left[Z_1(x_1, x_2), Z_2(x_1, x_2) \right],$$

$$Z_1(x_1, x_2) = 5x_1 - 2x_2,$$

$$Z_2(x_1, x_2) = -x_1 + 4x_2,$$

subject to:

$$-x_1 + x_2 \leq 3,$$

$$x_1 + x_2 \leq 8,$$

$$x_1 \leq 6,$$

$$x_2 \leq 4,$$

$$x_1, x_2 \geq 0.$$

Sketch for this problem the decision-space and the objective space. Identify the set of non-inferior solutions and use the Utopia point method to find the best trade-off solution.

2. A manufacturing line makes two products. Production and demand data are shown in Table 20.1.

Table 20.1: Demand data.

	Product 1	Product 2
Set-up time (hrs)	6	11
Set-up cost ($)	250	400
Production time/unit (h)	0.5	0.75
Production cost/unit ($)	9	14
Inventory holding cost/unit	3	3
Penalty cost for unsatisfied demand/unit ($)	15	20
Selling price ($/unit)	25	35

	Wk 1	Wk 2	Wk 3	Wk 4
Product 1	75	95	60	90
Product 2	20	30	45	30

The total time available (for production and set-up) in each week is 80 hours. Starting inventory is zero, and inventory at the end of week 4 must be zero. Only one product can be produced in any week, and the line must be shut down and cleaned at the end of each week. Hence the set-up time and cost are incurred for a product in any week in which that product is made. No production can take place while the line is being set up. Formulate and solve this problem as an MILP, maximizing total net profit over all products and periods.

20.7 Take-away message

A plant or factory is often a single node in a larger network of plants, raw material suppliers, product markets, intermediate production facilities and storage sites. In EWO the design is evaluated at this higher spatial level. In EWO the strategic decision where a new factory should be built, what should be its capacity and how and from whom raw materials should be purchased as well as to whom and where the products should be sold are made. Often EWO models are large-scale optimization models in the form of MILP and MINLP models. These models are optimized to minimize cost, maximize profit or minimize environmental impact.

20.8 Further reading

Grossmann, I. (2005). Enterprise-wide optimization: A new frontier in process systems engineering. *AIChE Journal*, 51 (7), pp. 1846–1857.

Garcia, D. J. and You, F. (2015). Supply chain design and optimization: Challenges and opportunities. *Computers and Chemical Engineering*, 81, pp. 153–170.

Grossmann, I. E. (2012). Advances in mathematical programming models for enterprise-wide optimization. *Computers and Chemical Engineering*, 47, pp. 2–18.

Shah, N. K., Li, Z. and Ierapetritou M. G. (2011). Petroleum refining operations: Key issues, advances, and opportunities. *Industrial and Engineering Chemistry Research*, 50(3), pp. 1161–1170.

Shashihara, S. (2011). *The Optimization Edge – Reinventing decision making to maximize all your company's assents*. McGraw Hill.

van Elzakker, M. (2013). *Enterprise-wide optimization for the fast moving consumer goods industry*. PhD thesis, Eindhoven University, the Netherlands.

A Appendix

Table A.1: PDCA example: the hospital code cart problem.

Step	Description
PLAN	A code cart is basically a miniature warehouse of inventory that is used by medical staff when a patient is in direct need of help. Pharmaceuticals, supplies, implements, and oxygen are some of the basic components that are kept on these carts. The hospital had numerous complaints by medical staff that the cart contained improperly cleaned implements and wrong or inadequate drugs, and that oxygen and supplies were often missing or not functioning properly. As a result, medical personnel would bring extra supplies to medical emergencies because they were concerned the carts were not being stocked properly on a consistent basis. The current process was flowcharted and measured, and several areas were cited for improvement. One floor of the hospital was selected as the pilot floor.
DO	Medical personnel on the floors was taking items from the cart for floor use instead of just for medical emergencies. It was not uncommon to have the oxygen tank taken for another use. Central supply was not aware of what items needed to be on the cart and how clean these needed to be. Medical staff had not defined what items should be kept on the cart, so the internal departments that supply the carts were trying to make assumptions about what should be kept on the cart, and those assumptions were not always correct. The pharmacy department was in charge of restocking the oxygen, but they had no mechanical information about how to assemble the oxygen tank. Appropriate procedures and training were developed and implemented to remedy this problem.
CHECK	The hospital code cart team designed two basic measurements in the do phase. These measurements included whether the carts were being stocked within the two-hour agreed-upon turnaround, and the other measurement was to check to see whether all the items were included on the cart. This team showed amazing success with their implementation of the appropriate countermeasures.
ACT	The implementation was so successful that the one floor that had been the pilot for this team became the model for implementation throughout the organisation. The team trained the balance of the organisation in both the new processes and procedures as well as the measurements to ensure the process was working correctly.

https://doi.org/10.1515/9783110570137-021

Index

www.ingramcontent.com/pod-product-compliance
Lightning Source LLC
Chambersburg PA
CBHW060955210326
41598CB00031B/4838